BOTANICAL RESEARCH AND PRACTICES

PLANT EXTRACTS

ROLE IN AGRICULTURE, HEALTH EFFECTS AND MEDICAL APPLICATIONS

BOTANICAL RESEARCH AND PRACTICES

Additional books in this series can be found on Nova's website under the Series tab.

Additional e-books in this series can be found on Nova's website under the e-book tab.

BOTANICAL RESEARCH AND PRACTICES

PLANT EXTRACTS

ROLE IN AGRICULTURE, HEALTH EFFECTS AND MEDICAL APPLICATIONS

ADRIANA GIORDANO
AND
ALFIERI COSTS
EDITORS

NOVA BIOMEDICAL
New York

For permission to use material from this book please contact us:
Telephone 631-231-7269; Fax 631-231-8175
Web Site: http://www.novapublishers.com

NOTICE TO THE READER

Additional color graphics may be available in the e-book version of this book.

Library of Congress Cataloging-in-Publication Data
Plant extracts : role in agriculture, health effects and medical applications / editors: Adriana Giordano and Alfieri Costs.
p. cm.
Includes index.
ISBN: 978-1-62417-534-3 (hardcover)
1. Plant extracts. 2. Plant extracts--Therapeutic use. 3. Materia medica, Vegetable. 4. Botanical chemistry. I. Giordano, Adriana. II. Costs, Alfieri.
QK861.P533 2013
615.3'21--dc23
2012046630

Published by Nova Science Publishers, Inc. † New York

Contents

Preface

In this book, the authors present current research in the study of the role in agriculture, health effects and medical applications of plant extracts. Topics discussed include the risks and benefits of phytochemicals present in fruits and vegetables; medicinal plants and their pharmacological attributes; Moringa oleifera and its benefits in nutrition and health; the potential health benefits and applications of the agave species; herbal medicinal plants used in the treatment of diabetes in Africa; phenolic compounds extracted from medicinal plants as an alternative for cancer prevention and treatment; using the antifeedant and repellent effects of plant extracts in the protection of plants and human health; essential oils and extracts of medicinal plants in the control of diseases in plants; herbal therapies for dermatological disorders; polyphenols as a source of antimicrobial agents against human pathogens; and new trends in plant extract use for food-borne pathogenic bacterial control.

Chapter 1 – Epidemiologic evidence suggests that regular consumption of fruits and vegetables may reduce the risk of some diseases, including cancer. The clinical pharmacologic interest in the efficacy and safety of phytochemicals present in fruits and vegetable has grown during past years because of the realization that many people self-medicate using these agents.

Therefore, the main reason for devoting a major review to the risks and benefits of phytochemicals is the enormous importance of fruits and vegetables used for their beneficial effects as nutrients, and as components in folk medicine, but the authors also raise concerns about the safety of these agents, which have not been fully assessed. The present review is limited to some of the most widely used fruits and vegetables to give medical practitioners some ideas of the possible pharmacological and physiological effects, potential side effects, and drug interactions for common fruits and vegetables used by their patients.

The authors discuss the cultural uses, phytochemistry, pharmacological, and toxicological properties of phytochemicals present in the following fruit and vegetables: grapefruit, orange, tangerine, grapes, cranberry, pomegranate, mango, guava, black raspberry, black mulberry, apple, broccoli, cauliflower, watercress, spinach, tomato, carrot, and avocado. The authors also examine the literature on their safety to raise a warning flag whenever potential hazardous risks are possible.

The authors conclude that their knowledge of the therapeutic benefits and risks of some fruits and vegetables or their components is still limited and efforts to elucidate them should be intensified.

Chapter 2 – Drugs derived from natural resources represent a significant segment of the pharmaceutical market. Traditional systems of medicines like Ayurveda, Traditional Chinese Medicine (TCM) or the European phytotherapy assumed synergistic approach wherein all ingredients of the plants could bring about maximum therapeutic efficacy. Due to the escalated cost of development of pre-clinical synthetic drugs, there has been a Worldwide demand for phytopharmaceuticals. The modern therapy applies more and more combinatorial approach in treatment of diseases like cancer, cardiovascular or rheumatic diseases. High-throughput methods are available to investigate complex mixtures of whole plant extracts and establish synergistic effects for multi-therapeutic target treatments. Demands are increasing to establish open access data banks in a multidisciplinary effort for medicinal plant research with the aim to develop a common standard molecular/therapeutic "bar codes" for characterization of medicinal plants. In the present chapter an attempt has been made to compile information on medicinal plant extracts with respect to their pharmacological attributes to promote wider acceptance and use of these plant-based drugs in main stream of medicine. The mode of action of phytopharmaceuticals as assessed by up / down-regulation of therapeutic target molecules with respect to anti-cancer, anti-inflammatory, anti-hypertensive, anti-viral, anti-obesity, antidiabetic and pro-gynecological properties complying with multi-target therapy will be described. This approach further helps scientific validation of plant extracts / bioactive molecules that are already being used in traditional medicine.

Chapter 3 – *Moringa oleifera* Lam. (Moringaceae), also known as the horseradish or drumstick tree, is a native plant of pan-tropical areas such as sub-Himalayan regions of India, Pakistan, Bangladesh and Afghanistan. However, other locations such as Africa, Ceylon and Thailand have started cultivating this plant. Due to the use of almost every part of the plant, its resistance to drought, pests and disease, as well as its acclimatization to different soil types and high biomass per annum yields it has become known as a 'miracle tree'. Apart from its use as a nutritional aid and medicinal remedy in both humans and animals, it has also been utilized as a source of biofuel. These effects are attributed to the many different macro- and micronutrients stored within the Moringa plant, as well as the presence of bioactive phytochemicals. Leaves contain high levels of protein and carbohydrates, vitamin A and C, and minerals such as calcium, supporting its use as meal replacement in malnourished individuals or as animal fodder. Medicinally it is used to treat various disorders, such as ascites, rheumatism, inflammation, infections, diabetes and cancer. It is also used in the prevention of hepatic, renal and cardiovascular diseases. These uses are supported by the bioactivities reported for the plant, including antioxidant, anti-inflammatory, hepatoprotective, nephroprotective, antimicrobial, larvacidal, antidiabetic and radioprotective activities. Toxicological studies have shown that it is well-tolerated at high doses in several animal models. It is apparent that due to the beneficial properties and low toxicity of Moringa that it could become an inexpensive and safe alternative to various allopathic medicines, as well as providing nutrition to people in developing countries.

Chapter 4 – An analysis and a summary of the reported uses of the genus *Agave,* mainly related to their medicinal applications, are briefly outlined. Based on years of experience from traditional medicine to the present time, different extracts, as well as pure isolated chemical compounds, have been obtained from the *Agave* genus. This chapter attempts to combine the existing information on uses, distribution, ethnobotany, and medicinal and social importance of the different *Agave* species. Over the last decade, ethnobotanical studies of medicinal plants of Yucatán, Mexico revealed that some *Agave* species were already used by the Mayans since the pre-hispanic times to treat illnesses with cancer-like symptoms.

Research on natural products with biological activity has led to a boom of commercially significant compounds with potential health benefits and applications in the pharmaceutical industry. Hauad-Marroquin (2010) referred to the Phytochemistry as a branch of science that has allowed for the detection and subsequent identification of active principles responsible for some property attributed to a plant. On those bases, several research groups have been focusing on the study of Agave extract applications in the treatment of different diseases such as kidney cancer, diabetes, gonorrhea and diseases of the excretory system. Agave has also been used as a diuretic, as poultices or agents to cure skin infections, and in the treatment of snake bites, among others. Recently, Ramachandran et al. (2011) pointed out that various parts of *Agave cantula* were used as a laxative, emmenagogue, in the treatment of scurvy, syphilis, scrutula, swelling, retention of urine, cancer, citotoxic or diuretic conditions, and even as an aphrodisiac and antisyphilitic agent. Similarly, several reports describe cytostatic activity in aqueous extracts of *Agave* species. There are also reports describing antitumor activity from *Agave* species. Furthermore, the medical applications of steroidal saponins from Agave have also been described, including the treatment of cognitive, as well as for anti-fertility treatment. Also, the antimicrobiological properties of Agave extracts have been known for a long time.

In general, monocotyledonous families contain compounds with potential industrial applications as sources of alkaloids, non-protein amino acids, anthraquinones, cyanogens, mono- and sesquiterpenoids, and saponins. In this regard, the steroidal saponins are mainly found in monocotyledonous families such as *Agavaceae*, *Dioscoreaceae* and *Liliaceae*. Therefore, several *Agave* species can be a good source of saponins (mainly heconin and tigonin) and sapogenins, from which derivative compounds can be obtained, such as contraceptives, anti-inflammatories, sexual hormones, etc. The cytotoxicity of Agave saponins has been well documented.

Polyphenols are also bioactive compounds with potential health benefits present in *Agave* species, based mostly on their antioxidant properties.

In summary, there are many investigations directed to the use of Agave extracts, as well as to the isolation, purification, characterization and structural elucidation of bioactive compounds with potential health benefits.

On the other hand, the presence of fructans in *Agave* species and their health applications have been well documented. These properties stem from the fact that prebiotic food ingredients (e.g., inulin-type fructans) reach the colon undigested, and once there, they are fermented by colonic microflora to short chain fatty acids (e.g., lactate, acetate, propionic and butyric acid). It is now believed that the beneficial effect on the colonic crypts, together with the antiproliferative neoplastic properties and the prevention of colon cancer (via enhanced apoptosis of colon cancer cells), are mostly due to the production of butyric acid. Additionally, Ulloa et al. (2010) list several diseases and pathological conditions that can be prevented or reduced by inulin-type fructans, including constipation and diarrhea, especially those associated with intestinal infections, reducing the risk of osteoporosis, and cardiovascular atherosclerosis associated with dyslipidemia, hypertriglyceridemia and insulin resistance, obesity, diabetes type 2 associated with insulin resistance, irritable bowel syndrome, and chemically induced colon cancer.

Chapter 5 – Diabetes mellitus, a metabolic disorder of the pancreas in which the blood glucose levels are abnormally high (hyperglycaemia), is a growing threat to public health in modern society. The number of people suffering from diabetes is increasing due to urbanization, population growth, aging and increasing prevalence of obesity and physical inactivity. While some Non Insulin Dependent Diabetic Mellitus (NIDDM) patients can be managed by diet alone, others require hypoglycaemic therapy and/or insulin. Although insulin therapy affords an effective glycaemic control, drawbacks such as oral ineffectiveness, short shelf life, requirement of constant refrigeration, parenteral therapy with its attendant abscesses and fatal hypoglycaemia in the event of excess dosage has limited its usage. On the other hand, pharmacotherapy with sulphonylureas, biguanides and thiazolidones is also associated with side effects. In recent years, the popularity of complementary medicine has increased considerably and the WHO has suggested the evaluation of the potentials of plants as effective therapeutic agents, especially in areas where the modern drugs are not readily available. Due to their relative safety and low costs, herbal medicinal plants are prescribed for the treatment and prevention of a variety of diseases. These medicinal plants are used in their crude forms as herbal teas, syrups, infusions, ointments, liniments and powders. There are many herbs available having antidiabetic activity and they have been shown to act by different mechanisms. Some have been shown to have stimulating or regenerating effects on β cell or possess extra pancreatic effect for hypoglycemic activity. This article examines some of the herbal medicinal plants that have been used in the folkloric treatment of diabetes in Africa and the experimental evidences to support the antidiabetic properties.

Chapter 6 – Medicinal plants have provided therapeutic agents for thousands of years and continue being a great source of novel pharmaceutical products. Two-thirds of the world population still relies to some extent on natural medicines for primary health care. In addition, the demand for medicinal plant products is now increasing in both developing and developed countries as a consequence of a growing aging population and increasing consumer awareness about general health. Nevertheless, there is a widespread concern regarding safety, efficacy and quality of herbal medicines. Currently, many countries are conscious about the need of regulatory guidelines for medicinal herbs formulation, production and use. Phytomedicines represent a wide range of product types: from raw herbs (dried or fresh) to others with different processing degrees (such as tinctures and extracts). The production of herbal medicines involves adequate cultivation and post-harvest methodologies as well as botanical identification of the vegetable drugs. For herbal medicine end-products with added value, appropriate extraction procedures (i.e., in terms of bioactivity), standardization of extracts, selection and tuning of the drying processes operating conditions and design of pharmaceutical formulations are also necessary.

In this chapter the current global scenario of phytomedicine is first presented, evaluating its impact on public health and markets. The challenges associated to safety, efficacy and quality of herbal medicines due to the extracts complexity are then introduced. Understanding the great importance of providing accurate dosages of herbal therapeutic agents, the production of dry plant extracts is next discussed focusing on: the drying step including the formulation of the solution to be dried (which comprises the herbal extract) and the dried extract quality. Regarding the available technologies for preparing dry medicinal plants extracts, spray drying is particularly described. This technique is the most commonly used in the herbal processing industry due to its ability to generate particulate products with precise specifications in continuous operations. Several key studies about the effect of the operating conditions and feed composition on the yield of the spray drying process and the particles properties are subsequently summarized. For the production of tablets by direct compression, stable herbal dried powders with good flow properties are required. Consequently, the effects of different processing variables on the stability and flowability of dried extracts are also presented. The formulation of solid dosage forms, based on spray-dried plant extracts, for conventional and modulated delivery is then reviewed. Besides, existing strategies to modulate the drug delivery of plant extracts are also described. Since modified-release phytomedicines are still in an exploratory stage, the prospective for the development of novel products is finally discussed.

Chapter 7 – Cancer is a chronic degenerative disease that involves multiple environmental, chemical, physical and metabolic factors as well as genetic predisposition to this condition. It is the principal cause of mortality in the world and worldwide, 7.6 million of deaths in 2008 were attributed to cancer. However, recent epidemiological evidence indicates that a diet based on fruits and vegetables rich in antioxidants significantly reduces risk of developing several types of cancer. This suggests that some antioxidants such as polyphenols, which are found in different foods, may be effective for cancer prevention and treatment. There are reports that this type of phytochemicals are involved in prevention mechanisms against development of tumors through inhibition of genotoxic effects, increasing anti-inflammatory activity, inhibition of proteases and cell proliferation, and protection of transduction pathway signals that modulate apoptosis process. In addition, some studies claim that many natural chemo-preventive agents induce cell cycle arrest or apoptosis by activating tumor suppressors such as p53 and its target genes. To elucidate the mechanisms involved in the action mode of these phyto-chemicals, extracts of some plants like strawberry, raspberry, blueberry, and pomegranate had been evaluated. Rich-polyphenol extracts were tested on different cancer cells including, MCF-7 (breast cancer), HT-29, HCT-116 (colon cancer), LNCaP and DU-145 (prostate cancer) cell lines causing tumor growth inhibition. Similarly, it has been demonstrated that pecan shell extracts reduce growth of kidney cancer cells (A-498, and 769-P). Moreover, recent studies using semi-desert plants (*Turnera diffusa* and *Origanum vulgare*) extracts reported high toxicity against leukemia cells (K562) and different cervical cancer cell lines (HeLa and SiHa), increasing levels of cell apoptosis. However, it is necessary to know the mechanism of action of these compounds against tumor cells and how affect their functionality. Hence, this chapter is intended to inform about cytotoxic effect of phenolic extracts obtained from plants and fruits used in traditional medicine, as well as the mechanisms involved in such activity, in addition, information about how natural compounds generally decrease the adverse effects caused by current therapies in treating cancer is presented too.

Chapter 8 – Mutual plant-insect interactions have developed in the course of many million years. Thanks to the high ability of plants to produce numerous various secondary metabolites that may lead to significant reduction of pest population dynamics, new and efficient extracts can be obtained today, which can be used in the production of botanical insecticides. Botanical insecticides (BI) play a significant role not only in plant protection against pests, but also in the protection of humans and domestic animals against significant vectors or against troubling insects.

Besides substances causing acute or chronic insect toxicity, a large group of compounds can be found in the plant kingdom, which is classified as allomones, thus substances exerting repellent, antiovipositional and antifeeding effects. The effects of some of these substances are so good that they can be used for the development of new BIs safe for the environment and for the health. This chapter provides a critical summary of the most recent knowledge on plant allomones and their possible use in plant and human protection against pests and important vectors.

Chapter 9 – In recent years medicinal plants have been widely studied because they have bioactive compounds that are able to inhibit or retard the growth of microorganisms. Currently, with an increasing demand from society for healthy food and sustainable agriculture, there have been more studies with the use of oils and extracts from medicinal plants for disease control in field and post harvest, to replace pesticides. Bioactive compounds of plants are generally extracted in the aqueous form, with ethanol or other organic solvents, or by steam, when present in the essential oils. These substances are mostly secondary metabolites which have biological effects on microorganisms. In fungi, the bioactive compounds are able to inhibit and/or delay mycelial growth, cause morphological changes in the hyphae, vacuolization and disorganization of the cytoplasm, rupture of plasma and mitochondrial membranes and other organelles, as well as changes in cell permeability and the flow of its constituents. In bacteria, these compounds, besides impairing pathogenicity, cause irreversible damage to cellular structure with wall degradation; changes in cytoplasm density (loss of electro-dense material); cytoplasmic vacuolation; alterations in membrane permeability with negative effects on the metabolic activity, the inhibition of specific enzymatic processes; and protein denaturation that leads to breakage, leakage of the cytoplasm and cell death. The compounds can also act indirectly on the pathogens by activating the plant defense system, which recognizes the bioactive substances present in extracts or essential oils through membrane-anchored receptors and activates the transcription of genes responsible for production of pathogenesis-related protein defense (PRPs - chitinase, β-1,3-glucanases, peroxidases, among others) and synthesis of phenolic compounds (phenol and lignin) and phytoalexins. Such effects have been confirmed in several pathosystems by molecular tools (gene expression), light microscopy and scanning electron and transmission and biochemical analyses of the activity of defense enzymes.

Chapter 10 – Dermatological disorders (DD) have been projected to amount to as high as 34% of all occupational diseases. Amplification of chronic, degenerative and malignant diseases, economic factors and limitations of conventional medicines such as antibiotics resistance have encouraged renewed public interest in herbal therapies (phytotherapies), which is classified as complementary and alternative medicines (CAM) to treat and/or manage common DD. It is of no denying that herbal extracts have proved to be key sources of novel templates for the pharmaceutical armamentarium. Interestingly, herbal extracts geared towards skin disorders, are currently gaining popularity among patients and to a lesser degree among physicians. For example, ancient Egyptians used the fruits and leaves of the bishop's weed (*Ammi majus*) to treat vitiligo, a skin condition characterised by a loss of pigment. Recently, β-methoxypsoralen, a drug produced from this plant which showed promising potential to treat psoriasis and other skin disorders, as well as T-cell lymphoma has been developed. Additionally, herbal medicines for DD have proved to be of great importance in the treatment and/or management of HIV-associated dermatological problems. Since such patients already require panoply of conventional medicines, topical herbal therapies have proved to be of great significance in avoiding unnecessary drug-drug interactions which can set up harmful allergic reactions. Consequently, an advanced investigation should be geared towards traditionally used medicinal plants with the aim of identifying any novel molecule that can be used as alternative remedies against common DD. However, appropriate standardization and exact mechanism(s) of action must be considered for effective and safe uses of such remedies. This present book chapter endeavors to present an overview of the prevalence of common DD and highlights the potential of plant extracts in the management and/or treatment of common human DD.

Chapter 11 – Polyphenols or phenolic compounds are groups of secondary metabolites that are frequently found in extracts of medicine plants. Many medicine plants have been tested in clinical and experimental studies as antimicrobials. As least for some species, this biological activity can be partially explained by the presence of phenolic compounds. These compounds comprise a wide variety of molecules that are usually divided into flavonoids and nonflavonoids structures and sequentially into several sub-groups of compounds depending on the oxidation state. Apart from this clear chemical classification, the literature is rather confusing in presenting the health effects of polyphenols. The major aim of this review is to present an updated material on the effects of polyphenols acting as antimicrobial agents against human pathogens. There is a wide variety of proposals for using phenolic compounds as antimicrobial agents but very little is discussed about the real potential of these molecules when incorporated in medicines and surface cleaning agents. The antibacterial activity of these molecules against pathogens of the oral cavity is particularly evidenced since limitations on the use of phenolic compounds in the oral cavity can exist (e.g. stains of the teeth and restorations). Examples of polyphenol-rich medicine plants that can show antibiotic activity against resistant bacteria are presented. The synergism and antagonism effects of the combination of phenolic compounds and other drugs are also discussed due to the fact that this is of upmost interest for health professionals. Finally, some medicine plants are presented as source of antibacterial phenolic compounds but framing these plants in the perspective of advantages and disadvantages for promoting health in humans.

Chapter 12 – Food-borne pathogenic bacteria are one of the major public concerns worldwide. Prevention of pathogenic and spoilage microorganisms in food is usually achieved by using synthetic chemical preservatives, but some of them are responsible for many carcinogenic and teratogenic damages, as well as residual toxicity, together with growing concern of microbial resistance toward conventional preservatives. All of the above are concerns that have originated pressure on the food industry, for progressive removal of chemical preservatives and adoption of natural alternatives, to obtain its goals concerning safe food with longer shelf live. Many plant derived products such as spices, fruit preparations, vegetal preparations or extracts have been used for centuries for preservation and extension of shelf life of foods. In this chapter, trends on food-borne bacteria control using plant extract are mentioned. Today, scientific research reveals that chemicals from plants not only have an effect against a particular disease, but also antioxidant property from plant extracts gives a beneficial effect to human health. Considerable research on antimicrobial plant extracts has been reported, moreover, different mixtures of plant or by-product extracts have been used for food-borne bacteria control. Currently, there is a tendency to create natural products and environmentally friendly. In addition, the use of different organic solvents allowed in food, which are in concordance with the context of organic production systems, as well as the use of green technologies for plant and by-product extracts obtaining is also encouraged.

Chapter 13 – The use of medicinal and aromatic plants has an aged history in many cultures; numerous varieties of these plants are listed in the historic manuscripts and religious books.

Many of the recurrent used spices and herbs were applied by the ancient nations throughout the world, and they were valued for their medicinal and preservative potentialities in addition to their flavor enhancement properties. The dependence on natural agent like plant extracts, as powerful alternatives to chemical and synthetic antimicrobials, is greatly desired from modern consumers.

While many ethno-pharmacological plants already exhibited promising characteristics for treating endemic and emergent diseases, however, the door is still opening for the discovery of other useful properties from phytomedicine.

The antimicrobial properties of plant extracts have been discussed by numerous investigators and everyday novel results are presented concerning their potentialities and prospects applications.

The present chapter will take us in a voyage, to indicate the apparent used antimicrobial plant extracts, in human therapy and as food biopreservatives, as well as the potential applications of these extracts to improve human health and safety.

Chapter 14 – This chapter reports the application of microwave technology for extraction of antioxidants compounds from natural product. This novel method based on an electromagnetic heating was named by the authors' research group as "Microwave Integrated Soxhlet" abbreviated as the MIS method. This technique was compared to the traditional method, Soxhlet extraction, which has been used as the reference method for evaluating the performance of the proposed method. Rosemary (*Rosmarinus officinalis L.*) was selected as natural source used for extracting antioxidants compounds by MIS process, owing to its antioxidants properties, as well as it is easily grown in the Mediterranean area. Response Surface Methodology (RSM), was applied to the MIS extraction of the antioxidants components of the rosemary plant, to determine the most influential factors in the process in order to optimize it and therefore, find the combination of factors which would result in obtaining the extract with the highest antioxidant content. After executing a set of designed experiments the MIS extraction parameters including extraction time, leaching time and microwave irradiation power were optimized.

In: Plant Extracts
Editors: Adriana Giordano and Alfieri Costs
ISBN: 978-1-62417-534-3

Chapter 1

Risks and Benefits of Phytochemicals Present in Fruits and Vegetables

Lourdes Rodríguez-Fragoso**, *Elizabeth Alvarez-Ayala, Ana Isabel Gonzaga-Morales, Lourdes Anzures-Cortés and Jorge Reyes-Esparza
Universidad Autónoma del Estado de Morelos,
Facultad de Farmacia, Cuernavaca, México

Abstract

Epidemiologic evidence suggests that regular consumption of fruits and vegetables may reduce the risk of some diseases, including cancer. The clinical pharmacologic interest in the efficacy and safety of phytochemicals present in fruits and vegetable has grown during past years because of the realization that many people self-medicate using these agents.

Therefore, the main reason for devoting a major review to the risks and benefits of phytochemicals is the enormous importance of fruits and vegetables used for their beneficial effects as nutrients, and as components in folk medicine, but we also raise concerns about the safety of these agents, which have not been fully assessed. The present review is limited to some of the most widely used fruits and vegetables to give medical practitioners some ideas of the possible pharmacological and physiological effects, potential side effects, and drug interactions for common fruits and vegetables used by their patients.

We discuss the cultural uses, phytochemistry, pharmacological, and toxicological properties of phytochemicals present in the following fruit and vegetables: grapefruit, orange, tangerine, grapes, cranberry, pomegranate, mango, guava, black raspberry, black mulberry, apple, broccoli, cauliflower, watercress, spinach, tomato, carrot, and avocado. We also examine the literature on their safety to raise a warning flag whenever potential hazardous risks are possible.

* Correspondence: Lourdes Rodríguez-Fragoso M.D., Ph.D. Fax: 52 777 329-7089 Email: mrodriguezf@uaem.mx.

We conclude that our knowledge of the therapeutic benefits and risks of some fruits and vegetables or their components is still limited and efforts to elucidate them should be intensified.

Introduction

Dietary habits are an important modifiable environmental factor influencing human health and disease. Epidemiologic evidence suggests that regular consumption of fruits and vegetables may reduce the risk of some diseases, including cancer (Liu, 2004). These properties have been attributed to foods that are rich sources of numerous bioactive compounds such as phytochemicals. Modifying the intake of specific foods and/or their bioactive components seems to be a prudent, non-invasive, and cost-effective strategy for preventing some diseases in people who appear to be "healthy" (Liu, 2003).

The clinical pharmacologic interest in the efficacy and safety of fruits and vegetables has grown during past years because of the realization that many people self-medicate using these foods.

There is a limited knowledge for healthcare workers (such as doctors, pharmacists, nurses, and social workers) about the pharmacology and toxicology for the most commonly used fruits and vegetable in their patients. Here we review some of the most widely used fruits and vegetables to give medical practitioners some ideas of the possible pharmacological and physiological effects, potential side effects, and drug interactions for common fruits and vegetables used by their patients.

We also examine the literature on the safety of these foods to raise a warning flag whenever potential hazardous risks are possible. As will be discussed in this chapter, potential problems occur when patients taking medicines regularly also consume certain fruits or vegetables.

Although many scientific articles have been published on fruits and vegetables and their diverse effects, each one has several different natural constituents, the majority of which have not been studied.

The present chapter is limited to species such as grapefruit (*Citrus paradisi),* orange (*Citrus sinensis),* tangerine (*Citrus reticulata),* grapes (*Vitis vinífera),* cranberry (V*accinium macrocarpon),* pomegranate (*Punica granatum),* mango (*Mangifera indica),* guava (*Psidium guajava),* black raspberry (*Rubus coreanus),* black mulberry (*Morus nigra),* apple (*Malus domestica),* broccoli (*Brassica oleracea),* cauliflower (*Brassica oleracea),* watercress (*Nasturtium officinale),* spinach (*Spinacia oleracea),* tomato (*Lycopersicum esculentum),* carrot (*Dactus carrota),* and avocado (*Persea americana).* Based on our experience, these are the most frequent fruits and vegetables sought by people who have health problems. Many of these fruits and vegetables have no popular name that is consistently used in English; in other cases, the English name may refer to two or more botanically distinct species (see Table 1 and 2).

Presently, there are few studies that combine a phytochemical and a detailed pharmacological approach (Heinrich, 2003) or studies that systematically explore the risk and benefits of fruit and vegetables. It is not possible at this point to list all of the fruit and vegetables commonly used by healthy and non-healthy people.

Table 1. Potential benefits of commonly consumed fruits

Fruit	Cultural uses	Clinical trials
Grapefuit *Citrus paradisi*	Antioxidant, antiseptic, cardiotonic, detoxicant, hypocholesterolemic, sedative, antimicrobian, insomnia, anxiety or nervousness (Kiani, 2007).	Reduces atherosclerotic plaque formation and inhibits breast cancer (Kiani, 2007). Provides resistance to common cold and wound infections (Meera and Kalidhar, 2008).
Orange *Citrus sinensis*	Nervous system depressant disorders of nervous origin, and as a sleep aid, gastrointestinal disorders, inflammatory ailments in respiratory tract, and arthritis. (Liu, 2003)	Decreases diastolic blood pressure. (Morand and others, 2011). Reduced oxidative DNA damage in circulating blood mononuclear cells (Milenkovic and others, 2011)
Tangerine *Citrus tangerine* *Citrus reticulate* *Citrus deliciosa*	Respiratory and gastrointestinal tract ailments and arthritis (Lyle, 2006).	Non documented
Grape *Vitus vinifera*	Antianemic, inflammatory ailments in respiratory tract and others (Yadav and others, 2009).	Beneficial effects against coronary heart disease and cancer preventive activity (Stakleff, 2012). Beneficial effects in diabetes, obesity and antihypertensive (Agostini, 2012).
Cranberry *Vaccinium macrocarpon*	Urinary tract infections nephrolithiasis wound healing (Rossi and others, 2010) Antioxidant (Vinson and others, 2001).	Antioxidants (Ruel and others, 2007; Steinberg, 2009). Suppress the inflammatory symptoms induced by *E. coli* infections (Howell, 2002; Huang and others, 2009).
Pomegranate *Punica granatum*	Inflammatory, respiratory and gastrointestinal tract ailments. (Jindal and Sharma 2004; Shabtay and others, 2008). Hemorrhoids, cataracts, contraceptive and abortifacient (Murali, 2002).	Heart disease and antihypertensive (Aviram and others, 2004; Aviram and Dornfeld, 2001). Anticancer (Mustafa and others, 2009)
Mango *Mangifera indica*	Heart diseases, laxative and diuretic, tuberculosis and respiratory related ailments (Chimponda and Mukanganyama, 2010; Shah and others, 2010).	Non documented
Guava *Psidium guajava L*	Stomachache, diarrhea, Genitourinary ailments, hypertension and diabetes (De Wet and others, 2010)	Non documented
Berries *Rubus coreanus* *Rubus idaeus* *Rubus fruticosus* *Rubus leucodermis*	Obesity, cancer, and cardiovascular and neurodegenerative diseases (Brownmiller and others, 2008). Chronic diseases (Abd El-Mawla and others, 2011)	Safety/tolerability raspberries (Stoner and others, 2005).
Apple *Malus domestica*	Diuretic, genitourinary ailments, inflammatory ailments in respiratory tract, and reduced risk of cancers (Mahbubeh and others, 2009).	Non documented
Papaya *Carica papaya*	Abdominal discomfort, pain, malaria, diabetes, obesity, infections and oral drug poisonings (Adeneye and Olagunju, 2009).	Non documented

Table 2. Potential benefits of commonly consumed vegetables

Vegetables	Cultural uses	Clinical trials
Tomato *Lycopersicum esculentum*	Treatment urinary tract disease symptoms and may have anticancer properties (Polívková and others, 2010).	Anticancerigen, beneficial effect on neurodegenerative and cardiovascular diseases (Zhang and others, 2009; Rao and Balachandran, 2002).
Carrot *Dactus carrota*	Constipation and antioxidant activity (Surles and others, 2004).	Antioxidant capacity, reduce oxidative stress, in women previously treated for breast cancer (Butalla, 2012). Reduce the risk of chronic diseases (Mech-Nowak and others, 2012).
Avocado *Persea americana*	Treatment of diabetes in Latin America and Africa (Plaza and others, 2009).	Non documented
Nopal *Opuntia ficus-indica*	Diabetes, headache, atherosclerosis, alcohol hangover, hypertension, asthma, burns, hypoglycemic, anti-allergic and gastrointestinal tract ailments (De Smet, 2002; Lee and others, 2000).	Antidiabetic and antidepressant (Godard and others, 2010).
Red pepper *Capsicum annuum L.*	Enhancing the palatability of food and drug such as counterirritant on stomach (Watanabe and others, 2001). Used in scarlatina, putrid sore throat, hoarsenes, dispepsia, yellow fever, piles and snakebite (Materska and Perucka, 2005).	Antimicrobial activity (Cichewicz, 1996). Antioxidant properties and potent antimutagenic and anticarcinogenic activities (Materska and Perucka, 2005).
Broccoli *Brassica oleracea*	Protective effects against various types of cancers (Juge and others, 2007).	Non documented
Cauliflower *Brassica oleracea*	Breast engorgement (Nikodem and others, 1993).	Pain and engorgement (Arora and others, 2009).
Watercress *Nasturtium officinale*	Antioxidant, hydratant, hypocholesterolemic (Getahun and Chung 1999).	Non documented
Spinach *Spinacia oleracea*	Antioxidant vegetable usually consumed after boiling the fresh or frozen leaves (Schirrmacher and others, 2010).	Non documented

Pharmacognosy

According to traditional medicine, there are diverse healing strategies, as well as different preferences for the plant parts used for various diseases. It is generally accepted that the beneficial effects of medicinal plants can be obtained from active constituents present in the whole plants, parts of plant (as flowers, fruits, roots or leaves), or plant materials or combinations thereof, whether in crude or processed state (De Smet, 2002).

Many herbalists believe that isolated ingredients have weaker clinical effects than whole plant extracts, a notion that would obviously require proof in each case. The multiconstituent character of crude fruit and vegetable can render efficacy testing more complex for purified constituents. However, it is necessary to chemically characterize whole fruit or vegetables and extracts to determine in preclinical studies the efficacy as well as toxicity and to determine if they have biological activity in human clinical trials.

The concept of several active principle ingredients acting in a synergistic manner in natural remedies may be somewhat unusual to pharmaceutical scientists who are more accustomed to monotherapy using specific therapeutic agents. An example of this phenomenon may be seen in the fruits and vegetables that have different active constituents. The most commonly used fruits and vegetables are reported to contain between 7-12 constituents, this is the case of grapefruit, black mulberry, and mango or tomato, broccoli, and watercress.

Potential Benefits of Fruits and Vegetables on Health

For the past several decades, public health policy statements have called for an increase in fruit and vegetable consumption by adults and children. The rationale for promoting greater intake includes a potential reduction in risk for cancer, hypertension, cardiovascular disease, reduced inflammatory symptoms in those diagnosed with rheumatoid disease, and possibly a reduction in body weight or adiposity (Thomson and Ravia, 2011). Health food is, simply, food that is claimed to be healthy, either because it has a healthy image or because it contains ingredients that are commonly assumed to be healthy such as fiber, antioxidants and polyunsaturated fatty acids (Table 3 and 4).

Fruits and vegetables are known to be important components in a healthy diet, since they have low energy density and are sources of micronutrients, fiber, and other components with functional properties, called phytochemicals. Increased fruit and vegetable consumption can also help displace food high in saturated fats, sugar or salt. Low fruit and vegetable intake is among the top 10 risk factors contributing to mortality. According to the World Health Organization (WHO), increased daily fruit and vegetable intake could help prevent major chronic, non-communicable diseases (WHO, 2003).

In addition to providing fiber, vitamins, and minerals, fruits and vegetables are rich in phytochemicals made by plants to protect them from the damaging effects of their environment. The presence of different plant pigments signify that content of phytochemicals is different in each fruit and vegetable. Therefore, eating from a rainbow of colors ensures the greatest number of benefits. Examples of pigment colors, foods in which they occur, and the name of the associated phytochemical are: yellow/tangerines/β-cryptoxanthin; orange /cantaloupe and carrots/β-carotene; pink/red/tomatoes and grapefruit/lycopene; green/broccoli, brussels sprouts/indoles; and blue/purple/blueberries, pomegranate, eggplant /anthocyanins (Myklebust, 2006).

Table 3. Phytochemical and ethnopharmacology of commonly consumed fruits

Fruit	Active components	Pharmacological activity
Grapefruit	Bergamottin, nobilein, tangeretin, quercetin, diosmin, naringenin, naringin, kaempferol, and furanocoumarins (Hanley and others, 2011). The peel has five compounds friedelin, β-sitosterol, limonin and cordialin B (Meera, 2008).	Cardiometabolic benefits (Silver and others, 2011).
Orange	Flavonoids as tangeretin, nobiletin, diosmin, hesperetin, naringenin, isohesperidina and aurantiamarina (Morand and others, 2011)	Antioxidant, antiinflammatory, antiallergic, antiproliferative activities, antibacterial, antifungal, and antiviral activities (Yañez and others, 2004; Seeram and others, 2008).
Tangerine	Tangeretine, nobiletin, limolene, alpha-pinene, myrcene, gamma-terpinene, citronellal, linalool, nerol, neryl acetate, geranyl acetate, geraniol, thymol, and carvone (Lyle, 2006).	Reduces triglycerides and cholesterol levels
Grape	Resverestrol, viniferin, ß-carotene, lutein, neochrome, neoxanthin, violaxanthin, luteoxanthin, flavoxanthin, zeaxanthin and norisoprenoid compounds (Bunea, 2012). Grape seeds oil: tocopherols, tannins, gallic acid, catechin, epicatechin, procyanidins, linoleic acid and α-tocopherol (Agostini and others, 2012).	Antitumoral, anti-inflammatory, antioxidant, antibacterial, antiobesity, antidiabetic (Stakleff and others, 2012; Zhu and others, 2012, Agostini and others, 2012).
Cranberry	Phenolic acids, flavonoids, (anthocyanins and proanthocyanidins), flavonols (Katsargyris and others, 2012).	Antioxidant and anticancer activities (Vinson and others, 2001). Anti-inflamatory, antibacterial, and anticancer (Ruel and others, 2007).
Pomegranate	Polyphenolic compound, anthocyanins, hydrolyzable tannins, ellagitannins and flavonoids (Schubert and others, 1999).	Non documented
Mango	Phenolic acids, flavonoids, carotenoids, essential oils, fatty acids, lectins, phenols, saponins, alkaloids, and triterpenes, terpenoids, ascorbic acid (Kim, 2011). Quercetin and glycosylated xanthones such as mangiferin (Berardini and others, 2005).	Analgesic, antitumor, antiproliferative, antiplasmodial, and antiviral (Engels and others, 2009; Shah and others, 2010; Khan and Islam, 2012)
Guava	Phenolics acids, flavonoid, carotenoid, terpenoid and triterpene (Quian and Nihorimbere, 2004).	Anti-inflammatory, antioxidant, antibacterial activity, antidiabetic (Sunagawa and others 2004). Antidiarrhoeal, antigiardial and antirotaviral activities (Tannaz and others, 2011; Barbalho and others, 2012).
Black raspberry	Vitamins A, C, E and folic acid, β-sitosterol, ellagic acid, ferulic acid, quercitin, anthocyanins and cyaniding (Tian and others, 2005a).	Chemopreventive, anticancer, anti-inflammatory and antiangiogenic (Katsuzaki and others, 2003; Li-Shu and others, 2011).
Apple	Quercetin-3-galactoside, quercetin-3-glucoside, quercetin-3-rhamnoside, catechin, epicatechin, procyanidin, cyanidin-3-galactoside, coumaric acid, chlorogenic acid, gallic acid, and phloridzin (Guyot and others, 2003).	Antiviral activity (Lee and others, 2003)
Papaya	Proteolytic enzymes (papain, chymopapain), alkaloids (carpain, carpasemine), sulfurous compounds (benzyl isothiocyanate), flavonoids, tannins, triterpenes, anhocyaninsorganic acids and oils (Mahmood and others, 2005).	Antioxidant, antihypertensive, antifungal, antihelminthic (Eno and others, 2000; Mehdipour and others, 2006).

Table 4. Phytochemical and ethnopharmacology of commonly consumed vegetables

Vegetable	Active components	Pharmacological activity
Tomato	Carotenoids phytofluene, phytoene, neurosporene, γ -carotene, lycopene, phytoene, phytofluene, quercetin, polyphenols and kaempferol (Mourvaki and others, 2005).	Antioxidant, hydratant and hypocholesterolemic (Freedman and others, 2008).
Carrot	Polyphenols, α- and β-carotene, quercetin, myrecetin and panaxynol (Surles and others, 2004).	Decreased systolic pressure, antioxidant capacity, protects the cardiovascular system decreasing lipid peroxidation (Potter and others, 2011).
Avocado	Monounsaturated fatty acids and sterols (alkanols, aliphatic acetogenins, terpenoid) glycosides, flavonoids and coumarin (Duester, 2001).	Antiinflammatory, antioxidant, hypocholesterolemic, vasorelaxant, anticonvulsant activity (Plaza and others, 2009). Hypoglicemic activity (De Almeida and others, 1998). Induction cell cycle arrest, inhibits growth, and induces apoptosis in precancerous and cancer cell lines (Lima and others, 2012).
Nopal	Pectin, mucilage, aromadendrin, taxifolin, dihydroquercetin, isorhamnetin, vitexin, kaempferol, quercetin, derivatives of pyrone, ascorbic acid, betalains, betacyanins and flavonoids (Stintzing and Carle, 2005; Galati and others, 2005).	Antioxidant, anticarcinogenic, hypoglycemic, hypolipidemic, hypocholesterolemic and analgesic (Park and others, 2001; Guevara-Arauza, 2009).
Red pepper	Capsaicinoids: dihydrocapsaicin and capsaicin (Materska and Perucka, 2005).	Non documented
Broccoli	Aliphatics glucosinolate and glucoraphanin (Cartea and others, 2010).	Anticancer effect (Juge and others, 2007; Russo and others, 2010).
Cauliflower	Isothiocyanates, glucosinolate, indole-3-carbinol, sulforaphane and indol (Tian and others, 2005b).	Antioxidant, anticancer activities (Nikodem and others, 1993).
Watercress	Isothiocyanates, flavonoids such as quercetin and hydroxycinnamic acids (Palaniswamy and others, 2003). β-carotene, lutein and glucosinolates (Getahun and Chung, 1999).	Non documented
Spinach	Flavonoid glycosides (Bergquist and others, 2005), glucuronides and acylated di-and triglycosides of methylated and methylene dioxide derivatives of 6-oxygenated flavonols (Lomnitski and others, 2003).	Decrease bioaccessibility and cellular uptake of Cd. (Yang and others, 2012).

Evidence is emerging that specific combinations of phytochemicals may be far more effective in protecting against some diseases than isolated compounds (Table 1 and 2). The protective effects are due to different mechanisms such as an increased intake of essential micronutrients as well as different non-nutritive phytochemicals.

A dietary pattern characterized by a high intake of fruits and vegetables has a much higher protective effect than the intake of pharmacological doses of single nutrients or phytochemicals (Craig, 2010).

It has also been proposed that the additives and synergistic effects of phytochemicals in fruits and vegetables are responsible for their antioxidant and anti-cancer activities, and that the benefit of plant-based diets is attributed in part to the complex mixture of phytochemicals present in whole food (Liu, 2003). Clearly, no single antioxidant can replace the natural combination of thousands of phytochemicals that exist in whole foods. Given the history of the diverse intake of plant foods by mankind, it is sensible to encourage a diverse intake. The exact amounts of fruits and vegetables needed each day to minimize cancer risk are not known and will require a great deal of additional research. However, the data of the benefits of fruits and vegetables suggests that it is not premature to advise an increased intake of a variety of colorful fruits and vegetables.

Several epidemiologic, clinical, and experimental studies have established that certain types of diet may have beneficial effects on health. For example, the traditional Mediterranean diet has been shown to reduce overall mortality and coronary heart disease events (Pauwels, 2011). Increasing evidence suggests that a Mediterranean-style diet rich in fruits, vegetables, nuts, fish and oils with monounsaturated fat and low in meat promotes cardiovascular health and aids in cancer prevention because of its positive effects on lipid profile, endothelial function, vascular inflammation, insulin resistance, and its antioxidant properties (Nadtochiy and Redman, 2011). Vegetarians, on the other hand, exhibit a wide diversity of dietary practices often described by what is omitted from their diet. When a vegetarian diet is appropriately planned and includes fortified foods, it can be nutritionally suitable for adults and children and can promote health and lower the risk of major chronic diseases (Craig, 2010). The vegan dietary category may be more comparable across countries and cultures because avoiding all animal products leaves little choice but to include large quantities of vegetables, fruit, nuts, and grains for nutritional adequacy. Admittedly, vegetable and fruit variety may also vary widely according to location (Li, 2011). Due to their high content of fruits and vegetables, all of these diets contain a large proportion of antioxidant vitamins, flavonoids, and polyphenols (Halliwell, 2007).

Potential Risks of Fruits and Vegetables on Health

Although the public and some healthcare professionals believe that fruits and vegetables are relatively safe because they are "natural", there is remarkably little data to support this assumption. Recent publications have highlighted the severe consequences from side effects from certain food products (Bush et al, 2007). Many dangerous and lethal side effects have been reported from the use of food products. Most of this information has been obtained from health centers and emergency rooms. These side effects may occur through several different mechanisms, including direct toxic effects of the plants, effects of contaminants, and interactions with drugs or other herbs.

At the present time, it is easier to determine which food remedies might induce direct toxicity, because it is known which chemical compounds are present in these plants and which of these compounds are associated with a number of side effects in a significant fraction of users. For example, flavonoids may be natural components which come from many vegetables and fruits. A variety of health effects of these compounds were found. However, a few safety studies indicated that certain adverse effects of these compounds could influence on human health, which could involve genotoxic/carcinogenic effects or interaction with thyroid hormone biosynthesis.

Table 5. Potential risk of commonly consumed fruits

Fruit	Experimental evidence	Clinical evidence
Grapefruit	Effects on CYP450 drug oxidation and transportation (Hanley and others, 2011). Inhibition of P-glycoprotein and MRP2-mediated drug efflux (Yoo and others, 2007). Inhibit the OATP-B function *in vitro* (Satoh and others, 2005; Greenblatt, 2009).	Increases plasma levels of several drugs (Pawelczyk and Kloszewska, 2008; Dresser and others, 2005; Amory and Amory, 2005).
Orange	Affects the pharmacokinetics of CYP3A4 substrates (Ho and others, 2000). Inhibits P-glycoprotein (Takanaga and others, 2000). Inhibits the function of human OATP-A (OATP1A2, gene symbol *SLC21A3/SLCO1A2*) *in vitro* (Dresser and others, 2002).	Reduces the intestinal absorption of substrates of OATP-B and decreases the oral bioavailability of several drugs (Dresser and others, 2005; Satoh and others, 2005; Saito and others, 2005; Greenblatt, 2009).
Tangerine	Inhibits P450 1A2 and P450 3A4 activity (Obermeier and others, 1995). Increase the absorption co-administered drugs able to serve as P-gp substrates (Yoo and others, 2007).	Non documented
Grape	Inhibitor of CYP3A4 and CYP2E1 in microsomes (Piver and others, 2001).	Non documented
Cranberry	Inhibits diclofenac metabolism in human liver microsomes (Ushijima and others, 2009). Inhibits nifedipine metabolism (Van den Bout-Van den Beukel and others, 2006).	Drug interaction with warfarin by inhibiting CYP3A4 and/or CYP2C9 enzymes (Izzo, 2005). Increase the bioavailability of CYP3A4 substrates (e.g., calcium antagonists or calcineurin inhibitors) (Van den Bout-Van den Beukel and others, 2006).
Pomegranate	Inhibition of the enteric CYP3A activity (Hidaka and others, 2005). Inhibited human CYP2C9 activity (Nagata and others, 2007).	Non documented
Mango	Reduces CYP1A2 activity (Rodeiro and others, 2009). Affects the activity of the multidrug transporter P-gp ABCB1 (Chieli and others, 2009).	Non documented
Guava	Inhibitory effect on P-gp mediated efflux (Junyaprasert and others, 2006).	Non documented
Black raspberry	Inhibit the human CYP3A (Hidaka and others, 2005; Kim and others, 2006). Inhibits OATP-B function (Satoh and others, 2005).	Non documented
Apple	Inhibition of the OATP transporter family (Pohl and others, 2006).	Non documented
Papaya	LD_{50} of 325.2 mg/kg in mice administered i.p. (Eno and others, 2000). Abortifacient and foetal toxicity (Oderinde and others, 2002). Antifertility effects (Lohiya and others, 1992, 2002). Inhibition of CYP3A activity (Hidaka and others, 2004).	Hipoglicemic effect (Danese and others, 2006).

Table 6. Potential risk of commonly consumed vegetables

Vegetable	Experimental evidence	Clinical evidence
Tomato	Contain solanine and tomatin be toxic alkaloids (Barceloux, 2009). Inhibits: CYP1A1, CYP1B1, UGP (Wang and Leung, 2010).	*In vitro* system: diethylnitrosamine, N-methyl-N-nitrosourea, and 1,2-dimethylhydrazine (Wang and Leung, 2010).
Carrot	Carrots induce phenol sulfotransferase activity (Yeh and Yen, 2005) and decrease CYP1A2 activity (Bradfield and others, 1985).	Non documented
Avocado	Inhibited the effect of warfarin (Wells and others, 1994)	Non documented
Nopal	Non documented	Non documented
Red pepper	Inhibit the constitutive enzymes CYP (Zhang, 1993). Induces alterations in intestinal brush border fluidity and passive permeability properties (Prakash and Srinivasan, 2010). Reduces oral salicylate bioavailability (Cruz and others, 1999).	A possible risk might occur of the interaction between red pepper and hypoglycemic drugs in diabetic patients (Bouraoui and others, 1995)
Broccolli	Inhibits: CYP1A1, CYP2B1/2, CYP3A 4, CYP2E1, hGSTA1/2, MRP-1, MRP-2, BCRP, UDP, glucorosytransferases, sulfotransferases, quinone reductases phenolsulfotransferases (Fimognari and others, 2008; Telang and others, 2009).	Non Documented
Cauliflower	Inhibits: CYP1A1, CYP2B1/2, CYP3A 4, CYP2E1, hGSTA1/2, MRP-1, MRP-2, BCRP, UDP, glucorosytransferases, sulfotransferases, quinone reductases phenolsulfotransferases (Fimognari and others 2008; Telang and others, 2009).	Non documented
Watercress	Interact with P-glycoprotein, MRP1, MRP2 and BCRP (Telang and others, 2009). Inhibits: CYP2E1, P-glycoprotein, MRP1, MRP2, and BCRP (Leclercq and others, 1998; Telang and others, 2009).	Inhibits the hydroxylation of chlorzoxazone, an in vivo probe for CYP2E1. Increase the enzymes SOD and GPX in blood cells *in vitro* and *in vivo* (Hofmann and others, 2009).
Spinach	Protective effect of spinach against the genotoxic effects of 2-amino-3-methylimidazo[4,5-f]quinoline (IQ) by interaction with CYP1A2 as a mechanism of anti-genotoxicity. (Platt and others, 2010).	Non documented

Isoflavones estrogenic activity could influence sexual development of young animals and children. Furthermore, flavonoids could inhibit the absorption of some nutrients, interact with certain drug pharmacokinetics and affect neurobehavioral development. In this paper, the potential toxicities of flavonoids from recently published papers were reviewed (Wang and Leung, 2010). Examples of this kind of risk have been demonstrated for some fruits and vegetable, including some in this review (Table 5 and 6). Besides that, all contain chemical compounds that are considered potentially toxic and they are cited in the Hazardous Substances Data Bank (National Library of Medicine, Bethesda, MD, January 2002).

Nowadays, food safety is receiving more attention than ever before by governments and policymakers, health professionals, the food industry, the biomedical community, and last but not least, the consumers. Side effects may also occur due to contaminants in fruits and vegetables, as heavy metals, including lead, mercury, or arsenic, and other undeclared pharmaceuticals, purposefully and illegally added to food products to produce a desired effect, generally in manufactured foods (Gagnier et al, 2006). Pesticide residues in fruits and vegetables are a major concern to consumers due to their negative health effects. They have been found in both raw and processed fresh produce (Keikotlhaile et al, 2010). Organic fruits and vegetables can be expected to contain fewer agrochemical residues than conventionally grown alternatives; yet, the significance of this difference is questionable, inasmuch as actual levels of contamination in both types of food are generally well below acceptable limits. Also, some leafy, root, and tuber organic vegetables appear to have lower nitrate content compared with conventional ones, but whether or not dietary nitrate indeed constitutes a threat to human health is a matter of debate. On the other hand, no differences can be identified for environmental contaminants (e.g. cadmium and other heavy metals), which are likely to be present in food from both origins (Magkos et al, 2006).

In addition, other factors exist that might also affect the content of active constituents in the herbal product (such as microorganisms, microbial toxins, and genetic factors). It is well known that as plants are sessile, they synthesize a vast array of secondary metabolites as defense mechanisms for protecting themselves against pathogen infections (bacteria, fungi, and viruses) (Liu, 2003). Nutritionally, fruit and vegetables represent a good source of vitamins, minerals, and dietary fiber, and fresh-cut produce satisfies consumer demand for freshly prepared, convenient, healthy food. However, fresh-cut produce deteriorates faster than corresponding intact produce, as a result of damage caused by minimal processing, which accelerates many physiological changes that lead to a reduction in produce quality and shelf-life (Palumbo et al, 2007). The symptoms of produce deterioration include discoloration, increased oxidative browning at cut surfaces, flaccidity as a result of loss of water, and decreased nutritional value. Damaged plant tissues also represent a better substrate for growth of microorganisms, including spoilage microorganisms and foodborne pathogens. The risk of pathogen contamination and growth is one of the main safety concerns associated with fresh-cut produce, as highlighted by the increasing number of produce-linked foodborne outbreaks in recent years. The pathogens of major concern in fresh-cut produce are *Listeria monocytogenes*, pathogenic *Escherichia coli* mainly O157:H7, and *Salmonella spp.* (Francis et al, 2012).

Fruits and vegetables behave as authentic medicines because the chemical substances of which they are formed can have a biological activity in humans. For this reason, their joint administration with "conventional medicines" can produce variations in the magnitude of the effect. Thousands of drugs are commercially available and a great percentage of the population takes at least one pharmacologically active agent on a regular basis. Given this magnitude of use and variability in individual nutritional status, dietary habits and food composition, there is a high potential for drug-nutrient interactions. However, there is a relatively short list of documented fruit-drug or vegetable-drug interactions, necessitating further and extensive clinical evaluation. There are numerous patients who encounter increased risks of adverse events associated with drug-nutrient interactions. These include elderly patients, patients with cancer and/or malnutrition, gastrointestinal tract dysfunctions, acquired immunodeficiency syndrome and chronic diseases that require the use of multiple drugs, as well as those receiving enteral nutrition or transplants. There are currently few studies that combine a nutrient-based and detailed pharmacological approach (Liu, 2004), or studies that systematically explore the risk and benefits of fruit and vegetables (Dangour et al, 2010). Nutritional status and diet can affect drug action by altering metabolism and function. In addition, various dietary components can have pharmacological activity under certain circumstances (Schmidt and Dalhoff, 2002). For healthy-treatment intervention, it is necessary to understand how these drug-food interactions can induce a beneficial result or lead to detrimental therapeutic conditions (less therapeutic action or more toxicity). Equal attention must be paid to food-drug interactions. An interaction is considered significant from a clinical perspective if it alters the therapeutic response. Food-drug interactions can result in two main clinical effects: the decreased bioavailability of a drug, which predisposes to treatment failure, or an increased bioavailability, which increases the risk of adverse events and may even precipitate toxicities (Singh et al, 2009; Custodio et al, 2008). Drug-drug interactions are widely recognized and evaluated as part of the drug-approval process, whether pharmaceutical, pharmacokinetic, or pharmacodynamic in nature. Therefore, equal attention must be paid to food-drug interactions.

Until recently, little regard was given to the possibility that food and food components could cause significant changes to the extent of drug absorption via effects on intestinal and liver transporters. It is now well known that drug-food interactions might affect the pharmacokinetics of prescribed drugs when co-administered with food (Huang et al, 2008). We know little about the processes through which phytochemicals (and/or their metabolites) are absorbed into the body, reach their biological target, and are eliminated. Recent studies show that some of phytochemicals are substrates and modulators of specific members of the superfamily of ABC-transporting proteins (Telang et al, 2009). Indeed, *in vitro* and pre-clinical data in rats suggest that a variety of foodstuffs (Huang and others 2008; Tomlinson et al, 2008), including herbal teas (Dresser et al, 2002; Meijerman et al, 2006) and vegetables and herbs (Zhou et al, 2004; Alvarez et al, 2010) can modulate the activity of drug transporters. It is not yet known whether these effects are predictive of what will be observed clinically. Often, influence on drug metabolism by compounds that occur in the environment, most remarkably foodstuffs, is bypassed. Dietary changes can alter the expression and activity of hepatic drug metabolizing enzymes. Although this can lead to alterations in the systemic elimination kinetics of drugs metabolized by these enzymes, the magnitude of the change is generally small (Santos and Boullata, 2005). Metabolic food-drug interactions occur when a certain food alters the activity of a drug-metabolizing enzyme, leading to a modulation of the pharmacokinetics of drugs metabolized by the enzyme (Schmidt and Dalhoff, 2002). Foods, such as fruits, vegetables, alcoholic beverages, teas, and herbs, which consist of complex chemical mixtures, can inhibit or induce the activity of drug-metabolizing enzymes (Mandlekar et al, 2006). The observed induction and inhibition of CYP enzymes by natural products in the presence of a prescribed drug has (among other reasons) led to the general acceptance that natural therapies can have adverse effects, contrary to popular beliefs in countries with active ethnomedicinal practices. Herbal medicines such as St. John's Wort, garlic, piperine, ginseng, and gingko, which are freely available over the counter, have given rise to serious clinical interactions when co-administered with prescription medicines (Nahrstedt and Butterweck, 2010). Such adversities have spurred various pre-clinical and *in vitro* investigations on a series of other herbal remedies, with their clinical relevance yet to be established. The CYP3A4-related interaction based on food componentis the best known; it might be related to the high level of expression of CYP3A4 in the small intestine, as well as its broad substrate specificity. If we consider that CYP3A4 is responsible for the metabolism of more than 50% of clinical pharmaceuticals, all nutrient-drug interactions should be considered clinically relevant, in which case all clinical studies of drugs should include a food-drug interaction screening (Kimura et al, 2010).

Pharmacodynamic interactions have not been well described in the literature. However, according to the pharmacological effects reported, there may be examples where herbal medicines might produce synergistic effects when they are used along with conventional medicines (e.g., nopal-hypoglycemiant). Pregnant women frequently use fruits and vegetables, but information about their safety during pregnancy and breast-feeding is scarce. There are case reports and epidemiologic studies to suggest that certain herbs (e.g., those that are rich in unsaturated pyrrolizidine alkaloids) are associated with embryotoxic or fetotoxic effects.

Fruits and Vegetables: Examples with Clinical Relevance

Grapefruit (*Citrus paradisi*)

Grapefruit juice is rich in a number of phytochemicals, including flavonoids and furanocoumarins (Hanley et al, 2011). The peel of grapefruit has five isolated compounds; friedelin, β-sitosterol, 7(3',7',11',14'-tetramethy)-pentadec-2',6',10'-trienyloxycoumarin, limonin and cordialin B (Meera and Kalidhar, 2008). Traditionally, grapefruit juice has been found to contain antioxidant, antinitrosaminic, antiseptic, aperitif, cardiotonic, detoxicant, hypocholesterolemic, sedative and stomachic activities. Medical data has suggested that grapefruit juice reduces atherosclerotic plaque formation and inhibits breast cancer cell proliferation and mammary cell tumorigenesis (Kiani and Imam, 2007).

Epidemiologic studies reveal that approximately 2% of the population in the United States consumes at least one glass of regular strength grapefruit juice per day. This becomes pertinent if we consider that, many people suffer from chronic metabolic diseases (including hypertension, hyperlipidemia, and cardiovascular diseases) and receive calcium channel antagonis therapy and HMG-CoA reductase inhibitors. The juice of the grapefruit can help build resistance to the common cold and wound infections (Meera and Kalidhar, 2008). Because of this, it has been traditionally prescribed for anorexia, bacteria, benign prostatic hypertrophy, cancers (breast, colon, prostate, lung, skin and throat), candida, colds, diabetes, dysuria, high cholesterol, infections, insomnia, mycobacteria, mycosis, nervousness, pseudomonas, rheumatism, staphylococcus and yeast (Kiani and Imam, 2007). Consumption of grapefruit and grapefruit juice may result in cardiometabolic benefits, since trials have shown it can increase serum HDL-cholesterol concentrations (Silver et al, 2011).

The interaction of grapefruit with certain drugs was unintentionally discovered two decades ago (Flanagan, 2005). Since then, there have been numerous reports on the effects of grapefruit and its components on CYP450 drug oxidation and transportation (Hanley et al, 2011). Several findings showed that grapefruit juice had a major effect on the intestinal CYP system with a minor effect at the hepatic level (Bressler, 2006). The predominant mechanism for this interaction is the inhibition of cytochrome P-450 3A4 in the small intestine, which results in a significant reduction of drug pre-systemic metabolism. Grapefruit juice intake has been found to decrease CYP3A4 mRNA activity through a post-transcriptional activity, possibly by facilitating degradation of the enzyme (Kim et al, 2006). An additional mechanism may be the inhibition of P-glycoprotein and MRP2-mediated drug efflux, transporters that carry drugs from enterocytes back to the gut lumen, all of which results in a further increase in the fraction of drug absorbed and increased systemic drug bioavailability (Yoo et al, 2007). It has also been reported that the major constituents of grapefruit significantly inhibit the OATP-B function *in vitro* (Greenblatt, 2009).

The interaction between grapefruit juice and drugs has been potentially ascribed to a number of constituents. It has been suggested that flavonoids such as naringin, naringenin, quercetin, and kaempferol, the major components in grapefruit, are responsible for drug interaction. Some of these chemicals are also found in other fruit juices. Pomegranate, for example, shares certain properties with grapefruit, suggesting that both could modify the bioavailability of drugs (Hidaka et al, 2005). Another group of compounds that has been detected in grapefruit juice are the furanocoumarins (psoralens), which are known to be mechanism-based inactivators of CYP450. The major furanocoumarin present in grapefruit is bergamottin, which demonstrated a time- and concentration-dependent inactivation of CYP enzymes *in vitro* (Kim et al, 2006). One interesting characteristic of this interaction is that grapefruit juice does not need to be taken simultaneously with the medication in order to produce the interaction. The bioavailability of drugs has been reported being doubled by grapefruit juice, even when taken 12 h after ingestion. Colored grapefruit juice and white grapefruit juice are equally effective in producing drug interactions.

This inhibitory interaction should be kept in mind when prescribing drugs metabolized by CYP3A4. Examples of drugs affected by grapefruit or its components include: calcium channel antagonists such as felodipine, nisoldipine, amlodipines, verapamil, and diltiazem (Sica, 2006); central nervous system modulators, including diazepam, triazolam, midazolam, alprazolam, carbamazepine, buspurone and sertraline (Pawelczyk and Kloszewska, 2008); HMG-CoA reductase inhibitors, such as simvastatin, lovastatin, atorvastatin, and pravastatin (Reamy and Stephens, 2007); immunosuppressants such as cyclosporine (Paine et al, 2008); anti-virals such as saquinavir (Van den Bout-van den Beukel et al, 2006); a phosphodiesterases-5 inhibitor such as sildenafil (Bailey and Dresser, 2004); antihistamines, including terfenadine and fexofenadine (Dresser et al, 2005); anti-arhythmics such as amiodarone (Bailey and Dresser, 2004); and antibiotics such as eritromicine (Amory and Amory, 2005).

In the case of many drugs, an increase in serum drug concentration has been associated with increased frequency of dose-dependent adverse effects (Kiani and Imam, 2007; Pillai et al, 2009). In light of the wide-ranging effects of grapefruit juice on the pharmacokinetics of various drugs, physicians need to be aware of these interactions and should make an attempt to warn and educate patients regarding the potential consequences of concomitant ingestion of these agents.

Orange (*Citrus sinensis*)

Orange juice consumption is increasing worldwide, probably due to its public perception as a healthy, natural source of nutrients and increased public interest in health issues. Orange leaves, flowers and peel share medicinal properties with an essential oil composition and concentration variable. The bitter glycosides in the leaves, flowers and fruit of bitter orange, are hesperidin, isohesperidine and aurantiamarine. In traditional medicine, the orange plant is used as a nervous system depressant and is quite useful in disorders of nervous origin, as well as a sleeping aid. This property increases when using the flower, and the fruit still with the peel and leaves. The plant, especially the peel, is used to calm digestive discomforts such as flatulence, dyspepsia, digestive sluggishness and lack of appetite, along with inflammatory ailments in the respiratory tract, and arthritis (Liu, 2003).

It has been observed that the consumption of orange juice decreases diastolic blood pressure and significantly improves post-prandial microvascular endothelial reactivity in healthy, moderately overweight men. This effect is attributed mainly to hesperidin, the juice's main flavonoid. The vascular protective effects of orange juice and the possible specific role of hesperidin in mediating these effects are supported by a recent prospective study that posits a convincing link between the dietary intake of flavanones and flavanone-rich foods and reduced risks of coronary heart disease (Morand et al, 2011). Other clinical studies in healthy subjects have shown that orange juice consumption reduces oxidative DNA damage and may prevent meal-induced oxidative and inflammatory stress in circulating blood mononuclear cells. The reduction of reactive oxygen species (ROS) generation and NF-κB binding following orange juice intake might be due to its flavonoid content, as suggested by *in vitro* results (Milenkovic et al, 2011).

Orange juice antagonizes oxidative and inflammatory stress, an effect at least partially attributed to the presence of hesperidin and naringenin, which are two flavonoids that suppress ROS generation *in vitro.* Although *in vitro* antioxidant potency does not prove *in vivo* biological activity, there is also consistent clinical evidence of antioxidant potency for the most potent beverages, including both pomelo juice and red wine (Seeram et al, 2008).

A new study claims that drinking orange juice with meals can counter the pro-inflammatory effects of high-carbohydrate, high-fat meals. High-carbohydrate, high-fat meals induce a protein that suppresses cytokine signaling 3 (SOCS-3), which interferes with the action of insulin promoting insulin resistance. These adverse changes can be avoided when orange juice is consumed with a meal. This implies that drinking orange juice could potentially help prevent insulin resistance, diabetes, CHD and strokes. This benefit might be attributed to the flavonoids in orange juice, which suppress reactive oxygen metabolites (McAllister and Bernard, 2011).

It was recently reported that polymethoxylated flavones (PMFs) are almost exclusively found in the citrus genus, particularly in the peels of mandarins (*Citrus reticulata*) and sweet oranges (*Citrus sinensis*) (Li et al, 2006). PMFs have recently raised interest due to their broad spectrum of biological activities, including antioxidant, anti-inflammatory, anti-allergic, and antiproliferative activities and effects on mammalian metabolism (Yañez et al, 2004). Recent studies report that PMFs are also potent inhibitors of microbial growth, possessing antibacterial, antifungal, and antiviral properties. As a result of increasing consumer preferences for organic, healthier products, scientific research has begun to focus on the screening of natural antimicrobial compounds as possible biopreservatives (Tripoli and others, 2007; Smejkal et al, 2008).

Consumption of most types of orange juice does not appear to alter CYP3A4 activity *in vivo* (Takanaga et al, 2000). However, orange juice made from Seville oranges appears to be somewhat similar to grapefruit juice and can affect the pharmacokinetics of CYP3A4 substrates (Ho et al, 2000). It has been previously shown that consumption of a single 240 mL serving of Sevilla orange juice resulted in a 76% increase in felodipine exposure, comparable to what is observed after grapefruit juice consumption (Malhotra et al, 2001).

Presumably, the mechanism of this effect is similar to that of grapefruit juice-mediated interactions, because Sevilla orange contains significant concentrations of flavonoids, mainly bergamottin and 6´,7´- dihydroxybergamottin (Kamath et al, 2005). Orange juice has also been shown to exert inhibitory effects on P-glycoprotein (P-gp)-mediated drug efflux. It has been reported that 3,3′,4′,5,6,7,8-heptamethoxyflavon and tangeretin were the major P-gp inhibitors present in orange juice and showed that another component, nobiletin, was also a P-gp inhibitor (Takanaga et al, 2000). Therefore, the intake of orange juice might inhibit the efflux transporters by P-gp, which could enhance the bioavailability of drugs and thus lead to an increase in the risk of adverse events (Yoo et al, 2007).

It has also been observed that components of orange juice-naringin, in particular, are *in vitro* inhibitors of OATP transport activity (Farkas and Greenblatt, 2008). It has been previously reported that orange juice inhibits the function of human OATP-A (OATP1A2, gene symbol *SLC21A3/SLCO1A2*) *in vitro* (Dresser et al, 2002). OATP-A, however, is predominantly expressed in the brain, but not in the intestine. On the other hand, it was reported that OATP-B-mediated uptake of glibenclamide as well as estrone-3-sulfate was significantly inhibited by 5% orange juice (Satoh et al, 2005). Orange juice might reduce the intestinal absorption of substrates of OATP-B (e.g., digoxin, benzylpenicillin, and hormone conjugates), resulting in a decrease in concentration in the blood.

Previous studies in humans using fexofenadine as a probe showed that oral co-administration with orange juice decreased the oral bioavailability of fexofenadine (Dresser et al, 2005). Orange juice and its constituents were shown to interact with members of the OATP transporter family by reducing their activities. The functional consequences of such an interaction are reflected in a significant reduction in the oral bioavailability of fexofenadine, possibly by preferential direct inhibition of intestinal OATP activity.

Other reports indicate that orange juice slightly reduced the absorption of ciprofloxacin, levofloxacin and celiprolol (Saito et al, 2005). A study of an interaction between orange juice and pravastatin showed an increase in AUC (Greenblatt, 2009). Orange juice also moderately reduces the bioavailability of atenolol, which may necessitate a dose adjustment (Lilja et al, 2005).

Tangerine (*Citrus reticulata*)

Tangerines are a good source of vitamin C, folate and beta-carotene. They also contain some potassium, magnesium, vitamins B_1, B_2 and B_3, lutein and zeaxanthin. Like all citrus oils, tangerine oil has limonene as its major constituent, along with alpha-pinene, myrcene, gamma-terpinene, citronellal, linalool, nerol, neryl acetate, geranyl acetate, geraniol, thymol, and carvone (Lyle, 2006). Diosmin is one of the main components of citrus fruits such as tangerine. Previous studies on hamsters fed supplements with mixtures of 1% polymethoxylated flavonoids that contained tangeretin and nobiletin, showed lower plasma concentrations of both triglycerides and cholesterol and reduced hepatic triglycerides (Kurowska and Manthey, 2004).

Early studies demonstrated the influence of tangeretin, a flavonoid found in high levels in tangerine juice, on drug metabolizing liver enzymes. It was demonstrated that tangeretin inhibits P450 1A2 and P450 3A4 activity in human liver microsomes (Obermeier et al, 1995). Tangeretin is a potent regioselective stimulator of midazolam 1'-hydroxylation by human liver microsomes CYP3A4. Although clinical studies have shown no influence on midazolam pharmacokinetics *in vivo,* further studies are needed to evaluate its effects on other drugs (Backman et al, 2000). Diosmin is one of the main components of citrus fruits, such as tangerine. Diosmin may increase the absorption or bioavailability of co-administered drugs able to serve as P-gp substrates. As a result, some caution may be required with its clinical use (Yoo et al, 2007). Studies *in vivo* indicate that prevention of the hepatic lipid load by nobiletin limits the availability lipid for hepatic storage, lipoprotein secretion and deposition in peripheral tissues (Mulvihill et al, 2011).

Grapes (*Vitis vinifera*)

The grape is considered a source of unique and potentially useful medicinal natural products; they are also used in the manufacturing of various industrial products (Yadav et al, 2009). ß-carotene and lutein are the most abundant carotenoids in grapes (85% of the total). The rest are neochrome, neoxanthin, violaxanthin, luteoxanthin, flavoxanthin, zeaxanthin and cis-isomers of lutein and ß-carotene. Carotenoids are directly involved in grape´s aroma because they can suffer degradation followed by the apparition of norisoprenoid compounds (Bunea, 2012). Grapes are consumed as fresh fruit, wine, juice or raisins and are cultivated in vast amounts for the wine industry. They contain a wide range of chemical substances such as sugars, organic acids, mineral salts, vitamins, enzymes, and also phytochemicals responsible for the sensory characteristics of wines and their health-related properties. Phytochemicals in grapes are mostly phenolic compounds.

The main biologically active and well-characterized constituent from the grape is resveratrol, which is known for various medicinal properties in treating human diseases (Yadav et al, 2009). The skin of grapes has a significant amount of resveratrol, the source behind the beneficial effects of red wines on cancer prevention and against coronary heart disease. In animal studies, resveratrol prevents or delays the development of skin, mammary, and prostate tumors, as well as esophageal, gastric, small intestinal, colonic pancreatic, and hepatic tumorigenesis. Resveratrol has also been shown to possess *in vitro* cytotoxic effects against a wide variety of human tumor cells, including lymphoid and myeloid cancer cells as well as skin, breast, ovary, cervix, prostate, stomach, colon, pancreas, liver and thyroid carcinoma. Resveratrol has a multitude of biochemical and molecular effects on the activities of cyclooxygenase (COX), iNOS, CYP-450 and PKC and estrogen receptors (Stakleff et al, 2012).

Many studies demonstrated that the phenolic compounds present in grapes could reduce the incidence of serious chronic problems such as cancer and cardiovascular diseases due to their antioxidant abilities (Zhu et al, 2012). The antioxidant activities of grapes are due to the presence of antioxidant components such as flavonoids, phenolic acids, anthocyanidins and carotenoids. Carotenoids play an important role in human nutrition through their pro-vitamin A activity, but also by acting as antioxidants for the prevention of age-related macular degeneration, or skin protection against UV radiation. *Vitis vinifera* extract might be a potential candidate in the prevention of UV-induced skin damage because they have reduced levels of IL-6 and TNF-α in SKH-1 in mice skin (Filip et al, 2011). On the other hand, grape seed extract had a considerable antibacterial activity against methicillin-resistant *Staphylococcus aureus* (Al-Habib, 2010). The antioxidant capacity of carotenoids has been proved for pure compounds as well as for plant extracts and food. Recent studies have examined the carotenoid profile in some Italian grape varieties and the effect of lightening and irrigation on carotenoid concentration (Bunea et al, 2012).

Grape seeds contain an important amount of oil with nutritional potential based on their high level of unsaturated and high content of tocopherols as well as tannins, gallic acid, catechin, epicatechin, and a large amount of procyanidins. They are also rich in linoleic acid and α-tocopherol (vitamin E), the latter of which reduces the risk of cardiovascular diseases, diabetes, and cancer, prevents sexual impotence and is not synthesized in the human body. In addition, this oil helps dissolve thrombi in arteries, reducing platelet aggregation and preventing heart attacks; it can prevent hypertension and normalization of injuries caused by poor circulation due to obesity and diabetes. It may be used to treat obesity, cellulite, and stretch marks, since it aids tissue elasticity, reduces swelling and edema, restores collagen and improves peripheral circulation (Agostini et al, 2012).

Resveratrol was shown to be an irreversible (mechanism-based) inhibitor of CYP3A4 and a non-competitive reversible inhibitor for CYP2E1 in microsomes from rat liver and human liver cells containing cDNA-expressed CYPs (Piver et al, 2001). Resveratrol is an electron-rich molecule with two aromatic benzene rings linked by an ethylene bridge. CYP3A-mediated aromatic hydroxylation and epoxidation of resveratrol are possible resulting in a reactive p-benzoquinone methide metabolite, which is capable of binding covalently to CYP3A4, leading to inactivation and potential drug interactions.

Cranberry (*Vaccinium macrocarpon*)

American cranberry is a fruit used as a prophylactic agent against urinary tract infections (Rossi et al, 2010). Cranberry contains a great amount of phenolics, including simple phenolic acids, flavonoids that include anthocyanidins and proanthocyanidins (PACs), and flavonols. These phenolics vary according to the degree of unsaturation, oxidation of the three-carbon segment, and polymerization, which may influence their biological activity. Only a small percentage of the total flavonol content in cranberry or cranberry juice exists as aglycones such as free myricetin, quercetin and kaempferol (Singh et al, 2009). When one thinks of cranberries, the color red comes to mind: this is due to the presence of anthocyanidins, the health benefits of which have been extensively studied. There are six aglycones in the anthocyanidin class, and the number of potential isomers is extremely large due to the different positions of sugar attachment and the large number of different mono- and disaccharides (glycosides), along with the less common but extant phenolic acids and acyl groups that can be attached.

Growing data points to the many potential pharmacological activities of cranberry and cranberry juice. Due to its high content of flavonoids and phenolic acids, cranberry ranks highly among the fruits in both for its antioxidant qualities and quantity (Vinson et al, 2008). In fact, it has the highest level of polyphenols among a group of 22 studied fruits. These antioxidant properties are likely to contribute to cranberry's disease-fighting ability. Cranberry compounds can cause an improvement in antioxidant status, which might be beneficial in the case of chronic diseases. It has been shown that when cranberry juice is spiked on human plasma, low-density lipoprotein (LDL) and very-low-density lipoprotein (VLDL) can be protected from oxidation in a dose-dependent manner. This second effect is directly related to atherosclerosis (Steinberg et al, 1989).

Cranberry juice (7 mL/kg) was given to healthy men (n = 21) for 14 days. Although there was no effect on lipids, the subjects oxidized LDL was significantly decreased (10%) and plasma antioxidant capacity was significantly increased (7%), even if inflammatory markers were not improved. In the same study, men with elevated LDL and body mass index (BMI) were given low-calorie cranberry juice at doses of 125, 250, and 500 mL/day for three successive 4-week periods. Significant decreases in oxidized LDL, systolic pressure, and the inflammatory markers ICAM-1 and VCAM-1 were recorded at the end of the study. All cranberry effects in this study were found to be beneficial in cases of heart disease (Ruel et al, 2005).

Cranberry's role as a potential chemopreventive agent is gradually emerging from *in vitro* model studies. Cranberry compounds may act against cancers by either inhibiting oxidative stress or other pathways. Over the past 10 years, numerous *in vitro* studies have addressed the effects of cranberry and its constituents against tumor cell proliferation and its possible mechanisms of action. Several recent reports of *in vivo* studies have lent support to the tumor-inhibiting potential of cranberry; these studies are currently under discussion. A summary of *in vitro* and *in vivo* studies on the anticancer properties of cranberry, the constituents contributing to these properties, and some possible mechanisms of action are found in Katsargyris et al (Katsargyris et al, 2012).

Water-soluble cranberry polyphenolic extracts from commercial cranberry powder were found to inhibit proliferation of several human tumor cell lines (Seeram et al, 2004) including two oral (CAL27 and KB), four colon (HT-29, HCT-116, SW480, and SW620), and three prostate (RWPE-1, RWPE-2, and 22Rv1) cancer cell lines. Although both anthocyanidin and proanthocyanidin subfractions inhibited proliferation, the total polyphenolic extract was found to be more effective in this study. The primary polyphenolic constituents of cranberry are flavonols, anthocyanidins, and proanthocyanidins.

Several individual constituents of cranberry may have anti-inflammatory properties. Inhibition of COX activity by anthocyanidins, including those found in cranberry, has been previously reported (Seeram et al, 2001). On the other hand, for many years, scientists and health practitioners believed that the antibacterial effects of cranberry juice were due to the acidification of urine by the hippuric acid produced by the metabolism of the quinic acid in cranberries. However, this claim was never substantiated. Studies correlating urinary pH with cranberry juice consumption show either no significant change in urine acidity or only a slight reduction in pH, which is insufficient to cause a bacteriostatic effect (Howell, 2002). Data from clinical trials suggests that extracts of cranberry or cranberry juice reduce the bacterial load of *E. coli* and also suppress the inflammatory symptoms induced by *E. coli* infections. A methanol extract prepared from 10 Kg of dehydrated cranberries did not directly inhibit the growth of *E coli* strains ATCC 700336 or ATCC 25922 in concentrations of up to 256 μg/mL *in vitro* (Huang et al, 2009).

Drug interactions with cranberry juice might be related to the fact that the juice is rich in flavonol glycosides, anthocyanins, proanthocyanidins, and organic and phenolic acids. Izzo described a total of eight cases of interaction between cranberry juice and warfarin, leading to changes in international normalized ratio (INR) values and bleeding (Izzo, 2005). The mechanism behind this interaction might be the inhibition by cranberry flavonoids of

CYP3A4 and/or CYP2C9 enzymes, which are responsible for warfarin metabolism (Pham and Pham, 2007).

It has also been shown that cranberry juice inhibits diclofenac metabolism in human liver microsomes, but this has not been demonstrated clinically in human subjects (Ushijima and others, 2009). Cranberry juice may increase the bioavailability of CYP3A4 substrates (e.g., calcium antagonists or calcineurin inhibitors) as discussed by Van den Bout-Van den Beukel et al, (2006). Uesawa and Mohri have demonstrated that nifedipine metabolism in rat intestinal and human hepatic microsomes are inhibited by preincubation with cranberry juice. Furthermore, cranberry juice increased the nifedipine concentration in rat plasma. These findings suggest that cranberry juice might affect the plasma concentration of nifedipine in humans as well (Uesawa and Mohri, 2006).

Pomegranate (*Punica granatum*)

Pomegranate is commonly eaten around the world and has been used in folk medicine for a wide variety of therapeutic purposes (Shabtay and others, 2008). The rind of the fruit and the bark of the pomegranate tree are used as traditional remedies against diarrhea, dysentery and intestinal parasites (Jindal and Sharma, 2004). The seeds and juice are considered tonics for the heart and throat. The astringent qualities of the flower juice, rind and tree bark are considered valuable for a variety of purposes, including stopping nose and gum bleeds, toning skin (after blending with mustard oil), firming-up sagging breasts, and treating hemorrhoids. Pomegranate is a rich source of several chemicals such as pectin, tannins, flavonoids, and anthocyanins.

Pomegranate juice (of specific fruit strains) is also used as eyedrops, since it is believed to slow the development of cataracts. Pomegranate has been used as a contraceptive and abortifacient, by means of consuming the seeds or rind, as well as by using the rind as a vaginal suppository. This practice is recorded in ancient Indian literature, medieval sources, and in modern folk medicine (Murali, 2002). Pomegranate aril juice provides about 16% of an adult's daily vitamin C requirement per 100 mL serving and is a good source of vitamin B_5 (pantothenic acid), potassium, and natural phenols such as ellagitannins and flavonoids. Pomegranates are listed as having a high fiber content in some nutritional value charts. That fiber, however, is entirely contained in the edible seeds that also supply unsaturated oils. People who choose to discard the seeds forfeit the nutritional benefits conveyed by the seed fiber, oils and micronutrients (Schubert et al, 1999).

Preliminary laboratory research and clinical trials show that the juice of the pomegranate may be effective in reducing heart disease risk factors such as LDL oxidation, macrophage oxidative status, and foam cell formation (Aviram and others, 2004). In a limited study of hypertensive patients, consumption of pomegranate juice for two weeks reduced systolic blood pressure by inhibiting the serum angiotensin-converting enzyme (Aviram and Dornfeld, 2001). Recent research has shown that pomegranate extracts selectively inhibit the growth of breast, prostate, colon and lung cancer cells in culture. In pre-clinical animal studies, oral consumption of pomegranate extract inhibited growth of lung, skin, colon, and prostate tumors. An initial phase II clinical trial of pomegranate juice in patients with prostate cancer reported significant prolongation of prostate-specific antigen doubling time (Mustafa and others, 2009).

It has been reported that pomegranate juice influenced the pharmacokinetics of carbamazepine in rats by inhibiting enteric CYP3A activity. Such inhibition of the enteric CYP3A activity by a single exposure to pomegranate juice appears to last for approximately 3 days (Hidaka et al, 2005). Nagata et al found that pomegranate juice inhibited human CYP2C9 activity and increased tolbutamide bioavailability in rats (Nagata et al, 2007). Recently, pomegranate juice was shown to potently inhibit the sulfoconjugation of 1-naphthol in Caco-2 cells. Saruwatari and others have suggested that some constituents of pomegranate juice, most probably punicalagin, may impair the metabolic functions of the intestine (specifically sulfoconjugation) and therefore might have effects upon the bioavailability of drugs (Saruwatari et al, 2008).

Mango (*Mangifera indica*)

Mangifera indica, also known as mango, has been an important plant in the Ayurvedic and indigenous medical systems for over 4000 years. Mangoes belong to the genus *Mangifera*, which consists of about 30 species of tropical fruiting trees in the flowering plant family, Anacardiaceae. Ayurveda attributes varied medicinal properties to different parts of the fruit (Shah et al, 2010). Previous studies of mango fruit have shown that it contains various classes of polyphenols, terpenoids, carotenoids, and ascorbic acid (Kim et al, 2011). Preliminary phytochemical screening revealed the presence of flavonoids, including quercetin and glycosylated xanthones such as mangiferin (Berardini et al, 2005). Quercetin has been shown to possess antioxidant, antimicrobial, anti-tumor, antihypertensive, anti-atherosclerosis, and anti-inflammatory properties (Bischoff, 2008). Plants have been screened for antimycobacterial activity given that some of them are known to have this property. Furthermore, some of the screened plants are used in traditional treatments for tuberculosis and other respiratory ailments (Chimponda and Mukanganyama, 2010).

In an analgesic test, *M. indica L.* extract provided protection at doses of 250 and 500 mg/Kg body weight in mice. The bark of *M. indica L.* has, therefore, a significant analgesic and cytotoxic activity and can be a choice in the treatment of pain and tumors (Khan and Islam, 2012). The significant cytotoxic activities of stem bark mango extract have been tested against the MCF 7, MDA-MB-435 and MDA-N breast cancer cell lines, as well as against a colon cancer cell line (SW-620) and a renal cancer cell line (786-0) (Shah et al, 2010). The ethanol/water (1:1) extract of dried aerial parts of mango administered intraperitoneally to mice at a dose of 250 mg/kg was inactive on Leuk-P388. In vitro, dose- and time-dependent administration of mangiferin inhibited the proliferation of K562 leukemia cells and induced apoptosis in the K563 cells line, probably through the down-regulation of BCR/ABL gene expression. These results suggest that mangiferin has a potential as a naturally-occurring chemopreventive agent (Shah et al, 2010). Mangiferin also mediates the downregulation of NF-$_{\kappa}$B, suppresses NF-$_{\kappa}$B activation induced by inflammatory agents including tumor nuclear factor (TNF), increases intracellular glutathione (GSH) levels, and potentiates chemotherapeutic agent-mediated cell death; this suggests a possible role in combination cancer therapy (Knodler and others, 2008).

In glucose-loaded normal rats, mangiferin induces a significant improvement in oral glucose tolerance but does not alter basal plasma glucose levels. These studies also showed that mangiferin (10 and 20 mg/kg, i.p.) exhibits potent anti-diabetic, antihyperlipidemic, antiatherogenic and antioxidant properties without causing hypoglycemia; mangiferin would then offer therapeutic benefits in the management of diabetes mellitus and diabetic complications associated with abnormalities in lipid profiles (Shah et al, 2010).

The anti-helminthic, antiplasmodial and antifungal activities of mango have been experimentally researched. The stem bark extract had a substantial effect on *P. falciparum* in vitro, with a growth inhibition of 50.4% at 20 μg/mL (Shah et al, 2010). Mangiferin at 100 mg/Kg had a similar inhibitory activity on *Cryptosporidium parvum* than the same dose (100 mg/Kg) of an active drug (paromomycin) in a neonatal mouse model (Shah et al, 2010). The *in vitro* effect of mangiferin was studied against *Herpes simplex* virus type 2; mangiferin does not directly inactivate HSV-2 but it inhibits a late event in HSV-2 replication. *In vitro* mangiferin was also able to inhibit HSV-1 virus replication in cells and antagonize the cytopathic effects of HIV (Engels et al, 2009).

In a series of studies, Rodeiro et al have shown the effects of mango on drug metabolizing enzymes and drug transporters (Rodeiro et al, 2009). They found that exposure of hepatocytes to mango extract produced a significant reduction (60%) in 7-methoxyresorufin-O-demethylase (MROD; CYP1A2) activity and an increase (50%) in 7-penthoxyresorufin-O-depentylase (PROD; CYP2B1) activity. This group also studied the effect of mangiferin on CYP enzymes and found that mangiferin reduced the activities of five P450s: POD (CYP1A2), midazolam 1'-hydroxylation (M1OH; CYP3A1), diclofenac 4'-hydroxylation (D4OH; CYP2C6), S-mephenytoin 4'-hydroxylation (SM4OH), and chlorzoxazone 6-hydroxyaltion (C6OH; CYP2E1). Recently, mango and mango-derived polyphenols have been shown to potentially affect the activity of the multidrug transporter P-gp ABCB1 (Chieli and others, 2009). These findings suggest that mango and its components inhibit the major human P450 enzymes involved in drug metabolism and some transporters. The potential for drug interactions with mango fruit should therefore be considered.

Guava (*Psidium guajava L*)

Guava is an important food crop and medicinal plant in tropical and subtropical countries; it is widely used as food and in folk medicine around the world (De Wet et al, 2010). *P. guajava* is a small medicinal tree that is native to South America. All parts of this tree, including fruits, leaves, bark, and roots, have been used for treating stomachache and diarrhea in many countries. A number of metabolites such as phenolics, flavonoid, carotenoid, terpenoid and triterpene have been found in this fruit. Extracts and metabolites of this plant, particularly those from the leaves and fruit, possess useful pharmacological activities (Quian and Nihorimbere, 2004). The extract of the whole guava plant excluding the roots is reportedly devoid of any antibacterial, antifungal, antiviral, anti-fertility, hypoglycemic, diuretic, cough sedative, antihypertensive, antiobesity or anti-inflammatory activities (Ojewole, 2006; Nair and Chanda, 2007). *P. guajava* leaf extract is known to improve symptoms of allergic disease and have potential anti-inflammatory activities (Keiichi et al, 2010). Guava is rich in dietary fiber associated with natural antioxidant compounds. The fruit contains a high percentage of vitamin C, carotene, vitamin B1, B2, B6, and pectin. Recently, the ripe fruit peel has been found to possess hyperglycemic activity in diabetic patients. Antidiabetic activity based on a higher concentration of Mg in the raw fruit peel of P. guajava has also been found. Guava leaves inhibited the increase of plasma sugar level in alloxan-induced diabetic rats during glucose tolerance testing. Flavonoid glycosides such as strictinin, isostrictinin, and pedunculagin are effective constituents that have been used in the clinical treatment of diabetes to improve insulin sensitivity (Perera and Li, 2012).

A recent study showed that *P. guajava* aqueous extract had antibacterial activity against *Salmonella typhi* and *Klebsiella pneumoniae*, but no effect on the growth of *Escherichia coli*, *Staphylococcus aureus*, and *Streptococcus fecalis* organisms (Sunagawa et al, 2004). The antidiarrheal activity of the aqueous extract of the leaves has been studied and various mechanisms of action (namely, antimicrobial activity, antispasmodic activity, inhibition of acetylcholine release, inhibition of bacterial colonization, and production/action of bacterial enterotoxins) have been proposed. Anti-giardial and anti-rotaviral studies with *P. guajava* leaves from different geographical locations have been previously reported. Because of known differences in the biological efficacy of medicinal plants due to differences in geographical locations, we screened an Indian variety of *P. guajava* leaves for their *in vitro* anti-giardial and anti-rotaviral activity (Barbalho et al, 2012). A hot aqueous extract (decoction) was used for the assays based on its field use as revealed during an ethnobotanical survey carried out by us. The decoction affected the viability of *G. lamblia* trophozoites at 5% and 10% concentrations, and decreased cell death in virus infected MA-104 cells at 5% and 10% concentrations, indicating antirotaviral activity. The decoction at these concentrations was not cytotoxic to MA-104 cells. Results of the study thus showed that the decoction of *P. guajava* leaves has antigiardial and antirotaviral activity (Tannaz et al, 2011).

There is only one report about the effect of guava extracts on drug transport: guava extract showed a potent inhibitory effect on P-gp mediated efflux in Caco-2 cells. It was also found to inhibit efflux transport from serosal to mucosal surfaces in the rat ileum (Junyaprasert et al, 2006). This means that guava could interact with P-gp substrates such as digoxin, fexofenadine, indinavir, vincristine, colchicine, topotecan, and paclitaxel in the small intestine. For this reason, this fruit should be consumed with caution by patients taking medicines.

Berries (*Rubus Spp.*)

Like other fruits, raspberries contain micro- and macronutrients such as vitamins, minerals, and fiber. Their biological properties, however, have been largely attributed to high levels of various phenolic compounds, as well as the interactive synergies among their natural phytochemical components (e.g., ellagic acid, quercetin, gallic acid, anthocyanidins, cyanidins, pelargonidins, catechins, kaempferol and salicylic acid). Berries also contain vitamins A, C, E and folic acid, calcium, and selenium (Tian et al, 2005a). They have been shown to have a positive impact on several chronic conditions, including obesity, cancer, and cardiovascular and neurodegenerative diseases (Brownmiller et al, 2008).

Cyanidin and delphinidin are potent inhibitors of epidermal growth factor receptor (EGFR) resulting in a significant decrease in the proliferation of A431 human vulva carcinoma cells, which over express EGFR (Meiers et al, 2001). In addition, cyanidin reduced the growth of human colon cancer cell lines HT29 and HCT116, and cyanidin 3-*O*-beta-D glucoside-induced apoptosis; it also inhibited the growth of human lymphoid leukemia Molt 4B cells (Katsuzaki et al, 2003). An ethanol extract of black raspberry was shown to suppress cell proliferation, inhibit the translation of VEGF, suppress inducible nitric oxide synthase activity and induce apoptosis in human oral squamous cell carcinoma cells (Rodrigo et al, 2006). This extract was also shown to inhibit the growth of malignant human oral cells by targeting cell cycle regulatory proteins (Han et al, 2005). Hecht and others found that ethanolic and water extracts of black raspberry inhibit benzo(*a*)pyrenediol-epoxide induction of AP-1 and NF-κB in mouse epidermal JB6 cells; they identified the most active constituents of the subfractions to be the anthocyanidins cyanidin-3-*O*-glucoside, cyanidin-3-*O*-rutinoside and cyanidin-3-*O*xylosylrutinoside (Hecht et al, 2006). Ethanolic extract and the anthocyanidins (cyanidin-3-O-glucoside and cyanidin-3-Orutinoside) selectively caused significant growth inhibition and induction of apoptosis in a highly tumorigenic cell line (RE-149 DHD) but not in a weakly tumorigenic line (RE-149).

Black raspberries have a selective effect on the growth and apoptosis of highly tumorigenic rat esophageal epithelial cells *in vitro*; this may be due to preferential uptake and retention of its component anthocyanidins, which may also be responsible for the greater inhibitory effects of freeze-dried whole berries on tumor cells *in vivo* (Zikri et al, 2009). On the other hand, in studies using a rat model of nitrosamine-induced esophageal squamous cell carcinoma, black raspberries were found to reduce proliferation, inflammation and angiogenesis, and stimulate apoptosis and differentiation of premalignant cells and tissues, resulting in reduced tumor development. Genes associated with these cellular functions were also protectively modulated by black raspberry diets (Stoner et al, 2007). Clinical trials have determined the safety/tolerability of freeze-dried black raspberries and measured specific anthocyanidins-cyanidin-3-glucoside, cyanidin-3-sambubioside, cyanidin-3-rutinoside, and cyanidin-3-xylosylrutinoside, as well as ellagic acid, in plasma and urine. It was found that 45 g of freeze-dried black raspberries a day are well tolerated and result in a quantifiable increase in anthocyanidins and ellagic acid in plasma and urine (Stoner et al, 2007). The effects of black raspberries on biomarkers of tumor development in the human colon and rectum, including demethylation of relevant tumor suppressor genes, cell proliferation, apoptosis, angiogenesis and expression of Wnt pathway genes has also been evaluated. Black raspberry has the ability to demethylate tumor suppressor genes like *SFRP2, SFRP5*, and *WIF1* and modulate other biomarkers of tumor development in the human colon and rectum. While demethylation of genes did not occur in colorectal tissues from all treated patients, the positive results with the secondary endpoints suggest that additional studies of black raspberry for the prevention of colorectal cancer in humans are warranted (Li-Shu et al, 2011).

Raspberry or raspberry constituents have also antioxidant and anti-inflammatory properties, and inhibit cancer cell growth (Juranic and Zixack, 2005; Seeram et al, 2006). Black raspberries (*Rubus coreanus)* have been called the "king of berries" for their superior health benefits, whereas black mulberry (*Morus nigra)* is most commonly used for its antioxidants properties and for its high bioactive content of phenolics, anthocyanins, and gallic acid.

Traditional herbal medicine has been used for centuries to treat various chronic diseases. The *Morus* species has been traditionally employed as an anti-diabetic herbal medication. The mulberry belongs to the genus *Morus* of the family Moraceae. There are 24 species of *Morus* and one subspecies, with at least 100 known varieties. Biologically active secondary metabolites such as alkaloids, flavonoids, glycoproteins, benzofurans etc. have been isolated and identified across different *Morus* species (Abd El-Mawla et al, 2011). *Morus nigra* is widely used in Brazilian folk medicine for diabetes treatment. The effect of *Morus nigra* aqueous extract treatment on maternal lipid and oxidative stress profile, reproductive outcomes, and also fetal anomaly incidence from diabetic and non-diabetic rats has been evaluated. After treatment with *Morus nigra* extract, non-diabetic and diabetic rats presented no glycemic changes. Fetuses from diabetic dams were small for pregnancy age regardless of *Morus nigra* treatment. Treatment caused a reduction in MDA, cholesterol, triglycerides and VLDL levels in diabetic dams, and decreased placental index and weight as compared to the diabetic group. The fetuses from diabetic rats treated with *Morus nigra* extract had a lower frequency of skeletal and visceral anomalies as compared to the diabetic group. Thus, *Morus nigra* leaf aqueous extract failed to control hyperglycemia in diabetic rats. However, *Morus nigra* treatment had an antioxidant effect that helped reduce the incidence of internal anomalies in offspring of diabetic dams (Volpato et al, 2011).

It has been shown that black raspberry and black mulberry are able to inhibit the human CYP3A-catalyzed midazolam 1-hydroxylation activity in liver microsomes, and the inhibitory effects are somewhat greater than those of pomegranate (Hidaka et al, 2005; Kim et al, 2006). It has also been reported that black mulberry extract potently inhibits OATP-B function at concentrations that seem to be physiologically relevant *in vitro* (Satoh et al, 2005). These results suggest that black raspberry and black mulberry may decrease the plasma concentrations of concomitantly ingested OATP-B substrate drugs or increase the plasma concentration levels of concomitantly ingested CYP3A-substrate drugs. *In vivo* studies on the interaction between black mulberry and black raspberry and CYP3A substrates are needed to determine whether inhibition of CYP3A activity by fruit juices is clinically relevant.

Apple (*Malus domestica*)

Apple is one of those fruits that can play a role in decreasing the risk of chronic diseases due to its fiber content and chemical components such as flavonoids, polyphenols and carotenoids. Its importance can be explained by different factors, including their market availability throughout the year in a variety of forms and their reputation as a healthy food. The concentration of these phytochemicals may depend on many factors, such as cultivar, harvest, storage and processing of the fruit. The concentration of phytochemicals in the apple peel also varies greatly from that in the apple flesh. Some of the best-studied antioxidant compounds in apples include quercetin-3-galactoside, quercetin-3-glucoside, quercetin-3-rhamnoside, catechin, epicatechin, procyanidin, cyanidin-3-galactoside, coumaric acid, chlorogenic acid, gallic acid, and phloridzin (Guyot et al, 2003).

Researchers recently examined the average concentrations of the main phenolic compounds in six cultivars of apples. They found that average phenolic concentrations among the six cultivars were: quercetin glycosides, 13.2 mg/100 g fruit; vitamin C, 12.8 mg/100 g fruit; procyanidin B, 9.35 mg/100g fruit; chlorogenic acid, 9.02 mg/100 g fruit; epicatechin, 8.65 mg/100 g fruit; and phloretin glycosides, 5.59 mg/100 g fruit (Lee et al, 2003). Although freshly prepared juice is a more potent antiviral agent than the commercial product, it loses potency more quickly during storage. This instability may be due to oxidation of the tannins, which reduces their effective reaction to viruses. Freshly prepared apple juice was particularly effective as an antiviral, but its activity declined more readily than that of commercial juice in response to heat and storage. The component responsible for the activity was located both in the flesh and skin (Konowalchuk, 1978).

Apple and its products contain high amounts of polyphenols, which show diverse biological activities and may contribute to beneficial health effects such as protecting the intestine against inflammation due to chronic inflammatory bowel diseases (Gerhauser, 2008). Overall, animal studies and *in vitro* research have thrown light on the mechanisms by which apples may help prevent chronic disease. First, the strong antioxidant activity of apples may help prevent lipid and DNA oxidation. Cancer cell culture research has demonstrated that apples inhibit cell proliferation *in vitro*, which may contribute to the association of apple intake with decreased cancer risk. Apples significantly lowered lipid oxidation both in humans and rats, and lowered cholesterol in humans. These effects, which may be attributed to both the phenolics and the dietary fiber found in apples, may partially explain the inverse association of apple intake and risk of cardiovascular disease (Boyer, 2004). Epidemiological studies have linked the consumption of apples with reduced risk of some cancers, cardiovascular disease, asthma, diabetes, lipid oxidation, cholesterol, and atherosclerosis progression. In addition, quercetin has shown significant anti-inflammatory activity (Mahbubeh et al, 2009).

Apple juice markedly reduces the plasma concentrations and renin inhibiting effect of aliskiren, probably by reducing its absorption in the gastrointestinal tract. These interactions may be caused by the inhibition of the OATP2B1-mediated intestinal absorption of aliskiren by the constituents in apple juice. Pohl and others found that apple juice extract inhibits CYP1A1 at levels of CYP1A1 mRNA, protein, and enzymatic activity (Pohl et al, 2006).

On the other hand, it has also been reported that apple juice and its constituents can interact with members of the OATP transporter family (OATP-1, OATP-3 and NTCP) by reducing their activities *in vitro*. The functional consequence of such an interaction was a significant reduction in the oral bioavailability of fexofenadine in human plasma levels, possibly by preferential direct inhibition of intestinal OATP activity (Dresser et al, 2002). These findings suggest that apple might interact with OATP substrates (e.g., estrone-3-sulfate, deltorphin II, fexofenadine, vasopressin, and rosuvastatin).

Papaya (*Carican papaya L.*)

Carica papaya L. leaves and seeds are known to contain proteolytic enzymes (papain, hymopapain), alkaloids (carpain, carpasemine), sulfurous compounds (benzyl isothiocyanate), flavonoids, triterpenes, organic acids and oils (Cowan, 1999). Papaya is known worldwide for its flavor and nutritional properties. Ethno-botanical survey showed that papaya was commonly used in traditional medicine for the treatment of various human diseases including abdominal discomfort, pain, malaria, diabetes, obesity, infections and oral drug poisonings (Adeneyeand Olagunju, 2009). Papaya leaves and seeds are known to contain proteolytic enzymes (papain, chymopapain), alkaloids (carpain, carpasemine), sulfurous compounds (benzyl isothiocyanate), flavonoids, tannins, triterpenes, anhocyanins organic acids and oils (Mahmood et al, 2005).

Extracts from different papaya tissues have been shown to be bioactive. Aqueous extracts of the leaves and seeds are known to have antifungal activity against *Colletotrichum gloeosporioides* (Bautista-Baños et al, 2002), and aqueous and organic extracts of seeds have anti-helminthic activity against *Caenorhabditis elegans* (Adebiyi and Adaikan, 2005). Alcoholic extracts of the epicarp, endocarp, roots and seeds from ripe and unripe papaya fruit have anti-diarrheic, anti-dysenteric and antibacterial properties (Doughari et al, 2007). The leaves of *Carica papaya* are used for treating malaria and gastritis (Simbo, 2010).

Aqueous extracts of green papaya epicarp and ripe papaya epicarp have been applied on induced wounds in mice (Anuar et al, 2008). *In vitro* studies using isolated rabbit arteries concluded that the juice of *C. papaya* probably contains anti-hypertensive agent(s), mainly exhibiting alpha-adrenoceptor activity (Eno et al, 2000). Studies about the toxicological and antioxidant potential of dried *C. papaya* juice *in vitro* and *in vivo* indicate its safety and antioxidative stress potential, which was found to be comparable to the standard antioxidant compound alpha-tocopherol (Mehdipour et al, 2006). Lethality studies showed a dose-mortality relationship with an LD_{50} of 325.2 mg/Kg in mice administered i.p. injection (Eno and others, 2000).

Lycopene is another compound found in papaya. It has been linked to antioxidation, antiproliferation (growth inhibition, cell cycle arrest, apoptosis), anti-angiogenesis, anti-inflammation, and immunomodulation in prostate, lung, breast, gastric, liver, pancreas, colorectal, head, neck and skin cancer. Synergistic interaction with genistein, adriamycin, and cisplatin was also observed. Different targets include cyclin D1, Bcl-2, Bcl-xL, AKT, BAD, NF-KB, MMP-9, Sp-1, and IGF-BP3 (Amin et al, 2009).

A number of models have reported anti-fertility effects for papaya. Oral administration in langur monkeys (*Presbytis entellus entellus*) showed that the extract gradually decreased sperm concentration in days 30-60 of treatment, with a total inhibition of sperm motility, a decrease in sperm viability, and an increase in sperm abnormality. Azoospermia was observed after day 90 of treatment and continued during the whole treatment period (Lohiya et al, 1992). Furthermore, Lohiya et al, (1992) have reported the contraceptive effect of a crude chloroform extract of papaya seeds in male albino rats.

Papaya fruit is "generally recognized as safe" (GRAS) by the US Food and Drug Administration (Rasiya and Nayar, 2011). The traditional use of papaya to treat many diseases, especially skin conditions, and cautions against its consumption during pregnancy prompted us to determine whether papaya extracts from both green and ripe fruits improved wound healing and produced fetal toxicity. Experiments addressed the abortifacient potential of aqueous extract of *Carica papaya* seeds in female Sprague-Dawley rats and found that low doses of aqueous crude extract did not adversely affect prenatal development. The altered toxicological profile indicates that abortifacient properties involve a high-dose side effect (Oderinde et al, 2002).

It has been investigated the ability of some fruits to inhibit the midazolam 1-hydroxylase activity of CYP3A in human liver microsomes. They found that papaya produced an inhibition of CYP3A activity in human microsomes (Hidaka et al, 2004). So far, there has been no clinical report suggesting the adverse food-drug interaction caused by the intake of papaya. Accordingly, the inhibition of CYP3A by papaya may not be observed *in vivo*. However, the results obtained by others raised the hypothesis that the extracts of the papaya were capable of altering pharmacokinetics of therapeutic drugs co-administered via CYP3A inhibition, similar to the case with grapefruit. Thus, the possibility of the adverse food-drug interaction by papaya with medicine through CYP3A metabolism should be examined *in vivo*. The empirical experience of the wide use of fermented papaya preparation (FPP), especially by elderly people, has indicated an unknown collateral effect, i.e., blood sugar level dropping signs especially in the afternoon. Those findings have been corroborated by a clinical study where have been shown that FPP use can induce a significant decrease in plasma sugar levels in both healthy subjects and type 2 diabetic patients (Danese et al, 2006). Therefore, a potential drug-food interaction could suffer those patients that have taken papaya and antidiabetic therapy.

Tomatoes (*Lycopersicon esculentum*)

Tomatoes (*Lycopersicon esculentum)* and tomato-based products are a source of important nutrients and contain numerous phytochemicals, such as carotenoids, that may influence health (carotenoids such as phytofluene, phytoene, neurosporene, γ-carotene, and ζ-carotene) (Tan et al, 2010). Tomatoes are also a source of a vast array of flavonols (e.g., quercetin and kaempferol), phytosterols, and phenylpropanoids (Ellinger et al, 2006). Lycopene is the most important carotenoid present in tomatoes and tomato products, and their dietary intake has been linked to a decreased risk of chronic illnesses such as cancer and cardiovascular disease (Riccioni et al, 2008; Waliszewski and Blasco, 2010).

Many epidemiological and experimental studies have shown lycopene's protection against oxidative damage (Mourvaki et al, 2005). Tomatoes and tomato-derived products are the richest source of lycopene in the human diet (Mourvaki et al, 2005). Tomato consumption has been associated with a decreased risk of breast cancer (Zhang et al, 2009), head and neck cancers (Freedman et al, 2008) and might also offer strong protection against neurodegenerative diseases (Rao and Balachandran, 2002). Tomatoes and tomato sauces and puree are said to help reduce urinary tract disease symptoms and may have anti-cancer properties (Polívková et al, 2010). Tomato consumption might be beneficial for reducing the cardiovascular risk associated with type 2 diabetes (Shidfar et al, 2011). Tomato (*Lycopersicon esculentum*) plants emit a blend of volatile organic compounds that mainly consists of terpenes, two tomato monoterpene synthases, LeMTS1, and LeMTS2 (Van Schie et al, 2007).

Leaves, stems, and the unripe fruit of the tomato plant contain small amounts of the toxic alkaloid tomatine. They also contain solanine, a toxic alkaloid found in potato leaves and other plants in the nightshade family. Use of tomato leaves in tea (tisane) has been responsible for at least one death. However, levels of tomatine in foliage and green fruit are generally too small to be dangerous unless large amounts are consumed, for example, as greens. Small amounts of tomato foliage are sometimes used for flavoring without ill effect, and the green fruit is sometimes used for cooking, particularly as fried green tomatoes. Compared to potatoes, the amount of solanine in green or ripe tomatoes is low. As with potatoes, solanine poisoning resulting from dosages several times normal human consumption has been demonstrated. However, actual cases of poisoning resulting from excessive consumption of potatoes that have a high concentration of solanine are rare (Barceloux, 2009).

Studies performed on human recombinant CYP1 showed that lycopene inhibits CYP1A1 and CYP1B1. Lycopene has also been shown to slightly reduce the induction of ethoxyresorufin-*O*-deethylase activity by 20% by DMBA in MCF-7 cells (Wang and Leung, 2010). It appears to inhibit bioactivation enzymes and induce detoxifying enzymes. It has been suggested that lycopene might have a potential advantage over other phytochemicals by facilitating the elimination of genotoxic chemicals and their metabolites (Wang and Leung, 2010). Recent *in vitro* studies suggest that high dose lycopene supplementation increases hepatic cytochrome P4502E1 protein and inflammation in alcohol-fed rats (Veeramachaneni et al, 2008).

Carrots (*Daucus carrota*)

Carrots are widely consumed as food. The active components of carrots, which include beta-carotene and panaxynol have been studied by many researchers (Surles et al, 2004). The carotenoids present in *Daucuscarota L.* are also antioxidants capable of reducing the risk of chronic diseases (cardiovascular diseases, cancer etc.). In Poland, carrots are grown on a large scale and have great economic importance (Mech-Nowak et al, 2012). Anthocyanidins are the main antioxidants in purple-yellow and purple-orange carrots; chlorogenic acid is a major antioxidant in all carrots. Carotenoids do not contribute to total antioxidant capacity, but correlate with the antioxidant capacity of hydrophobic extracts. Purple-yellow carrots have the highest antioxidant capacity, followed by purple-orange carrots; the other carrots did not significantly differ.

Evidence suggests that higher plasma carotenoid concentrations can protect from breast cancer recurrence. A simple randomized carrot juice intervention study was designed to test the hypothesis that daily intake of 8 ounces of fresh Beta Sweet (anthocyanidin-rich) or Balero orange carrot juice would increase plasma total carotenoid concentrations to levels previously shown to be associated with reduced breast cancer recurrence. It was hypothesized that regular carrot juice intake would be associated with reductions in oxidative stress and inflammation. The results suggest daily intake of fresh carrot juice is a simple and effective approach to increasing total plasma carotenoids and, in turn, reducing oxidative stress (but not inflammatory markers) in women previously treated for breast cancer (Butalla et al, 2012).

Drinking carrot juice did not affect plasma cholesterol, triglycerides, Apo A, Apo B, LDL, HDL, body fat percentage, insulin, leptin, interleukin-1α, or C-reactive protein; carrot juice decreased systolic pressure, but did not influence diastolic pressure. Carrot juice significantly increased total plasma antioxidant capacity and decreased plasma malondialdehyde production; it may also protect the cardiovascular system by increasing total antioxidant status and decreasing lipid peroxidation independent of any of the cardiovascular risk markers (Potter et al, 2011).

Carrots induce phenol sulfotransferase activity (Yeh and Yen, 2005) and decrease CYP1A2 activity (Nugon-Budon et al, 1996). Bradfield and others have reported that a carrot diet increased the activity of ethoxycoumarin O-deethylase ECD activity in a mouse model (Bradfield et al, 1985).

Avocado (*Persea americana*)

Avocado is a rich source of nutrients and phytochemicals. Although some of the individual phytochemicals found in it have been well characterized, many new uncharacterized ones with potential activity are being discovered. Avocado is a good source of bioactive compounds such as monounsaturated fatty acids and sterols (Duester, 2001). These compound classes may be divided into alkanols (also sometimes termed "aliphatic acetogenins"), terpenoid glycosides, various furan ring-containing derivatives, flavonoids, and coumarin. Growing data on the health benefits of avocadoes have led to increased consumption and research on potential health benefits (Whiley and Schaffer, 2002). Phytochemicals extracted from avocado can selectively induce several biological functions (Plaza et al, 2009).

Oberlies et al, isolated 1,2,4-trihydroxyheptadec-16-ene, 1,2,4-trihydroxyheptadec-16-yne, and 1,2,4-trihydroxynonadecane from the unripe fruits of *P. americana*, and found these substances to be moderately cytotoxic when evaluated against a small panel of cancer cell lines (Oberlies et al, 1998). An ethanol-soluble extract of the dried leaves of avocado showed anti-inflammatory activity in a carrageenan-induced edema protocol (Adeyemi et al, 2002). The aqueous extract of *P. americana* (0.01-12.8 mg/mL) produced a concentration-related vasorelaxation response in rat aortic rings with intact endothelium precontracted with noradrenaline. The vasorelaxant action of the aqueous extract of *P. americana* was significantly attenuated in the endothelium-denuded rings (Adeyemi et al, 2002).

The aqueous leaf extract of *P. americana* (800 mg/Kg) significantly inhibits the swelling caused by carrageenan. This effect was similar to that produced by indomethacin during the same time lapse (Adeboye et al, 1999). On the other hand, the aqueous leaf extract of *P. americana* has been reported to have anticonvulsant activity in mice. The effectiveness of the extract in the experimental convulsion paradigm probably suggests that the plant could be used in both *petit* and *grand mal* types of epilepsy (De Almeida et al, 1998). The leaves of *P. americana* have been popularly used to treat diabetes in countries of Latin America and Africa. The hydroalcoholic extract of the leaves of *P. americana* reduces blood glucose levels and improves metabolic state in animals. It possibly acts to regulate glucose uptake in liver and muscles by way of PKB/Akt activation, restoring intracellular energy balance (Lima et al, 2012).

Avocado seeds contain elevated levels of phenolic compounds and exhibit antioxidant properties. Avocado seeds reduced total cholesterol and LDL-C levels, as well as the prediction of the atherogenic index. Therefore, the antioxidant activity of phenolic compounds and dietary fiber in avocado seeds may be responsible for their hypocholesterolemic activity in a hyperlipidemic model of mice (Pahua-Ramos et al, 2012). On the other hand, studies have shown that phytochemicals extracted from the avocado fruit selectively induce cell cycle arrest, inhibit growth, and induce apoptosis in precancerous and cancer cell lines. Our recent studies indicate that phytochemicals extracted with 50% methanol from avocado fruits aid the proliferation of human lymphocyte cells and decrease chromosomal aberrations induced by cyclophosphamide (Paul et al, 2011)

Two papers published in the 1990's reported avocados interact with warfarin, stating that the fruit inhibited the effect of warfarin. They, however, did not establish the cause of such inhibition (Blickstein et al, 1991; Wells et al, 1994).

Nopal (*Opuntia ficus-Indica*)

In Mexico, the *Opuntia ficus indica* cactus (nopal) has been employed since pre-Columbian times as an important dietary and economic component (Betancourt-Dominguez et al, 2006). In Sicilian folk medicine it is used for treating gastric ulcers (Galati et al, 2001). *Opuntia* prickly pads are an important source of nutritional elements like pectin, mucilage and minerals. Several studies have also reported high levels of amino acids, especially proline, taurine and serine (Piga, 2004; Stintzing and Carle, 2005). The presence of total phenolic compounds (free and conjugated) with concentrations of 80-90 mg/100 g dried weight include aromadendrin, taxifolin or dihydroquercetin, isorhamnetin, vitexin, kaempferol, quercetin, and derivatives like myricitin, orientin and some derivatives of pyrone (Stintzing and Carle, 2005). Phenolic compounds from nopal have antioxidant, anti-cancer, anti-inflammatory, analgesic, anti-ulcerogenic, hypoglycemic, hypolipidemic and hypocholesterolemic properties (Guevara-Arauza, 2009). Nopal fruit juice contains many phenol compounds, ascorbic acid, betalains, betacyanins, and a flavonoid fraction that consists mainly of rutin and isorhamnetin derivatives. Hepatoprotection may be related to the flavonoid fraction of the juice, but other compounds, such as vitamin C and betalains, could synergistically counteract many degenerative processes by means of their antioxidant activity (Galati et al, 2005).

Opuntiaficus indica var. *saboten* is used in the manufacture of health foods such as tea, jam and juice. Its fruits and stems have been traditionally used in Asian folk medicine to treat diabetes, hypertension, asthma, burns, edema and indigestion (Ahn, 1998). The extracts of fruits or stems exhibit hypoglycemic (Trejo-Gonzalez et al, 1996), anti-ulcer (Galati et al, 2005), and anti-allergic activities (Lee et al, 2000). In addition, Park and others reported analgesic and anti-inflammatory activity for fruit and stem extracts, and isolated β-sitosterol as an active anti-inflammatory principle from the stem extract (Park and others, 2001). A clinical trial evaluated the acute and chronic effects of nopal in obese, pre-diabetic men and women and found acute blood glucose lowering effects and the long-term safety of the nopal, supporting its traditional use for blood glucose management (Godard et al, 2010).

Nopal dietary fiber is composed of several chemical components that are resistant to digestive enzymes such as cellulose, hemicelluloses, pectin, lignin, gums, *etc.* (Periago et al, 1993). The benefits associated with fiber content are well known, especially for the prevention of illnesses such as diabetes, treatment of gastrointestinal disorders, illnesses associated with low dietary fiber intake, reduction of glucose values in the blood, anti-hyperlipidemic and anti-hypercholesterolemic effects (Feugang et al, 2006). Age-related differences within the species could be attributed to the type of *Opuntia* and the climatic conditions where the plants grow (rain precipitation or available irrigation). Previous studies have shown young prickly pads are rich in calcium and this increases with age (Hernández-Urbiola et al, 2011). Most current literature addresses the usefulness of young pads but does not include data regarding the nutritional and mineral content of prickly pads in their advanced maturity stages (Rodríguez-Garcia et al, 2007).

Betanin isolated from nopal induced the release of cytochrome c into the cytosol poly (ADP) ribose polymerase (PARP) cleavage, as well as down-regulation of Bcl-2 and reduction in cell membrane potentials. Betatin also induced apoptosis in K562 cells through the intrinsic pathway and was mediated by the release of cytochrome c from mitochondria into the cytosol and PARP cleavage. The antiproliferative effects of betanin add further value to the nutritional characteristics of the nopal fruit (Sreekanth et al, 2007). *Opuntia ficus indica* polyphenolic compounds triggered an increase in intracellular [Ca^{2+}] in human Jurkat T-cell lines (Yoon et al, 2004).

The ethanol extracts of the nopal fruit and stem were pharmacologically evaluated. Both extracts inhibited the writhing syndrome induced by acetic acid, indicating that they contain an analgesic effect. Moreover, the extracts suppressed the release of beta-glucuronidase, a lysosomal enzyme in rat neutrophils. It was also noted that the extracts had a protective effect on gastric mucosal layers (Park et al, 1998). Another study has described the antidepressant effect of two flavonoids (kaempferol and quercitrin) isolated from the *Opuntia ficus-indica* var. *saboten* (Park et al, 2010).

Red Pepper (*Capsicum annuum L.*)

Red pepper is fruit used as a spice for enhancing the palatability of food and drug such as counter-irritant on stomach medicines in many countries (Watanabe et al, 2001). The pungent principle of red pepper is a group of compounds called capsaicinoids, which possess a variety of biological properties and provides its spicy flavor. Two major capsaicinoids, dihydrocapsaicin (DHC) and capsaicin (CAP) are responsible for up to 90% of the total pungency of pepper fruits. In traditional medicine, *Capsicum annuum* is used as a fruit stimulant, externally used as rubifacient. It is also useful in used in scarlatina, putrid sore throat, hoarsenes, dispepsia, yellow fever, piles and snakebite (Ishtiaq et al, 2007). Capsaicin (8-methyl-N-vanillyl-6-nonenamide) is a principal component of *Capsicum* fruits. Capsaicin is a known to have antioxidant properties and therefore has been associated with potent anti-mutagenic and anti-carcinogenic activities (Materska and Perucka, 2005).

A survey of Mayan pharmacopoeia revealed that the flesh of *Capsicum* species (*Solanaceae*) was included in a number of herbal remedies for a variety of ailments of possible microbial origin. Using a filter disk assay, plain and heated aqueous extracts from fresh *Capsicum annuum, Capsicum baccatum, Capsicum chinense, Capsicum frutescens, and Capsicum pubescens* varieties were tested for their antimicrobial effects with fifteen bacterial species and one yeast species. Two pungent compounds found in *Capsicum* species (*capsaicin and dihydrocapsaicin*) were also tested for their anti-microbial effects. The plain and heated extracts were found to exhibit varying degrees of inhibition against *Bacillus cereus, Bacillus subtilis, Clostridium sporogenes, Clostridium tetani*, and *Streptococcus pyogenes* (Cichewicz and Thorpe, 1996).

Chemically characterized compounds such as a saponin from pepper (*Capsicum frutescens L*), benzaldehyde, chitosan and 2-deoxy-D-glucose are being studied as natural fungicides. Resin glycosides for pathogen resistance in sweet potato and residues of semi-tropical leguminous plants for nematode control are also being researched (Duke et al, 2010).

Early studies have reported that capsaicin strongly inhibited the constitutive enzymes CYP 2A2, 3A1, 2C11, 2B1, 2B2 and 2C6 (Zhang et al, 1993). There is a report indicating that capsaicin is a substrate of CYP1A2. Pharmacokinetic studies in animals have shown that a single dose of *Capsicum* fruit could affect pharmacokinetic parameters of theophylline, while a repeated dose affected the metabolic pathway of xanthine oxidase (Bouraoui et al, 1995). Therefore, a potential interaction may occur when it is taken along with medicines. Recently, it has been studied that red pepper induces alterations in intestinal brush border fluidity and passive permeability properties, associated with the induction of an increased microvilli length and perimeter, resulting in an increased absorptive surface of the small intestine, providing for an increased bioavailability not only of micronutrients but also of drugs (Prakash and Srinivasan, 2010). Cruz et al have indicated that pepper ingestion reduces oral salicylate bioavailability, likely as a result of the gastrointestinal effects of capsaicin (Cruz et al, 1999).

On the other hand, Imaizumi et al have reported capsaicinoids-induced changes of glucose in rats. Therefore, a possible risk might occur of the interaction between red pepper and hypoglycemic drugs in diabetic patients (Imaizumi et al, 2011).

Broccoli (*Brassica oleracea*) and Cauliflower (*Brassica oleracea*)

These leafy vegetables are unique among the common cruciferous vegetables that contain high levels of the aliphatics glucosinolate and glucoraphanin (Cartea, et al, 2010). Upon hydrolysis, glucoraphanin produces several products that include the bioactive isothiocyanate sulforaphane. Cauliflower compounds present are glucosinolates (thioglucosides) and S-methyl cysteine sulfoxide. These compounds, which are derived in plant tissue by amino acid biosynthesis, show quite different toxicological effects and appear to possess anticarcinogenic properties.

Several epidemiological studies in Asia, the United States and Europe suggest that the consumption of vegetables from the Brassicaceae family, e.g., broccoli and cauliflower, may have protective effects against various types of cancers (Juge et al, 2007). Indole-3-carbinol is the main hydrolysis product of the glucosinolate glucobrassicin and can provide significant protection against cancer in animal models with a variety of chemical carcinogens, as well as in cultured human cancer cells. Initial clinical trials in women have shown that indole-3-carbinol may prove a promising agent against cervical and breast cancers. Also, 3,3'-diindolylmethane, an indole derivative produced in the stomach after the consumption of broccoli and other cruciferous vegetables, has been shown to exert anti-cancer effects in both *in vivo* and *in vitro* models. 3,3'-diindolylmethane (0 -30 μmol/L) inhibited the growth of HT-29 human colon cancer cells in a concentration-dependent fashion (Choi and others, 2009; Russo et al, 2010).

The percentage of isothiocyanate sulforaphanepresent in these vegetables may vary depending on conditions of hydrolysis, food handling, and preparation procedures (Tian et al, 2005b). In animal studies, dietary freeze-dried broccoli was found to offer protection against several cancers (Vasanthi et al, 2009). However, broccoli, cauliflower and their glucosinolate hydrolysis products have been shown to induce phase I and phase II drug-metabolizing enzymes in intact liver cells from both rats and humans. The isothiocyanate sulforaphane decreased the enzyme activities hepatocytes associated with CYP1A1 and 2B1/2, namely ethoxyresorufin-O-deethylase and pentoxyresorufin-O-dealkylase, respectively, in a dose-dependent manner (Anwar-Mohamed and El-Kadi, 2009). An increase in hGSTA1/2 mRNA has been observed in isothiocyanate sulforaphane-treated human hepatocytes, whereas the expression of CYP3A4, the major CYP in the human liver, markedly decreased at both mRNA and activity levels (Fimognari et al, 2008). Conversely, it was recently shown that sulforaphane induces mRNA levels of MRP1 and MRP2 in primary hepatocytes and Caco-2 cells (Harris and Jeffery, 2008). It has been additionally reported that broccoli is able to induce the activity of phenolsulfotransferases (Yeh and Yen, 2005). These results suggest that other vegetables with a high content of isothiocyanates, such as those of the family *Cruciferae* (e.g., cabbage, cauliflower, Brussels sprouts, watercress, broccoli, and kale) and the genus *Raphanus* (radishes and daikons) may have pharmacological and toxicological implications in humans.

Watercress (*Nasturtium officinale*)

Watercress (*Nasturtium officinale*) is another important member of the cruciferous vegetables and an excellent source for glucosinolates and other bioactive phytochemicals (Getahun and Chung, 1999). Watercress is an exceptionally rich dietary source of beta-phenylethylisothiocyanate (PEITC) (Palaniswamy et al, 2003), a compound that inhibits phase I enzymes, which are responsible for the activation of many carcinogens in animals, and induces phase II enzymes, which are associated with enhanced excretion of carcinogens.

Watercress supplementation was associated with reductions in basal DNA damage, in basal plus oxidative purine DNA damage, and in basal DNA damage in response to *ex vivo* hydrogen peroxide challenge. Beneficial changes seen after watercress intervention were greater and more significant in smokers than in non-smokers. Plasma lutein and beta-carotene increased significantly (100% and 33%, respectively) after watercress supplementation. These results support the theory that consumption of watercress can be linked to a reduced risk of cancer via decreased damage to DNA and possible modulation of antioxidant status by increasing carotenoid concentrations (Gill et al, 2007).

Previous studies by Leclercq and others have shown that a single ingestion of watercress inhibits the hydroxylation of chlorzoxazone, an *in vivo* probe for CYP2E1, in healthy volunteers (Leclercq et al, 1998). It has also been shown that watercress is a bifunctional agent with the ability to induce both phase I (CYP450) and II enzymes.

Adding watercress juice to human liver cells induced the activity of CYP4501A and ethoxyresorufin-O-deethylase and NAD(P)H-quinone reductase (Lhoste et al, 2004). According to reports, PEITC also has several anti-carcinogenic effects given that it can inhibit phase I enzymes and/or activate phase II enzymes. Watercress juice can increase the enzymes *SOD* and *GPX* in blood cells *in vitro* and *in vivo* (Hofmann et al, 2008). Isothiocyanates also interact with ATP-binding cassette (ABC) efflux transporters such as P-glycoprotein, MRP1, MRP2 and BCRP, and may influence the pharmacokinetics of substrates of these transporters (Telang et al, 2009). According to current data, watercress and isothiocyanate may have clinical repercussions by inducing changes in the bioavailability of some drugs.

Spinach (*Spinacia oleracea*)

Spinacia oleracea L., commonly called spinach or palang (vernacular name), is an annual plant (occasionally biennial). It is native to central and southwest Asia and may survive winters in temperate regions. The leaves are alternate, simple, ovate to triangular-based, variable in size, about 2-30 cm long and 1-15 cm broad, with larger leaves at the base of the plant and small leaves higher on the flowering stem. It is a favored food in India during the winter season and a dietary powerhouse, full of vitamins and minerals. However, there are no reports yet on the use of mucilage isolated from *S. oleracea L.* leaves (Nayak et al, 2010). Spinach is an important antioxidant vegetable usually consumed after boiling the fresh or frozen leaves (Schirrmacher et al, 2010). Freshly cut spinach leaves contain approximately 1 g of total flavonoids per kilogram, and the occurrence of at least 10 flavonoid glycosides has been reported. These are glucuronides and acylated di- and triglycosides of methylated and methylene dioxide derivatives of 6-oxygenated flavonols (Bergquist et als, 2005). Epidemiological and pre-clinical data supports the nutritional benefits of spinach and the safety of its consumption (Lomnitski et al, 2003). The protective effect of spinach has been evaluated for bioaccessibility and intestinal uptake of heavy metals (Arsenic (As), Cadmium (Cd), and Lead (Pb)) using an *in vitro* digestion model with Caco-2 cells. The leaf of water spinach proved the most effective for decreasing both bioaccessibility and cellular uptake of Cd (Yang et al, 2012). Little is currently known about the *in vivo* effects these compounds have on the bioavailability of xenobiotics the clearance and/or tissue distribution of which is determined by active transport and biotransformation. There are no publications about its effects on drug metabolizing enzymes and drug transporters. Platt et el reported the protective effect of spinach against the genotoxic effects of 2-amino-3-methylimidazo[4,5-f]quinoline (IQ) by interaction with CYP1A2 as a mechanism of anti-genotoxicity. Its high isothiocyanate and flavonoid content demands additional research to evaluate possible nutrient-drug interactions (Platt et al, 2010).

Other Vegetables

Various studies found cabbage leaves to be beneficial for reducing breast engorgement and pain regardless of temperature. However, a meta-analysis concluded that there is no good evidence that topical cabbage leaves were better than no treatment, because engorgement tends to improve over time regardless of treatment (Mangesi and Dowswell, 2010). Unrestricted nursing of the infant may be an important factor in reducing engorgement (Nikodem et al, 1993). Some low-quality evidence indicates that maternal cabbage ingestion might cause colic in their breastfed infants (Lust et al, 1996). A study compared the effectiveness of topical use of alternating cold and hot compresses to frozen cabbage leaves for the treatment of breast engorgement. The study was not randomized or blinded. Each treatment was applied for 30 minutes 3 times daily for 2 days. Outcomes were judged using breast engorgement and pain scales, although it is unclear who did the rating. Both treatments were effective in reducing pain and engorgement. The authors concluded that hot and cold compresses were more effective than cabbage leaves (Arora et al, 2009).

Vegetables such as cabbage, celery, onion and parsley are known to have a high content of polyphenols. It has been reported that polyphenols can potentially affect phase I metabolism either by direct inhibition of phase I enzymes or by regulating the expression of enzyme levels *via* their interactions with regulatory cascades. Several studies have directly and indirectly shown that dietary polyphenols can modulate phase II metabolism (Lambert et al, 2007). In addition, polyphenols have been shown to interact with ABC drug transporters involved in drug resistance and drug absorption, distribution and excretion (Alvarez et al, 2010).

It has been reported that asparagus, celery and eggplant induced significant phenol sulfotransferase-P (PST-P) activity, whereas asparagus, eggplant and potato induced PST-M activity (Yeh and Yen, 2005). Previous results have also reported that a diet supplemented with apiaceous vegetables (dillweed, celery, parsley, parsnip) resulted in a 13-15% decrease in CYP1A2 activity. The authors speculate that furanocumarins present in the apiaceous vegetables were responsible for the inhibitory effects on CYP1A2 (Harris and Jeffery, 2008).

Conclusion

WHO and the Food and Agriculture Organization of the United Nations (FAO) recommend a daily intake of at least 400 grams or five servings of fruits and vegetables to aid in the prevention of chronic illnesses such as heart disease, cancer, diabetes, and obesity. As a consequence, there is an increased global consumer demand for fruits and vegetables, and some consumers purchase organic foods with the understanding that they are healthy. The use of natural products for improving human health has evolved independently in different regions of the world and production, use, attitudes, and regulatory aspects vary globally. Although modern medicine may be available in most countries for the treatment of many chronic degenerative diseases, folk medicine (phytomedicine) has remained popular for historical and cultural reasons. This chapter shows evidence of beneficial influence of fruit, vegetables or their components (phytochemicals) for certain diseases; however, we also show their potential risks for the human population. In addition to all of this, we must account for the fact that we live in an era of very varied life styles. Some people are vegetarians, others take high doses of flavonoids or antioxidants as supplements, some ingest large amounts of bottled water from plastic bottles, or use chlorinated disinfectants. In industrialized countries, fruits and vegetables tend to be subjected to some sort of processing (e.g., refrigeration, acidification, fermentation, and thermal, high pressure, chemical, or physical processing) that might have an effect on the bioactive compounds. All of these factors could have an impact on quality of efficacy and safety of fruit, vegetables or their components. Most of the fruits and vegetables examined in this chapter contain a quite similar phytochemical mix, and just change the proportion of them. *In vitro* models and animal models have shown that many of these agents influence drug metabolizing enzymes and drug transporters. It is possible that other fruits and vegetables could have the same potential for risk and benefits on health, and this should be taken into account. Although the significance of interactions between drugs is widely appreciated, little attention has been given to interactions between drugs and nutrients.

However, the possibility that these effects could occur in humans should not be ignored. Despite its existence and continued use over many centuries, as well as its popularity and extensive use during the last decade, the use of fruit, vegetables or their components in traditional medicine have not been officially recognized in most countries. Consequently, education, training, and research in this area have not received due attention and support. Our knowledge regarding the potential risk and benefits of fruit, vegetables or their components is still limited. Therefore, efforts to elucidate potential risk and benefits should be intensified in order to prevent undesired and harmful clinical consequences.

References

Abd El-Mawla, A.M., Mohamed, K.M., Mostafa, A.M. (2011). Induction of biologically active flavonoids in cell cultures of *Morus nigra* and testing their hypoglycemic efficacy. *Sci. Pharm.*, 79, 951–961.

Adebiyi, A., Adaikan, P. (2005). Modulation of jejunal contractions by extract of *Carica papaya L.* seeds. *Phytother Res.*, 19, 628–632.

Adeboye, J.O., Fajonyomi, M.O., Makinde, J.M., Taiwo, O.B. (1999). A preliminary study on the hypotensive activity of *Persea americana* leaf extracts in anesthetized normotensive rats. *Fitoterapia*, 70, 15–20.

Adeneye, A.A., Olagunju, I.A. (2009). Preliminary hypoglycemic and hypolipidemic activities of the aqueous seed extract of *Carica papaya Linn.* in Wistar rats. *Biology and Medicine*, 1, 1–10.

Adeyemi, O.O., Okpo, S.O., Ogunti, O.O. (2002). Analgesic and anti-inflammatory effects of the aqueous extract of leaves of *Persea americana* Mill (Lauraceae). *Fitoterapia*, 73, 375–380.

Agostini, F., Bertussi, R.A., Agostini, G., Atti Dos Santos, A.C., Rossato, M., Vanderlinde, R. (2012). Supercritical extraction from vinification residues: fatty acids, alpha-tocopherol, and phenolic compounds in the oil seeds from different varieties of grape. *The Scientific World Journal*, 9 pages.

Ahn, D. K. (1998). Illustrated book of Korean medicinal herbs. Kyohak Publishing Co, Korea seoul. pp 497.

Al-Habib, A., Al-Saleh, E., Safer, A. M., Afzal, M. (2010). Bactericidal effect of grape seed extract on methicillin-resistant *Staphylococcus aureus* (MRSA). *J. Toxicol. Sci.*, 35, 357-364.

Alvarez, A.I., Real, R., Perez, M., Mendoza, G., Prieto, J.G., Merino, G. (2010). Modulation of the activity of ABC transporters (P-glycoprotein, MRP2, BCRP) by flavonoids and drug response. *Pharm. Sci.*, 99, 598-617.

Amin, A.R., Kucuk, O., Khuri, F.R., Shin, D.M. (2009). Perspectives for cancer prevention with natural compounds. *J. Clin. Oncol.*, 27, 2712-2725.

Amory, J.K., Amory, D.W. (2005). Oral erythromycin and the risk of sudden death. *N. Engl. J. Med.*, 352, 301-304.

Anuar, N.S., Zahari, S.S., Taib, I.A., Rahman, M.T. (2008). Effect of green and ripe *Carica papaya* epicarp extracts on wound healing and during pregnancy. *Food Chem. Toxicol.*, 46, 2384-2389.

Anwar-Mohamed, A., El-Kadi, A.O. (2009). Sulforaphane induces CYP1A1 mRNA, protein, and catalytic activity levels via an AhR-dependent pathway in murine hepatoma Hepa 1c1c7 and human HepG2 cells. *Cancer Lett.*, 275, 93-101.

Arora, S., Vatsa, M., Dadhwal, V. (2009). Cabbage leaves vs hot and cold compresses in the treatment of breast engorgement. *Nurs J. India*, 100, 52-54.

Aviram, M., Dornfeld, L. (2001). Pomegranate juice consumption inhibits serum angiotensin converting enzyme activity and reduces systolic blood pressure. *Atherosclerosis*, 158, 195–198.

Aviram, M., Rosenblat, M., Gaitini, D. (2004). Pomegranate juice consumption for 3 years by patients with carotid artery stenosis reduces common carotid intima-media thickness, blood pressure and LDL oxidation. *Clin. Nutr.*, 23, 423–433.

Backman, J.T., Maenppa, J., Belle, D.J., Wrighton, S.A., Kivisto, K.T., Neuvonen, P.J. (2000). Lack of correlation between *in vitro* and *in vivo* studies on the effects of tangeretin and tangerine juice on midazolam hydroxylation. *Clin. Pharmacol. Ther.*, 67, 382–390.

Bailey, D.G., Dresser, G.K. (2004). Interactions between grapefruit juice and cardiovascular drugs. *Am. J. Cardiovasc. Drugs*, 4, 281-297.

Barbalho, S.M., Farinazzi-Machado, F.M., Goulart, R., Brunnati, C.S., Bueno-Ottoboni, A.M., Teixeira-Nicolau, C.C. (2012). *Psidium Guajava* (Guava): a plant of multipurpose medicinal applications. *Med. Aromat. Plants*, 1, 4.

Barceloux, D.G. (2009). Potatoes, tomatoes, and solanine toxicity (*Solanum tuberosum L., Solanum lycopersicum L.*). *Disease a Month*, 55, 391–402.

Bautista-Baños, S., Barrera-Necha, L.L., Bravo-Luna, I., Bermudes-Torres, L. (2002). Antifungal activity of leaf and stem extracts from various plant species on the incidence of Colletotrichum gloesporoides of papaya and mango fruit after storage. *Rev. Mex Fitopatol.*, 20, 8–12.

Berardini, N., Fezer, R., Conrad, J., Beifuss, U., Carle, R., Schieber, A. (2005). Screening of mango (*Mangifera indica L.*) cultivars for their contents of flavonol O- and xanthone C-glycosides, anthocyanins, and pectin. *J. Agric. Food Chem.*, 53, 1563-1570.

Bergquist, S.A., Gertsson, U.E., Knuthsen, P., Olsson, M. E. (2005). Flavonoids in baby spinach (*Spinacia oleracea L.*): changes during plant growth and storage. *J. Agric. Food Chem.*, 53, 9459-9464.

Betancourt-Domínguez, M.A., Hernández-Pérez, T., García-Saucedo, P., Cruz-Hernández, A., Paredes-López, O. (2006). Physico-chemical changes in cladodes (nopalitos) from cultivated and wild cacti (*Opuntia spp.*). *Plant Foods Hu. Nutr.*, 61, 115–119.

Bischoff, S.C. (2008). Quercetin: potentials in the prevention and therapy of disease. *Curr. Opin. Clin. Nutr. Metab. Care*, 11, 733-740.

Blickstein, D., Shaklai, M., Inbal A. (1991). Warfarin antagonism by avocado. *Lancet*, 337, 914-915.

Bouraoui, A., Brazier, J.L., Zouaghi, H., Rousseau, M. (1995). Theophylline pharmacokinetics and metabolism in rabbits following single and repeated administration of Capsicum fruit. *Eur. J. Drug Metab. Pharmacokinet*, 20, 173-178.

Boyer, J., Liu, R.H. (2004). Apple phytochemicals and their health benefits. *Nutrition Journal*, 12, 5.

Bradfield, C.A., Chang, Y., Bjeldanes, L.F. (1985). Effects of commonly consumed vegetables on hepatic xenobiotic-metabolizing enzymes in the mouse. *Food Chem. Toxicol*, 23, 899-904.

Bressler, R. (2006). Grapefruit juice and drug interactions. Exploring mechanisms of this interaction and potential toxicity for certain drugs. *Geriatrics*, 61,12-18.

Brownmiller, C., Howard, L.R., Prior, R.L. (2008). Processing and storage effectson monomeric anthocyanins, percentpolymeric color, and antioxidant capacityof processed blueberry products. *J. Food Sci*, 73, H72-H79.

Bunea, C.I., Pop, N., Babe, A.C., Matea, C., Dulf, F.V., Bunea, A. (2012). Carotenoids, total polyphenols and antioxidant activity of grapes (*Vitis vinifera*) cultivated in organic and conventional systems. *Chem. Central J.*, 6, 66.

Bush, T.M., Rayburn, K.S., Holloway, S.W., Sanchez-Yamamoto, D.S., Allen, B.L., Lam, T., So, B.K., Tran, H., Greyber, E.R., Kantor, S., Roth, L.W. (2007). Adverse interactions between herbal and dietary substances and prescription medications: a clinical survey. *Altern Ther. Health Med.*, 13, 30–35.

Butalla, A.C., Crane, T.E., Patil, B., Wertheim, B.C., Thompson, P., Thomson, C.A. (2012). Effects of a carrot juice intervention on plasma carotenoids, oxidative stress, and inflammation in overweight breast cancer survivors. *Nutr. Cancer*, 64, 331-341.

Cartea, M.E., Francisco, M., Soengas, P., Velasco, P. (2010). Phenolic compounds in Brassica vegetables. *Molecules*, 16, 251-280.

Chieli, E., Romiti, N., Rodeiro, I., Garrido, G. (2009). *In vitro* effects of *Mangifera indica* and polyphenols derived on ABCB1/P-glycoprotein activity. *Food Chem. Toxicol.,* 47, 2703-2710.

Chimponda, T., Mukanganyama, S. (2010). Antimycobacterial activities of selected medicinal plants from Zimbabwe against *Mycobacterium aurum* and *Corynebacterium glutamicum*. *Tropical biomedicine*, 27, 595-610.

Choi, H.J., Lim, D.Y., Park, J.H. (2009). Induction of G1 and G2/M cell cycle arrests by the dietary compound 3,3'-diindolylmethane in HT-29 human colon cancer cells. *BMC Gastroenterology,*9, 39.

Cichewicz, R.H., Thorpe, P.A. (1996). The antimicrobial properties of chile peppers (Capsicum species) and their uses in Mayan medicine. *J. Ethnopharmacol.*, 52, 61-70.

Cowan, M.M. (1999). Plant products as antimicrobial agents. *Clin. Microbiol. Rev.*, 12, 564–582.

Craig, W.J. (2010). Nutrition concerns and health effects of vegetarian diets. *Nutr. Clin. Pract.*, 25, 613-620.

Cruz, L., Castañeda-Hernández, G., Navarrete, A. (1999). Ingestion of chili pepper (*Capsicum annuum*) reduces salicylate bioavailability after oral asprin administration in the rat. *Can. J. Physiol. Pharmacol.*, 77, 441-446.

Custodio, J.M., Wu, C.Y., Benet, L.Z. (2008). Predicting drug disposition, absorption/elimination/transporter interplay and the role of food on drug absorption. *Adv. Drug Deliv. Rev.*, 60, 717-733.

Danese, C., Esposito, D., D'Alfonso, V., Cirene, M., Ambrosino, M., Collotto, M. (2006). Plasma glucose level decreases as collateral effect of fermented papaya preparation use. *Clin. Ter.*, 157, 195–198.

Dangour, A.D., Lock, K., Hayter, A., Aikenhead, A., Allen, E., Uauy R. (2010). Nutrition-related health effects of organic foods: a systematic review. *Am. J. Clin. Nutr.*, 92, 203-210.

De Almeida, A.P., Miranda, M.M., Simoni, I.C., Wigg, M.D., Lagrota, M.H., Costa, S.S. (1998). Flavonol monoglycosides isolated from the antiviral fractions of *Persea americana* (Lauraceae) leaf infusion. *Phytother. Res.*, 12, 562–567.

De Smet, P.A. (2002). Herbal remedies. *N. Engl. J. Med.*, 347, 2046–2056.

De Wet, H., Nkwanyana, M.N., Van Vuuren, S.F. (2010). Medicinal plants used for the treatment of diarrhea in northern Maputaland, KwaZulu-Natal Province, South Africa. *J. Ethnopharmacol.*, 130, 284-289.

Doughari, J.H., Elmahmood, A.M., Manzara, S. (2007). Studies on the antibacterial activity of root extracts of *Carica papaya L. Afr J. Microbiol. Res*, 1, 37–41.

Dresser, G.K., Bailey, D.G., Leake, B.F., Schwarz, U.I., Dawson, P.A., Freeman, D.J., Kim, R.B. (2002). Fruit juices inhibit organic anion transporting polypeptide-mediated drug uptake. *Clin. Pharmacol. Ther.*, 71, 11-20.

Dresser, G.K., Kim, R.B., Bailey, D.G. (2005). Effect of grapefruit juice volume on the reduction of fexofenadine bioavailability: possible role of organic anion transporting polypeptides. *Clin. Pharmacol. Ther.*, 77, 170-177.

Duester, K.C. (2001). Avocado fruit is a rich source of beta-sitosterol. *J. Am. Diet Assoc.*, 101, 404-405.

Duke, S.O., Baerson, S.R., Dayan, F.E., Rimando, A.M., Scheffler, B.E., Tellez, M.R., Wedge, D.E., Schrader, K.K., Akey, D.H., Arthur, F.H., De Lucc, A.J., Gibson, D.M., Harriso, H.F., Peterso, J.K., Gealy, D.R., Tworkoski, T., Wilson, C.L., Hervert-Hernandez, D., Sayago-Ayerdi, S.G., Goni, I. (2010). Bioactive compounds of four hot pepper varieties (*Capsicum annuum L.*), antioxidant capacity, and intestinal bioaccessibility. *J. Agric. Food Chem.,58*, 3399-3406.

Ellinger, S., Ellinger, J., Stehle, P. (2006). Tomatoes, tomato products and lycopene in the prevention and treatment of prostate cancer: do we have the evidence from intervention studies ? *Curr. Opin. Clin. Nutr. Metab. Care*, 9, 722-727.

Engels, C., Knodler, M., Zhao, Y.Y., Carle, R., Ganzle, M.G., Schieber, A. (2009). Antimicrobial activity of gallotannins isolated from mango (*Mangifera indica L.*) kernels. *J. Agric. Food Chem.*, 57, 7712– 7718.

Eno, A.E., Owo, O.I., Itam, E.H., Konya, R.S. (2000). Blood pressure depression by the fruit juice of *Carica papaya* (*L.*) in renal and DOCA-induced hypertension in the rat. *Phytother. Res.*, 14, 235-239.

Farkas, D., Greenblatt, D.J. (2008). Influence of fruit juices on drug disposition: discrepancies between *in vitro* and clinical studies. *Expert. Opin. Drug Metab. Toxicol.*, 4, 381-393.

Feugang, J.M., Konarski, P., Zou, D., Stintzing, F.C., Zou, C. (2006). Nutritional and medicinal use of cactus pear (*Opuntia spp.)* cladodes and fruits. *Front. Biosc.*, 11, 2574–2589.

Filip, A., Clichici, S., Daicoviciu, D., Catoi, C., Bolfa, P., Postescu, I.D., Gal, A., Baldea, I., Gherman, C., Muresan, A. (2011.) Chemopreventive effects of *Calluna vulgaris* and *Vitis vinifera* extracts on UVB-induced skin damage in SKH-1 hairless mice. *J. Physiol. Pharmacol.*, 62, 385-392.

Fimognari, C., Lenzi, M., Hrelia, P. (2008). Interaction of the isothiocyanate sulforaphane with drug disposition and metabolism: pharmacological and toxicological implications. *Curr. Drug Metab.*, 9, 668-678.

Flanagan, D. (2005). Understanding the grapefruit-drug interaction. *Gen. Dent.*, 53, 282-285.

Francis, G.A., Gallone, A., Nychas, G.J., Sofos, J.N., Colelli, G., Amodio, M.L., Spano, G. (2012). Factors affecting quality and safety of fresh-cut produce. *Crit. Rev. Food Sci. Nutr.*, 52, 595-610.

Freedman, N.D., Park, Y., Subar, A.F., Hollenbeck, A.R., Leitzmann, M.F., Schatzkin, A., Abnet, C.C. (2008). Fruit and vegetable intake and head and neck cancer risk in a large United States prospective cohort study. *Intl. J. Cancer,* 122, 2330–2336.

Gagnier, J.J., DeMelo, J., Boon, H., Rochon, P., Bombardier, C. (2006). Quality of reporting of randomized controlled trials of herbal medicine interventions. *Am. J. Med.*, 119, 1–11.

Galati, E.M., Mondello, M.R., Lauriano, E.R., Taviano, M.F., Galluzzo, M., Miceli, N. (2005). *Opuntia ficus indica* (L.) Mill fruit juice protects liver from carbon tetrachloride-induced injury. *Phytother. Res.*, 19, 796-800.

Gerhauser, C. (2008). Cancer chemopreventive potential of apples, apple juice, and apple components. *Planta Med.*, 74,1608-1624.

Getahun, S.M., Chung, F.L. (1999). Conversion of glucosinolates to isothiocyanates in humans after ingestion of cooked watercress, cancer epidemiology, biomarkers and prevention. *Am. Soc. Preventive Oncol.*, 8, 447-451.

Gill, C.I., Haldar, S., Boyd, L.A., Bennett, R., Whiteford, J., Butler, M., Pearson, J.R., Bradbury, I., Rowland, I.R. (2007). Watercress supplementation in diet reduces lymphocyte DNA damage and alters blood antioxidant status in healthy adults. *Am. J. Clinical Nutr.*, 85, 504-510.

Godard, M.P., Ewing, B.A., Pischel, I., Ziegler, A., Benedek, B., Feistel, B. (2010). Acute blood glucose lowering effects and long-term safety of OpunDia supplementation in pre-diabetic males and females. *J. Ethnopharmacol.*, 130, 631-634.

Greenblatt, D.J. (2009). Analysis of drug interactions involving fruit beverages and organic anion-transporting polypeptides. *J. Clin. Pharmacol.*, 49, 1403-1407.

Guevara-Arauza, J.C. (2009). Efectos biofuncionales del nopal y la tuna. *Horticultura Intl*, 71, 18-24.

Guyot, S., Marnet, N., Sanoner, P., Drilleau, J. (2003). Variability of the polyphenolic composition of cider apple (Malus domestica) fruits and juices. *J. Agric. Food Chem.*, 51, 6240-6247.

Halliwell, B. (2007). Dietary polyphenols: good, bad, or indifferent for your health? *Cardiovascular Res.*, 73, 341-347.

Han, C., Ding, H., Casto, B., Stoner, G., D'Ambrosio, S. (2005). Inhibition of growth of pre-malignant human oral cell lines by extracts and components of black raspberries. *Nutr. Cancer*, 52, 207–217.

Hanley, M.J., Cancalon, P., Widmer, W.W., Greenblatt, D.J. (2011). The effect of grapefruit juice on drug disposition. *Expert Opin. Drug Metab. Toxic*ol, 7, 267-286.

Harris, K.E., Jeffery, E.H. (2008). Sulforaphane and erucin increase MRP1 and MRP2 in human carcinoma cell lines. *J. Nutr. Biochem.*, 19, 246-254.

Hecht, S., Huang, C., Stoner, G., Li, J., Kenney, P. (2006). Identification of cyanidin glycosides as constituents of freeze-dried black raspberries which inhibit anti-benzo[a]pyrene-7,8 diol-9,10- epoxide induced NF-κB and AP-1 activity. *Carcinogenesis*, 27, 1617–1626.

Heinrich, M. (2003). Ethnobotany and natural products: the search for new molecules, new treatments of old diseases or a better understanding of indigenous cultures. *Curr. Topics Med. Chem.*, 3, 141–154.

Hernández-Urbiola, M.I., Pérez-Torrero, E., Rodríguez-García, M.E. (2011). Chemical analysis of nutritional content of prickly pads (*Opuntia ficus indica*) at varied ages in an organic harvest. *Int. J. Environ. Res. Public Health*, 8, 1287-1295.

Hidaka, M., Fujita, K., Ogikubu, T., Yamasaki, K., Iwakiri, T., Okumura, M., Arimori, K. (2004). Potent inhibition by star fruit of human cytochrome p450 3a (cyp3a) activity. *Drug Metab. Dispos*, 32, 581-583.

Hidaka, M., Okumura, M., Fujita, K., Ogikubo, T., Yamasaki, K., Iwakiri, T., Setoguchi, N., Arimori, K. (2005). Effects of pomegranate juice on human cytochrome p450 3A (CYP3A) and carbamazepine pharmacokinetics in rats. *Drug Metab. Dispos.*, 33, 644-648.

Ho, P.C., Saville, D.J., Coville, P.F., Wanwimolruk, S. (2000). Content of CYP3A4 inhibitors, naringin, naringenin and bergapten in grapefruit and grapefruit juice products. *Pharm. Acta Helv.*, 74, 379-385.

Hofmann, T., Kuhnert, A., Schubert, A., Gill, C., Rowland, I.R., Pool-Zobel, B. L., Glei, M. (2009). Modulation of detoxification enzymes by watercress: *in vitro* and *in vivo* investigations in human peripheral blood cells. *Eur. J. Nutr*, 48, 483-491.

Howell, A.B. (2002). Cranberry proanthocyanidins and the maintenance of urinary tract health. *Crit. Rev. Food Sci. Nutr.*, 42, 273-278.

Huang, S.M., Strong, J.M., Zhang, L., Reynolds, K.S., Nallani, S., Temple, R., Abraham, S., Habet, S.A., Baweja, R.K., Burckart, G.J., Chung, S., Colangelo, P., Frucht, D., Green, M.D., Hepp, P., Karnaukhova, E., Ko, H.S., Lee, J.I., Marroum, P.J., Norden, J.M., Qiu, W., Rahman, A., Sobel, S., Stifano, T., Thummel, K., Wei, X.X., Yasuda, S., Zheng, J.H., Zhao, H., Lesko, L.J. (2008). New era in drug interaction evaluation: US Food and Drug Administration update on CYP enzymes, transporters, and the guidance process. *J. Clin. Pharmacol.*, 48, 662–670.

Huang, Y., Nikolic, D., Pendland, S., Doyle, B.J., Locklear, T.D., Mahady, G.B. (2009). Effects of cranberry extracts and ursolic acid derivatives on P-fimbriated *Escherichia coli*, COX-2 activity, pro-inflammatory cytokine release and the NF-kappabeta transcriptional response *in vitro*. *Pharm. Biol.,* 47, 18-25.

Imaizumi, K., Sato, S., Kumazawa, M., Arai, N., Aritoshi, S., Akimoto, S., Sakakibara, Y., Kawashima, Y., Tachiyahiki, K. (2011). Capsaicinoids-induced changes of plasma glucose, free fatty acids and glicerol concentrations in rats. *J. Toxicol. Sci.*, 36, 109-116.

Ishtiaq, M., Hanif, W., Khan, M.A., Ashraf, M., Butt, A.M. (2007). An ethnomedicinal survey and documentation of important medicinal folklore food phytonims of flora of Samahni Valley (Azad Kashmir) Pakistan. *Pakistan J. Biol. Sci.*, 10, 2241-2256.

Izzo, A.A. (2005). Herb–drug interactions: an overview of the clinical evidence. *Fundam Clin. Pharmacol.*, 19,1–16.

Jindal, K., Sharma, R.C. Recent trends in horticulture in the Himalayas. New Delhi. Indus Publishing ; 2004.

Juge, N., Mithen, R.F., Traka, M. (2007). Molecular basis for chemoprevention by sulforaphane: a comprehensive review. *Cell Mol. Life Sci., 64*, 1105–1127.

Junyaprasert, V.B., Soonthornchareonnon, N., Thongpraditchote, S., Murakami, T., Takano, M. (2006). Inhibitory effect of Thai plant extracts on P-glycoprotein mediated efflux. *Phytother. Res., 20*, 79-81.

Juranic, Z., Zixack, Z. (2005). Biological activities of berries: from antioxidant capacity to anti-cancer effects. *BioFactors*, 23, 207-211.

Kamath, A.V., Yao, M., Zhang, Y., Chon, S. (2005). Effect of fruit juices on the oral bioavailability of fexofenadine in rats. *J. Pharm. Sci.*, 94, 233-239.

Katsargyris, A., Tampaki, E.C., Giaginis, C. Theocharis, S. (2012).Cranberry as promising natural source of potential anti-cancer agents: current evidence and future perspectives. *Anti-cancer Agents,* 12, 619-630.

Katsuzaki, H., Hibasami, H., Ohwaski, S., Ishikawa, K., Imai, K., (2003). Cyanidin 3-O-beta-D-glucoside isolated from skin of black glycine max and other anthocyanins isolated from skin of red grape induce apoptosis in human lymphoid leukemia Molt 4B cells. *Oncol. Rep.*, 10, 297–300.

Keiichi, M., Rie I., Kaori, K., Susumu, K. (2010). New benzophenone and quercetin galloyl glycosides from Psidium guajava L. *J. Nat. Med.*, 64, 252–256.

Keikotlhaile, B.M., Spanoghe, P., Steurbaut, W. (2010). Effects of food processing on pesticide residues in fruits and vegetables: a meta-analysis approach. *Food Chem. Toxicol.*, 48, 1–6.

Khan, M.A., Islam, M.T. (2012). Analgesic and cytotoxic activity of *Acorus calamus L., Kigelia pinnata L., Mangifera indica L.* and *Tabernaemontana divaricata L. J. Pharm. Bioallied Sci.*, 4, 149-154.

Kiani, J., Imam, S. (2007). Medicinal importance of grapefruit juice and its interaction with various drugs. *Nutr. J.*, 6, 33-35.

Kim, H., Yoon, Y.J., Shon, J.H., Cha, I.J., Shin, J.G., Liu, K.H. (2006). Inhibitory effects of fruit juices on CYP3A activity. *Drug Metab. Dispos*, 34, 521-523.

Kim, S.H., S, K.C., Min, T.S., Kim, Y., Yang, S.O., Kim, H.S., Hyun, S.H., Kim, H., Kim, Y.S., Choi, H.K. (2011). Ameliorating effects of Mango (*Mangifera indica L.*) fruit on plasma ethanol level in a mouse model assessed with H-NMR based metabolic profiling, *J. Clin. Biochem. Nutr.,* 48, 214-221.

Kimura, Y., Ito, H., Ohnishi, R., Hatano, T. (2010). Inhibitory effects of polyphenols on human cytochrome P450 3A4 and 2C9 activity. *Food Chem. Toxicol.*, 48, 429–435.

Knodler, M., Conrad, J., Wenzig, E.M., Bauer, R., Lacorn, M., Beifuss, U., Carle, R., Schieber, A. (2008). Anti-inflammatory 5-(11'Z-heptadecenyl)- and 5-(8'Z,11'Z-heptadecadienyl)-resorcinols from mango (*Mangifera indica L.*) peels. *Phytochemistry*, 69, 988-993.

Konowalchuk, J.L. (1978). Anti-viral effect of apple beverages. *Appl. Environm Microbiol.*, 36, 6.

Kurowska, E., Manthey, J., (2004). Hypolipidemic effects and absorption of citrus polymethoxylated flavones in hamsters with diet-induced hypercholesterolemia. *J. Agric. Food Chem.*, 52, 2879–2886.

Lambert, J.D., Sang, S., Lu, A.Y., Yang, C.S. (2007). Metabolism of dietary polyphenols and possible interactions with drugs. *Curr. Drug Metab.*, 8, 499-507.

Leclercq, I., Desager, J.P., Horsmans, Y. (1998). Inhibition of chlorzoxazone metabolism, a clinical probe for CYP2E1, by a single ingestion of watercress. *Clin. Pharmacol. Ther.*, 64, 144-149.

Lee, K., Kim, Y., Kim, D., Lee, H., Lee, C. (2003). Major phenolics in apple and their contribution to the total antioxidant capacity. *J. Agric. Food Chem.*, 51, 6516-6520.

Lee, N.H., Yoon, J.S., Lee, B.H., Choi, B.W., Park, K.H. (2000). Screening of the radical scavenging effects, tyrosinase inhibition, and anti-allergic activities using *Opuntia ficus indica. Korean J. Pharmacog.*, 31, 412-415.

Lhoste, E.F., Gloux, K., De Waziers, I., Garrido, S., Lory, S., Philippe, C., Rabot, S., Knasmüller, S. (2004). The activities of several detoxication enzymes are differentially induced by juices of garden cress, water cress and mustard in human HepG2 cells. *Chem. Biol. Interact*, 150, 211-219.

Li, D. (2011). Chemistry behind Vegetarianism. *J. Agric. Food Chem*, 59, 777-784.

Li, S.M., Lo, C.Y., Ho, C.T. (2006). Hydroxylated polymethoxyflavones and methylated flavonoids in sweet orange (*Citrus sinensis*) peel. *J. Agric. Food Chem.*, 54, 4176– 4185.

Lilja, J.J., Raaska, K., Neuvonen, P.J. (2005). Effects of orange juice on the pharmacokinetics of atenolol. *Eur. J. Clin. Pharmacol.*, 61, 337-340.

Lima, C.R., Vasconcelos, C.F., Costa-Silva, J.H., Maranhão, C.A., Costa, J., Batista, T.M., Carneiro, E.M., Soares, L.A., Ferreira, F., Wanderley, A.G. (2012). Anti-diabetic activity of extract from *Persea americana* Mill leaf via the activation of protein kinase B (PKB/Akt) in streptozotocin-induced diabetic rats. *J. Ethnopharmacol.*, 141, 517-525.

Li-Shu, W., Mark, A., Yi-Wen, H., Christine, S., Claire, S., Edward, M., Tim, H.M., Ken, R., Steven, S., Wendy, F., Dennis, P., Yiqing, X., John, W., Guang-Yu, Y., Gary S. (2011). Modulation of genetic and epigenetic biomarkers of colorectal cancer in humans by black raspberries: a phase I pilot study. *Clin. Cancer Res.*, 17, 598-610.

Liu, R.H. (2003). Health benefits of fruit and vegetables are from additive and synergistic combinations of phytochemicals. *Am. J. Clin. Nutr.*, 78, 517S-520S.

Liu, R.H. (2004). Potential synergy of phytochemicals in cancer prevention: mechanism of action. *J. Nutr.*, 134, 3479S–3485S.

Lohiya, N.K., Ravi, B.G. (1992). Anti-fertility investigations on the crude chloroform extract of *Carica papaya* seeds in male albino rats. *Indian J. Exp. Biol.*, 30, 1051-1055.

Lomnitski, L., Bergman, M., Nyska, A., Ben-Shaul, V., Grossman, S. (2003). Composition, efficacy, and safety of spinach extracts. *Nutr. Cancer*, 46, 222-231.

Lust, K.D., Brown, J.E., Thomas, W. (1996). Maternal intake of cruciferous vegetables and other foods and colic symptoms in exclusively breast-fed infants. *J. Am. Diet Assoc.*, 96, 46-48.

Lyle, S. (2006). Fruit and nuts: a comprehensive guide to the cultivation, uses and health benefits of over 300 food-producing plants. *Timber Press*, 145.

Magkos, F., Arvaniti, F., Zampelas, A. (2006). Organic food: buying more safety or just peace of mind? A critical review of the literature. *Crit. Rev. Food Sci. Nutr.*, 46, 23–56.

Mahbubeh, S., Sedighe, A., Akram, E., Ali, H. R., Nafiseh, E. (2009). Effects of apple juice on risk factors of lipid profile, inflammation and coagulation, endothelial markers and atherosclerotic lesions in high cholesterolemic rabbits. *Lipids Health Dis.*, 8, 39.

Mahmood, A.A., Sidik, K., Salmah, I. (2005). Wound healing activity of *Carica papaya L.* aqueous leaf extract in rats. *Int. J. Mol. Med. Adv. Sci.*, 1, 398-401.

Malhotra, S., Bailey, D.G., Paine, M.F., Watkins, P.B. (2001). Seville orange juice-felodipine interaction: comparison with dilute grapefruit juice and involvement of furocoumarins. *Clin. Pharmacol. Ther.*, 69, 14-23.

Mandlekar, S., Hong, J.L., Kong, A.N. (2006). Modulation of metabolic enzymes by dietary phytochemicals: a review of mechanisms underlying beneficial versus unfavorable effects. *Curr. Drug Metab.*, 7, 661-675.

Mangesi, L., Dowswell, T. (2010). Treatments for breast engorgement during lactation. *Cochrane Database Syst. Rev.*, 9.

Materska, M., Perucka, I. (2005) Antioxidant activity of the main phenolic compounds isolated from hot pepper fruit (*Capsicum annuum L.*). *J. Agric. Food Chem.*, 53, 1750-1756.

McAllister, P.M., Bernard K.W. (2011). Efficacy of dietary behavior modification for preserving cardiovascular health and longevity. *Cardiol. Res. Pract.*, 820457.

Mech-Nowak, A., Swiderski, A., Kruczek, M., Luczak, I., Kostecka-Gugala, A. (2012). Content of carotenoids in roots of seventeen cultivars of *Daucus carota L. Acta biochimica Polonica*, 59, 139-141.

Meera, Kalidhar, S.B. (2008). A new coumarin from *Citrus paradisi* Macf. *Indian J. Pharm. Sci.*, 70, 517-519.

Mehdipour, S., Yasa, N., Dehghan, G., Khorasani, R., Mohammadirad, A., Rahimi, R., Abdollahi, M. (2006). Antioxidant potentials of Iranian *Carica papaya* juice *in vitro* and *in vivo* are comparable to alpha-tocopherol. *Phytother. Res.*, 20, 591-594.

Meiers, S., Kemeny, M., Weyand, U., Castpar, R., Von Angerer, E. (2001). The anthocyanidins cyaniding and delphinidin are potent inhibitors of the epidermal growth factor receptor. *J. Agric. Food Chem.*, 49, 958–962.

Meijerman, I., Beijnen, J.H., Schellensa, J.H. (2006). Herb–drug interactions in oncology: focus on mechanisms of induction. *Oncol.,* 11, 742–752.

Milenkovic, D., Deval, C., Dubray, C., Mazur, A., Morand, C. (2011). Hesperidin displays relevant role in the nutrigenomic effect of orange juice on blood leukocytes in human volunteers: a randomized controlled cross-over study. *Plos One*, 6, e26669.

Morand, C., Dubray, C., Milenkovic, D., Lioger, D., Martin, J.F. (2011). Hesperidin contributes to the vascular protective effects of orange juice: a randomized crossover study in healthy volunteers. *Am. J. Clin. Nutr.*, 93, 73–80.

Mourvaki, E., Gizzi, S., Rossi, R., Rufini, S. (2005). Passion flower fruit — A new source of lycopene? *J. Med. Food*, 8, 104–106.

Mulvihill, E., Assini, J., Lee, J. (2011). Nobiletin attenuates VLDL overproduction, dyslipidemia, and atherosclerosis in mice with diet-induced insulin resistance. *Diabetes*, 5, 1446–1457.

Murali, M. (2002). Ayurveda for all. *Pustak Mahal*, ISBN 81-223-0764-7.

Mustafa, V., Khan, N., Mukhtar, H. (2009). Cancer chemoprevention by pomegranate: laboratory and clinical evidence. *Nutr. Cancer*, 61, 811–815.

Myklebust, M. (2006). The healing foods pyramid: an integrative nutrition tool. *Explore*, 2, 352-356.

Nadtochiy, S.M., Redman, E.K. (2011). Mediterranean diet and cardioprotection: the role of nitrite, polyunsaturated fatty acids, and polyphenols. *Nutrition*, 27, 733-744.

Nagata, M., Hidaka, M., Sekiya, H., Kawano, Y., Yamasaki, K., Okumura, M., Arimori, K. (2007). Effects of pomegranate juice on human cytochrome P450 2C9 and tolbutamide pharmacokinetics in rats. *Drug Metab. Dispos.*, 35, 302-305.

Nahrstedt, A., Butterweck, V. (2010). Lessons learned from herbal medicinal products: the example of St. John's Wort (perpendicular). *J. Nat. Prod.*, 28, 1015-1021.

Nair, R., Chanda, S. (2007). *In-vitro* antimicrobial activity of *Psidium guajava L.* leaf extracts against clinically important pathogenic microbial strains. *Braz. J. Microbiol.*, 38, 452–458.

Nayak, A.K., Pal, D., Pany, D.R., Mohanty B. (2010). Evaluation of *Spinacia oleracea L.* leaves mucilage as an innovative suspending agent. *J. Adv Pharm Technol Res.*, 1, 338–341.

Nikodem, V.C., Danziger, D., Gebka, N. (1993). Do cabbage leaves prevent breast engorgement? A randomized, controlled study. *Birth*, 20, 61-64.

Nugon-Budon, L., Roland, N., Flãœnoãœs, J.P., Beauhe, P. (1996).Hepatic cytochrome P450 and CIDP-Glucuronosyl transferase areaffected by five sources of dietary fiber in germ-free rats. *J. Nutr.*, 126, 403-409.

Oberlies, N.H., Rogers, I.I., Martin, J.M., Mclaughlin, J.L. (1998). Cytotoxic and insecticidal constituents of the unripe fruit of *Persea americana*. *J. Nat. Prod.*, 61,781-785.

Obermeier, M.T., White, R.E., Yang, C.S. (1995). Effects of bioflavonoids on hepatic P450 activities. *Xenobiotica*, 25, 575-584.

Oderinde, O., Noronha, C., Oremosu, A., Kusemiju, T., Okanlawon, O.A. (2002). Abortifacient properties of *Carica papaya* (Linn) seeds in female Sprague-Dawley rats. *Niger Postgrad. Med. J.*, 9, 95–98.

Ojewole, J.A. (2006). Anti-Inflammatory and analgesic effects of *Psidium guajava* Linn. (Myrtaceae) leaf aqueous extracts in rats and mice. *Meth. Find Exp. Clin. Pharmacol.*, 28, 441–446.

Pahua-Ramos, M.E., Ortiz-Moreno, A., Chamorro-Cevallos, G., Hernández- Navarro, M.D., Garduño-Siciliano, L., Necoechea-Mondragón, H., Hernández-Ortega, M. (2012).

Hypolipidemic effect of avocado (*Persea americana* Mill) seed in a hypercholesterolemic mouse model. *Plant Foods Hum. Nutr.,* 67, 10-16.

Paine, M.F., Widmer, W.W., Pusek, S.N., Beavers, K.L., Criss, A.B., Snyder, J., Watkins, P.B. (2008). Further characterization of a furanocoumarin-free grapefruit juice on drug disposition: studies with cyclosporine. *Am. J. Clin. Nutr.*, 87, 863-871.

Palaniswamy, U.R., McAvoy, R.J., Bible, B.B., Stuar, J.D. (2003). Ontogenic variations of ascorbic acid and phenethyl isothiocyanate concentrations in watercress (*Nasturtium officinale* R.Br.) leaves. *J. Agric. Food Chem.*, 51, 5504-5509.

Palumbo, M.S., Gorny, J.R., Gombas, D.E., Beuchat, L.R., Bruhn, C.M., Cassens, B., Delaquis, P., Farber, J.M., Harris, L.J., Ito, K., Osterholm, M.T., Smith, M., Swanson, K.M.J. (2007). Recommendations for Handling Fresh-cut Leafy Green Salads by Consumers and Retail Food service Operators. *Food Prot. Trends*, 27, 892–898.

Park, E.H., Kahng, J.H., Paek, E.A. (1998). Studies on the pharmacological action of cactus: identification of its anti-inflammatory effect. *Arch. Pharm. Res.*, 21, 30-34.

Park, E.H., Kahng, J.H., Lee, S.H., Shin, K.H. (2001). An anti-inflammatory principle from cactus. *Fitoterapia*, 72, 288- 290.

Park, S.H., Sim, Y.B., Han, P.L., Lee, J.K., Suh, H.W. (2010). Antidepressant-like effect of kaempferol and quercitirin, isolated from *Opuntia ficus-indica* var. saboten. *Exp. Neurobiol.*, 19, 30-38.

Paul, R., Kulkarni, P., Ganesh, N. (2011). Avocado fruit (*Persea americana* Mill) exhibits chemo-protective potentiality against cyclophosphamide induced genotoxicity in human lymphocyte culture. *J. Exp. Ther. Oncol.,* 9, 221-230.

Pauwels, E.K. (2011). The protective effect of the Mediterranean diet: focus on cancer and cardiovascular risk. *Med. Princ. Pract.*, 20, 103-111.

Pawełczyk, T., Kłoszewska, I. (2008). Grapefruit juice interactions with psychotropic drugs: advantages and potential risk. *Przegl Lek*, 65, 92-5.

Perera, P.K., Li, Y. (2012). Functional herbal food ingredients used in type 2 diabetes mellitus. *Pharmacogn. Rev.*, 6, 37-45.

Periago, M.J., Ros, G., López, G., Martínez, M.C., Rincón, F. (1993). The dietary fiber components and their physiological effects. *Rev. Esp. Ciencia Tecnol. Aliment,* 33, 229-246.

Pham, D.Q., Pham, A.Q. (2007). Interaction potential between cranberry juice and warfarin. *Am. J. Health Syst. Pharm.*, 64, 490-494.

Piga, A. (2004). Cactus pear, a fruit of nutraceutical and functional importance. *J. Prof. Assoc. Cactus Dev.,* 6, 9-22.

Pillai, U., Muzaffar, J., Sen, S., Yancey, A. (2009). Grapefruit juice and verapamil: a toxic cocktail. *South Med. J.*, 102, 308-309.

Piver, B., Berthou, F., Dreano, Y., Lucas, D. (2001). Inhibition of CYP3A, CYP1A and CYP2E1 activities by resveratrol and other non volatile red wine components. *Toxicol. Let*, 125, 83–91.

Platt, K.L., Edenharder, R., Aderhold, S., Muckel, E., Glatt, H. (2010). Fruits and vegetables protect against the genotoxicity of heterocyclic aromatic amines activated by human xenobiotic-metabolizing enzymes expressed in immortal mammalian cells. *Mutat. Res.*, 13.

Plaza, L., Sánchez, M.C., Pascual, T.S., D'Ancos, B., Cano, M.P. (2009). Fatty acids, sterols, and antioxidant activity in minimally processed avocados during refrigerated storage. *J. Agric. Food Chem.*, 57, 3204-3209.

Pohl, C., Will, F., Dietrich, H., Schrenk, D. (2006). Cytochrome P450 1A1 expression and activity in Caco-2 cells: modulation by apple juice extract and certain apple polyphenols. *J. Agric. Food Chem.*, 54, 10262-10268.

Polívková, Z., Šmerák, P., Demová, H., Houška, M. (2010). Anti-mutagenic effects of lycopene and tomato purée. *J. Med. Food.*, 13, 1443–1450.

Potter, A.S., Foroudi, S., Stamatikos, A., Patil, B.S., Deyhim, F. (2011). Drinking carrot juice increases total antioxidant status and decreases lipid peroxidation in adults. *Nutr. J.*, 10, 96.

Prakash, U.N., Srinivasan, K. (2010). Gastrointestinal protective effect of dietary spices during ethanol-induced oxidant stress in experimental rats. *Appl. Physiol. Nutr. Metab.*, 35, 134-141.

Qian, H., Nihorimbere, V. (2004). Antioxidant power of phytochemicals from *Psidium guajava* leaf. *J. Zhejiang Univ. Sci.*, 5, 676-683.

Rao, A.V., and Balachandran, B. (2002). Role of oxidative stress and antioxidants in neurodegenerative diseases. *Nutr. Neurosci.*, 5, 291–309.

Rasiya, B.A., Nayar, T.S. (2011). Plants used for natal healthcare in folk medicine of Kerala, India. *Indian J. Tradit. Knowl.*, 10, 523-527.

Reamy, B.V., Stephens, M.B. (2007). The grapefruit-drug interaction debate: role of statins. *Am. Fam. Phys.*, 76, 190-192.

Riccioni, G., Mancini, B., Di Ilio, E., Bucciarelli, T., D'Orazio, N. (2008). Protective effect of lycopene in cardiovascular disease. *Eur. Rev. Med. Pharmacol. Sci.*, 12, 183-190.

Rodeiro, I., Donato, M.T., Jimenez, N., Garrido, G., Molina-Torres, J., Menendez, R., Castell, J.V., Gómez-Lechón, M.J. (2009). Inhibition of human P450 enzymes by natural extracts used in traditional medicine. *Phytother Res.*, 23, 279–282.

Rodrigo, K.A., Rawal, Y., Renner, R.J., Schwartz, S.J., Tian, Q., Larsen, P.E., Mallery, S.R. (2006). Suppression of the tumorigenic phenotype in human oral squamous cell carcinoma cells by an ethanol extract derived from freeze-dried black raspberries. *Nutr. Cancer*, 54, 58-68.

Rodríguez-García, M.E., De Lira, C., Hernández-Becerra, E., Cornejo-Villegas, M.A., Palacios-Fonseca, A.J., Rojas-Molina, I., Reynoso, R., Quintero, L.C., Del real, A., Zepeda, T.A. (2007). Physicochemical characterization of prickly pads (*Opuntia ficus indica*) and dry vacuum prickly pads powders as a function of the maturation. *Plant Foods Hum. Nutr.*, 62, 107–112.

Rossi, R., Porta, S., Canovi, B. (2010). Overview on cranberry and urinary tract infections in females. *J. Clin. Gastroenterol.*, 44, S61-S62.

Ruel, G., Pomerleau, S., Couture, P., Lamarche, B., Couillard, C. (2005). Changes in plasma antioxidant capacity and oxidized low-density lipoprotein levels in men after short-term cranberry juice consumption. *Metabolism*, 54, 856-861.

Russo, M., Spagnuolo, C., Tedesco, I., Russo, G.L. (2010). Phytochemicals in cancer prevention and therapy: truth or dare? *Toxins, 2*, 517-551.

Saito, M., Hirata-Koizumi, M., Matsumoto, M., Urano, T., Hasegawa, R. (2005). Undesirable effects of citrus juice on the pharmacokinetics of drugs: focus on recent studies. *Drug Saf.*, 28, 677-694.

Santos, C.A., Boullata, J.I. (2005). An approach to evaluating drug-nutrient interactions. *Pharmacother.*, 25, 1789-1800.

Saruwatari, A., Okamura, S., Nakajima, Y., Narukawa, Y., Takeda, T., Tamura, H. (2008). Pomegranate juice inhibits sulfoconjugation in Caco-2 human colon carcinoma cells. *J. Med. Food*, 11, 623-628.

Satoh, H., Yamashita, F., Tsujimoto, M., Murakami, H., Koyabu, N., Ohtani, H., Sawada, Y. (2005). Citrus juices inhibit the function of human organic anion-transporting polypeptide OATP-B. *Drug Metab. Dispos.*, 33, 518-523.

Schirrmacher, G., Skurk, T., Hauner, H., Grassmann, J. (2010). Effect of *Spinacia oleraceae L.* and *Perilla frutescens L.* on antioxidants and lipid peroxidation in an intervention study in healthy individuals. *Plant Foods Hum. Nutr.*, 65, 71-76.

Schmidt, L.E., Dalhoff, K. (2002). Food-drug interactions. *Drugs*, 62, 1481-1502.

Schubert, S.Y., Lansky E.P., Neeman I. (1999). Antioxidant and eicosanoid enzyme inhibition properties of pomegranate seed oil and fermented juice flavonoids. *J. Ethnopharmacol.*, 66, 11–17.

Seeram, N.P. (2008). Berry fruits: compositional elements, biochemical activities, and the impact of their intake on human health, performance, and disease. *Agric. Food Chem.*, 56, 627-629.

Seeram, N.P., Adams, L.S., Hardy, M.L., Heber, D. (2004). Total cranberry extract versus its phytochemical constituents: antiproliferative and synergistic effects against human tumor cell lines. *J. Agric. Food Chem.*, 5, 2512-2517.

Seeram, N.P., Adams, L.S., Zhang, Y., Lee, R., Sand, D., Scheuller, H.S., Heber, D. (2006). Blackberry, black raspberry, blueberry, cranberry, red raspberry, and strawberry extracts inhibit growth and stimulate apoptosis of human cancer cells in vitro. *J. Agric. Food Chem.*, 54, 9329-9339.

Seeram, N.P., Momin, R.A., Nair, M.G., Bourquin, L.D. (2001). Cyclooxygenase inhibitory and antioxidant cyanidin glycosides in cherries and berries. *Phytomedicine,* 8, 362-369.

Shabtay, A., Eitam, H., Tadmor, Y., Orlov, A., Meir, A., Weinberg, P., Weinberg, Z.G., Chen, Y., Brosh, A., Izhaki, I., Kerem, Z. (2008). Nutritive and antioxidative potential of fresh and stored pomegranate industrial by product as a novel beef cattle feed. *J. Agric. Food Chem.*, 56, 10063-10070.

Shah, K.A., Patel, M.B., Patel, R.J., Parmar, P.K. (2010). *Mangifera indica* (mango), *Pharmacognosy Rev.*, 4, 42-48.

Shidfar, F., Froghifar, N., Vafa, M., Rajab, A., Hosseini, S., Shidfar, S., Gohari, M. (2011). The effects of tomato consumption on serum glucose, apolipoprotein B, apolipoprotein A-I, homocysteine and blood pressure in type 2 diabetic patients. *Intl. J. Food Scis. Nutr., 62,* 289–294.

Sica, D.A. (2006). Interaction of grapefruit juice and calcium channel blockers. *Am. J. Hypertens*, 19, 768-773.

Silver, H.J., Dietrich, M.S., Niswender, K.D. (2011). Effects of grapefruit, grapefruit juice and water preloads on energy balance, weight loss, body composition, and cardiometabolic risk in free-living obese adults. *Nutr. Metab. (Lond),* 8, 8.

Simbo, D. (2010). An ethnobotanical survey of medicinal plants in Babungo, Northwest Region, Cameroon. *J. Ethnobiol. Ethnomed.*, 6, 8.

Singh, A.P., Wilson, T., Kalk, A.J., Cheong, J., Vorsa, N. (2009). Isolation of specific cranberry flavonoids for biological activity. *Assess. Food Chem.,* 15, 963–968.

Smejkal, K., Chudik, S., Kloucek, P., Marek, R., Cvacka, J., Urbanova, M., Julinek, O., Kokoska, L., Slapetova, T., Holubova, P., Zima, A., Dvorska, M. (2008). Antibacterial C-geranylflavonoids from *Paulownia tomentosa* fruits. *J. Nat. Prod.*, 71, 706-709.

Sreekanth, D., Arunasree, M.K., Roy, K.R., Chandramohan, T, Reddy, GV, Reddanna P. (2007). Betanin a betacyanin pigment purified from fruits of Opuntia ficus-indica induces apoptosis in human chronic myeloid leukemia cell line-K562. *Phytomedicine*, 14, 739-746.

Stakleff, K.S., Sloan, T., Blanco, D., Marcanthony, S., Booth, T.D., Bishayee, A. (2012). Resveratrol exerts differential effects *in vitro* and *in vivo* against ovarian cancer cells. *Asian Pac. J. Cancer Prev.*, 13, 1333-1340.

Steinberg, D., Parthasarathy, S., Carew, T.E., Khoo, J.C., Witztum, J.L. (1989). Beyond cholesterol Modifications of low-density lipoprotein that increase its atherogenicity. *N. Engl. J. Med.*, 320, 915–924.

Stintzing, F.C., Carle, R. (2005). Cactus stems (*Opuntia spp.*): a review on their chemistry, technology, and uses. *Mol. Nutr. Food Res.*, 49, 175–194.

Stoner, G.D., Wang, L.S., Zikri, N. (2007). Cancer prevention with freeze-dried berries and berry components. *Semin. Cancer Biol.*, 17, 403–410.

Sunagawa, M., Shimada, S., Zhang, Z., Oonishi, A., Nakamura, M., Kosugi, T. (2004). Plasma insulin concentration was increased by long-term ingestion of guava juice in spontaneous non-insulin-dependant diabetes mellitus (NIDDM) rats. *J. Health Sci.*, 50, 674–678.

Surles, R.L., Weng, N., Simon, P.W., Tanumihardjo, S.A. (2004). Carotenoid profiles and consumer sensory evaluation of specialty carrots (*Daucus carota, L.*) of various colors. *J. Agric. Food Chem.*, 52, 3417-3421.

Takanaga, H., Ohnishi, A., Yamada, S., Matsuo, H., Morimoto S, Shoyama, Y., Ohtani, H., Sawada, Y. (2000). Polymethoxylated flavones in orange juice are inhibitors of P-glycoprotein but not cytochrome P450 3A4. *J. Pharmacol. Exp. Ther.*, 293, 230-236.

Tan, H.L., Thomas-Ahner, J.M., Grainger, E.M., Wan, L., Francis, D.M., Schwartz, S.J., Erdman. J.W.Jr., Clinton, S.K. (2010). Tomato-based food products for prostate cancer prevention: what have we learned? *Cancer Metastasis Rev.*, 29, 553-568.

Tannaz, J.B., Poonam, G.D., Brijesh, S., Tetali, P. (2011). *In vitro* antigiardial and antirotaviral activity of *Psidium guajava L.* Leaves. *Indian J. Pharmacol.*, 43, 616–617.

Telang, U., Ji, Y., Morris, M.E. (2009). ABC transporters and isothiocyanates: potential for pharmacokinetic diet–drug interactions. *Biopharm. Drug Dispos*, 30, 335–344.

Thomson, C.A., Ravia, J. (2011). A systematic review of behavioral interventions to promote intake of fruit and vegetables. *J. Am. Diet Assoc.*, 111, 1523-1535.

Tian, Q., Aziz, R.M., Stoner, G.D., Schwartz, S.J. (2005a). Anthocyanins determination in black raspberry (Rubus occidentalis) and biological specimens using liquid chromatography-electrospray ionization tandem mass spectrometry. *J. Food Sci.*, 70, 43–47.

Tian, Q., Rosselot, R.A., Schwartz, S.J. (2005b). Quantitative determination of intact glucosinolates in broccoli, broccoli sprouts, brussels sprouts, and cauliflower by high-performance liquid chromatography-electrospray ionization-tandem mass spectrometry. *Anal. Biochem.*, 343, 93-99.

Tomlinson, B., Hu, M., Lee, V.W. (2008). *In vivo* assessment of herb-drug interactions: possible utility of a pharmacogenetic approach*? Mol. Nutr. Food Res.*, 52, 799–809.

Trejo-Gonzalez, A., Gabriel-Ortiz, G., Puebla-Perez, A.M., Huizar-Contreras, M.D., Munguia-Mazarigos, M.R., Mejia-Arreguin, S., Calva, E. (1996). A purified extract from prickly pear cactus (*Opuntia fulginosa*) controls experimentally-induced diabetes in rats. *J. Ethnopharmacol.,* 55, 27-33.

Tripoli, E., La Guardia, M., Giammanco, S., Di Majo, D., Giammanco, M. (2007). Citrus flavonoids: molecular structure, biological activity and nutritional properties: a review. *Food Chem.*, 104, 466– 479.

Uesawa, Y., Mohri, K. (2006). Effects of cranberry juice on nifedipine pharmacokinetics in rats. *J. Pharm. Pharmacol.*, 58, 1067-1072.

Ushijima, K., Tsuruoka, S., Tsuda, H., Hasegawa, G., Obi, Y., Kaneda, T., Takahashi, M., Maekawa, T., Sasaki, T., Koshimizu, T.A., Fujimura, A. (2009). Cranberry juice suppressed the diclofenac metabolism by human liver microsomes, but not in healthy human subjects. *Br. J. Clin. Pharmacol.*, 68,194-200.

Van den Bout-Van den Beukel, C.J., Koopmans, P.P., Van der Ven, A.J., De Smet, P.A., Burger, D.M. (2006). Possible drug-metabolism interactions of medicinal herbs with anti-retroviral agents. *Drug Metab. Rev.*, 38, 477-514.

Van Schie, C.N., Haring, M., Schuurink, R. (2007). Tomato linalool synthase is induced in trichomes by jasmonic acid. *Plant Mol. Biol.*, 64, 251–263.

Vasanthi, H.R., Mukherjee, S., Das, D.K. (2009). Potential health benefits of broccoli- a chemico-biological overview. *Mini Rev. Med. Chem.*, 9, 749-759.

Veeramachaneni, S., Ausman, L.M., Choi, S.W., Russell, R.M., Wang, X.D. (2008). High-dose lycopene supplementation increases hepatic cytochrome P4502E1 protein and inflammation in alcohol-fed rats. *J. Nutr.*, 138, 1329-1335.

Vinson, J.A., Bose, P., Proch, J., AlKharrat, H., Samman, N. (2008). Cranberries and cranberry products: powerful *in vitro*, *ex vivo*, and *in vivo* sources of antioxidants. *J. Agric. Food Chem.,* 56, 5884-5891.

Volpato, G.T., Calderon, I.M., Sinzato, S., Campos, K.E., Rudge, M.V., Damasceno, D.C. (2011). Effect of *Morus nigra* aqueous extract treatment on the maternal-fetal outcome, oxidative stress status and lipid profile of streptozotocin-induced diabetic rats. *J. Ethnopharm.*, 138, 691-696.

Waliszewski, K.N., Blasco, G. (2010). Nutraceutical properties of lycopen. *Salud Publica Mex.*, 52, 254-265.

Wang, H., Leung, L.K. (2010). The carotenoid lycopene differentially regulates phase I and II enzymes in dimethylbenz[a]anthracene-induced MCF-7 cells. *Nutrition*, 26, 1181-1187.

Watanabe, T., Sakurada, N., Kobata, K. (2001). Capsaici-, resiniferatoxin-, and olvanil-induced adrenaline secretion in rats via the vanilloid receptor. *Biosci. Biotechnol. Biochem.*, 65, 2443-2447.

Wells, P.S., Holbrook, A.M., Crowther, N.R., Hirsh, J. (1994). Interactions of warfarin with drugs and food. *Ann. Intern Med.*, 121, 676-683.

Whiley, A.W., Schaffer, B. (2002). History, distribution and uses. In the avocado: botany, production, and uses. CABI Publishing, 1, 1.

World Health Organization. (2003). Diet, nutrition, and the prevalence of chronic diseases. Geneva, Switzerland: World Health Organization. Contract No.: WHO Technical Report Series No. 916.

Yadav, M., Jain, S., Bhardwaj, A., Nagpal, R., Puniya, M., Tomar, R., Singh, V., Parkash, O., Prasad, G.B., Marotta, F., Yadav, H. (2009). Biological and medicinal properties of grapes and their bioactive constituents: an update. *J. Med. Food.*, 12, 473-484.

Yang, U.J., Yoon, S.R., Chung, J.H., Kim, Y.J., Park, K.H., Park, T.S., Shim, S.M. (2012). Water spinach (Ipomoea aquatic Forsk.) reduced the absorption of heavy metals in an in vitro bio-mimicking model system. *Food Chem. Toxicol.*, Jul 25.

Yañez, J., Vicente, V., Alcaraz, M., Castillo, J., Benavente-Garcia, O., Canteras, J.A. (2004). Cytotoxicity and antiproliferative activities of several phenolic compounds against three melanocytes cell lines: relationship between structure and activity. *Nutr. Cancer*, 2004, 49, 191– 199.

Yeh, C.T., Yen, G.C. (2005). Effect of vegetables on human phenolsulfotransferases in relation to their antioxidant activity and total phenolics. *Free Radic. Res.*, 39, 893-904.

Yoo, H.H., Lee, M., Chung, H.J., Lee, S.K., Kim, D.H. (2007). Effects of diosmin, a flavonoid glycoside in citrus fruits, on P-glycoprotein-mediated drug efflux in human intestinal Caco-2 cells. *J. Agric. Food Chem.*, 55, 7620-7625.

Yoon, H.J., Won, C.H., Moon, S.E. (2004). Allergic contact dermatitis due to *Opuntia ficus-indica* var. saboten. *Contact Dermatitis*, 51, 311-312.

Zhang, C.X., Ho, S.C., Chen, Y.M., Fu, J.H., Cheng, S.Z., Lin, F.Y. (2009). Greater vegetable and fruit intake is associated with a lower risk of breast cancer among Chinese women. *Intl. J. Cancer*, 125, 181–188.

Zhang, Z., Hamilton, S.M., Stewart, C., Starter, A., Teel, R.W. (1993). Inhibition of liver microsomal cytochrome P450 activity and metabolism of the tobacco-specific nitrosamine NNK by capsaicin and ellagic acid. *Anticancer Res.*, 13, 2341-2346.

Zhou, S., Lim, L.Y., Chowbay, B. (2004). Herbal modulation of P-glycoprotein. *Drug Metab. Rev.*, 36, 57-104.

Zhu, L., Zhang, Y., Lu, J. (2012). Phenolic contents and compositions in skins of red wine grape cultivars among various genetic backgrounds and originations. *Intl. J. Mol. Sci.*, 13, 3492-3510.

Zikri, N.N., Riedl, K.M., Wang, L., Lechner,J.F., Schwartz, S.J., Stoner, G.D. (2009). Black raspberry components inhibit proliferation, induce apoptosis and modulate gene expression in rat esophageal epithelial cells. *Nutr. Cancer*, 61, 816–826.

In: Plant Extracts
Editors: Adriana Giordano and Alfieri Costs
ISBN: 978-1-62417-534-3

Chapter 2

Therapeutic Targets of Medicinal Plants of Western Ghats, India

S. Shailasree and H.S. Prakash*
Department of Studies in Biotechnology, University of Mysore,
Manasagangotri, Mysore, India

Financial Support

The work is supported under Institution of Excellence programme of the University of Mysore funded by Government of India - Ministry of Human Resource Development.

Abstract

Drugs derived from natural resources represent a significant segment of the pharmaceutical market. Traditional systems of medicines like Ayurveda, Traditional Chinese Medicine (TCM) or the European phytotherapy assumed synergistic approach wherein all ingredients of the plants could bring about maximum therapeutic efficacy. Due to the escalated cost of development of pre-clinical synthetic drugs, there has been a Worldwide demand for phytopharmaceuticals. The modern therapy applies more and more combinatorial approach in treatment of diseases like cancer, cardiovascular or rheumatic diseases. High-throughput methods are available to investigate complex mixtures of whole plant extracts and establish synergistic effects for multi-therapeutic target treatments.

Demands are increasing to establish open access data banks in a multidisciplinary effort for medicinal plant research with the aim to develop a common standard molecular / therapeutic "bar codes" for characterization of medicinal plants. In the present chapter an attempt has been made to compile information on medicinal plant extracts with respect to their pharmacological attributes to promote wider acceptance and use of these plant-

* Corresponding author: Prof. H. S. Prakash; Ph: +91 0821 2419877, Fax/Ph: +91 0821 2414450; Email: hsp@appbot.uni-mysore.ac.in, hasriprakash@gmail.com.

based drugs in main stream of medicine. The mode of action of phytopharmaceuticals as assessed by up / down-regulation of therapeutic target molecules with respect to anti-cancer, anti-inflammatory, anti-hypertensive, anti-viral, anti-obesity, antidiabetic and pro-gynecological properties complying with multi-target therapy will be described. This approach further helps scientific validation of plant extracts / bioactive molecules that are already being used in traditional medicine.

1. Introduction

One of the major areas of present-day research and development in life sciences include Drug Discovery research. Plants are being exhaustively evaluated worldwide as sources of lead molecules in drug discovery and development and evaluation of the potential of natural products as biopesticides and bio-insecticides. Basically the pharmaceutical research includes studies of pathways in understanding a disease for identifying a safe and effective new treatment. Several pieces of information on the basic cause of the disease are studied at the level of genes, proteins and cells with ultimately identifying and understanding the 'targets'.

Plants form the basis of sophisticated traditional medicine system with its existence for thousands of years. They have been used by several societies throughout the world for treatment of innumerable varieties of acute and chronic conditions. Traditionally they have played a major role in health care, further with evolving technology natural product discovery has become a technology-intensive activity.

The use of herbal drugs has escalated in the form of complementary and alternative medicine (CAM) (NCCAM, 2007). The World Health Organisation (WHO) estimates that approximately 80% of the world's population relies mainly on traditional medicine, predominantly originated from plants. Ancient sytem of medicine viz., Ayurveda, Chinese system of medicine and European phytotherapy applied a combination of standardized complex plant mixtures. They have provided potentials for the development of Western medicine.

Creation of a library of synthetic molecules for validation of its activity constructed with computational strategy is an ongoing process. A lack of lead compounds from synthetic libraries for some therapeutic targets viz., anti-infectives, immunosuppression, oncology and metabolic diseases is attributed to the difference in their chemical interaction with biomolecules (Piggott and Karuso, 2004).

Added to this draw back, the toxic and adverse side effects of synthetic drugs as well as failure of conventional treatments has given a scope for herbal medicine as source of bioactive molecules generally identified from traditional medicinal uses (Harun-ur *et al.*, 2002). Other reason for prompting researchers to investigate natural products could be the higher cost of synthetic compounds and in addition culturally plant-based natural products are more acceptable as a consumer equates natural-with safe.

The interesting aspect of natural products with drug discovery is their structural diversity. The natural products have provided scientists some unique chemical structures with an inspiration to persue new chemical entities which are beyond human imagination and completely different structures. These are sterically complex than synthetic compounds and differ in their function. About 2,00,000 natural compounds are currently reported for natural resources, with only about 5-15% of the plants and nearly ~2,50,000 higher plants still untapped (Yong, 1997). It has been estimated that 40% of medicines have their origin from natural products (Harvey, 2000; Newman *et al.*, 2000). Some of the first molecules identified in the early 1970s were clavulanic acid; a β-lactamase inhibitor from *Streptomyces clavuligerus* (Brown *et al.*, 1976); mevastatin, ML236B, a HMG-CoA reductase inhibitor from *Penicillium citrinum* (Endo *et al.*, 1976; Endo, 1992). The future of natural products appears to be a tale of justifiable hope in drug development bringing relief to millions of patients.

Pharmacology, molecular biology and clinical studies unravealed a synergistic mechanism (Ulrich-Merzenich *et al.*, 2010) for plant extracts. Natural products could act by improving solubility, resorption and bioavailability of drug molecules (Anand *et al.*, 2007) and in combination with a synthetic molecule could exhibit better antimicrobial potential thereby decreasing the minimal effective dose of antibiotics (Rosato *et al.*, 2007).

Modern sciences apply several methods to convert medicinal plants into drugs with a basic simple extraction and then its concentration. Though unreliable and inaccurate, two general methods are in use. They apply the ethanomedicinal knowledge of age-old herbal remedies used in many households and/ or go to the wild and gather the samples from all manner of plant life and trees based on the word of mouth handed down over generations from very old cultures/ folk remedy. In the recent years a renewed interest in plants is also witnessed in Western world.

Ethanopharmacological leads have resulted in introduction of novel single molecule drugs. The World Health Organization (WHO) and United Nations Educational, Scientific and Cultural Organization (UNESCO) have recognized the importance of Tradtional botanical knowledge, its utilization and practice as a heritage. But, Health authorities are reluctant to accept traditional drug preparations from cultural areas without well documented data on quality, safety and efficacy. There is a general feel to integrate the science of ethnobothany with pharmacology and science. Documentation of scientific information of drugs from natural resources with supporting scientific literature has regained interest (Aggarwal *et al.*, 2011; Rates, 2011).

India is a Centre or origin of many traditional systems of medicine viz., Ayurveda, Unani, Siddha, Tibetian, Yoga etc., and published wealth of information but is not organized to cater to the needs of the public health care. In India, Western Ghats locally known as the Sahyadri, is one of the global biodiversity hotspots running along the western coast with flora identified to have potential medicinal value (Gadgil, 1996; Daniel, 1997; Myers *et al.*, 2000). Plants like lemon grass, patchouli, *Dioscorea* spp. and the *Vetiver* spp. have originated in this area. *Rauvolfia serpentine, Saraca asoca, Gymnema sylvestre, Gloriosa superb, Strycnos nux-vomica, Myristica malabarica, Garcinia indica, Utleria salicifolia, Coscinium fenestratum* and *Vateria indica* rich in their medicinal value are included in the IUCN Red List (IUCN, 1994) due to over exploitation. The present chapter is a comprehensive review related to 53 medicinal plants, their extracts, bioactive molecules with their biological activity and patented technologies relevant to the targets they act in chronic diseases viz., anti-inflammation, anti-cancer, cardioprotective, anti-viral, anti-obesity and antidiabetic activities.

With questions raised on the efficacy of monotarget therapies considering multitarget causes for the chronic diseases, modern medical therapy presently applies combination therapies. With this in view analytical information on the medicinal plants with the target they act on is compiled (Table 1).

2. Anti-Inflammatory Activity

Inflammation is a self-defense reaction in its first phase. Under specific circumstances it could turn into chronic state and subsequently become a causative factor in the pathogenesis. Hence in many chronic and disabling diseases, inflammation is regarded as the main therapeutic target and often, the best choice to treat the disease and alleviate the symptoms (Ajmone-Cat *et al.*, 2010). The pro-inflammatory nature of arthritis involves cell signaling network of molecules tumor necrosis factor alpha (TNF-α), interleukin-1-beta (IL-1β), interleukin-6 (IL-6), chemokines, nitric oxide (NO), superoxide (O_2^-), matrix metallo-protease-9 (MMP-9), adhesion molecules, inducible nitric oxide (iNOS), inflammatory enzymes cyclooxygenase (COX)-2, 5-lipoxygenase (LOX) and 15-LOX regulated by master regulator, nuclear factor-kappaB (NF-κB) (Kaibori *et al.*, 1999; Khanna *et al.*, 2007; Ma *et al.*, 2010). Prostaglandins (PG) E2, products of enzyme catalyzed reaction of the COX isoforms 1 and 2; leucotrienes, the second main family of arachidonic acid derivatives, synthesized from the activity of 5- LOX play a major role in the inflammatory process. The suppression of their synthesis at sites of inflammation has been regarded as primarily responsible beneficial anti-inflammatory property. The current treatment for arthritis using non-steroidal anti-inflammatory drugs (NSAIDS) results in unwanted side effects opening up novel opportunities for naturally derived products as potential anti-arthritic treatment.

Practitioners of traditional Indian medicine have been using 'Dashamoola' (combination of roots of ten plants) as a standard Ayurvedic medicine for inflammatory diseases. Guggulsterone [4,17(20)-pregnadiene-3,16-dione], a plant sterol derived from the gum resin (guggulu) of the tree *Commiphora mukul* (Shishodia and Aggarwal, 2004), boswellic acid (a pentacyclic triterpenic acid) and its derivatives (acetyl-β-boswellic acid, 11-keto-β-boswellic acid and acetyl-11-keto-β-boswellic acid) (Safayhi *et al.*, 1995; Takada *et al.*, 2006), curcumin from turmeric, resveratrol (red grape seeds, cranberries and peanuts), tea poly-phenols, genistein (Soy), quercetin (onions), silymarin (artichoke) and withenolides (Ashwagandha) are termed as 'gold molecules' exhibiting anti-inflammatory activity inhibiting NF-κB activation and down-regulating the expression of COX-2, 5-LOX and MMP-9 (Khanna *et al.*, 2007). Other vast resource of medicinal plants with anti-inflammatory activities are reported as discussed in the following section.

The experimental models used in the search of anti-inflammatory metabolites are broadly classified into two types: acute inflammatory models and chronic inflammatory models (Lewis, 1989). Acute models are designed to test drugs modulating erythema, changes in vascular permeability, leukocyte migration, measurement of local pain, local analgesic action and rat paw edema (Barbosa-Filho *et al.*, 2006). Chronic models are designed to find drugs modulating disease process induced by sponge, pellet implants, granuloma pouches and adjuvant induced arthritis (Lewis, 1989).

In vitro studies involving inhibition of NF-κB activation thereby down regulation of inflammatory enzymes (LOX, COX-1, -2, iNOS) were employed to screen medicinal plant extracts for anti-inflammatory activity. Most often, the production of inflammatory mediators viz., PGE_2, iNOS, NO are among the several parameters looked into. Macrophage cell lines RAW264.7 stimulated with proinflammatory stimuli [lipopolysaccharide (LPS), interferon gamma (IF-γ) or calcium ionophore, A23187] most commonly were used for *in vivo* studies.

Aphanamixis polystachya (Wall.) Parker bark paste has been used traditionally to reduce swelling at the site of wound. Studies using bark extract was assessed for anti-inflammatory activity for its capacity to inhibit potato 5-LOX that has a role in onset of rheumatoid arthritis, periodontal disease, asthma, bowel disease and others such as free radical mediated disorders viz., Alzheimer's, Parkinson's and cardiovascular disease. Of the various preparations bioenriched fraction of R20 resin column recorded IC_{50} of 0.80 mg/ml for 5-LOX inhibition. *In vivo* studies in Freund complete adjuvant induced arthritis model of Sprague Dawley rats potent pro-inflammatory biomarkers in serum such as nitrite, TNF-α, IL-1β were reduced by 31.6%, 20.6% and 21% respectively. Further *A. polystachya* treated group showed 23.4% and 19.3% inhibitions of liver malondialdehye (a marker compound for lipid peroxidation) and glutathione when compared to control group. Compositions including neutraceutical, cosmeceutical and pharmaceutical formulations with therapeutic amounts of *A. polystachya* extracts in combination with known anti-inflammatory agents (acetyl-11-keto-β-boswellic acid enriched *B. serrata* extracts, curcumin mixture, demethylated curcumin mixture, glucosamine, resveratrol, garlic extract, ginger extract, chondroitin, methylsulfonylmethane, bromelain, serration peptidase or quercitine) for oral, parenteral and topical administration are reported (Gokaraju *et al.*, 2010).

Aristolochia acuminata Lam. is traditionally used for treating rheumatic pains and fever. Several extracts were investigated both *in vivo* and *in vitro* involving carrageenan-induced hind paw edema in rats and macrophage cell line RAW 267.4 stimulated with proinflammatory stimuli (LPS, IF-γ or calcium ionophore A23187) to determine release of PGE_2 or LTB_4. The ethyl acetate extract at 200 and 400 mg/kg p.o., reduced the paw edema in rats. It inhibited the induction of PGE_2 and LTB_4 with IC_{50} of 39.1 and 29.5 mg/mL respectively (Battu *et al.*, 2011).

Artocarpus hirsutus Lam., is endemic to Western Ghats and native to India is a source of several phytochemicals. Another species *A. heterophyllus* is a source of phytochemicals *viz.*, artocarpesin [5,7,2',4'-tetrahydroxy-6-(3-methylbut-3-enyl) flavone], norartocarpetin (5,7,2',4'-tetrahydroxyflavone) and oxyresveratrol [trans-2,4,3',5'-tetrahydroxystilbene]. They suppressed the LPS-induced production of NO and PGE2 through down-regulation of iNOS and COX-2 in RAW 264.7 murine macrophage cells (Fang *et al.*, 2008a, b; Jagtap and Bapat, 2010).

Buchanania lanzan Spreng., is traditionally used in the treatment of skin diseases. The methanolic extract of kernel significantly decreased paw volume in carrageenan-induced edema of rats at a 200mg/kg dose. Oral administration of this extract also reduced pathogenesis of inflammation when the the rats were subjected to carragenan or formaldehyde injection (Warokar *et al.*, 2010). *In vitro* studies with methanolic extract of *B. lanzan* bark further augmented evidence for anti-inflammatory potential of this plant. It could inhibit 15-LOX and human COX-2 with IC_{50} 32μg and 23.45μg (Shailasree *et al.*, 2012). The wound healing effect of alcoholic extract of *B. lanzan* was studied in Albino rats. Three wound models viz. incision, excision and dead space wounds were looked into. Coadministration of *B. lanzan* extract with dexamethasone significantly increased the breaking strength of dexamethasone treated group, increase wound contraction with reduced time required for epithelialization.

Cinnamomum verum Presl., traditionally used to treat diabetis complications. However, a study involving bark extract was tested for anti-inflammatory property. When tested in LPS-induced RAW264.7 cells at a concentration of 200μg/mL it could inhibit PGE_2 production and NO formation inhibiting iNOS (Ho and Tsai, 2004).

Garcinol from *Garcinia* spp. offers protection against several pathophysiological disorders like protection to oxidative damage, inflammation and cancer. Garcinol, tri-isoprenylated chalcone isolated from the dried fruit rind of *Garcinia indica* blocked LPS activation of NF-κB further interfering 5-LOX when assayed in cell-free system (Padhye *et al.*, 2009). It modulated arachidonic acid (AA) metabolism by retarding the phosphorylation of cytosolic PLA2 ($cPLA_2$) through inhibiting the extracellular ERK1/2 kinase activation. It interfered with NO production by suppression of iNOS modulating (JAK)/STAT-1 signaling pathway and inhibited expression of COX-2 in LPS activated macrophages (Liao *et al.*, 2005; Pandey *et al.*, 2009; Padhye *et al.*, 2009). Gambogic acid, polyprenylated xanthone found abundant in the resin of *Garcinia morella* was isolated and studied for its role to mitigate the expression of pro-inflammatory genes. This molecule inhibited LPS-mediated expression of NF-κB, decreased nuclear p65 levels in RAW 264.7 macrophages. *In vitro* kinase assays with interaction studies using biotinylated gambogic acid with proteomic analysis demonstrated covalent modification of IKK-β [IkappaB (inhibitory kappaB) kinase-beta] a key kinase in NF-κB signaling at Cys-179 resulting in its inhibition (Palempalli *et al.*, 2009).

Myristica fragrans nut finds use as herbal remedies and is used in form of herbal tea / infusion. Myristicin (1-allyl-5-methoxy-3, 4-methylenedioxybenzene) an aromatic compound found in its seeds have been reported for cholinergic, antibacterial and hepatoprotective effects. It was evaluated for anti-inflammtory property in virus stimulated macrophages. It did not reduce cell viability of RAW 264.7 mouse macrophages. It significantly reduced the production of calcium, NO, IL-6, interferon inducible protein-10, monocyte chemotactic protein (MCP)-1, MCP-3, granulocyte-macrophage colony stimulating factor, macrophage inflammatory protein (MIP)-1α, MIP1β and leukemia inhibitory factor in dsRNA [polyinosinic-polycytidylic acid]-induced RAW264.7 cells. The anti-inflammatory property was attributed to work via calcium pathway (Lee and Park, 2011).

Macelignan was isolated from dried seed kernels of *M. fragrans*. The antioxidant and anti-inflammatory activities were studied in murine hippocampal HT22 cell line. It attenuated the ROS production and neurotoxicity induced by glutamate. In microglials activated by LPS it potentially suppressed expression of iNOS and COX-2. It suppressed secretion/generation of pro-inflammatory cytokine TNF-α and IL-6 in these cells when induced by LPS (Jin *et al.*, 2005). Hence macelignan possesses therapeutic potentials against neurodegenerative diseases with oxidative stress and neuroinflammation.

Rubia cordifolia traditionally used for coughs and joint pains. The methanolic extract at a dose of 10 and 20mg/kg showed reduction in paw edema when rats were injected with carrageenan. It was comparable to that of standard, phenylbutazone (100mg/kg) used in the study. It could inhibit lipoxygenase enzyme pathway with decreased cumene hydroperoxides and leukotrienes production. A decreased NO with concomitant decrease in iNOS was recorded in LPS-induced J774.1 macrophages was recorded (Deshkar *et al.*, 2008).

Alkaloids like valesamine and scholaricine were identified from *Alstonia scholaris*. They could reduce acetic acid-induced writhing response and xylene-induced ear edema in mice. They significantly increased superoxide dismutase (SOD) activity and decreased the levels of NO, PGE_2 and malondialdehyde (MDA) levels in air pouch mice model. They inhibited inflammatory mediators (COX-1, COX-2 and 5-LOX) (Shang *et al.*, 2010).

3. Anti-Cancer Activity

Cancer, a hyperproliferative disorder affecting millions people around the world as diagnosed by the International Agency for Research on Cancer. It involves tumour initiation, promotion and progression involving dysregulation of apoptosis, unregulated proliferation, invasion, angiogenesis and metastasis. An agent with carcinogenic property should possess growth-inhibition power and the ability to induce cell death. Cell death on the basis of main morphologic criteria of cellular shrinkage, condensation and margination of nuclear chromatin, DNA fragmentation, cytoplasmic vacuolization, cell lysis is termed apoptosis (Kerr *et al.*, 1972). Necrosis is associated traditionally with inflammation that results from additive effect of a number of independent biochemical events activated by severe depletion of cell energy. Inhibition of cell cycle progression in cancer cells targeting epidermal growth factor, DNA Topo I and II, iNOS, telomerases, COX-1, -2, NF-κB, IL-1, IL-6, chemokines, cytokines, cyclin D1, cyclin E cell cycle proteins, vascular endothelial growth factor angiogenic factor, up regulating anti-apoptotic molecules Bcl-2, Bcl-X_L, XIAP, survivin, FLIP, MMP and cellular proliferation molecules e.g., c-myc, AP-1 by phytochemicals has been reviewed (Aggarwal *et al.*, 2006).

In vitro and *in vivo* assay systems are applied to evaluate the cytotoxicity of a candidate anti-tumor agent are experiments on neoplastic animal cell lines and using animal systems bearing certain types of cancer for primary screening of drug molecules (Kintzios, 2006). The biological parameters affected by the medicinal plant extracts / bioactives isolated have been detailed below. The pathophysiological response and the possible anti-cancer activity at the therapeutic targets and site of action has been indicated (Figure 1).

Aeginetia indica L., a plant parasite found on roots of pampa grasses or sugar canes, has been used as a tonic and an antiinflammatory herbal agent. A 55-kDa protein, AILb-A was isolated from its seeds. In *in vitro* studies AILb-A induced cytokine production (Okamoto *et al.*, 2004). It markedly induced helper T cell 1 (Th1)-type cytokines. The cytokines activated apoptosis-inducing factors such as TNF-α, TNF-β, Fas ligand, TNF-related apoptosis-inducing ligand and perforin on the cell lines used for the study, the human peripheral blood mononuclear cells *in vitro* (Okamoto *et al.*, 2000). AILb-A induced a Th1-dominant state and elicited marked antitumor effects in syngeneic Meth-A tumor-bearing *BALB/c* mice, in which the Th2 response was genetically dominant. Toll like receptor (TLR)-4 mediated NF-κB activation induced by AILb-A was observed in their experiments. As TLR combined with carcinogens finds application in a wide variety of melanoma immunotherapeutic strategies tt has been strongly suggested that AILb-A may be a useful immunotherapeutic agent for patients with malignancies (Ohe *et al.*, 2001; Okamoto *et al.*, 2004).

Callophyllum polyanthum Wall. Ex. Choisy crude methanol and chloroform extracts were screened for antiproliferative activity using human cancer cell lines B16F10 (melanoma), HeLa (cervical carcinoma) and KB (epidermoid carcinoma) as studied by the sulfur-hodamine-B (SRB) binding assay. It exhibited this activity to some extent (Manosroi *et al.*, 2007). In search of anti-proliferative agents traditionally used *C. inophyllum* L. was studied for anti-cancer properties. Ten 4-phenylcoumarins previously isolated were examined for their possible inhibitory effects on Epstein--Barr virus early antigen (EBV-EA) activation induced by 12-O-tetradecanoylphorbol-13-acetate in Raji cells (human Burkitts lymphoma).

These compounds tested showed inhibitory activity against EBV, without showing any cytotoxicity. Calocoumarin-A showed more potent activity than any of the other compounds tested. It exhibited a marked inhibitory effect on (skin) tumor promotion in an *in vivo* two-stage carcinogenesis test (Itoigawa *et al.*, 2001).

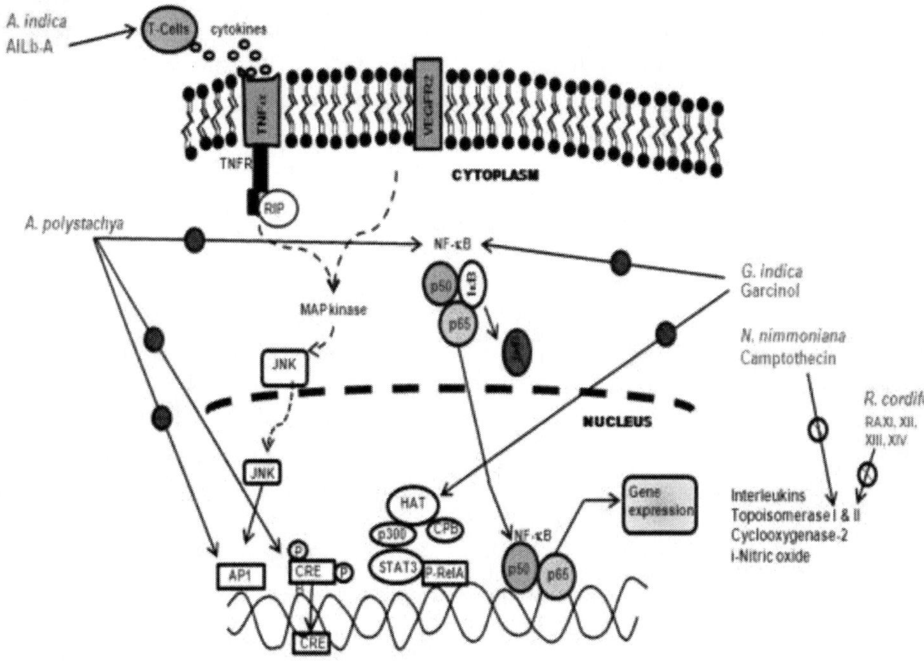

Represents the inhibitory targets of medicinal plants and bioactives.

Figure 1. Pathophysiology of anti-cancer response indicating medicinal plants / bioactivies with possible anti-cancer activities at the therapeutic targets and site of action.

Dioscorea spp. form the natives of Western Ghats. Anti-tumor promoting effect of *D. bulbifera* was reported using the neoplastic transformation assay of mouse epidermal JB6 cell lines. Of the several extract preparations, 75% ethanol extracts ethanolic and butanolic extracts exhibited activities against tumor promotion when induced by promoter of 12-*O*-tetradecanoylphorbol-13-acetate. Rhizome extracts of *Dioscorea oppositifolia* L. exhibited cytotoxicity to JB6 cells (Gao *et al.*, 2007).

A crucial factor of human diseases such as cancer and age-related mascular degeneration is the formation of new blood vessels known as angiogenesis. Vascular endothelial growth factor (VEGF)-2 has been identified as a key in signalling pathway which stimulates endothelial cell survival and proliferation and thus an attractive target for anti-angiogenesis treatment. Lu *et al.* (2010) when screening for natural inhibitors derived from diet, their studies identified *C. zeylanicum* water-based extract had capacity to inhibit VEGFR2 kinase activity, mitogen-activated protein (MAP) kinase- and STAT3-mediated signaling pathway in endothelial cells. If activated, STAT3 interacts with p65 of NF-κB, RelA to recruit the p300 HAT complex mediating STAT3-NF-κB gene transcription resulting in survival, proliferation and tube-formation aggreviating cancerous condition. Further experiments identified Procyanidin type A trimer (molecular weight, 864) and a tetramer (molecular weight, 1152) purified from the *C. zeylanicum* extract were bioactives inhibiting kinase activity of purified VEGFR2 with the capacity to inhibit angiogenesis (Lu *et al.*, 2010).

The principal antioxidant substance of *G. indica* (Thouars) Choisy, garcinol exhibited good anti-tumor activity against human leukemia (HL-60) cells and cyclin D1-positive cells. It inhibited histone acetyltransferases preventing DNA transcription (HATs, $IC_{50} = 7\ \mu M$) and p300/CPB-associated factor (PCAF, $IC_{50} = 5\ \mu M$), both of which are known to modulate gene expression in cancer conditions (Padhye *et al.*, 2009). Gambogic acid from *G. morella* was tested for anti-tumor activity. It could prevent the pathogenesis of this chronic disease by ligation with transferrin receptor-1 (TfR1). It is a carrier protein needed to import iron into the cell and maintain cellular iron homeostasis. By this interaction the loss of homeostasis could have resulted in death of cancer cells (Palempalli *et al.*, 2009).

Nothapodytes nimmoniana (Grah.) Mabb., an endangered medicinal plant contains alkaloid camptothecin and 9-methoxy camptothecin. Camptothecin exhibited anti-tumor action due to its inhibitory activity to DNA Topoisomerase-I. Two derivatives of camptothecin, irinotecan and topotecan, were developed to improve the water-solubility and were approved for use against breast cancer by U.S. Food and Drug Administration (FDA) in 1996. Novel water soluble alkaloids from *N. nimmoniana*, foetidine 1 and 2, having anti-cancer and anti-viral properties suitable for the treatment of the patients is currently in used (Bombardelli *et al.*, 2006).

Rubia cordifolia, has a series of bicyclic hexapeptides (RA-XI, -XII, X-III and XIV) with potent anti-tumor activities against P-388 and HepG2 cell lines. It also possessed Topo II inhibitory activities, but clinical trials reported undesirable side effects (Yoshida *et al.*, 1994; Deshkar *et al.*, 2008).

Sida rhombifolia ssp. retusa is a well established drug. Anti-cancer activity of the total ethyl acetate extract showed potent cytotoxicity with LC_{50} values (5.41 ppm) comparable to the reference standard, gallic acid (Islam *et al.*, 2003). In another study aqueous extracts of known medicinal plants were combined and its anti-proliferative effect was assessed on HeLa cancer cell line. A combination of *Nymphoides indica* L. Kuntze, *Sida rhombifolia* L., and *Xylinbaria minutiflora* Pierre. (N040) exhibited the highest anti-proliferative. It inhibited MMP-2 at 30 and threefolds of cisplatin, respectively (Manosroi *et al.*, 2012).

Sterospermum suaveolens medicinal uses include use in coughs, fevers and as tonic. Of the several bioactives reported, lapachol, 2-hydroxy-3-(3-methyl-2-butenyl)-1,4-naptho-quinone, a vitamin K antagonist has anti-tumor activity. Its effect was studied on experimental metastasis of murine B16BL6 melanoma cells. A single oral administration of high toxic dose of lapachol (80-100 mg/kg) promoted metastasis in T-cell deficient mice and NK-suppressed mice. *In vitro* treatment of B16BL6 cells with could promote metastasis to a limited extent. When tested against Walker 256 carcinosarcome in animals by intraperitoneal, subcutaneous, intramuscular and oral routes, the best activity was recorded when the drug was given twice daily by oral route (Rao *et al.*, 1968).

It is reported that 102 plant genera with anti-cancer properties belonging to 61 families have been reported but, so far only 15 species have been utilized in cancer chemotherapy at clinical level. It is anticipated that the next breakthrough in this therapy will be from plant mixtures rather than a single characterized molecule (Kintzios, 2006).

4. Cardioprotective Activity

Cardiovascular system comprises of the circulatory system, the heart and blood vessels carrying nutrients and oxygen to the tissues of the body and removing CO_2 and other wastes from them. Cardiovascular diseases affect the heart and blood vessels, peripheral vascular system causing congenital heart disease and heart failure. Increased ROS, hypertriglyceridaemia and increased levels of cholesterol ester in serum might also be responsible for altered cardiovascular functions (Freedman *et al.*, 1988). Cardioprotective drugs include antioxidant molecules, ionotropic agents, β blockers, calcium antagonists, organic nitrates, anti-arrhythmics, Na^+, K^+-ATPase modulators, ACE inhibitors, diuretics, cholesterol lowering drugs, clot-busters, anti-coagulants and anti-platelet drugs.

The serum enzymes creatine kinase (CK), lactate dehydrogenase (LDH), aspartate transaminase (AST), alanine transaminase (ALT) and cardiac specific proteins like troponins serve as sensitive indices to assess the severity of myocardial infarction (Nigam, 2007). These cardio-specific marker enzymes are released from the heart into the blood during myocardial damage due to myofibril degeneration and myocyte necrosis (Senthil *et al.*, 2007; Velavan *et al.*, 2009). Many plants are reported in Ayurveda with cardiotonic properties attributed to their antioxidant and ROS scavenging activities (Tilak-Jain and Devasagayam, 2006).

Premna serratifolia (Verbenaceae) contains alkaloids and iridoid glycoside decreased the activity of Na^+K^+ATPase and Mg^{2+}ATPase and increased Ca^{2+}ATPase increasing intra-cellular Ca^{2+} concentrations through Na^+/Ca^{2+} exchange. The intracellular Ca^{2+} plays a significant role in the regulation of normal physiology and biochemistry of cardiac and smooth muscles. (Goto *et al.*, 1992; McGarry and Williams, 1993; Wang *et al.*, 2001). The extract did not alter the activities of cardiac markers such as ALT, AST, LDH and CK indicating no changes in physiological conditions of the heart crediting positive ionotropic and negative chronotropic actions for its cardiotonic properties (Rajendran *et al.*, 2008).

Ayurvedic/Siddha formulation with cardioprotective activities was reviewed by Tilak-Jain and Devasagayam (2006). *T. arjuna* reported to have cardioprotective activity is available as formulations *viz.*, Arjunsal, Arjunaristam, Arjunaghrita, Arjuna Siddha Dugdha (Tilak-Jain and Devasagayam, 2006). Caps-HT2, herbal Ayurvedic medicine contains methanolic extracts of selected parts of plants, *Commiphora mukul, Allium sativum, Plumbago indica, Semecarpus anacardium, Hemidesmus indicus, T. arjuna, Tinospora cordifolia, Withania somnifera* Dunal., *O. sanctum* with reported cardioprotective function (Mary *et al.*, 2003).

5. Anti-Viral Activity

A (HIV) virus infects a susceptible host cell by binding to cell surface receptors CD4. Following fusion, the genetic material (RNA) enters the cell and undergoes reverse transcription into DNA catalyzed by viral associated reverse transcriptase (RT). The viral DNA then enters host cell nucleus and integrates into host genetic material catalyzed by enzyme integrase causing viral infection. HIV infection in a cell can lead to the formation of "syncytia" due to fusion of cells to form large cells with many nuclei. Infection could stay in latent state within the host DNA for years making its eradication difficult and the patients are required to be on anti-virus drugs for life (Yang *et al.*, 2001; Singh *et al.*, 2005). Natural products from medicinal plants have a moderate to good anti-viral activity offering an alternative source of medicine. They are reported to work both at an early stage of HIV life cycle by inhibiting RT as well as in later stages by inhibiting cellular fusion and syncytium formation (Singh *et al.*, 2005). *Callophyllum inophyllum* Linn. commonly known as Indian laurel or Alexandrian laurel is rich in coumarins and xanthones with biological activities. *Callophyllum* sp. are gaining importance as a source of anti-HIV medicines. Coumarins *viz.*, calanolides [(+)-calanolide, (–)-calanolide and its dihydro-derivative, (–)-7, 8-dihydrocalanolide B] with non-nucleoside-specific inhibitory property of HIV-reverse transcriptase (RT) significantly inhibited the cytopathic effects of HIV-1 in T-cell lines (CEM-SS cells and MT-2 cells), laboratory-adapted HIV-1 variants, the clinical viral isolates, syncytium-inducing and non-syncytium inducing isolates, the T-tropic and monocyte-tropic isolates (Singh *et al.*, 2005; Khare, 2007).

Pharmaceutical companies currently have started phase II clinical trials of calanolide-A from *C. lanigerum* for assessing its long-term anti-HIV activity in combination with other anti-HIV agents (Singh *et al.*, 2005). A comprehensive catalog of 90 anti-HIV compounds isolated and characterized by the Laboratory for Drug Discovery Research and Development (LDDRD) chemists from natural product sources is open to the public with the goal of promoting improved health through drug discovery and development.

6. Anti-Obesity Activity

Lipid metabolism is elegantly balanced to maintain homeostasis (Hofbauer, 2002). Obesity or hyperlipidemia is traditionally believed to be associated with life styles, dietary patterns and physical activity levels and this condition leads to variety of serious diseases *viz.*, atherosclerosis, hypertension, diabetes and functional depression of certain organs.

A few medicinal plants such as *Garcinia gummi-gutta* (Syn. *G. cambogia*) are mentioned in Ayurvedic texts with anti-obesity properties (Mangal and Sharma, 2009). Human subjects involving 135 overweight men and women were subjected to 12-week randomized, double-blind, placebo-conmtrolled trial to evaluate the efficacy of this hydroxycitric acid for body weight and fat-mass loss in overweight patients. Although no significant weight loss was recored (Heymsfield *et al.*, 1998), it is reported that hydroxycitric acid competitively inhibited the extramitochondrial enzyme adenosine triphosphate–citrate (pro-3S)-lyase. As a citrate cleavage enzyme that may play an essential role in *de novo* lipogenesis inhibition, *G cambogia* is claimed to lower body weight and reduce fat mass in humans (Girola *et al.*, 1996). A preliminary clinical study with a herbal drug powder *obeloos* containing one part each of *Achyranthus aspera*, *Cyprus rotundus*, *C. multiforum* and *G. gummigutta* fruits decoction was tested on 60 clinically diagnosed patients and followed up for 30 to 60 days. Amongst the tested medicinal plants, *G. gummigutta* was reported to be effective in inhibiting lipogenesis, lowering production of cholesterol and fatty acids, increasing production of glycogen in liver, suppresses appetite, increases body's heat production by activating thermogenesis (Mangal and Sharma, 2009).

Salacia spp. roots have been used in Ayurvedic medicine for diabetes and obesity since antiquity and have been extensively consumed in Japan, the United States and other countries as a food supplement for the prevention of obesity and diabetes. They exhibited multi-target actions contributing improvement of type 2 diabetes and obesity-associated hyperglycemia, dyslipidemia and related cardiovascular complications seen in humans and rodents. The bioassay-guided identification indicate that mangiferin, salacinol, kotalanol and kotalagenin 16-acetate are in part responsible for these multi-target regulatory activities of *Salacia* roots (Li *et al.*, 2008). Inhibition of triglyceride digestion and absorption via inhibition of pancreatic lipase (PL) forms a new approach for the reduction of fat absorption and obesity. Hot water extract of *Salacia* spp., *S. reticulata* roots has been reported to inhibit PL in a concentration-dependent manner with an IC_{50} of 264 mg/ml (Birari and Bhutani, 2007).

7. Antidiabetic Activity

Diabetes is a syndrome identified as chronic disorder of carbohydrate metabolism resulting in abnormally high blood sugar level (hyperglycemia). It could be caused by many factors including absolute or relative deficiency of insulin secretion with/without varying degree of insulin resistance, hereditary, increasing age, poor diet, imperfect digestion, obesity, sedentary lifestyle, stress, drug-induced, infection in pancreas, hypertension, high serum lipid, lipoproteins and less glucose utilization. Therapeutic targets to counter diabetics include inhibition of β-glucosidase, upregulation of glucose trasporter (GLUT), glycolytic enzymes, insulin receptor substrate (IRS) and down regulation of gluconeogenic enzymes. Alloxan/ streptozoin-induced diabetic animal studies and anti-hyperglycemic attributes are the commonly analyzed parameter for antidiabetic properties. The treatment of diabetes with synthetic drugs includes use of second-generation sulfonylureas, biguanides, thiazolidinediones, meglitinides, and alpha-glucosidase inhibitors with the drawback being expensive and side effects like abdominal upset, headache, retention of fluid in the body and hypersensitivity (Bolen *et al.*, 2007). Ayurvedic drugs and traditional medicinal system are reported for treatment of diabetes.

Alstonia scholaris aqueous methanol extract was found to have potent β-glucosidase inhibitory activity. Phytochemicals, quercetin 3-O- β-D-xylopyranosyl (1'''-2'')- β-D-galactopyranoside and (-)-lyoniresinol 3-O-β -D-glucopyranoside were isolated and identified to inhibit β-glucosidase. The former inhibited maltase with IC_{50} value of 1.95mM and the latter acted against both sucrase and maltase with IC_{50} values of 1.95 and 1.43mM. These bioactives could slow starch digestion in the small intestine and decreasing glucose release into bloodstream overcoming type 2 diabetic (Anurakkun *et al.*, 2007).

Cinnamomum verum, leaf is reported to be antidiabetic (Khare, 2007). Intake of 1, 3, or 6 g of cinnamon per day was recommended to reduce serum glucose, triglyceride, LDL cholesterol, and total cholesterol in people with type 2 diabetes. Hydroxychalcone purified from cinnamon had the capacity of mimicking insulin like activity that needs further research attention (Jarvill-Taylor *et al.*, 2001).

Coscinium fenestratum has been mainly used to treat diabetes mellitus reported in Ayurvedic and Siddha systems of medicine. Alcoholic stem extract of this medicinal plant was studied for its carbohydrate metabolism and antioxidant status in streptozotocin-nicotinamide induced type-2 diabetic rats. Oral administration of the extract in graded doses caused a significant increase in enzymatic antioxidants viz., catalase, SOD, glutathione synthetase, peroxidase and glutathione peroxidase and the nonenzymatic antioxidants viz., ascorbic acid, ceruloplasmin and tocopherol. A significant increase in the expression of glycolytic enzymes such as glucose-6-phosphate dehydrogenase, lactate dehydrogenase and hexokinase was recorded. A significantly decreased levels of gluconeogenic enzymes viz., glucose-6-phosphatase and alanine aminotransferase was recorded in diabetic rats reducing glucose levels in blood (Punitha *et al.*, 2005; Tushar *et al.*, 2008).

Salacia roots have been used in Ayurvedic medicine for diabetes and obesity and used as food supplement in many countries. It has the capacity to modulate multiple targets: peroxisome proliferator-activated receptor-alpha-mediated lipogenic gene transcription, AGN II/AGN II type 1 receptor, β-glucosidase, aldose reductase and pancreatic lipase improving type 2 diabetes and obesity-associated hyperglycemia and dyslipidemia (Masayuki *et al.*, 2003; Li *et al.*, 2008; Singh and Duggal, 2010). These multi-target actions may mainly contribute to improvement of diabetes type-2, obesity associated hyperglycemia, dyslipidemia and related cardiovascular complications. Bio-assay guided fractionation identified mangiferin, salacinol, kotalanol and kotalagenin-16-acetase in part responsible for multi-target activity (Li *et al.*, 2008). Peroxisome proliferatory-activated receptor (PPAR) α/γ dual agonists have the potential for use as therapeutic targets in the treatment of type-2 diabetes. Macelignan from *M. fragrans*, reduced serum glucose, insulin, triglycerides, free fatty acid levels and triglycerides levels in the skeletal muscle and liver of *db/db* mice (*db/db* mouse - a model of obesity, diabetes and dyslipidemia. Leptin receptor activity is deficient as the mice are homozygous for a point mutation in the gene for the leptin receptor). It significantly improved glucose and insulin tolerance without altering food intake, their body weights were slightly reduced in comparison with standards (troglitazone-) treated mice. An increased levels of adiponectin expression and decreased TNF-α and IL-6 was recorded in adipose tissue and serum. It reduced endoplasmic reticulum stress and c-Jun NH2-terminal kinase activation in the liver and adipose tissue with subsequent increased insulin signaling alleviating symptoms of diabetis (Han *et al.*, 2008). Upto 600 traditional plant medicine formulations with 3 to 25 herbs are reportedly used as antidiabetic drugs. They include hypoid tablets, Madhuhari powder, Madhumeha churna, Diagon tablets, Glucolev capsule, Diasulin, Glucocare, Glucomap tablets (Jarald *et al.*, 2008).

Conclusion

Western Ghats plants as a source of new drug targets looks increasingly promising. The challenges are formidable: ethnopharmacologists must identify a medicine, its uses, and active components. These efforts are urgent, as traditional knowledge—and traditional plant species—are being lost at an alarming rate. Taking a lead from traditional and folk medicine it should be possible to look for a new bioactive molecule and link it to the modern medicinal practice. This review has cited medicinal plants as source of natural products from the perspective of pharmaceutical industry, research and teaching Institutions. Continued effort to extrapolate *in vitro* and *in vivo* results to human situation through proper designed clinical trials should help to realize the potential of this class of compounds (Kim *et al.*, 2010). It is hoped that this review will be able to address the recent contributions of natural products research to overall drug discovery, development and the role that this field will play in future.

Table 1. Medicinal plants from Western Ghats: Analytical information of medicinal plants at species level and the biological parameters affected

	Plant name	Therapeutic activity (TA)/ Indication	Biochemical parameters
1.	*Aeginetia indica* Linn. **Synonym-** **Family-** Orobanchaceae **CN/AN-** Forest ghost flower/-	TA: tonic and anti-inflammatory	**Anticancer**-Polysaccharides- B cell mitogenic, thymocyte co-stimulatory effects *in vitro*; proteins- 50–60 kDa protein with non-carbohydrate epitopes - antitumor activity *in vivo* (Chai *et al.*, 1995) **AILb-A- 55kDa protein** - ↑Th1-type (induced helper T cell 1) cytokines and mediates potent anticancer immunity *in vitro* and *in vivo* (Okamoto *et al.*, 2004)
2.	*Alangium salvifolium* (L.f.) Wang **Synonyms -** *Grewia salvifolia* L. f., **Family -** Alangiaceae	TA : External use – acute inflammation, rabies, leprosy, (Porchezhian *et al.*, 2001)	**Anti-inflammatory**: Salviifoside ↓NO, ↓ PGE_2, ↓TNF-α (Hung *et al.*, 2009) **Stem bark extract**: ↑Abortifacient (Murugan *et al.*, 2000); ↓Cholesterol (Kshirsagar *et al.*, 2010)
3.	*Alstonia scholaris* (L.) R. Br. **Synonym-** *Echites scholaris* (Linn.). **Family-** Apocynaceae **CN/*AN*** - Devil's tree/ *Saptaparna*	TA : Antidiabetic, hypoglycemic; alstonoside –anti-inflammatory, anti-cancer; Corialstonine and corialstonidine- anti-malarial	**Leaf extract–** bronchovasodilatory, antidiarrhoeal and spasmolytic activity (Channa *et al.*, 2005; Shah *et al.*, 2010) **Anti-inflammatory**: Vallesamine & scholaricine - ↓COX-1, ↓COX-2, ↓ 5-LOX (Shang *et al.*, 2010) **Antidiabetic**: 3-O-β-D-xylopyranosyl (1''2'')-b-D-galactopyranoside and (_)-lyoniresinol 3-O- β -D-glucopyranoside and (-)-lyoniresinol 3-O- β -D-glucopyranoside - ↓ α-Glucosidase (Anurakkun *et al.*, 2007)
4.	*Aphanamixis polystachya* Parker **Synonyms-** ***Aglaia*** *polystachya* Wall. *(*basionyn*);* **Family – Meliaceae**	TA : Anticancer, anti-inflammatory	**Anti-inflammatory: Extract-**↓5-LOX mediated diseases (Gokaraju *et al.*, 2010) **Anti-cancer**, ↓NF κB, ↓AP-1, ↓CREB (Aggarwal *et al.*, 2006)

	Plant name	Therapeutic activity (TA)/ Indication	Biochemical parameters
	CN/*AN* - -/ ***Rohitaka***		**Antioxidant**:↓SOD, ↓CAT, ↓GST (Sen *et al*., 2010)
5.	*Aristolochia tagala* Cham. **Synonym-** *Aristolochia acuminata* Lam. **Family**-Aristolochiaceae **CN/*AN***-Indian Birthwort/*Sunanda*	TA: helping child birth, rheumatic pains and fever	**Anti-inflammatory-** ethylacetate extract-↓PGE($_2$) from RAW264.7 cells (Battu *et al*., 2011)
6.	***Artocarpus hirsutus*** Lam.* **Synonym -** *Artocarpus hirsuta* Lam. **Family** – Moraceae **CN/*AN*-**	TA : Skin care, anti-inflammatory, dysentery, tuberculosis; bark-toothache	**Anti-inflammatory**: ↓iNOS (Jagtap and Bapat, 2010)
7.	*Bombax ceiba* Linn. **Synonyms -** *Bombax malabaricum* Sch. & Endl.; *Salmalia malabaricum* DC **Family –** Bombacaceae **CN/*AN*-**	TA : Analgesic, gonorrhea, diabetes	**Urolithiasis**: Fruit extracts (Gadge and Jalalpure (2012) **Anti-hypertensive**: Xanthone C - glucoside - (Patent Appl No. 814/98, 1998); http://www.iccs.edu/patentsIccs.html
8.	*Buchanania lanzan* Spreng. **Synonyms -** *B. latifolia* Roxb **Family-** Anacardiaceae **CN/*AN*-**	TA:	**Anti-inflammatory**: kernel methanolic extract (Warokar *et al*., 2010); bark methanolic extract - ↓15-LOX, ↓human COX-2, ↓ hyaluronidase, ↓elastase (Shailasree *et al*., 2012)
9.	*Calophyllum inophyllum* Linn. **Synonym -** *Calophyllum bintagor* Roxb **Family –** Clusiaceae **CN/*AN* -Indian Laurel/ *Punnaaga***	**TA :** Friedelane – Anti-cancer Callophylloide – Anti-inflammatory (Shah *et al*., 2006)	**Anti-viral**: Pyranocoumarins, calophyllolide, inophyllums B, C, G_1, G_2 and P - ↓HIV-1; Calanolide - ↓HIV-1 / ↓HIV-2 (Buckheit, 1999; Singh *et al*., 2005); ↓RT (Boyd *et al*., 1999)
10.	*Calophyllum polyanthum* Wall. Ex Choisy – **Synonym :** *Calophyllum elatum* Bedd. **Family** - Clusiaceae **CN/*AN*-**		**Anti-proliferative** – Methanol, chloroform extracts-↓B16F10 (melanoma), ↓HeLa (cervical carcinoma) and ↓KB (epidermoid carcinoma) (Manosroi *et al*., 2007)

Table 1. (Continued)

	Plant name	Therapeutic activity (TA)/ Indication	Biochemical parameters
			Anti-cancer: calocoumarin A- ↓ tumor promotion (Itoigawa *et al*., 2001)
11.	*Canarium strictum* Roxb. **Synonyms -** *Canarium sikkimense King* **Family –** Burseraceae **CN/*AN*** -Black Dammar/ *Raal Dhuup*	TA : Anti-inflammatory, analgesic	**Extract** – antimicrobial, anti-inflammatory (Goel *et al*., 2002)
12.	*Cinnamomum verum* Presl. **Synonym-***Cinnamomum zeylanicum* Bl. **Family-** Lauraceae **CN/*AN*** – Cinnamom/ *Tvak*	TA : Cold, liver problems. Hydroxychalcone - antidiabetic (Jarvill-Taylor, 2010)	**Anti-cancer** : water extract of bark and procyanidin Type-A trimer - ↓VEGFR2; (Lu *et al*., 2010) **Anti-inflammatory-** ↓PG-E_2, ↓NO inhibiting iNOS (Ho and Tsai, 2004) **Antidiabetic**: ↓serum glucose, LDL, cholesterol; hydroxychalcone –Insulin like activity (Jarvill-Taylor *et al*., 2001)
13.	*Cinnamomum malabathrum* Burm. f. (Bl.) **Synonyms -** *Cinnamomum iners* auct. non Reinw. ex Burm. f. **Family-** Lauraceae **CN/*AN*** – Tejpata/*tamālapattram*	TA: essential oils- antimicrobial (Kumar *et al*., 2010)	**Anti-peroxidative effect (**Gupta and Sharma, 2010)
14.	*Coscinium fenestratum* Coleb.* **Family** – Menispermaceae **CN/*AN***-	TA : Antidiabetic, hepatoprotective	**Antidiabetic**: Alcoholic stem extract- ↑G-6-PDH, ↑ LDH, ↑ HK, ↓ALT, ↓G6P (Punitha *et al*., 2005; Tushar *et al*., 2008)
15.	*Costus speciosus* (Koenig) Sm. **Family** - Costaceae **CN/*AN*** – Canereed/ *Kebuka*	TA : Hepatoprotective, antistress. Eremanthin – Antidiabetic, anti-lipidemic (Bavarva and Narasimhacharya, 2008)	**Hepatoprotective**: Extract-↓SGOT, ↓SGPT, ↓TB, ↓ALP, ↓dopamine, ↓MAO (Verma and Khosa, 2009) **Neuroprotective**: Total alkaloids - ↓AChE (Bhaltacharya *et al*., 1972)

	Plant name	Therapeutic activity (TA)/ Indication	Biochemical parameters
16.	*Dioscorea bulbifera* Linn. **Synonyms -** *Dioscorea sativa* Thumb auct.non L. **Family-** Dioscoreaceae **CN/*AN*** - Patoto Yam/ *Vaaraahi*	TA : Cure osteoposoris – proliferation and differentiation of bone marrow (Prasad, 2004); relieves pain and anti-inflammatory (Nguelefack *et al*., 2010)	**Anti-cancer** : Dihydroresveratrol (*trans*-3,5,4'–trihydroxybibenzylstilbene) (Gao *et al*., 2002) **Anti-tumor promoting-** Rhizome extract- ↓ JB6 cells (Gao *et al*., 2007)
17.	*Dioscorea oppositifolia* L **Synonyms -** *Dioscorea batatas* Dcne., **Family-** Dioscoreaceae **CN/*AN*** - **Chinese yam/** *Amlardraka*	TA: asthma and diabetes	**Anti-atherosclerosis** -hypolipidemic - ↓8-hydroxy-2'-deoxyguanosine (Chang *et al*., 2005)
18.	*Embelia basaal* (R. & S.) A. DC **Synonyms-** **Family**-Myrsinaceae **CN/*AN***-Malabar Embelia/ *Bidanga*	TA: Dry fruit powder - oral contraceptive; fresh fruits - rheumatoid inflammation (Karuppusamy *et al*., 2009).	**Acetone extract** –anti-microbial (Rahul *et al*., 2011)
19.	*Embelia ribes* Burm. f. **Family** - Myrsinaceae **CN/*AN*** – Embelia/ *Vidanga*	TA : Anthelmetic	**Anti-inflammatory**: Embelin and its 2, 5-isobutylmine salts –↓ COX-2, ↓NF-□B, ↑apoptosis (Ahn *et al*., 2007)
20.	***Elaeocarpus munronii* (Wl.) Masters** **Synonym-** *Monocera munronii* Wt. **Family-** Elaeocarpaceae **CN/*AN*-**		**Antimicrobial** (Karuppusamy and Rajasekaran, 2010)
21.	*Elaeocarpus serratus* Linn. **Synonym-** **Family- Elaeocarpaceae** **CN/*AN*** -	TA: leaf ethanolic extract – diuretic, cardiovascular stimulant, used in rheumatism; fruits - diarrhea and dysentery. (Ghani, 1998).	**Cytotoxicity-** Brime shrimp lethality (Parvin *et al*., 2009)
22.	*Elaeocarpus tuberculatus* Roxb. **Synonym-Family-Elaeocarpaceae CN/*AN*-** Deccan olive/-	TA: rheumatism, typhoid and epilepsy (Pullaiah, 2006).	**Antioxidant** (Surveswaran *et al*., 2007)

Table 1. (Continued)

	Plant name	Therapeutic activity (TA)/ Indication	Biochemical parameters
23.	***Ervatamia heyneana*** (Wall.) T. Cooke **Synonyms -** *Tabernaemontana heyneana* **Family-** Apocynaceae **CN/*AN* –Nagarkuda/-**	TA: **Leaf extracts is applied externally on mumps. Bark juice along with oil, cumin, boiled and cooled is applied on cuts and wounds (Prakasha and Krishnappa 2006)**	**Anti tumor** (Chitnis *et al.*, 1979)
24.	*Garcinia gummi-guta* (L.) N. Robson* **Synonym -** *Garcinia cambogia* Desr. **Family -** Clusiaceae **CN/*AN*** - Gamboge tree/ *Vrkshaamla*	TA : (−)-Hydroxycitric acid : anti-obese (Mangal and Sharma 2009); Garcinol : anti-ulcer (Girola *et al.*, 1996; Heymsfield *et al.*, 1998)	**Decoction –rheumatism** (Orwa, 2009) **Neuroprotective**: Garcinol, camboginol : ↓AChE, ↓butylcholinesterase, ↓ lipase (Lenta *et al.*, 2007)
25.	*Garcinia indica* (Thouars) Choisy* **Synonym -** *Garcinia purpurea* Roxb. **Family -** Clusiaceae **CN/*AN*** - Kokam Butter tree, *Vrkshaamla*	TA : Hypolipidemic, anti-inflammatory, anti-cancer (Pandey *et al.*, 2009)	**Anti-cancer, anti-inflammatory** : Garcinol derivatives- ↓cPLA2, ↓ERK1/2 kinase; ↓iNOS, ↓JAK / STAT-1, ↓NF□B, ↓p38 MAPK, ↓COX-2, ↓5LOX, ↓HAT (Liao *et al.*, 2005; Pandey *et al.*, 2009; Padhye *et al.*, 2009)
26.	*Garcinia morella* (Gaertn.) Desr. **Synonym :** *Mangostana morella* Gaertn.; *Garcinia elliptica* Wall.; *G. gutta* Wt. **Family -** Clusiaceae **CN/*AN*** – Indian gamboges/*Vrikshamla*	TA: dysentery and stomach-ache (Boruah *et al.*, 2012)	**Anti-inflammatory -** Gambogic acid- ↓TfR1 (transferrin receptor-1); LPS-dependent expression of NF-κB; ↓nuclear p65; ↓IKKβ (Palempalli *et al.*, 2009)
27.	*Garcinia xanthochymus* Hook. F Synonyms: ***Garcinia pictoria*** (Roxb.) Engl. ***Garcinia tinctoria*** (DC.) W. Wight, orth. var. **Family** - Clusiaceae **CN/*AN*** –	TA: Cardiotonic	**Cytotoxicity**: ↓breast cancer (MDA-MB-435S) and ↓lung adenocarcinoma (A549) (Han *et al.*, 2007)

	Plant name	Therapeutic activity (TA)/ Indication	Biochemical parameters
28.	*Iphigenia indica* (L.) A. Gray ex Kunth **Synonyms-** *Melanthium indicum* L. (basionym), *Anguillaria indica* (L.) R. Br. Cent **Family-**Colchicaceae **CN/*AN*-** Indian grass lilly/Kadu belluli	TA: infection and cancer	-
29.	*Madhuca indica* J. F. Gmel **Synonyms-** *Madhuca longifolia* (L.) J. F. Macbr. var. *latifolia* (Roxb.) A. Chev **CN/*AN*:** Mahuwa, Indian butter tree /*Madhuka*	TA: rheumatism, ulcers, tonsillitis, diabetis milletus	**Antihyperglycemic**: Bark methanolic extract-(Dahake *et al*., 2009) **Anticonvulsant:** heart wood extract - GABAergic and opioid systems (Patel *et al*., 2011) **Antioxidant**: Madhucosides A and B- ↓superoxide (Pawar and Bhutani, 2004)
30.	*Memecylon malabaricum* (C. B. Clarke.) Cogn.* **Synonym-***Memecylon amplexicaule* Roxb. var. *malabarica* Cl.; *Memecylon randeriana* SM & MR Almeida **Family-** Melastomataceae **CN/*AN*-**	TA: bacterial infections, inflammation and skin diseases including herpes, chickenpox (Prakasha *et al*., 2010).	**Antimicrobial** (Hullati and Rai, 2004); **Antisproriatic** (Dhanabala *et al*., 2012)
31.	*Memecylon umbellatum* Burm. f. **Synonym-** *M. edule* Roxb.; *M. tinctorium* Koen. Ex Wight & Arn **Family**: Melastomataceae **CN/*AN*:** Iron wood tree	TA: Antidiabetic (Ayyanar *et al*., 2008)	**Hepatoprotective**: ↓SGOT, ↓SGPT and ↓ALP (Shetty *et al*., 2011)
32.	*Mimusops elangii* **Synonym-** **Family-**Sapotaceae	TA: specific cure for teeth and gum diseases	Cytotoxic (Bhujbal *et al*., 2011)

Table 1. (Continued)

	Plant name	Therapeutic activity (TA)/ Indication	Biochemical parameters
33.	*Myristica dactyloides* Gaertn. (Red listed IUCN) **CN/*AN***-bitter nut meg		**Bioactives-** lignans- rel-(8*S*,8′*R*)-dimethyl-(7*S*,7′*R*)-bis(3,4-methylenedioxyphenyl) tetrahydrofuran and rel-(8*R*,8′*R*)- dimethyl-(7*S*,7′*R*)-bis (3,4-methylenedioxy phenyl) tetrahydrofuran, diaryl alkanone - 1-(2,6-dihydroxyphenyl)-9-(4-hydroxy-3-methoxyphenyl)nonan-1-one (Herath and Priyadarshani, 1996) **Antimicrobial** (Koperuncholan and John, 2011)
34.	*Myristica fragrans* Houtt **Family** - Myristicaceae **CN/*AN*** – Nutmeg/ *Jaatiphala*	TA : Aphrodisiac, hepatoprotective (Morita *et al.*, 2003)	**Hepatoprotective, anti-inflammatory, anti-cancer**: Myristicin -↓Ca, ↓NO, IL-6, ↓MCP-1, -3, Macelignan –↓MAPK, ↓NF-κB, ↓MIP 1β, ↓iNOS, ↓COX-2, ↓IL-6, ↓TNF- α (Morita *et al.*, 2003; Jin *et al.*, 2005; Ma *et al.*, 2009) **Antidiabetic**-PPAR α/γ, ↓TNF- α, IL-6 (Han *et al.*, 2008)
35.	*Myristica malabarica* Lamark **Family** - Myristicaceae **CN/*AN***-Rampatri/-	-	**Hepatoprotective** (Manjunath *et al.*, 2011)
36.	*Nothapodytes nimmoniana* (Grah.) Mabb. **Synonym** - *Mappia foetida* (Wt.) Miers **Family-** Icacinaceae **CN/*AN***-	TA : Campothecine – induces apoptosis	**Anti-cancer**: ↓Topo I, ↑apoptosis (Sharma *et al.*, 2009) **Antiviral and anticancer**: Foetidine 1, 2 (derivatives of camptothecin and of 9-methoxy-camptothecin) (Bombardelli *et al.*, 1996)

	Plant name	Therapeutic activity (TA)/ Indication	Biochemical parameters
37.	***Persea macrantha* (Nees) Kosterm.-** **Synonym :** *Machilus macrantha* Nees **Family-***Lauraceae* **CN/*AN*- Machils/*Picchilataru, Kshudramra***	TA: asthma and rhematism	**Anti-inflammatory-** petroleum ether extract and aqueous extract - (Kulkarni *et al*., 2009)
38.	*Premna obtusifolia* R.Br. **Synonyms -***Premna serratifolia* Linn.; *P. integrifolia* L., *P. corymbosa* auct. non Rottl. & Willd. **Family** – Verbenaceae **CN/*AN*** – Headache tree/*Agnimantha*	TA : Cardiac stimulant, anti-inflammatory	**Cardiotonic** :↓$Na^{+}K^{+}$ATPase and ↓Mg^{2+}ATPase ↑Ca^{2+}ATPase (Rajendran *et al.*, 2008; Goto *et al.*, 1992; McGarry and Williams, 1993; Wang *et al.*, 2001)
39.	*Premna latifolia* Roxb. **Synonym –** **Family-** Verbenaceae **CN/*AN*-**Dusky fire brand bark/*Agnimantha*	TA: Rheumatism	**Anti-inflammatory** -Methanolic leaf extract - Iridoid glycosides, diterpenes and saponins (Mahire *et al*., 2009)
40.	*Rauwolfia serpentine* Benth. ex Kurz. **Synonym-** **Family-** Apocynaceae **CN/*AN*:** Indian snake root /*Sarpagandha*	TA: reduces blood pressure	**Hypotensive-Reserpine -** ↓Monoamine oxidase (McQueen *et al*., 1954)
41.	*Rubia cordifolia* Linn. **Synonym -** *Rubia munjesta* Roxb. **Family -** Rubiaceae **CN/*AN*** - Indian Madder/ *Manjishthaa*	TA : Hepatoprotective	**Hepatoprotective**: ↓SGOT, ↓SGPT, ↓ALP (Deshkar *et al*., 2008) **Anti-inflammatory**: ↓iNOS in J774.1 cell line (Deshkar *et al*., 2008) **Anti-cancer**: RAXI, XII, XIII and XVI- ↓Topo- I, ↓Topo-II (Deshkar *et al*., 2008)

Table 1. (Continued)

	Plant name	Therapeutic activity (TA)/ Indication	Biochemical parameters
42.	*Salacia chinensis* Linn. **Synonyms-** *Salacia prinoides* DC. **Family-** Celastraceae **CN/*AN*** - Chinese salacia/ *saptarangi*	TA : Anti-obesity and antidiabetic (Masayuki *et al.*, 2003); type 2 diabetes, hyperglycemia, dyslipidemia (Li *et al.*, 2008)	**Antidiabetic**:↓α -glucosidase (Masayuki *et al.*, 2003) **Anti-obesity**: ↓ α -lipogenic gene, ↓ANGII, ↓ α -glucosidase, ↓ aldose reductase, ↓pancreatic lipase (Li *et al.*, 2008)
43.	*Salacia oblonga* Wall. **Synonym-***Comocladia serrata* Blanco, *Sethia oblonga* Wight. **Family-** Celastraceae **CN/ *AN*-** -/*Saptarangi*	TA:antidiabetic (Lia *et al.*, 2004)	**Antidiabetic:** kotalagenin 16-acetate- α -glucosidase and aldose reductase (Matsuda *et al.*, 1999) Root extract and Mangiferin -↑mRNA levels of collagen I, collagen IV, fibronectin, angiotensin II type 1 receptor, transforming growth factor (TGF)-β1, plasminogen activator inhibitor; ↓angiotensin II- (He *et al.*, 2009) **Antidiabetic-** ↑glucose transportor 4, ↓ α -glucosidase (Giron *et al.*, 2009)
44.	*Saraca asoca* (Roxb.) De Wilde **Synonym** - *Saraca indica* auct. non L. **Family** - Caesalpiniaceae **CN/*AN*** - Ashoka tree/ *Ashoka*	TA : Progynecological activity (Chunekar 1969; Pradhan *et al.*, 2009)	**Anti-inflammatory**: ↓NF- κB, ↓AP-1, ↓CREB (Lampronti *et al.*, 2008)
45.	*Sida cordifolia* Linn **Family-** Malvaceae **CN/*AN*** - Country Mallow/ *Balaa*	TA : Tranquilizer, anti-inflammatory, (Kanth and Diwan, 1999) CAP Pain Massage Oil	**Anti-inflammatory**: ↓COX-2 (Sutradhar *et al.*, 2006) **Anti-hypertensive**: ↓ACE (Hansen, 1995)
46.	*Sida rhombifolia* **Family-** Malvaceae **CN/AN-**Common bala/*Mahaabalaa*	TA: pulmonary tuberculosis, nervous diseases and rheumatism (Khare, 2007)	**Antiarthritic-**ethanol and aqueous extracts (Gupta *et al.*, 2009) **Cytotoxicity** -(Islam *et al.*, 2003; Manosroi *et al.*, 2012)

	Plant name	Therapeutic activity (TA)/ Indication	Biochemical parameters
47.	*Sida accuta* Burm. f **Family-** Malvaceae **CN/*AN*-**	TA: fever and skin ailments (Karou *et al*., 2007)	**Antiplasmodic, antimicrobial,** ↓lethal effects of *Bothrops atrox* venom (Karou *et al*., 2007) **Scopoletin-**↑lipoprotein lipase in 3T3 L1 cells (Karou *et al*., 2007)
48.	*Stereospermum colais* Mabb. **Synonyms –** *Stereospermum chelonoides* (Linn. f.); **Family** - Bignoniaceae **CN/*AN*** - Trumpet Flower/ *Paatalaa*	TA : Analgesic, anti-pyretic, anti-inflammatory, cytotoxic (Haque *et al*., 2006)	-
49.	*Stereospermum personatum* (Hassk.) Chatterjee **Synonym -** *Stereospermum colais* (Buch.-Ham. ex Dillwyn) Mabb. **Family**–Bignoniaceae **CN/*AN***-Padar/-		**Antioxidant:** (+)-Cycloolivil – antioxidant (Rao *et al*., 2004)
50.	*Stereospermum suaveolens* DC. **Family** - Bignoniaceae **CN/*AN***- Padiri	TA: intermittent and puerperal fevers	**Antidiabetic-** bark extract-↓glutathione, ↓ catalase, ↓superoxide dismutase (Balasubramaniam *et al*., 2012) **Anti-cancer-** lapachol, 2-hydroxy-3-(3-methyl-2-butenyl)-1,4-naphthoquinone- ↓Walker 256 carcinosarcoma (Rao *et al*., 1968)
51.	*Symplocos racemosa* Roxb. **Synonym :** *Symplocos beddomei* Cl.; *Symplocos candolleana* Brand.; *Symplocos racemosa* Roxb. var. *racemosa* **Family-** Symplocaceae **CN/AN**-Lodh tree/*Lodhra*	TA:bark for progynecological activity (Khare, 2007)	Hepatoprotective- ethanolic extract ↓AST, ↓ALT (Wakchaure *et al*., 2011)

Table 1. (Continued)

	Plant name	Therapeutic activity (TA)/ Indication	Biochemical parameters
52.	*Symplocos cochinchinensis* (Lour.) S.Moore ssp. *laurina* (Retz.) Noot. **Family-**Symplocaceae **Synonym :** *Symplocos spicata* Roxb. **CN/*AN*-**Pachotti/-	TA: antidiabetic (Sunil *et al.*, 2011)	**Antidiabetic-**leaf hexane extract-↓plasma insulin, ↓plasma and hepatic total cholesterol, triglycerides and ↓free fatty acids (Sunil *et al.*, 2011) **Hypolipidemic -**↓total cholesterol, ↓triglycerides, ↓HDL-C, ↓ LDL-C (Sunil *et al.*, 2012)
53.	*Vateria indica* Linn. **Synonym -** *Vateria malabarica* Bl. **Family -** Dipterocarpaceae **CN/*AN* -** White Damar/ *Sarja*	TA : Anti-tumor (Mishima *et al.*, 2003); CAP : GonoRem (Indmal)®	-

* = Endemic to Western Ghats; CN/*AN* = Common names/ *Ayurvedic names.*

↑ = up regulated/ higher concentration.

↓ = down regulated/ lower concentration.

Acknowledgments

The authors acknowledge the recognition of University of Mysore as Institution of Excellence and financial support from Ministry of HRD and the University Grants Commission, New Delhi. SS acknowledges the receipt of Senior Research Associate (Scientists' Pool Scheme) from Council of Scientific and Industrial Research, New Delhi, India.

References

Aggarwal, B. B., Ichikawa, H., Garodia, P., Weerasinghe, P., Sethi, G., Bhatt, I. D., Pandey, M. K., Shishodia, S. and Nair, M. G. (2006) From traditional ayurvedic medicine to modern medicine: identification of therapeutic targets for suppression of inflammation and cancer. *Expt. Opin. Ther. Targets* 10:87-118.

Aggarwal, B. B., Prasad, S., Reuter, S., Kannappan, R., Yadev, V. R., Park, B., Kim, J. H., Gupta, S. C., Phromnoi, K., Chitra, S., Prasad, S., Chaturvedi, M. M. and Sung, B. (2011) Identification of Novel Anti-inflammatory Agents from Ayurvedic Medicine for Prevention of Chronic Diseases: "Reverse Pharmacology" and "Bedside to Bench" Approach. *Curr. Drug. Targets* 12:1595–1653.

Ahn, K. S., Sethi, G. and Aggarwal, B. B. (2007) Embelin, an inhibitor of X chromosome-linked inhibitor-of-apoptosis protein, blocks Nuclear Factor-kB (NF-□B) signaling pathway leading to suppression of NF-□B-regulated antiapoptotic and metastatic gene products. *Mol. Pharm.* 71:209–219.

Ajmone-Cat, M. A., Bernardo, A., Greco, A. and Minghetti, L. (2010) Non-steroidal anti-inflammatory drugs and brain inflammation: effects on microglial functions. *Pharma* 3:1949-1964.

Anand, P., Kunnumakkara, A. B., Newman, R. A. and Aggarwal, B. B. (2007). Bioavailability of curcumin: problems and promises. *Mol. Pharm.* 4: 807–818.

Anurakkun, N. J., Bhandari, M. R. and Kawabata, J. (2007) α -Glucosidase inhibitors from Devil tree (*Alstonia scholaris*). *Food Chem* 103:1319–1323.

Ayyanar, M., Sankarasivaraman, K. and Ignacimuthu, S. (2008) Traditional herbal medicines used for the treatment of diabetes among two major tribal groups in South Tamil Nadu, India. *Ethnobotanical Leaflets* 12: 276-280.

Balasubramanian, T., Chatterjee, T. K., Senthilkumar, G. P. and Mani, T. (2012) Effect of potent ethyl acetate fraction of *Stereospermum suaveolens* extract in streptozotocin-induced diabetic rats. *Scientific World J.* 2012: 413196.

Barbosa-Filho, J. M., Piuvezam, M. R., Moura, M. D., Silva, M. S., Lima, K. V. B., Leitão da-Cunha, E. V., Fechine, I. M. and Takemura, O. S. (2006) Anti-inflammatory activities of alkaloids. A twenty-century review. *Revista Brasil Farmaco.* 16:109-122.

Battu, G. R., Parimi, R. and Chandra Shekar, K. B. (2011) *In vivo* and *in vitro* pharmacological activity of *Aristolochia tagala* (syn: *Aristolochia acuminata*) root extracts. *Pharm. Biol.* 49:1210-1214.

Bavarva, J. H. and Narasimhacharya, A. V. (2008) Antihyperglycemic and hypolipidemic effects of *Costus speciosus* in alloxan induced diabetic rats. *Phytother. Res.* 22:620-626.

Bhaltacharya, S. K., Parikh, A. K., Debnath, P. K., Pandey, V. B. and Neogy, N. C. (1972) Anticholinesterase activity of *Costus speciosus* alkaloids. *Indian J. Pharmacol.* 4:178-179.

Bhujbal, S. S., Deshmukh, R. P., Bidkar, J. S., Thatte, V. A., Awasare, S. S. and Garg, P. P. (2011) Evaluation of cytotoxic activity of barks of *Mimusops elengi*. *Eurasia J. Biosci.* 5: 73-79.

Birari, R. B. and Bhutani, K. K. (2007) Pancreatic lipase inhibitors from natural sources: unexplored potential. *Drug Discovery Today* 12:879-889.

Bolen, S., Feldman, L., Vassy, J., Wilson, L., Yeh, H. C., Marinopoulos, S., Wiley, C., Selvin, E., Wilson, R., Bass, E. B. and Brancati, F. L. (2007) Systematic review: comparative effectiveness and safety of oral medications for type 2 diabetes mellitus. *Ann. Intern. Med.* 147:386-399.

Bombardelli, E., Mustich, G. and Verotta, L. Alkaloids from *Mappia foetida*, the use thereof and formulations containing them. (1996) Patent US5525609.

Boruah, M. P., Mahanta, D. and Handique (2012) Comparative evaluation of antioxidant activities of some fresh and preserved herbal products of North East India by cyclic voltammetry. *Indian J of Nat. Prod. Resour.* 3: 40-47.

Boyd, M. R., Cardellina, J. H., II, Gustafson, K. R., McMahon, J. B., Fuller, R. W., Cragg, G. M., Kashman, Y. and Soejarto, D. (1999) Calanolide and related antiviral compounds, compositions, and uses thereof. US patent 5859049.

Brown, A. G., Butterworth, D., Cole, M., Hanscomb, G., Hood, J. D., Reading, C. and Rolinson, G. N. (1976) Naturally-occuring beta-lactamase inhibitors with antibacterial activity. *J. Antibiot.* 29: 668.

Buckheit, R.W. Jr, White, E. L., Fliakas-Boltz, V., Russell, J., Stup, T. L., Kinjerski, T. L., Osterling, M. C., Weigand, A. and Bader, J. P. (1999) Unique anti-human immunodeficiency virus activity of the nonnucleoside reverse transcriptase inhibitors calanolide A, costatolide and dihydrocostatolide. *Antimicrob. Agents Chemother.* 43:1827-1834.

Chai, J-G., Okamoto, M., Bando, T., Nagasawa, H., Hisaeda, H., Sakai, T., Himeno, K., Sato, M. and Ohkubo, S. (1995) Dissociation between the mitogenic effect and antitumor activity of seed extract from *Aeginetia indica* L. *Immunopharmacol.* 30: 209–215

Chang, W-C., Yu, Y-M., Wu, C-H., Tseng, Y-H. and Wu, K-Y. (2005) Reduction of oxidative stress and atherosclerosis in hyperlipidemic rabbits by *Dioscorea* rhizome. *Can. J. Physiol. Pharmacol.* 83:423-430

Channa, S., Dar, A., Ahmed, S. and Atta-Ur-Rahman (2005) Evaluation of *Alstonia scholaris* leaves for broncho-vasodilatory activity. *J. Ethnopharmacol.* 97:469-476

Chitnis, M. P., Bhatia, K. G. and Phatak, M. K. (1979) Anti-tumour activity of the methanol extract *Ervatamia heyneana* (NSC B668619). *Indian J. Exp. Biol.* 17:212.

Chunekar, K. C. (1969) Bharvaprakasa Nighantu (Indian Meteria Medica of Sri Bhavamisra. Chaukhamba Bharati Academy, Varanasi.

Dahake, A. P., Chakma, C. S., Chakma, R. C. and Bagherwal (2009) Antihyperglycemic activity of methanolic extract of *Madhuca longifolia* bark. *Diabetologia Croatia* 39:3-8

Daniel, R. J. R. (1997) Taxonomic uncertainties and conservation assessment of the Western Ghats. *Curr. Sci.* 73: 169-170.

Deshkar, N., Tillo, S. and Pande, V. (2008) A comprehensive review of *Rubia cordifolia* Linn. *Pharmacogn. Rev.* 2: 124-134.

Dhanabala SP, Muruganathama N, Hanumanthappa K, Wadhwanib A and Shamasundard NM (2012) Antipsoriatic activity of extracts and fractions obtained from *Memecylon malabaricum* leaves. *J. Pharm. Pharmacol.* DOI: 10.1111/j.2042-7158.2012.01528.x

Endo A, Kuroda M and Tsujita Y (1976) ML-236A, ML-236B and ML236-C, new inhibitors of cholesterogenesis produced by *Penicillium citrinum. J. Antibiot* 29: 1346.

Endo, A. (1992) The discovery and development of HMG-CoA reductase inhibitors. *J. Lipid Res.* 33:1569.

Fang, S. C., Hsu, C. L., and Yen, G. C. (2008a) Anti-inflammatory effects of phenolic compounds isolated from the fruits of *Artocarpus heterophyllus. J Agri Food Chem* 56: 4463–4468.

Fang, S. C., Hsu, C. L., Yu, Y. S. and Yen, G. C. (2008b) Cytotoxic effects of new geranyl chalcone derivatives isolated from the leaves of *Artocarpus communis* in SW 872 human liposarcoma cells. *J Agri. Food Chem.* 56:8859–8868.

Freedman, D. S., Gruchow, H. W., Anderson, A. J., Rimm, A. A. and Barboriak, J. J. (1988) Relation of triglyceride levels to coronary artery disease: the Milwaukee Cardiovascular Data Registry. *Am J Epidemiol* 127:1118-1130.

Gadge, N. B. and Jalalpure, S. S. (2012) Curative treatment with extracts of *Bombax ceiba* fruit reduces risk of calcium oxalate urolithiasis in rats. *Pharm. Biol.* 50: 310-317

Gadgil, M. (1996) Western Ghats: A Lifescape. *J. Indian Inst. Sci.* 76: 495-504.

Gao, H., Hou, B., Kuroyanagi, M. and Wu, L. (2007) Constituents from anti-tumor-promoting active part of *Dioscorea bulbifera* L. in JB6 mouse epidermal cells. *Asian J Traditional Med.* 2:104-109.

Gao, H., Kuroyanagi, M., Wu, L., Kawahara, N., Yasuno, T. and Nakamura, Y. (2002) Antitumor-promoting constituents from *Dioscorea bulbifera* L. in JB6 mouse epidermal cells. *Biol. Pharm. Bull.* 25: 1241-1243.

Gao, H., Yi-Na, H., Pei-Yu, X. and Kawabata, J. (2007) Inhibitory effect on α-glucosidase by the fruits of *Terminalia chebula* Retz. *Food Chem.* 105: 628-634.

Ghani A. (1998) *Medicinal Plants of Bangladesh*, 2nd Ed. pp. 167-168 The Asiatic Society of Bangladesh, Dhaka

Girola, M., De Bernardi, M., Contos, S., Tripodi, S., Ventura, P., Guarino, C. and Marletta, M. (1996) Dose effect in lipid-lowering activity of a new dietary intetrator (chitosan, *Garcinia cambogia* extract, and chrome). *Acta Toxicol. Ther.* 17:25-40.

Girón, M. D., Sevillano, N., Salto, R., Haidour, A., Manzano, M., Jiménez, M. L., Rueda, R. and López-Pedrosa, J. M. (2009) *Salacia oblonga* extract increases glucose transporter 4-mediated glucose uptake in L6 rat myotubes: role of mangiferin.*Clin Nutr.* 28:565-574.

Goel, A. K., Kulshreshtha, D. K., Dubey, M. P. and Rajendran, S. M. (2002) Screening of Indian plants for biological activity: Part XVI. *Indian J Exp. Biol.* 40:812-827

Gokaraju, G. R., Gokaraju, R. R., Golakoti, T., Rao, V., Chirravuri, Alluri, V. K. R. and Bhupathiraju, K. Use of *Aphanamixis polystacha* extracts or fractions against 5-lipoxygenase mediated diseases (2010) Patent US20100178288

Goto, A., Yamada, K., Yagi, N., Yoshioka, M. and Sugimoto, T. (1992) Physiology and pharmacology of endogenous digitalis-like factors. *Pharmacol. Rev.* 44: 377-99.

Gupta, S. R., Nirmal, S. A., Patil, R. Y. and Asane, G. S. (2009) Anti-arthritic activity of various extracts of *Sida rhombifolia* aerial parts. *Nat. Prod. Res.* 23:689-695.

Gupta, V. and Sharma, M. (2010) Protective effect of *Cinnamomum tejpata* on lipid peroxide formation in isolated rat liver homogenate. *Current Res. J Biol. Sci.* 2: 246-249.

Han, K. L., Choi, J. S., Lee, J. Y., Song, J., Joe, M. K., Jung, M. H., and Hwang J-K. (2008) Therapeutic potential of peroxisome proliferators–activated receptor - α/γ dual agonist with alleviation of endoplasmic reticulum stress for the treatment of diabetes. *Diabetes* 57:737–745.

Han, Q. B., Qiao, C. F., Song, J. Z., Yang, N. Y., Cao, X. W., Peng, Y., Yang, D. J., Chen, S. L. and Xu, H. X. (2007) Cytotoxic prenylated phenolic compounds from the twig bark of Garcinia xanthochymus. *Chem. Biodivers.* 4:940-946.

Hansen, K., Nyman, U., Smitt, U. W., Adsersen, A., Gudiksen, L., Rajasekharan, S. and Pushpangadan, P. (1995) *In vitro* screening of traditional medicines for anti-hypertensive effect based on inhibition of the angiotensin converting enzyme (ACE). *J. Ethnopharmacol* 48:43-51.

Haque, M R., Rahman, K. M., Hasan, C. M. and Rashid, M. A. (2006) Antimicrobial and cytotoxic activities of *Stereospermum chelonoides. Dhaka Univ. J. Pharma. Sci.* 5: 71-72.

Harun-ur R. M., Gafur, M. A., Sadik, G. M. and Rahman, M. A. A (2002) Biological activities of a new acrylamide derivative from *Ipomoea turpithum. Pak. J. Biol. Sci.* 5: 968-969.

Harvey, A. (2000) Strategies for discovering drugs from previously unexplored natural products. *Drug Discovery Today* 5: 294.

He, L., Qi, Y., Rong, X., Jiang, J., Yang, Q., Yamahara, J., Murray, M. and Li, Y. (2009) The Ayurvedic medicine Salacia oblonga attenuates diabetic renal fibrosis in rats: suppression of angiotensin II/AT1 signaling. *Evidence-based Complementary Altern. Med.: eCAM.* doi:10.1093/ecam/nep:1-13.

Herath, H. M. T. B. and Priyadarshani, A. M. A. (1996) Two lignans and an aryl alkanone from *Myristica dactyloides. Phytochemistry* 42: 1439-1442.

Heymsfield, S. B., Allison, D. B., Vasselli, J. R., Pietrobelli, A., Greenfield, D. and Nunez, C. (1998) *Garcinia cambogia* (hydroxycitric acid) as a potential antiobesity agent: a randomized controlled trial. *JAMA* 280:1596-1600.

Ho, S. C. and Tsai, P-J. (2004) Comparison of the effects of "hot" and "cold" Chinese medicinal plants on the production of inflammatory mediators by RAW 264.7 Cells. *J. Food Drug Anal.* 12: 120-127.

Hullatti, K. K. and Rai, V. R. 2004. Antimicrobial activity of *Memecylon malabaricum* leaves. *Fitoterapia* 75:409-11.

Hung, T. M., Dang, N. H., Cheol, K. J., Sue, C. J., Kyu, L. H. and Sun, M. B. (2009) Phenolic glycosides from *Alangium salvifolium* leaves with inhibitory activity on LPS-indeced NO, PGE2 and TNF-α production. *Bioorg. Med. Chem. Lett.* 19: 4389-4393

Islam, M. E., Haque, M. E. and Mosaddik, M. A. (2003) Cytotoxicity and antibacterial activity of *Sida rhombifolia* (Malvaceae) grown in Bangladesh. *Phytotherapy Res.* 17:973-975.

Itoigawa, M., Ito, C., Tan, H. T., Kuchide, M., Tokuda, H., Nishino, H. and Furukawa, H. (2001) Cancer chemopreventive agents, 4-phenylcoumarins from *Calophyllum inophyllum. Cancer Lett.* 169:15-19.

Jagtap, U. B. and Bapat, V. A. (2010) Artocarpus: A review of its traditional uses, phytochemistry and pharmacology. *J. Ethnopharmacol.* 129: 142–166.

Jarald, E., Joshi, B. S. and Jain, D. C. (2008) Diabetes and herbal medicines. *Iran J. Pharmacol. Ther.*7: 97-106.

Jarvill-Taylor, K. J., Anderson, R. A. and Graves, D. J. (2001) A hydroxychalcone derived from cinnamon functions as a mimetic for insulin in 3T3-L1 adipocytes. *J. Am. Coll. Nutr.* 20:327-36.

Jin, D. Q., Lim, C. S., Hwang, J. K., Ha, I. and Han, J. S. (2005) Anti-oxidant and anti-inflammatory activities of macelignan in murine hippocampal cell line and primary culture of rat microglial cells. *Biochem. Biophys. Res. Commun.* 331:1264–1269.

Kaibori, M., Sakitani, K., Oda, M., Kamiyama, Y., Masu, Y. and Okumura, T. (1999) Immunosuppressant FK506 inhibits inducible nitric oxide synthase gene expression at a step of NF-κB activation in rat hepatocytes. *J. Hepatology* 30: 1138–1145.

Kanth, V. R. and Diwan, P. V. (1999) Analgesic, anti-inflammatory and hypoglycaemic activities of *Sida cordifolia*. *Phytotherapy Res.* 13:75-77.

Karou, S. D., Nadembega, W. M. C., Ilboudo, D. P., Ouermi, D., Gbeassor, M., De-Souza, C. and Simpore, J. (2007) *Sida acuta* Burm. f.: a medicinal plant with numerous potencies. *African J. Biotechnol.* 6:2953-2959

Karuppusamy, S. and Rajasekaran, K. M. (2010) A new approach to high throughput antibacterial screening of plant extracts by resazurin redox with special reference to medicinal plants of Western Ghats *Adv. Biol. Res.* 4 : 201-206

Karuppusamy, S., Muthuraja, G. and Rajasekaran, K. M. (2009) Lesser known ethnomedicinal plants of Alagar hills, Madurai district of Tamil Nadu, India *Ethnobotanical Leaflets* 13: 1426-1433.

Kerr, J. F., Wyllie, A. H. and Currie, A. R. (1972) Apoptosis: a basic biological phenomenon with wide-ranging implications in tissue kinetics. *Br. J. Cancer* 26:239-257.

Khanna, D., Sethi, G., Ahn, K. S., Pandey, M. K., Ajaikumar, B., Kunnumakkara, Sung, B. Aggarwal, A. and Aggarwal, B. B. (2007) Natural products as a gold mine for arthritis treatment. *Curr. Opinion Pharmacol.* 7:344–351.

Khare, C. P. (2007) Indian medicinal plant: An illustrated dictionary (eds. Khare CP) *Springer*.

Kim, J., Lee, H. J., and Lee, K. W. (2010) Naturally occurring phytochemicals for the prevention of Alzheimer's disease. *J. Neurochem* 112: 1415-1430.

Kintzios, S. E. (2006) Terrestrial plant-derived anticancer agents and plant species used in anticancer research. *Crit. Rev. Plant Sci.* 25: 79-113.

Koperuncholan, M. and John, S. A. (2011) Antimicrobial and phytochemical screening in *Myristica dactyloides Gaertn. J. Pharm. Res.* 4: 398-400.

Kshirsagar, R. P., Darade, S. S. and Takale, V. (2010) Effect of *Alangium salvifolium* (Alangiaceae) on dexamethasone induced insulin resistance in rats. *J. Pharm. Res.* 3: 2714-2716.

Kulkarni, Y. A., Gokhale, S. B., Veeranjaneyulu, A., Surana, S. J. and Tatiya, A. U. (2009) Effect of *Persea macrantha* against acute inflammation and adjuvant-induced arthritis in rats. *Pharma. Biol.* 47: 304-308

Kumar, H. B., Basheer, S. and Haseena (2010) Antioxidant Potential And Antimicrobial activity of *Cinnamomum malabathrum* (Batka) *Oriental J. Chem.* 26: 1449-1453.

Lampronti, I., Khan, M. T. H., Borgatti, M., Bianchi, N. and Gambari, R. (2008) Inhibitory effects of Bangladeshi medicinal plant extracts on interactions between transcription factors and target DNA sequences. *eCAM* 5:303–312.

Lee, J. Y. and Park, W. (2011) Anti-Inflammatory effect of myristicin on RAW 264.7 macrophages stimulated with polyinosinic-polycytidylic acid. *Molecules* 16: 7132-7142

Lenta, B. N., Vonthron-Senecheau, C., Weniger, B., Devkota, K. P., Ngoupayo, J., Kaiser, M., Naz, Q., Choudhary, M. I., Tsamo, E. and Sewald, N. (2007) Leishmanicidal and cholinesterase inhibiting activities of phenolic compounds from *Allanblackia monticola* and *Symphonia globulifera*. *Molecules* 12:1548-1557.

Lewis DA (1989) Anti-inflammatory drugs from plant and marine sources. Birkhause Verlag, Basel. Wiley and Sons. Chichester, England.

Li, Y., Huang, T., Hsun-Wei and Yamahara, J. (2008) *Salacia* root, a unique Ayurvedic medicine, meets multiple targets in diabetes and obesity. *Life Sci* 82: 1045-1049.

Lia, Y., Penga, G., Lia, Q., Wenb, S., Huanga, T. H-W., Roufogalisa, B. D. and Yamaharac, J. (2004) *Salacia oblonga* improves cardiac fibrosis and inhibits postprandial hyperglycemia in obese zucker rats. *Life Sci.* 75:1735–1746

Liao, C. H., Ho, C. T. and Lin, J. K. (2005) Effects of garcinol on free radical generation and NO production in embryonic rat cortical neurons and astrocytes. *Biochem. Biophys. Res. Commun.* 329:1306-1314.

Lu, J., Zhang, K., Nam, S., Anderson, R. A., Jove, R. and Wen, W. (2010) Novel angiogenesis inhibitory activity in cinnamon extract blocks VEGFR2 kinase and downstream signaling. *Carcinogenesis* 31: 481-488.

Ma, J. S., Kim, W. J., Kim, J. J., Kim, T. J., Ye, S. K., Song, M. D., Kang, H., Kim, D. W., Moon, W. K. and Lee, K. H. (2010) Gold nanoparticles attenuate LPS-induced NO production through the inhibition of NF-κB and IFN-β/STAT1 pathways in RAW264.7 cells. *Nitric Oxide* 23: 214-219.

Ma, J., Hwang, Y. K., Cho, W. H., Han, S. H., Hwang, J. K. and Han, J. S. (2009) Macelignan attenuates activations of mitogen-activated protein kinases and nuclear factor kappa B induced by lipopolysaccharide in microglial cells. *Biol. Pharm. Bull.* 32: 1085–1090.

Mahire, N. B., Tote, M. V., Jain, A. P., Undale, V. R. and Bhosle, A. V. (2009) Anti-inflammatory effect of *Premna latifolia* leaves. *Pharmacologyonline* 3: 929-937

Mangal, A. and Sharma, M. C. (2009) Evaluation of certain medicinal plants for anti obesity properties. *Indian J. Traditional Knowl.* 8: 602-605.

Manjunath, B. K., Mankani, K. L., Mukunda, S. K., Divakara, R., Sridar, B. K. and Paul, K. (2011) Antioxidant and hepato protective effect of *Myristica malabarica s*eed aril extracts on carbon tetrachloride induced hepatic damage. *Global J. Biotechnol. Biochem.* 6: 25-30

Manosroi, J., Boonpisuttinant, K., Manosroi, W. and Manosroi, A. (2012) Anti-proliferative activities on HeLa cancer cell line of Thai medicinal plant recipes selected from MANOSROI II database. *J. Ethnopharmacol.* 142:422-431

Manosroi, J., Wilairat, R. and Manosroi, A. (2007) Anti-Proliferative activity of extracts from Thai plantsin Guttiferae and Schisandraceae families on human cancer cell lines. *Pharm. Biol.* 45: 255-258.

Mary, N. K., Babu, B. H. and Padikkala, J. (2003) Antiatherogenic effect of *Caps HT2*, a herbal Ayurvedic medicine formulation. *Phytomedicine* 10: 474-482.

Masayuki, Y., Yutana, P., Akinobu, K., Tadashi, K., Wang, T., Toshio, M. and Hisashi, M. (2003) Biological activities of *Salacia chinensis* originating in Thailand: The quality evaluation guided by α-glucosidase inhibitory activity. *Yakugaku zasshi* 123: 871-880.

Matsuda, H., Murakami, T., Yashiro, K., Yamahara, J. and Yoshikawa, M. (1999) Antidiabetic principles of natural medicines. IV.1) aldose reductase and □-glucosidase inhibitors from the roots of *Salacia oblonga* Wall. (Celastraceae): Structure of a new friedelane-type triterpene, kotalagenin 16-Acetate. *Chem. Pharm. Bull.* 47: 1725—1729

McGarry, S. J. and Williams, A. J. (1993) Digoxin activates sarcoplasmic reticulum Ca^{2+} release channels: a possible role in cardiac ionotropy. *Br. J. Pharmacol.* 108: 1043-50.

Mcqueen, E. G., Doyle, A. E. and Smirk, H. (1954) Mechanism of Hypotensive Action of Reserpine, an Alkaloid of *Rauwolfia serpentine. Nature* 174: 1015 doi:10.1038/1741015b0.

Mishima, S., Matsumoto, K., Futamura, Y., Araki, Y., Ito, T., Tanaka, T., Iinuma, M., Nozawa, Y. and Akao, Y. (2003) Antitumor effect of stilbenoids from *Vateria indica* against allografted sarcoma S-180 in animal model. *J. Exp. Ther. Oncol.* 3:283-288.

Morita, T., Jinno, K., Kawagishi, H., Arimoto, Y., Suganuma, H., Inakuma, T. and Sugiyama, K, (2003) Hepatoprotective effect of myristicin from nutmeg (*Myristica fragrans*) on lipopolysaccharide/D-galactosamine-induced liver injury. *J. Agri. Food Chem. 51:* 1560–1565.

Murugan, V., Shareef, H., Rama-Sarma G. V. S., Ramanathan, M. and Suresh, B. (2000) Anti-fertility activity of the Stem bark of *Alangium salviifolium* (Linn.f) Wang in Wistar female rats. *Indian J. Pharmacol.* 32: 388-389.

Myers, N., Mittermeier, R. A., Mittermeier, C. G., da Fonseca G. A. B. and Kent, J. (2000) Biodiversity hotspots for conservation priorities. *Nature* 403: 853-858.

National Center for Complementary and Alternative Medicine (NCCAM). *National Institutes of Health.* August 3. Available at: http://nccam.nih.gov/. Accessed August 7, 2007.

Newman, D. J., Cragg, G. M. and Snader, K. M. (2000) The influence of natural products upon drug discovery. *Nat. Prod. Reports* 17: 215

Nguelefack, T. B., Dutra, R. C., Paszcuk, A. F., Andrade, E. L., Tapondjou, L. A. and Calixto, J. B. (2010) Antinociceptive activities of the methanol extract of the bulbs of *Dioscorea bulbifera* L. var sativa in mice is dependent of NO–cGMP–ATP-sensitive-K^+ channel activation. *J. Ethnopharmacol.* 128: 567-574.

Nigam, P. K. (2007) Biochemical markers of myocardial infarction. *Indian J. Clin. Biochem.* 22: 10-17.

Ohe, G. M., Okamoto, T., Oshikawa, S., Furuichi, H., Nishikawa, T., Tano, K., Uyama, T., Bando, H., Yoshida, T., Sakai, K., Himeno, K., Sato, M. and Ohkubo, S. (2001) Th1-cytokine induction and anti-tumor effect of 55 kDa protein isolated from *Aeginetia indica* L., a parasitic plant. *Cancer Immunol. Immunother.* 50:251-259.

Okamoto, M., Oh-E, G., Oshikawa, T., Furuichi, S., Tano, T., Ahmed, S. U., Akashi, S., Miyake, K., Takeuchi, O., Akira, S., Himeno, K., Sato, M. and Ohkubo, S. (2004) Toll-like receptor 4 mediates the antitumor host response induced by a 55-kilodalton protein isolated from *Aeginetia indica* L., a parasitic plant. *Clin Diagn Lab Immunol.* 11:483-95.

Okamoto, M., Ohe, G., Oshikawa, T., Nishikawa, H., Furuichi, S., Bando, T., Yoshida, H., Sakai, T., Himeno, K., Sato, M. and Ohkubo, S. (2000) Purification and characterization of cytokine-inducing protein of seed extract from *Aeginetia indica* L., a parasitic plant. *Immunopharmacol.* 49:377-389.

Orwa G. (2009) *Agroforestry Database 4.0* 1-5.

Padhye, S., Ahmad, A., Oswal, N. and Sarkar, F. H. (2009) Emerging role of Garcinol, the antioxidant chalcone from *Garcinia indica* Choisy and its synthetic analogs. *J. Hematol. Oncol.* 2:2-38.

Palempalli, U. D., Gandhi, U., Kalantari, P., Vunta, H., Arner, R. J., Narayan, V., Ravindran, A. and Prabhu, K. S. (2009) Gambogic acid covalently modifies IκB kinase-β subunit to mediate suppression of lipopolysaccharide-induced activation of NF-κB in macrophages *Biochem. J.* 419: 401–409.

Pandey, M. K., Singh, S. K., Patki, P. S. (2010) Efficacy and safety of bresol. *Indian J. Clin. Pract.* 20: 707-713.

Parvin, M. N., Sarwar, S. and Chowdhury, S. A. (2009) *In-vitro* cytotoxicity and antioxidant studies of *Elaeocarpus serratus. S. J. Pharm. Sci.* 2: 86-90.

Patel, S., Patel, S. and Patel, V. (2011) Investigation into the Mechanism of Action of *Madhuca longifolia* for its anti-epileptic activity. *Pharmacogn Commun.* 1:18-22

Pawar, R. S. and Bhutani, K. K. (2004) Madhucosides A and B, protobassic acid glycosides from *Madhuca indica* with inhibitory activity on free radical release from phagocytes. *J. Nat. Prod.* 67:668-671.

Piggott, A. M., and Karuso, P. (2004) Quality, not quantity: the role of natural products and chemical proteomics in modern drug discovery. *Comb. Chem. High Throughput Screen.* 7: 607-630.

Porchezhian, E., Ansari, S. H., and Ahmad, S. (2001) Analgesic and anti-inflammatory effects of *Alangium salvifolium. Pharm Biol.* 39: 65-66.

Pradhan, P., Joseph, L., Gupta, V., Chulet, R., Arya, H., Verma, R. and Bajpai, A. (2009) *Saraca asoca* (Ashoka): A review. *J. Chem. Pharm. Res.* 1:62-71.

Prakasha, H. M. and Krishnappa, M. (2006) People's knowledge on medicinal plants in Srigeri taluk, Karnataka. *Indian J Traditional Knowl.* 5: 353-357.

Prasad, D. G. (2004) Herbal preparation for management of cardiovascular and neurologic disorders. US patent US 2006/0034950 A1

Pullaiah, T. (2006) Encyclopedia of world medicinal plants Volume-2, Regency Publication, New Delhi. pp 852-853.

Punitha, S. R., Rajendran, K., Shirwaikar, A. and Shirwaikar, A. (2005) Alcoholic stem extract of *Coscinium fenestratum* regulates carbohydrate metabolism and improves antioxidant status in streptozotocin–nicotinamide induced diabetic rats. *eCAM* 2:375–381

Rajendran, R., Suseela, L., Meenakshi, S. R. and Saleem, B. N. (2008) Cardiac stimulant activity of bark and wood of *Premna serratifolia. Bangladesh J. Pharmacol.* 3: 107-113.

Rao, J. M., Tiwari, A. K., Kumar, U. S., Yadav, J. S. and Raghavan, K. V. (2004) (+)-Cycloolivil as antioxidant obtained from *Stereospermum personatum.* European patent EP1411910.

Rao, K.V., McBride, T. J. and Oleson, J. J. (1968) Recognition and evaluation of Lapachol as an antitumor agentâ. *Cancer Res.* 28: 1952-1954.

Rates, S. M. K. (2011) Plants as sources of drugs. *Toxicon* 39: 603-613.

Rosato, A., Vitali, C., Laurentis, N., Armanise, D. and Nulillo, M. A. (2007) Antibacterial effect of some essential oils administered alone or in combination with Norfloxacin. *Phytomedicine* 14:727

Safayhi, H., Sailer, E. R. and Ammon, H. P. (1995) Mechanism of 5-lipoxygenase inhibition by acetyl-11-keto-beta-boswellic acid. *Mol. Pharmacol.* 47:1212-1216.

Sen, S., Chakraborty, R., Sridhar, C., Reddy, Y. S. R., De, B. (2010) Free radicals, antioxidants, diseases and phytomedicines: Current status and future prospect. *Int. J. Pharm. Sci. Rev. Res.* 3:91-100.

Senthil, S., Sridevi, M. and Pugalendi, K. V. (2007) Protective effect of ursolic acid against myocardial ischemia induced by isoproterenol in rats. *Toxicology Mechm. Methd.* 17: 57-65.

Shah, A. J., Gowani, S. A., Zuberi, A. J., Ghayur, M. N. and Gilani, A. H. (2010) Antidiarrhoeal and spasmolytic activities of the methanolic crude extract of *Alstonia scholaris* L. are mediated through calcium channel blockade. *Phytother Res.* 24:28-32.

Shah, B. N., Nayak, B. S., Seth, A. K., Jalalpure, S. S., Patel, K. N., Patel, M. A. and Mishra, A. D. (2006) Search for medicinal plants as a source of anti-inflammatory and anti-arthritic agents - A review. *Pharmacogn. Mag.* 2: 77-86.

Shailasree, S., Ruma, K. and Prakash, H. S. (2012) Curative properties of *Buchanania lanzan*: an evidence for traditional knowledge as evaluated by its capacity to inhibit inflammatory enzymes. J Natural Pharmaceuticals (In Press)

Shang, J. H., Cai, X. H., Feng, T., Zhao, Y. L., Wang, J. K., Zhang, L. Y., Yan, M. and Luo, X. D. (2010) Pharmacological evaluation of *Alstonia scholaris*: anti-inflammatory and analgesic effects. *J. Ethnopharmacol.* 129:174-81.

Shang, J. H., Cai, X. H., Feng, T., Zhao, Y. L., Wang, J. K., Zhang, L. Y., Yan, M. and Luo, X. D. (2010) Pharmacological evaluation of *Alstonia scholaris*: anti-inflammatory and analgesic effects. *J. Ethnopharmacol.* 129:174-81.

Sharma, A., Namdeo, A. G. and Mahadik, K. R. (2009) Pharmacognostic studies on *Nothapodytes nimmoniana* (j. Graham) mabberly. *Int J Pharma Res Dev* – Online Publication Ref No.: IJPRD/2009/PUB/ARTI/VOV-1/ISSUE-9/NOV/008 ISSN 0974 – 9446 pp1-10

Shetty, P., Vijayanarayana, K., D'Souza, U. P., Shama, P. K. and Shastry, C. S. (2011) *Hepatoprotective properties of Memecylon umbellatum leaves. Int. J. Pharma. Sci. Biotech.* 1:313-315.

Shishodia, S. and Aggarwal, B. B. (2004) Guggulsterone inhibits NF- κB and IαBα kinase activation, suppresses expression of anti-apoptotic gene products, and enhances apoptosis. *J. Biol. Chem.* 279:47148-47158.

Singh, A. and Duggal, S. (2010) *Salacia* spp.: Hypoglycemic principles and possible role in diabetes management. *Integr Med* 9:40-44.

Singh, I. P., Bharate, S. B. and Bhutani, K. K. (2005) Anti-HIV natural products. *Curr. Sci.* 89: 269-290.

Sunil, C., Ignacimuthu, S. and Agastian, P. (2011) Antidiabetic effect of *Symplocos cochinchinensis* (Lour.) S. Moore. in type 2 diabetic rats. *J. Ethnopharmacol.* 134:298-304.

Sunil, C., Ignacimuthu, S. and Kumarappan, C. (2012) Hypolipidemic activity of *Symplocos cochinchinensis* S. Moore leaves in hyperlipidemic rats. *J. Nat. Med.* 66:32-38.

Surveswaran, S., Cai, Y-Z., Corke, H. and Sun, M. (2007) Systematic evaluation of natural phenolic antioxidants from 133 Indian medicinal plants. *Food Chem.* 102: 938–953

Sutradhar, R. K., Rahman, A. M., Ahmad, M., Bachar, S. C., Saha, A. and Guha, S. K. (2006) Bioactive alkaloid from *Sida cordifolia* Linn. with analgesic and anti-Inflammatory activities. *Iran J. Pharmacol. Ther.* 5:175-178.

Takada, Y., Ichikawa, H., Badmaev, V. and Aggarwal, B. B. (2006) Acetyl-11-ketobeta-boswellic acid potentiates apoptosis, inhibits invasion, and abolishes osteoclastogenesis by suppressing NF-κB and NF-κB-regulated gene expression. *J. Immunol.* 176:3127-3140.

Tilak-Jain, J. A. and Devasagayam, T. P. A. (2006) Cardioprotective and other beneficinal effects of some Indian medicinal plants. *J. Clin. Biochem. Nutr.* 38: 9-18.

Tushar, K. V., George, S., Remashree, A. B. and Balachandran, I. (2008). *Coscinium fenestratum* (Gaertn.) Colebr.-A review on this rare, critically endangered and highly-traded medicinal species. *J. Plant Sci.* 3: 133-145.

Ulrich-Merzenich, G., Panek, D., Zeitler, H., Vetter, H. and Wagner, H. (2010) Drug development from natural products : *Exploiting synergistic effects.* 48: 208-219.

Velavan, S., Selvarani, S. and Adhithan, A. (2009) Cardioprotective effect of *Trichopus zeylanicus* against myocardial ischemia induced by isoproterenol in rats. *Bangladesh J. Pharmacol.* 4: 88-91.

Wakchaure, D., Jain, D., Singhai, A. K. and Somani, R. (2011) Hepatoprotective activity of *Symplocos racemosa* bark on carbon tetrachloride-induced hepatic damage in rats. *J. Ayurveda Integr. Med.* 2:137-143.

Wang, S. Q., Song, L. S., Lakatta, E. G. and Cheng, H. (2001) Ca^{2+} signaling between single L-type Ca2+ channels and rynodine receptors in heart cells. *Nature* 410: 592-596.

Warokar, A. S., Ghante, M. H., Duragkar, N. J. and Bhusari, K. P. (2010) Anti-inflammatory and Antioxidant Activities of Methanolic extract of *Buchanania Lanzan* Kernel. *Indian J. Pharm. Educ. Res.* 44: 363-368

Yang, S. S., Cragg, G. M., Newman, D. J. and Bader, J. P. (2001) Natural Product-Based Anti-HIV Drug Discovery and Development Facilitated by the NCI Developmental Therapeutics Program. *J. Nat. Prod.* 64: 265-277.

Yong, P. (1997) Major microbial diversity initiative recommended. *ASM News* 63: 417.

Yoshida, F., Asai, R., Majima, H., Tsukagoshi, S., Furue, H., Suminaga, M., Sakamoto, K., Niitani, H., Murata, A. and Kurihara, M. (1994) Study of cardiac function in first stage examination of RA-700. *Gan To Kagaku Ryoho* 21:199-207.

In: Plant Extracts
Editors: Adriana Giordano and Alfieri Costs
ISBN: 978-1-62417-534-3

Chapter 3

Moringa Oleifera: Benefits in Nutrition and Health

W. Cordier[1], E. S. du Toit[2] and V. Steenkamp[1]
[1]Department of Pharmacology, Faculty of Health Sciences, University of Pretoria, Pretoria, South Africa
[2]Department of Plant Production and Soil Sciences, Faculty of Natural and Agricultural Sciences, University of Pretoria, Pretoria, South Africa

Abstract

Moringa oleifera Lam. (Moringaceae), also known as the horseradish or drumstick tree, is a native plant of pan-tropical areas such as sub-Himalayan regions of India, Pakistan, Bangladesh and Afghanistan. However, other locations such as Africa, Ceylon and Thailand have started cultivating this plant. Due to the use of almost every part of the plant, its resistance to drought, pests and disease, as well as its acclimatization to different soil types and high biomass per annum yields it has become known as a 'miracle tree'. Apart from its use as a nutritional aid and medicinal remedy in both humans and animals, it has also been utilized as a source of biofuel. These effects are attributed to the many different macro- and micronutrients stored within the Moringa plant, as well as the presence of bioactive phytochemicals. Leaves contain high levels of protein and carbohydrates, vitamin A and C, and minerals such as calcium, supporting its use as meal replacement in malnourished individuals or as animal fodder. Medicinally it is used to treat various disorders, such as ascites, rheumatism, inflammation, infections, diabetes and cancer. It is also used in the prevention of hepatic, renal and cardiovascular diseases. These uses are supported by the bioactivities reported for the plant, including antioxidant, anti-inflammatory, hepatoprotective, nephroprotective, antimicrobial, larvacidal, antidiabetic and radioprotective activities. Toxicological studies have shown that it is well-tolerated at high doses in several animal models. It is apparent that due to the beneficial properties and low toxicity of Moringa that it could become an inexpensive and safe alternative to various allopathic medicines, as well as providing nutrition to people in developing countries.

Keywords: Horseradish tree, *Moringa oleifera*, nutrition, pharmacological activity, toxicity

1. Introduction

Approximately 80% of the population of Africa and Asia rely on alternative medicines, such as herbal remedies, as primary health care. [1] The popularity of herbal remedies is growing due to their relative low-cost and easy procurement. Various cultures have established their own pharmacopeia of locally grown plants for the treatment of numerous ailments. Although vegetables are ingested as food for their nutritional value it has come to light that they could provide protection against the progression of various diseases. [2] Supplementation containing these herbal preparations and vegetables are becoming more popular and marketed as nutraceuticals. [3] The potential of these plants to decrease or reverse the progression of disease is being investigated.

Moringa oleifera Lam. (Moringaceae) (synonym *M. pterygosperma* Gaertner) is known as the horseradish tree, benzolive, drumstick tree, kelor, marango, mlonge, mulangay, nébéday, saijhan, magic tree, sajna and miracle tree. [4-7] This plant has been used since ancient times by the Greeks, Romans and Egyptians. [8] It is one of the best known members of the Moringaceae family and is cultivated as a vegetable, phytomedicine or vegetable oil source. [9] It is native to pan-tropical areas such as sub-Himalayan regions of India, Pakistan, Bangladesh and Afghanistan. It's popularity has caused it to be incorporated into other tropical locations, including Africa, Ceylon and Thailand. [5] *M. oleifera* is used to treat various diseases and disorders, including inflammatory, cancerous, hepatorenal and cardiovascular pathologies. [9]

M. oleifera is considered to be highly versatile with almost every part being used for either a food source or other beneficial quality. [10] *M. oleifera* is placed in the same market segment as "green superfoods" such as spirulina, green barley, wheat and alfalfa sprouts. [11] In South Africa, there are currently two start-up farming companies developing *M. oleifera* for its leaves, however internationally there are a number of companies already utilising these plant products. The leaves are beneficial for human nutrition. In some areas the value of the seeds in water treatment and the seed oil for cooking and cosmetic purposes are also important. This review broadly summarises reports on *M. oleifera* with emphasis on its nutritional and pharmacological value. Since safety of use is essential, studies regarding the toxicity of the plant are also included in this review.

2. Taxonomy and Morphology

M. oleifera belongs to the family Moringaceae, which is a monogeneric family of shrubs and trees. [12] The small genus comprised of ten species is distributed across Africa and Asia. This genus can be divided into three groups: slender trees, bottle trees and tuberous shrubs. *M. oleifera* is categorized as a slender tree. [12]

The tree is medium-sized and reaches about 10 m in height. The trunk is straight and approximately 10-30 cm thick, with the bark being whitish grey, corky, with longitudinal cracks. The tuberous taproot helps the plant to tolerate drought conditions.

Figure 1. *M. oleifera* (photo taken by ES du Toit).

The umbrella shaped tree comes with a loose crown of feathery foliage. [13] The foliage is evergreen or deciduous depending on the environment. In season the tree is covered with creamy white, honey-scented flowers arranged in drooping panicles. Flowers are insect pollinated and require a large number of insect visitations.

Flowers and fruits (pods) occur once or twice a year depending on the climate, however, in tropical areas these events may occur year-round. [13] The fruits are initially light green, slim, and tender, but turn dark green and firm to brown at maturity stage. Although most fruit shapes are straight, a few are wavy or curly. Fully mature dried seeds are covered by a lightly wooded shell with three papery wings. [13] Seeds have no dormancy period, so they can be planted as soon as they are mature and they will retain the ability to germinate for up to one year. [12]

3. Distribution

M. oleifera is indigenous to northwest India. [14] It grows wild in the sub-Himalayan tract from the river Chenab eastwards to the Sarda and in the tarai tract of Uttar Pradesh in India. The trees are also widely distributed in Egypt and are naturalized in Sri Lanka, Malaysia and the Philippines. It has been introduced to the West Indies, tropical America (from Mexico to Peru, Paraguay, and Brazil) and Africa. [13] In Africa, although widespread, *M. oleifera* is seldom cultivated intensively and in only selected areas is it put to use in an optimal manner. [13]

4. Propagation

Commercial plant production nurseries produce mainly container grown *M. oleifera*, planted either by seed, or by using hardwood stem cuttings. Seeds can be sown directly to the field or in seedbeds. [11,12]

Direct seeding is preferred when labour is limited and plenty of seed is available. Transplanting seedlings from seedbeds and containers also allows flexibility and a choice to the farmer to plant superior selections during field planting. To ensure good returns on investment, seeds should be tested for viability. A good seed should also be clean and disease free. Seeds should not be stored over long periods as they lose their germination capacity after approximately a year or two. [11,12]

Container grown stem cuttings are used when labour is available and seed is scarce. An advantage of stem cuttings is the multiplication of uniform clonal material which is preferred to predict growth performance and yields. [11,12]

Compared to trees planted from seed, trees from stem cuttings grow faster, but develop a shallow root system that makes them more sensitive to drought conditions. Hardwood stem cuttings are usually made from branches of a tree that are at least 10 to 12 months old. [11,12]

M. oleifera is a drought- and disease-resistant plant. [15] The yield is generally low in the first two years, while from the third year it generates a high biomass yield per hectare, with approximately 600 fruits per tree. There may be two crops per year in areas such as south India. [16]

5. Uses

5.1. Ethnomedicinal Uses

Almost all parts of *M. oleifera* have ethnomedicinal uses. Leaves are used for inflammatory disorders, microbial infections, cardiovascular pathologies, hepatorenal disorders, gastrointestinal concerns, cancers, sores, headaches, diabetes, hypercholestraemia, hypertension, haematological disorders and wound healing. [9,17] Seeds are used as an enhancer of sexual virility in both males and females. [8] Root, bark and fruits are used to treat piles and urinary discharge in Maharashtra, India. [18] Moringa seeds and leaves are used to treat diabetes mellitus in Limpopo, South Africa. [19] In the Philippines it is used to increase woman's milk production and to treat anemia. [20]

5.2. Other Uses

Seeds yield 38 - 40% non-drying oil, known as Ben oil, used for lighting, in arts and for lubricating watches and other delicate machinery. Ben oil is clear, sweet and odourless. As it turns rancid slowly, it is edible and useful in the manufacture of perfumes and cosmetics. [14] This oil resembles olive oil and may be of some value as a constituent of non-yellowing, non-drying plasticising alkyds. It is highly valued by perfumers as it absorbs and retains fugitive odours. [21]

The corky, grey bark yields a coarse fibre which is utilised in making mats and paper. The stem of the tree exudes a gum when injured, which is initially white but changes to reddish-brown or brownish-black on exposure. Although sparingly soluble in water, it swells when it comes into contact with water, making it a highly viscous solution. It is a polyuronide consisting of arabinose, galactose and glucuronic acid, as well as trace amounts of rhamnose. The gum is used in calico printing. The dark colour of the gum is much valued in the European market. [21]

Other uses include plant growth enhancers (fertilizer) from leaves, honey collected from bees, biogas from all tree parts, animal fodder from leaves and twigs, branches and seed press cake, fuel wood, sheltering/fencing from trees, spices (roots), agroforestry (shade for vegetables during intercropping) and for water purification (charged proteins in seed tissue). [12-14] As the moisture content in leaves is low, microbial infection is unlikely. [5] Due to its ability to froth, the leaves are used as a soap. [22]

6. Nutritional Value

Nutritional values of the plant parts vary between geographic location depending on agroclimatic conditions and maturity. [4] This tree is mainly valued for the tender pods which are cut into slices and used in cooking preparations or pickled. The soft seed extracted from immature pods are boiled and eaten like peas. [13] Seeds are consumed after frying and taste like peanuts. [12,14] Fresh flowers are eaten as vegetables [12,14], whereas both fresh and dried flowers are used for making teas. [23]

Leaves are eaten in various forms, including as a constituent of salads or vegetable dishes. They can be stored for many months without need for refrigeration with little loss of nutritional value. [5] Leaves provide an inexpensive source of proteins, vitamins and minerals. The leaves contain more beta-carotene than carrots, more protein than peas, more vitamin C than oranges, more calcium than milk, more potassium than bananas, and more iron than spinach. [23] Dried and milled leaves are stored and the powder added to daily meals. The leaf powder is sometimes used by food businesses to enrich their products in nutrients. The spicy roots are used as a garnish or pickled. For the latter they are peeled, dried, ground and stored in vinegar. [13,14] The drumsticks contain β-carotene as well as vitamin A and C. [9]

Crude protein concentrations are highest in meal and kernels, followed by leaves (Table 1). High quantities of sugars are present in the stems and roots, whereas higher fiber and cellulose content is found in pods than leaves. Gross energy appears to be similar between leaves, twigs, stems and meal, with higher quantities present in shells and kernels.

Younger leaves possess a lower ash content which increases with age as minerals accumulate. With aging the carbohydrate content decreases and protein content increases. [5] Moringa leaves provide energy, while having a low content of fats. [5,26] Fat content does however allow for better palatability by retaining flavours, and allows for a healthy percentage of the caloric energy (1 – 2%). [10] High levels of essential amino acids (Table 2), especially glutamic acid, is present which form the building blocks of proteins. [4] Various minerals are present as well, including calcium, phosphorous, potassium, magnesium, sodium and iron. [10]

Table 1. Nutritional values reported in literature for *M. oleifera*

Parameter	Leaves[a]	Twigs[b]	Stems[c]	Roots[d]	Shells[e]	Seeds[f]	Pods[g]	Kernels[h]	Meal[i]
Crude protein (%)	23.67	7.2	5.60	3.80	6.95	11.00	17.10	36.70	61.40
Crude fibre (%)	11.82	-	-	-	-	-	36.7	-	-
Fat (%)	5.07	-	-	-	2.20	-	3.9	-	0.00
Sugars (%)	5.18	-	24.57	25.89	0.89	3.37	-	-	-
Cellulose (%)	6.00	-	-	-	-	-	39	-	-
Gross energy (kcal/kg)	4069	4426	4529	-	5167	-	3540	6376	4634
Ash (%)	10.64	9.3	6.9	-	2.27	-	9.47	3.18	5.65

Values are averages from the following references:
a – [4,5,10,15,24-32]; b – [4]; c – [4,29]; d – [29]; e – [4,29]; f – [29]; g – [32,33] ; h – [4]; i – [4].

Table 2. Amino acid composition of different parts of *M. oleifera* [4]

Amino acids	g/16 g nitrogen				
	Leaves	Kernels	Meal	Extracted kernels	Extracted meal
Lysine	5.6	1.41	1.47	1.62	1.48
Leucine	8.7	5.11	5.27	5.85	5.84
Isoleucine	4.5	2.99	3.05	3.45	3.49
Methionine	1.98	1.93	1.9	2.23	2.13
Cystine	1.35	4.13	4.22	4.82	4.72
Phenylalanine	6.18	3.83	3.97	4.3	4.29
Tyrosine	3.87	1.44	1.5	1.52	1.41
Valine	5.68	3.4	3.47	3.72	3.63
Histidine	2.99	2.2	2.27	2.38	2.28
Threonine	4.66	2.15	2.25	2.41	2.28
Serine	4.12	2.64	2.75	3.05	2.85
Glutamic acid	10.22	19.46	19.35	20.7	19.63
Aspartic acid	8.83	3.8	3.97	3.96	3.76
Proline	5.43	5.27	5.52	6.37	6.04
Glycine	5.47	4.76	4.9	5	4.4
Alanine	7.32	3.67	3.77	4.15	4.05
Arginine	6.23	11.41	11.63	13.87	16.68
Tryptophan	2.1	ND	ND	ND	ND

Seeds contain various fatty acids, and appear to correlate well between different geographical areas and studies. [34,35] The seeds contain very few polyunsaturated fatty acids and has a high oxidative stability [36], while the leaves contain more polyunsaturated fatty acids. [26] Oleic acid is the major fatty acid found in seed oil (72.83%), with behenic (6.09%), palmitic (5.89%) and stearic acid (5.67%) are also found in high proportions. [34-36] Additionally, arachidic (3.80%), eicosenoic (2.22%), palmitoleic (1.27%), lignoceric

(1.08), linoleic (0.76%), α-linoleic (0.21%), erucic (0.12%) and myristic acid (0.12%) [34,35,37] are present.

Moringa does not contain high quantities of anti-nutritional factors. Although leaves, kernels and meal contain substantial saponin content, this does not appear to induce saponin-related toxicity. [4] However, phytate levels in kernels and meal might reduce the bioavailability of minerals such as zinc and calcium in monogastrics. [4] Cyanogenic glycoside concentrations are not considered near toxic levels. Although glucosinolate content might contribute to flavour, too high levels might be considered harmful and anti-nutritional. The kernels generally have a bitter taste, which could be attributed to alkaloids, cyanogenic glycosides, saponins and glucosinolates. [4]

7. Phytochemical Composition

Leaves have been shown to contain carbohydrates, alkaloids, glycosides, terpenoids, saponins, flavonoids, phenolic acids, tannins and fats. [4,38-42] Leaves contain approximately 93 volatile compounds, including those of aldehydes, alcohols, ketones, hydrocarbons, esters, terpenoids and acids. [27] Acetic acid and hexanoic acid appear to be the major volatile components. [27] Flowers contain saponins and anthraquinones [41], while seeds contain saponins, lectins and alkaloids. [38,41,43] Stem-bark contains sterols, glycosides, alkaloids, triterpenoids, flavonoids and tannins. [22] Kernels contain cyanogenic glycosides, alkaloids, saponins and glucosinolates. [4] Pods contain glucosinolates, isothiocyanates, thiocarbamates and flavonoids. [9] Various phenolic and flavonoid compounds are present in the leaves [44], seeds [45], stem-bark [46] and flowers [47], including catechin, caffeic acid, epicatechin, rutin, ferulic acid, ellagic acid, chlorogenic acid, gallic acid, quercetin, myricetin, kaempferol and vanillin. [7,44-48] The alkaloid, trigonelline, is present in high levels in the pods, but also found in the leaves, roots, seeds and flowers. [49] Numerous glycosides are present in the leaves, including niazirin, niaziridin [50], niazarin (4-(α-L-rhamnopyranosyloxy) phenylacetonitrile, pyrrolemarumine 4”-*O*-α-L-rhamnopyranoside, marumoside A, marumoside B, methyl 4-(α-L-rhamnopyranosyloxy)-benzylcarbamate, benzyl β-D-glucopyranoside, benzyl β-D-xylopyranosyl-(1→6)-β-D-gluco-pyranoside, kaempferol 3-*O*-β-D-glucopyranoside and quercetin 3-*O*-β-D-glucopy-ranoside. [51] The nitrile glycoside niazarin has also been isolated from the pods. [33] Fruits contain niazicin A, niazinin A, niazirin, 4-[(4’-*O*-acetyl-α-L-rhamnosyloxy)benzyl] isothio-cyanate, 4-[(α-L-rhamnosyloxy)benzyl]isothiocyanate, 4-[(2’-*O*-acetyl-α-L-rhamnosyl-oxy) benzyl] isothiocyanate, 4-[(3’-*O*-acetyl-α-L-rhamnosyloxy)benzyl] isothiocyanate and *S*-methyl-*N*-{4-[(α-L-rhamnosyloxy)benzyl]}thiocarbamate. [52]

Flowers contain the triterpene, β-amyrin, and sterol, β-sterol. [47] Two new caffeoyl quinic acid α-glucosides, 4-O-(4′-O-α-d-glucopyranosyl)-caffeoyl quinic acid and 4-*O*-(3′-*O*-α-d-glucopyranosyl)-caffeoyl quinic acid, have been isolated from moringa leaves. [53] Fumaric acid has been isolated from the stem-bark. [46] Methyl 2-[4-(α-L-rhamnopyranosol)phenyl]acetate, *N*-[4-(β-L-rhamnopyranosol)benzyl]-1-*O*-α-D-glucopyranosylthiocarboxamide, 1-*O*-phenyl-α-L-rhamnopyranoside, 4-[(β-D-glucopyranosol)-(1→3)-(α-L-rhamnopyranosol)] phenylacetonitrile, niazicin A, methyl *N*-{4-[(α-L-rhamnopyranosol) benzyl]}carbamate and methyl *N*-{4-[(4’-*O*-acetyl-α-L-rhamnopyranosol)benzyl])carbamate has been isolated from the fruits. [54] The aromatic monoamine, benzylamine or moringine, is present in various parts of the plant. [55] The leaves contain the glucosides 4-O-(4′-O-α-D-glucopyranosyl)-caffeoyl quinic acid and 4-*O*-(3′-*O*-α-D-glucopyranosyl)-caffeoyl quinic acid. [56] Aurantiamide acetate and 1,3-dibenzyl urea have been isolated from the roots. [57] Protease inhibitors have been isolated from the bark, roots, flowers, seeds and leaves. [47,58,59] The leaves and seeds yield the highest quantity of the protease inhibitor. [58,59]

Seeds contains numerous fixed oils, including 9-hexadecanoic acid, hexadecanoic acid, eicosanoic acid, 9-octadecanoic acid, docosanoic acid, pentadecanoic acid, tetracosanoic acid, hepatadecanoic acid, 1,3,5-cycloheptatrine, pentacosane and octacosane. [60]

8. Pharmacological Activity

8.1. Antioxidant Activity and Organoprotective Properties

Oxidative stress, or the overburdening of the body with free radicals has been coupled to the aetiology of various diseases, including cardiovascular complications, diabetes, neurodegeneration, inflammatory disorders and cancers. [44,61,62] Free radicals result in the oxidation and degradation of various biomolecules, such as proteins and lipids, which leads to lipid peroxidation, DNA damage, mitochondrial membrane destabilization and cellular death. [44,62] These can result in the progression of diseases. Antioxidants have gained popularity in literature due to their ability to neutralize free radicals, and elicit organoprotective functions. [44,61,62]

Although fruit peels do not possess any significant antioxidant activity [63], the seeds [45,64], leaves [44,65,66], fruit [45], stalks [28], bark [22] and stems [28] show varied antioxidant activity. Antioxidant activity appears to be more potent in leaves than fruits and seeds. [45] Antioxidant activity has been shown to be more pronounced in organic extracts, such as diethyl ether and ethyl acetate, than aqueous extracts due to higher extraction of polyphenolic constituents. [44] Also the antioxidants in leaves appear to have high thermal stability. [67]

Polyphenols are known to contribute greatly to antioxidant activity. Polyphenols isolated from *M. oleifera*, include gallic acid, ellagic acid, ferulic acid, kaempferol and quercetin. [44] The polyphenolic content in leaves constitutes approximately 3.42% [4,7,27,31,66,68,69], although values as high as 25% [44] and 51% [22] have been reported. Polyphenolic content in seeds [70], twigs and stems [4] total 4%, 1% and 0.5%, respectively. The meal and kernels appear to contain insignificant amounts of polyphenols. [4]

Aqueous extracts of leaves, fruit and seeds (5 to 25 μg/ml) have been reported to attenuate oxidative damage inflicted upon DNA plasmids. [45] Additionally, synergistic protection has been noted when extracts were combined with the synthetic antioxidant, Trolox. [45] Aqueous and ethanol extracts of leaves and pods have been shown to protect erythrocytes and lipids against oxidative stress, due to increased concentrations of reduced glutathione. [71] Aqueous leaf extracts at 50 mg/ml were found to decrease hydrogen-peroxide induced oxidative stress in KB tumour cells by reducing cytotoxicity (20%), DNA damage (50%) and lipid peroxidation (50%), as well as increasing the concentration of superoxide dismutase and catalase (near-normal levels). [72] The authors attributed this to the ability of polyphenolic constituents, such as quercetin and kaempferol, to attenuate hydrogen peroxide-induced damage. [72] Isolated compounds from the fruits; niazicin A, *N*-[4-(β-L-rhamnopyranosol) benzyl]-1-*O*α-D-glucopyranosylthiocarboxamide,4-[(β-D-glucopyranosol) -(1→3)-(α-L-rhamnopyranosol)]phenylacetonitrile and methyl *N*-{4-[(4'-*O*-acetyl-α-L-rhamnopyranosol)benzyl])carbamate have been found to decrease *in vitro* lipid peroxidation at 100 μg/ml. [54]

Hydro-ethanol leaf extracts have been shown to decrease lipid peroxidation induced in mice, rats and rabbits by gamma-irradiation, [7] streptozocin [73] and gentamicin [74], respectively. Ethanol and hexane leaf extracts increased antioxidant status in streptozocin-induced diabetic rats [73] and monocrotaline-induced pulmonary hypertensive rats. [50] Alcoholic extracts of pods increased antioxidant status and decreased lipid peroxidation induced by streptozocin [75] and 7,12-dimethylbenz(a)anthracene. [9,76]

Hexane extracts of leaves decreased pulmonary hypertension and arterial wall thickness in monocrotaline-induced pulmonary hypertensive rats. [50] A flavonoid-enriched fraction of leaves (2.5 mg/kg) decreased sodium-selenite induced cataractogenesis in rats by 79%, which could be attributed to the increased antioxidant status and reduced oxidative stress to near-normal levels. [77] Fluoride toxicity in rabbits was decreased with oral administration of aqueous seed extracts, which was evident from decreased serum hepatic enzyme levels, biochemical parameters (creatinine, urea and fluoride) and increased cortical indices. [78] This is thought to be due to decreased gastrointestinal absorption of fluoride due to the presence of a cationic polyelectrolyte binding to fluoride. [78] Administration of seed powder, post-exposure of arsenic, reduced chronic toxicity in rats. [79] Treatment resulted in reduced lipid peroxidation in the liver and kidneys, as well as increased reduced glutathione (blood) and superoxide dismutase (liver, kidneys and brain). Furthermore, a marginal reduction in arsenic levels was noted in the kidneys, liver and brain. [79] Aqueous stem-bark extracts (250 and 500 mg/kg) reduced isoproterenol-induced cardiotoxicity in rats, as was observed in histopathological examination. [80] Treated hearts were shown to have less oedema and inflammation than isoproterenol-exposed hearts. Furthermore, the antioxidant status of the heart was increased and lipid peroxidation decreased. Aspartate transaminase, alanine transaminase, lactate dehydrogenase and creatine kinase leakage from the heart was decreased. [80] A higher concentration of reduced glutathione and superoxide dismutase were present in Xhosa lop-eared goats fed leaves compared to grass hay, as well as having a reduced frequency of lipid peroxidation. [66]

Various *in vivo* studies have been carried out to determine the protective action of *M. oleifera* against hepatotoxicity. An ethyl acetate fraction of leaves reduced carbon tetrachloride-induced lipid peroxidation in the liver and kidney of rats at concentrations as low as 50 mg/kg. [44] Furthermore, concentrations of reduced glutathione, catalase and superoxide dismutase were elevated during treatment. [44] Ethanol extracts of leaves decreased hepatotoxicity induced by paracetamol by increasing antioxidant status, decreasing release of hepatic enzymes and protecting against histopathological changes. [81,82] Seed oil decreased carbon tetrachloride-induced hepatotoxicity through the alleviation of lipid peroxidation, protein degradation and elevation of serum hepatic enzyme levels. [83] Histopathological analysis indicated decreased structural damage to hepatocytes. This was thought to be through the antioxidant activity of tocopherols present in the oil, as well as fatty acid components. [83]

8.2. Antimicrobial Activity

Microbial infections form an immense problem worldwide, especially due to the presence of antimicrobial resistance. These infections can result in respiratory disorders [84], worsening of wounds [8], fungal contaminations and gastric irritation [85] amongst others. Activity has been shown against commonly found microbes (Table 3 and Table 4), including those causing respiratory infections (*Staphylococcus aureus*) [84], food poisoning (*Bacillus cereus*) [84] and gastric distress (*Escherichia coli*). [85] As the seeds are used in the coagulation and treatment of sewage water, the presence of antimicrobial activity is of considerable interest in eliminating water-borne microbes. Gram-negative bacteria appear to be more susceptible to the effects of *M. oleifera* than Gram-positive bacteria. In the majority of cases antimicrobial activity was only found at concentrations higher than 2 mg/ml.

Aqueous seed extracts have been shown to decrease growth of spread inoculums of *Aspergillus fumigates* [86], *Candida albicans* [87], *Aspergillus niger, Fusarium solani* and *Metarhizium aniscoplae* strains, while only slight activity has been noted against *Rhizopus solani.* [88] Microscopic evaluation indicated that antifungal activity resulted in structural damage to hyphae and conidia. [88] Antibacterial activity has been attributed to a variety of phytochemicals, including three lectins (SLL-1, SLL-2 and SLL-3) [43], hydroxyphenyl acetic acid, niazicin A, 4-(α-L-rhamnopyranosyloxyl)-benzyl isothiocyanate, 4-(β-D-gluco-pyranosyl-1→4-α-L-rhamnopyranosyloxyl)-benzyl isothiocyanate and 4-O-α-L-rhamno-py-ranosyloxy-N-glucopyranosyl-1→2-fructopyranosyloxy phenylmethanamine. [8] The lectins were found to inhibit *Echerichia coli*, *Shigella dysenteriae* and *Staphylococcus aureus* growth. [43] This finding is supported by a study showing that a water-soluble lectin extracted from seeds displayed a minimum inhibitory concentration (MIC) of 7.8 μg/ml and 250 μg/ml against *S. aureus* and *E. coli*, respectively. [100]

8.3. Antiviral Activity

Viral infections prove to be a constant problem worldwide, especially in developing countries. These include infections of herpes simplex (topical and genital infections) [101] and HIV/AIDS (immunosuppression). [96] Due to resistance to allopathic treatment, it is thus necessary to search for novel treatment methods.

Ethanol leaf extracts have been reported to display antiviral activity against herpes simplex virus-1 in Vero cells [101] and equine herpes virus-1. [102] No activity against polio or measles has been found. [101] Aqueous seed extracts decreased Newcastle Disease Virus *in ovo*. [103] Methanol extracts of bark decreased HIV reverse transcriptase activity by 15% at 66 μg/ml. [96]

Ethanol leaf extracts have been shown to decrease mortality and skin lesions associated with herpes simplex-1 viral infection in BALB/c mice. [101] The mechanism was thought to be due to immunomodulatory activity and other protective measures, as well as antiviral activity. [101]

Table 3. Qualitative antimicrobial activity against micro-organisms exposed to *M. oleifera* extracts

C	Extract	Micro-organism
Activity against spread inoculum		
Leaves	Acetone	GP: *Bacillus subtilis, Enterobacter aerogenes, Micrococcus luteus, Pseudomonas aeruginosa, Staphylococcus aureus [89], Proteus mirabilis [39]*
		GN: *Salmonella typhi, Salmonella paratyphi A [89]*
	Aqueous	GP: *Bacillus subtilis, Staphylococcus aureus [90]*
		GN: *Campylobacter jejuni [42], Escherichia coli [90]*
	Chloroform	GP: *Pseudomonas aeruginosa, Staphylococcus aureus, Staphyloccocus pyogenes[87]*
		GN: *Escherichia coli [87]*
		Fungus: *Aspergillus niger, Candida albicans [87]*
	Ethanol	GN: *Escherichia coli [62,90,91], Aeromonas caviae, Vibrio parahaemolyticus [92]*
		GP: *Bacillus subtilis [90], Enterococcus faecalis [92], Klebsiella pneumonia [62], Staphylococcus aureus [62,90,92,93]*
	Ethyl alcohol	GP: *Bacillus licheniformis, Bacillus megaterium, Staphylococcus aureus, Lactobacillus acidophilus [86]*
		Fungus: *Aspergillus niger, Aspergillus fumigates[86]*
	Isopropyl alcohol	GP: *Proteus mirabilis [39]*
	Methanol	GP: *Enterobacter aerogenes, Micrococcus luteus, Pseudomonas aeruginosa[89], Bacillus subtilis [90], Staphylococcus aureus [89,90]*
		GN: *Escherichia coli [90], Salmonella typhi, Salmonella paratyphi A[89]*
	Petroleum ether	GP: *Proteus mirabilis[39]*
Seeds	Aqueous	GP: *Staphyloccocus aureus, Bacillus subtilis[88,94], Pseudomonas aeruginosa[94]*
		GN: *Escherichia coli[88,94], Pasteurella multocida[88]*
		Fungus: *Aspergillus niger[88,94], Fusarium solani, Metarhizium aniscoplae[88]*
	Fixed oils	GP: *Staphyloccocus aureus, Bacillus stermophilus[95]*
	Hexane	GN: *Escherichia coli, Salmonella typhi, Vibrio cholera[85]*
	Methanol	GP: *Bacillus stermophilus[95], Staphyloccocus aureus[94,95], Bacillus subtilis, Pseudomonas aeruginosa[94]*
		GN: *Escherichia coli[85,94], Salmonella typhi, Vibrio cholera[85]*
Little to no activity against spread inoculum		
Bark	Methanol	GP: *Bacillus megaterium[96]*
		GN: *Escherichia coli[96]*
		Fungus: *Microbotryum violaceum[96]*

Table 3. (Continued)

C	Extract	Micro-organism
		Green algae: *Chlorella fusca[96]*
Leaves	Acetone	GP: *Klebsiella pneumonia[89]*
		GN: *Salmonella typhi[97]*
	Aqueous	GP: *Bacillus cereus, Listeria monocytogenes[42], Staphylococcus aureus[42,62], Klebsiella pneumoniae, Staphylococcus aureus[62]*
		GN: *Escherichia coli[42,62], Salmonella typhi[97], Salmonella typhimurium[42]*
	Ethanol	GP: *Bacillus subtilis, Klebsiella pneumonia[91], Pseudomonas aeruginosa[92]*
		GN: *Escherichia coli[91,92], Vibrio cholera[91], Salmonella typhi[97], Salmonella enteritidis[92]*
	Methanol	GP: *Klebsiella pneumonia[89]*
Seeds	Aqueous	Fungus: *Rhizophus solani[88], Candida albicans[94]*
	Fixed oils	GP: *Pseudomonas aeruginosa[95]*
		GN: *Escherichia coli[95]*
	Methanol	GP: *Pseudomonas aeruginosa[95]*
		GN: *Escherichia coli[95]*
		Fungus: *Aspergillus niger, Candida albicans[94]*

GP = Gram-positive bacteria; GN = Gram-negative bacteria.

Table 4. Quantitative antimicrobial activity against micro-organisms treated with *M. oleifera* extracts

Plant part	Extract (MIC category)	Micro-organism (MIC in mg/ml)[a]
Leaves	Aqueous (NA)	GN: *Escherichia coli (5), Pseudomonas aeruginosa (5), Klebsiella pneumoniae (5)[84]*
		GP: *Bacillus cereus (5), Bacillus pumilus (5), Enterobacter cloacae (5), Proteus vulgaris (5), Staphylococcus aureus (5), Streptococcus faecelis (5)[84]*
	Acetone (<2 mg/ml)	GP: *Micrococcus kristinae (0.5)[84]*
	Acetone (2-10 mg/ml)	GN: *Escherichia coli (5), Enterobacter cloacae (5), Proteus vulgaris (5)[84]*
		GP: *Staphylococcus aureus (5)[84]*
	Acetone (>10 mg/ml)	GN: *Proteus mirabilis (12.5)[39]*
	Acetone (NA)	GN: *Pseudomonas aeruginosa (5), Klebsiella pneumoniae (5)[84]*
		GP: *Bacillus cereus (5), Bacillus pumilus (5), Streptococcus faecelis (5)[84]*
	Ethanol (<2 mg/ml)	GN: *Campylobacter jejuni (1)[42]*

Plant part	Extract (MIC category)	Micro-organism (MIC in mg/ml)[a]
	Ethanol (2-10 mg/ml)	GN: *Salmonella typhi (8)[97]*
	Ethanol (NA)	GN: *Escherichia coli (4), Salmonella sp. (4)42, Shigella shinga (0.9), Shigella sonei (0.9), Pseudomonas aeruginosa (0.9), Pseudomonas sp. (0.9)[98]*
		GP: *Bacillus cereus (4), Listeria monocytogenes (4), Staphylococcus aureus (4)[42], Bacillus cereus (0.9), Bacillus subtilis (0.9), Staphylococcus aureus (0.9), Streptococcus B-haemolytica (0.9), Sarcina lutea (0.9), Bacillus megaterium (0.9)[98]*
	Methanol (2-10 mg/ml)	GN: *Enterobacter aerogenes (2), Shigella sp. (2)[38], Escherichia coli (2.4)[90]*
		GP: *Staphylococcus aureus (2)[38], Staphylococcus aureus (7.4), Bacillus subtilis (2.4)[90]*
	Methanol (NA)	GN: *Escherichia coli (4), Salmonella typhimurium (4), Salmonella typhi (4)[38]*
		GP: *Pseudomonas aeruginosa (4)[38]*
	Petroleum ether (>10 mg/ml)	GN: *Proteus mirabilis (6.25)[39]*
	Chloroform (<2 mg/ml)	GN: *Salmonella typhimurium (1)[38]*
	Chloroform (2-10 mg/ml)	GN: *Enterobacter aerogenes (2), Escherichia coli (2), Salmonella typhi (2)[38]*
	Chloroform (NA)	GN: *Enterobacter aerogenes (4), Shigella sp. (4)[38]*
		GP: *Pseudomonas aeruginosa (4), Staphylococcus aureus (4)[38]*
	Isopropyl alcohol (>10 mg/ml)	GN: *Proteus mirabilis (12.5)[39]*
	Powdered fresh leaf juice (<2 mg/ml)	GN: *Shigella shinga (0.5), Shigella sonei (0.2), Pseudomonas aeruginosa (0.2), Pseudomonas sp. (0.5)[98], Escherichia coli (0.2), Shigella dysenteriae (0.4), Salmonella sp. (0.2), Enterobacter sp. (0.2), Klebsiella pneumoniae (0.4), Serratia marcescens (0.4)[99]*
		GP: *Bacillus cereus (0.5), Bacillus subtilis (0.5), Sarcina lutea (0.5), Bacillus megaterium (0.5), Staphylococcus aureus (0.5), Streptococcus B-haemolytica (0.5)[98]*
Freshly-picked leaves	Ethanol (<2 mg/ml)	GN: *Shigella shinga (0.9), Shigella sonei (0.5), Pseudomonas aeruginosa (0.5), Pseudomonas sp. (0.5)[98], Escherichia coli (0.4), Shigella dysenteriae (0.8), Salmonella sp. (0.4), Enterobacter sp. (0.4), Klebsiella pneumoniae (0.8), Serratia marcescens (0.8)[99]*
		GP: *Bacillus cereus (0.5), Bacillus subtilis (0.5), Sarcina lutea (0.9), Bacillus megaterium (0.5)[98]*
	Ethanol (NA)	GP: *Staphylococcus aureus (0.9), Streptococcus B-haemolytica (0.9)[98]*
Seeds	Aqueous (<2 mg/ml)	GN: *Yersinia enterocolitica (0.9), Proteus mirabilis (1.75)[8]*
	Aqueous (2-10 mg/ml)	GN: *Proteus vulgaris (5), Escherichia coli (2.5), Acinetobacter baumanii (5), Klebsiella oxycota (2.5)[8]*
		GP: *Pseudomonas fluorescens (5)[8]*
	Aqueous (NA)	GN: *Klebsiella pneumoniae (10), Providencia stuartii (10), Burkholderia cepacia (10), Serratia rubidae (10), Salmonella pullorum (10)[8]*
		GP: *Streptococcus sp. (10)[8]*

Table 4. (Continued)

Plant part	Extract (MIC category)	Micro-organism (MIC in mg/ml)[a]
	Methanol (<2 mg/ml)	GN: *Acinetobacter baumanii (0.9), Proteus vulgaris (1.8)[8]*
	Methanol (2-10 mg/ml)	GN: *Burkholderia cepacia (2.5), Proteus mirabilis (2.5)(8), Enterobacter aerogenes (2)[38]*
	Methanol (2-10 mg/ml)	GP: *Pseudomonas fluorescens (2.5)[8], Staphylococcus aureus (4)[38]*
	Methanol (NA)	GN: *Escherichia coli (10), Klebsiella oxycota (10), Klebsiella pneumoniae (10), Providencia stuartii (10), Serratia rubidae (10), Salmonella pullorum (10), Yersinia enterocolitica (0.9)[8],Shigella sp. (4), Escherichia coli (4), Salmonella typhimurium (4), Salmonella typhi (4)[38]*
	Methanol (NA)	GP: *Streptococcus sp. (10)8, Pseudomonas aeruginosa (4)38*
	Chloroform (<2 mg/ml)	GN: *Salmonella typhimurium (1)[38]*
	Chloroform (2-10 mg/ml)	GN: *Shigella sp. (4), Salmonella typhi (2)[38]*
	Chloroform (NA)	GN: *Enterobacter aerogenes (4), Escherichia coli (4)[38]*
	Chloroform (NA)	GP: *Pseudomonas aeruginosa (4), Staphylococcus aureus (4)[38]*

GP = Gram-positive bacteria; GN = Gram-negative bacteria; NA = no activity observed at highest concentration tested; [a] – in regards to NA groups, this was the highest concentration tested.

8.4. Antihelminthic and Larvacidal Activity

In developing countries parasitic infections can result in decreased quality of life and mortality due to the presence of helminthes [104], Dengue fever [47] and malaria [105], amongst others.

Methanol and chloroform extracts of leaves decreased viability of the helminth, *Pheretima posthuma,* at 50 mg/ml by increasing paralysis and death. [40] Aqueous extracts from seeds decreased helminth egg counts and decreased turbidity in water through coagulation. [104]

Aqueous extracts of flowers and methanol extracts of seeds displayed larvacidal activity against *Aedes aegypti* [47] and *Anopheles stephensi* [105], respectively. Larvacidal activity was more potent against the late stages of development (instar L4) for *Aedes aegypti* [47], while the reverse was seen with *Anopheles stephensi* (instar L1). [105] The activity found in the flower extracts is thought to be due to the presence of a proteinaceous trypsin inhibitor. [47]

A water-soluble lectin isolated from seeds decreased *Aedes aegypti* viability, which is thought to be due to its haemagluttination activity and potential digestive enzyme inhibitory activity. [106]

8.5. Anti-Diabetic Activity and Attenuation of Metabolic Disorders

Diabetes remains one of the major diseases resulting in deaths worldwide due to inadequate control of blood glucose levels. [19] Various parameters have been targeted in the search for new anti-diabetic treatment, including enzyme inhibitors (α-amalyse and α-glucosidase) and insulin sensitizers. [69]

In vitro analysis have shown that methanol extracts of leaves inhibit both α-amylase and α-glucosidase. [69] Raw leaves have been found to be less effective in inhibiting the enzymes than cooked and blanched counterparts. Activity is thought to be due to antioxidant activity and polyphenolic content. [69] Additionally, leaf extracts have also been reported to inhibit cholesterol esterase enzymes and decrease cholesterol micelle formation, but only at high concentrations (>4 mg/ml). However, no inhibition of pancreatic lipase has been reported. [107] Methanol extracts of leaves have been shown to decrease uptake of cholesterol in Caco-2 colon carcinoma cells and inhibit pancreatic lipase 3-hydroxy-3-methylglutaryl-coenzyme A reductase activity *in vitro.* [108] These activities could result in the decreased severity of metabolic disorders due to a hypocholesterolaemic effect. [108]

Niazicin A, 1-*O*-phenyl-α-L-rhamnopyranoside, 4-[(β-D-glucopyranosol)-(1→3)-(α-L-rhamnopyranosol)]phenylacetonitrile, methyl *N*-{4-[(α-L-rhamnopyranosol) benzyl]} carbamate and methyl *N*-{4-[(4'-*O*-acetyl-α-L-rhamnopyranosol)benzyl])carbamate (all isolated from fruits) promoted release of insulin from INS-1 rodent pancreatic cells. [54]

Aqueous leaf extracts (100, 200 and 300 mg/kg) were reported to decrease blood glucose levels (21.1%, 29.9% and 27.9%, respectively) in rats. [109] A dose of 200 mg/kg reduced fasting and post-prandial blood glucose levels by 69% and 51%, respectively, in streptozocin-induced diabetic rats. [109] Aqueous leaf extracts (200 mg/kg) administration for 60 days prevented hyperglycaemia, weight changes, insulin fluxuations and hyperlipidaemia in type I diabetes and insulin resistant diabetic rat models. [110] Similarly, ethanol leaf extracts decreased hyperglycaemia in streptozocin-induced diabetic rats at 250 and 500 mg/kg by 75% and 81%, respectively. This decrease was not seen in normoglycaemic controls. [111] Leaf powder (200 mg/kg) has been shown to reduce glucose intolerance and gastric motility in Goto-kakizaki and Wistar rats, possibly through a hypoglycaemic effect, insulin-like action or decreased glucose absorption. [48] Ethanol extracts of the stem-bark (250 mg/kg) have been reported to decrease blood glucose levels in alloxan-induced diabetic mice by 81% after one week of daily oral administration. [112] Methanol extracts of pods (150 and 300 mg/kg) decreased streptozocin-induced diabetes by reducing hyperglycaemia and nitric oxide, as well as increasing serum insulin and protein levels. [75] Additionally, degeneration of pancreatic cells was decreased in histopathological evaluation. [75]

Chronic, 17 day-administration of benzylamine-supplemented drinking water has been shown to improve glucose tolerance in overweight, pre-diabetic mice, while decreasing total plasma cholesterol levels. The latter is thought to be due to an insulin-like action. [55]

Leaf and stem-bark extracts thus appear to be potential candidates for anti-diabetic treatment, though further isolation will need to be done to determine which bioactive phytochemicals are responsible for this activity.

8.6. Immunomodulatory and Anti-Nociceptive Activity

Inflammation and immune responses have been implicated in numerous diseases states, including asthma [113], arthritis [114] and chronic pain. [115] Modulation and inhibition of inflammatory responses have been reported by various herbal remedies, and as such form a potential for the treatment of inflammatory and pain disorders.

Methanol extracts of leaves (250 mg/kg and 750 mg/kg) displayed immunomodulatory activity in Wistar rats through the enhancement of immune responses. [116] Effects included:

enhancement of neutrophil adhesion in the presence of nylon fibers, decreased mortality induced by *Pasteurella multocida*, attenuation of cyclophosphamide-induced neutropaenia, higher clearance of carbon particles by phagocytes, increased serum immunoglobulin levels and reduced haemagglutination. The lower dose was more effective in enhancing the immune system than the higher dose – this was thought to be due to the presence of more toxic compounds at higher concentrations inducing added stress in the assays. [116] Ethanol extracts of leaves decreased cyclophosphamide-induced immunosuppression by increasing cellular and humoral response in terms of white blood cell count, neutrophil percentage, thymus and spleen weight, as well as phagocytosis. [117]

Anti-inflammatory activity has been ascribed to *M. oleifera*. Aqueous leaf [118], petroleum ether and methanol stem-bark [115], as well as methanol root-bark extracts [119] decreased carrageenan-induced paw-oedema. Furthermore, petroleum ether and methanol extracts of stem-bark decreased cotton pellet-induced granuloma formation. The anti-inflammatory action was thought to be due to an antagonistic effect against serotonin, histamine, bradykinin and prostaglandin. [115] Hydro-ethanol extracts of flowers (100 and 200 mg/kg) reduced Freund's adjuvant-induced arthritis in rats with regards to alterations in bodyweight, paw oedema volume, erythrocyte sedimentation rate, serum parameters (rheumatoid factor, tumour necrosis factor-α and interleukin-1) and histopathological changes in tissue. [114]

Aurantiamide acetate and 1,3-dibenzyl urea, isolated from the roots, have been shown to decrease cytokine production in whole blood. [57] Aurantiamide inhibited tumour necrosis factor α and interleukin-2, but had no activity against interleukin-6. 1,3-Dibenzyl urea inhibited interleukin-2. [57] Four phenolic glycosides isolated from from fruits: 4-[(2'-*O*-acetyl-α-L-rhamnosyloxy)benzyl]isothiocyanate (1.7 μM), 4-[(3'-*O*-acetyl-α-L-rhamno-syloxy)benzyl]isothiocyanate (2.7 μM), 4-[(4'-*O*-acetyl-α-L-rhamnosyloxy)benzyl] isothiocyanate (2.7 μM) and 4-[(α-L-rhamnosyloxy)benzyl]isothiocyanate (14.4 μM), had nitric oxide inhibitory activity in RAW 264.7 cells, and reduced the expression of inducible nitric oxide synthase. [52] It was concluded that an isothiocyanate group and acetylation of several carbon atoms were needed to potentiate inhibitory action. [52]

Furthermore, 4-[(2'-*O*-acetyl-α-L-rhamnosyloxy)benzyl]isothiocyanate displayed anti-inflam-atory properties by decreasing expression of lipopolysaccharide-induced cyclo-oxygenase-2 and inducible nitric oxide synthase in these macrophage cells. This activity was due to blockage of the ERK1/2, SAPK/JNK and NF-κB pathways. [120] 1-*O*-phenyl-α-L-rhamnopyranoside and 4-[(β-D-glucopyranosol)-(1→3)-(α-L-rhamnopyranosol)] phenyl-acetonitrile, isolated from fruits, inhibited cyclo-oxygenase 1 and 2 enzymes at 83 μg/ml. [54] These results support its ability to decrease inflammatory responses *in vivo*.

Pre-treatment of guinea pigs with β-sitosterol (2.5 mg/kg) isolated from seeds decreased ovalbumin-induced airway inflammation. [113] Clinically, respiratory rate was decreased and tidal volume increased. A reduction in inflammatory cell count was seen (eosinophils, monocytes, lymphocytes and neutrophils). β-sitosterol decreased cytokine (tumour necrosis factor-α, interleukin 4 and 5) levels both in serum and bronchoalveolar fluid, though little effect was seen on interlukin 6. [113] Furthermore, histamine levels in the lung tissues were normalized, as well as histopathological changes. The suggested mechanism of action is the attenuation of Th2-cell-mediated inflammation, inflammatory mediator release and counteraction of acetylcholine-induced bronchoconstriction. As such β-sterol could possess anti-asthmatic properties. [113]

Alcoholic extracts of leaves were found to decrease thermal hyperalgesia and mechanical allodynia in both chronic constriction injury-inflicted diabetic mice [73] and arthritic Wistar rats. [121]

Furthermore, juice and ethanol extracts of leaves (25, 50 and 100 mg/kg) decreased pain involved in the acetic acid-induced writhing, formalin-induced paw licking and heat-induced tail flicking assays in Swiss albino mice. [122] This activity was antagonized by naloxone, indicating both a peripheral and central activity from leaves. [122] In another study aqueous leaf extracts (10, 30 and 100 mg/kg) had similar effects, though the mechanism of action was thought to be due to central opiod receptor action, and no peripheral action was noted. [118] Although the root extracts decreased thermal hyperalgesia, it did not reverse mechanical allodynia. A mixture of leaf and root extracts had similar results, and also displayed a prophylactic anti-nociceptive effect. [121] Petroleum ether and methanol extracts of stem-bark [115], as well as methanol extracts of root-bark [119] decreased pain involved in thermal hyperalgesia and acetic acid-induced writhing models. The methanol extracts were more effective than the petroleum ether extracts, suggesting higher antinociceptive effects of polar compounds in *M. oleifera*. Local anaesthetic activity has been ascribed to methanol root-bark extracts. [119] Anti-nociceptive activity of *M. oleifera* is thought to be due to anti-inflammatory (inhibition of cyclo-oxygenases and lipoxygenases) [73,115], calcium channel inhibitory action [73] and antioxidant properties. [121]

Aurantiamide acetate and 1,3-dibenzyl urea have been shown to decrease thermal hyperalgesia in mice. [57] The latter compound has greater antinociceptive potency, which could be due to its close relationship to 1,3-dibenzyl thiourea, a compound which has been shown to decrease pain and inhibit calcium uptake. [57]

8.7. Anticonvulsant Activity

In mice administered methanol leaf extracts, decreased convulsions were noted after subjection to the maximal electroshock and pentylenetetrazole tests. [123] Complete inhibition was present only at 400 mg/kg treatment. Inhibition of maximal electroshock-induced hind limb tonic extension indicates the ability of the extract to prevent the spread of seizure discharges from the focal point. The decreased pentylenetetrazole-induced activity is thought to be due to the flavonoids modulating gamma-aminobutyric acid. [123]

8.8. Anti-Ulcerogenic Activity

Oral administration of methanol root extract (400 mg/kg) decreased ulcerogenic activity of ethanol in rats, which was ascribed to antioxidant activity. [124] Petroleum ether, acetone and methanol extracts of leaves decreased the gastrotoxic effects of acetic acid and cold stress in rats, while only acetone and methanol extracts were able to reduce pyloris-ligation-, ethanol-, indomethacin- and cysteamine-induced ulceration. Protection was seen with regards to ulcer index, ulcer area, regeneration of glandular epithelium, collagen and mucin content, as well as acidity of the stomach. [125] Aqueous extracts of leaves decreased ulceration in albino rats, possibly through the modulation of serotonin secretion in mucosal EC cells. This is suggested to increase secretion of protective prostaglandins, and as such offer increased protection against gastric irritation. [126]

8.9. Anticancer Activity

Data on anticancer activity for *M. oleifera* is limited. Studies are confined to a small selection of cancer cell lines. Potent anticancer activity against leukemic (CEM, acute lymphoid and acute myeloid) cells has been ascribed to leaf [65] and root extracts. [127] Leaf extracts displayed anticancer activity against HpG2 hepatocarcinoma cells [65], while root extracts inhibited B-16 melanoma cell growth. [127] Aqueous extracts were cytotoxic to HeLa epithelial cancer cells. [128] Minimal anticancer activity has been reported in the KB HeLa contaminant (aqueous leaf extract) [129], COLO320 colon carcinoma (ethanol bark extract) [130], acute myelocytic leukemia and HpG2 hepatocarcinoma (aqueous and ethanol seed extracts) [131], HL-60 pro-myelocytic and HCT-8 adenocarcinoma (ethanol root extract) cell lines. [127] Although aqueous leaf extracts did not induce much cytotoxicity in KB Hela contaminant cells, induction of reactive oxygen species and apoptosis, with membrane blebbing, chromatin condensation and DNA fragmentation was observed at 200 μg/ml. [129] This anticancer activity seems to be through the induction of oxidative stress in carcinomas, with resultant apoptotic signaling.

In vitro analysis has shown that 4(α-L-rhamnosyloxy)-benzyl isothiocyanate, niazimicin, 3-O-(6'-O-oleyol-β-D-glucopyranosyl-β-sitosterol and β-sitosterol-3-O-β-D-glucopyranoside inhibit 12-O-tetradecanoyl-phorbol-13-acetate-induced Epstein-Barr virus-early antigen activation in Raji cells, thus providing anticancer activity. [132] Niazirin, niazicin A, methyl 2-[4-(α-L-rhamnopyranosyl) phenyl]acetate, *N*-[4-(β-L-rhamnopyranosyl)benzyl]-1-*O*-α-D-glucopyransylthiocarboxamide 1-*O*-phenyl-α-L-rhamnopyranoside, 4-[(β-D-glu-copy-ranosol)-(1→3)-(α-L-rhamnopyranosol)]phenylacetonitrile, methyl *N*-{4-[(α-L-rhamnopyranosol) benzyl]}carbamate and methyl *N*-{4-[(4'-*O*-acetyl-α-L-rhamnopyranosol) benzyl]}carbamate (30 μg/ml) did not induce any cell death in MCF-7 breast carcinoma, SF-268 central nervous system, NCI-H460 lung and HCT-116 colon carcinoma cell lines. [54]

An *in vivo* study on melanoma-bearing mice found that methanol extracts of leaves and fruits had the potential to decrease cancer progression at 500 mg/kg body weight. [133] Potential candidates responsible for the activity include niazimicin, niaziminin, quercetin, nicotinic acid and other antioxidants. [133] Aqueous extracts of pods were able to decrease the clastogenic activity of mitomycin C and 7,12-dimethylbenz(a)anthracene in mice at 8.5 g/kg body weight by 37.5% and 33%, respectively. [134] Niazimicin displayed anticancer promoting activity in mice subjected to a 7,12-dimethylbenz(a)anthracene carcinogenesis model. [132]

Anticancer activity of *M. oleifera* is mostly ascribed to the leaves, roots and pods, with limited activity in the bark and seeds. Due to the limited information available, it cannot be concluded with certainty which solvent extraction method is most effective though a tendency for polar anticancer compounds seems to be evident.

8.10. Protease Inhibition

A high yield of a protease inhibitor has been reported in the leaves and seeds, while bark, roots and flowers had a much lower content. [58,59] The high percentage in the leaves is thought to be due to their exposure to more stressors and damage. [58] This inhibitor was shown to be approximately 23.6 kDa in size. [59] Inhibitory activity was present against elastase, thrombin, cathepsin B, chymotrypsin, papain and collagenase, all of which have a direct association with the progression of diseases, such as neurodegeneration and inflammation. [58] Furthermore, high inhibition of the commercially-available proteases obtained from *Bacillus licheniformis*, *Bacillus* sp, *Aspergillus oryza* and *Engyodontium album* was found. Ultimately, the protease inhibitor decreased proteolysis and spoilage in shrimp, which indicates its potential commercial value in nutrient protection. [58] Thus extracts could prove useful in limiting proteolytic degeneration of biomolecules and foodstuffs.

9. Toxicity

9.1. *In Vitro* Toxicity

Aqueous leaf extracts have been found to be toxic to peripheral mononuclear blood cells, but only at supra-supplementary concentrations (>20 mg/ml). [3] Alcoholic extracts of the seeds [95] and leaves [101] were not toxic to Vero green monkey kidney cells. Ethanol root extracts have been reported to have no haemolytic effect, though inhibition of sea urchin cell cycle has been noted. [127] Similarly, organic extracts of roots [127] and stem-bark [22] were minimally toxic to brine shrimp. Contrary, three seed-derived lectins, SLL-1, SLL-2 and SLL-3, have shown toxicity using the latter assay (LC50 values of 15.8, 17.8 and 14.1 μg/ml). [43]

Genotoxicity has been noted in the Ames test for the aqueous seed extracts at concentrations higher than 0.2 μg/μl. [135] However, no deleterious effects during the Kado and DNA plasmid breakage tests were found for these extracts. [135]

9.2. *In Vivo* Toxicity

Several acute *in vivo* toxicity studies have reported no toxicity after administration of organic extracts of fruit (5 g/kg) [125], as well as aqueous (3 g/kg to 20 g/kg) [3,6,68], methanol (≤1.5 g/kg) [116] and various organic (2 g/kg to 5 g/kg) [125,136] extracts of leaves in mice and rats. However, authors have reported that slight dullness occurs at high concentrations (1.6 g/kg to 3.2 g/kg) [6,136], though this dissipated within 2 to 5 h. An LD_{50} of 1.585 g/kg was determined 24 h after intra-peritoneal injections of aqueous leaf extracts in mice. [6] No mortality or adverse effects were observed in rats after oral administration of aqueous stem-bark extracts between 2 and 5 g/kg. [80]

Chronic *in vivo* studies have indicated that oral administration of aqueous and organic extracts of fruits (100 mg/kg) for 7 days resulted in non-significant changes to hepatic transaminases in mice. [137] Mortality has not been observed in rodents after oral administration of aqueous leaf extracts for 15 days (3 g/kg) [3], 60 days (1.5 g/kg) [6] or 6 months (1 g/kg). [68] Neither was rabbits affected adversely when fed hydro-alcoholic extracts of leaves for 10 days (5 g/kg). [74]

Toxicity has been noted in mice administered hydro-ethanol extracts of pods via gastric intubation for 14 days, with the manifestation of convulsions, behavioral and locomotion changes, as well as mortality (LD_{50} of 1.8 g/kg). [9,76] Leaf extracts were toxic to monogastrics, such as broiler chicks [15] and Harco chicks. [93] Methanol root extracts displayed a three week LD_{50} of 223.61 mg/kg in guinea pigs. [138] Leaf extracts at 175 mg/kg caused high abortifacient activity in mice. [139] Root-bark extracts have been reported to result in potentially lethal hypotensive crises and uterine contractions. [140]

Administration of leaf extracts have been found to result in no or marginal histopathological changes in the heart, brain, kidneys, liver or testes of mice. [6,74] Slight congestion of hepatocytes and reduced sperm count has been observed in mice after administration of leaf extracts, though this was not deemed to be toxic. [6] Reversible hepato- and nephrotoxicity has been reported in guinea pigs administered methanol root extracts (3.6 mg/kg and 4.6 mg/kg, respectively). [138] Aqueous seed extracts have been noted to result in aggregation of bile canaliculi in rats. [141]

Haematological changes have been reported after fruit [137], methanol root-bark [142] and aqueous leaf extract administration [136] in acute and chronic toxicity models. Yet, other studies report no haematological changes with aqueous or methanol leaf [3,68,116] and ethanol root [127], extract administration.

Table 5. *In vitro* toxicity assessment of *M. oleifera* extracts

In vitro parameter	Plant extract	Finding
Brine shrimp toxicity	Root(E)[127], Stem bark(M, C, PE)[22]	Low toxicity (>800 μg/ml)
Cytotoxicity	Leaves(E,A)[3,101]	Low cytotoxicity (>1 mg/ml)
Genotoxicity	Seeds(A)[135]	Genotoxicity >0.2 mg/ml
Haemolysis	Root(E)[127]	Damage to erythrocytes
Cell cycle inhibition	Root(E)[127]	Inhibition of sea urchin cell cycle

E = ethanol; M = methanol; C = chloroform; PE = petroleum ether; A = aqueous.

Changes to liver toxicity parameters (including ASP, ALT, ACP, ALP) have been noted for fruit [137], leaf [136,137], root-bark [142] and seed extracts. [141] Alteration of biochemical parameters, such as urea, sodium, chloride [3,136] and potassium [68] has also been observed.

Acute genotoxicity measured as the increase in micronucleated polychromatic erythrocytes has been reported in Sprague-Dawley rats administered 3 g/kg aqueous leaf extracts. [3] Aqueous extracts of pods showed no clastogenic activity up to 8.5 mg/kg chronically. [134] *In vitro* and *in vivo* toxicity data is summarized in Table 5 and Table 6, respectively.

Table 6. *In vivo* acute and chronic toxicity assessment of *M. oleifera* extracts

In vivo parameter	Animal model (dosage regime)	Finding
Acute toxicity	Wistar rats (aqueous leaves, 0.4-2 g/kg, oral, 48 h)[136]	No mortality or adverse effects noted
	Wistar albino mice (aqueous leaves, oral, 0.4-6.4 g/kg, 24 h)[6]	
	ICR mice (aqueous leaves, oral, 10 g/kg, 24 h)[68]	
	Wistar rats (organic leaves/fruits, oral, 5 g/kg, 24 h)[125]	
	Wistar rats (methanol leaves, oral, <5 g/kg, 24 h) [116]	
	Wistar rats (aqueous root bark, oral, 2-5 g/kg, 24 h)[80]	
	Wistar albino mice (aqueous leaves, ip, 0.25-1.5 g/kg, 24 h) [6]	LD_{50} = 1.585 g/kg
Chronic toxicity	Sprague-Dawley rats (aqueous leaves, 1-3 g/kg, oral, 15 days)[3]	No mortality or adverse effects noted
	Wistar albino mice (aqueous leaves, oral, 0.25-1.5 g/kg, 60 days)[6]	
	ICR mice (aqueous leaves, oral, 0.01-1 g/kg, 6 months)[68]	
	Swiss albino mice (aqueous/ethanol fruit/leaves, oral, 0.1 g/kg, 7 days)[137]	
	Rabbits (hydroethanol leaves, ip, 5 g/kg, 10 days)[74]	
	Swiss albino mice (hydroethanol leaves, oral, 0.1-.45 g/kg, 15 days)[7]	
	Harco chicks (leaves, oral, 3 weeks)[93]	Toxicity to monogastrics at high supplementation
	Broiler chickens (leaves, oral, 7 days)[15]	
	Guinea pigs (methanol roots, ip)[138]	LD_{50} = 223.61 mg/kg
	Swiss albino mice (hydroethanol pods, gastric intubation, 0.05-1.6 g/kg, 14 days)[9,76]	LD_{50} = 1.8 g/kg
Genotoxic potential	Sprague-Dawley rats (aqueous leaves, oral, 3 g/kg, 48 h)[3]	Genotoxicity seen with erythrocyte micronuclei
	ICR mice (aqueous pods, oral, 2.1-8.5 mg/kg, 14 days)[134]	No clastogenic activity
	Swiss albino mice (ethanol leaves, oral, 0.25-2 g/kg, 7 days)[3]	No polychromatic erythrocytic micronuclei or DNA damage
Biochemical parameters	Wistar rats (aqueous leaves, 0.4-2 g/kg, oral, 48 h)[136]	Alterations in total protein, bilirubin, urea
	Sprague-Dawley rats (aqueous leaves, 1-3 g/kg, oral, 15 days)[3]	Reduction in urea, sodium, chloride, albumin
	Wistar albino mice (aqueous leaves, oral, 0.25-1.5 g/kg, 60 days)[6]	Increase in urea and creatinine
	ICR mice (aqueous leaves, oral, 0.01-1 g/kg, 6 months)[68]	Reduction in potassium
	Mice (methanol root bark, ip, 70 mg/kg, weekly)[142]	Changes to urea, bilirubin, plasma protein and non-protein nitrogen
	Guinea pigs (methanol roots, ip, 3-7 mg/kg, 3 weeks)[138]	Liver degeneration

Table 6. (Continued)

In vivo parameter	Animal model (dosage regime)	Finding
	Albino rats (aqueous seeds, oral, 2 mg/ml, 21 days)[141]	Aggregation of bile canaliculi
	Wistar rats (methanol seeds, 0.4-1.6 g/kg, oral, 21 days)[143]	Dose-dependent increase of cortical congestion, portal cellular infiltration and degeneration of hepatocytes
Hepatotoxicity makers	Wistar albino mice (aqueous leaves, oral, 0.25-1.5 g/kg, 60 days)[6]	Increase in serum liver enzymes (ALT, ASP, ALP)
	Swiss albino mice (aqueous/ethanol fruit/leaves, oral, 0.1 g/kg, 7 days)[137]	
	Albino rats (aqueous seeds, oral, 2 mg/ml, 21 days)[141]	
	Wistar rats (aqueous leaves, 0.4-2 g/kg, oral, 48 h)[136]	
	Wistar rats (methanol seeds, 0.4-1.6 g/kg, oral, 21 days)[143]	
	Mice (methanol root bark, ip, 46-70 mg/kg, weekly)[142]	
Nephrotoxicity histopathology	Guinea pigs (methanol roots, ip, 3-7 mg/kg, 3 weeks)[138]	Liver inflammation
Adverse effects	Wistar rats (aqueous leaves, 0.4-2 g/kg, oral, 48 h)[136]	Slight dullness observed
	Wistar albino mice (aqueous leaves, ip, 0.25-2 g/kg, 24 h)[6]	
	Swiss albino mice (hydroethanol pods, gastric intubation, 0.05-1.6 g/kg, 14 days)[9,76]	Behavioral changes, locomotor changes, convulsions
Haematological parameters	Wistar rats (aqueous leaves, 0.4-2 g/kg, oral, 48 h)[136]	Changes observed in haematological parameters (red and white blood cell count)
	Swiss albino mice (aqueous/ethanol fruit/leaves, oral, 0.1 g/kg, 7 days)[137]	
	Wistar rats (methanol leaves, oral, 1.5 g/kg, 24 h)[116]	
	Mice (methanol root bark, ip, 46-70 mg/kg, weekly)[142]	
	Mice (methanol root bark, ip, 7 mg/kg, daily)[142]	
Organ histopathology	Wistar albino mice (aqueous leaves, oral, 0.25-1.5 g/kg, 60 days)[6]	No changes to brain, heart, testes
Reproductive potential	Wistar albino mice (aqueous leaves, oral, 0.25 g/kg, 60 days)[6]	Reduction in sperm count

ALT = alanine transferase; ALP = alanine phosphatase; ASP = aspartate phosphatase; ip = intraperitoneal.

Conclusion

Diseases rampant in Africa, such as diabetes, cancer, as well as viral and bacterial infections require new avenues for pharmaceutical treatment. Pharmacological activity of considerable interest described in this article, especially in developing countries, would be the ability of extracts to attenuate oxidative stress-related toxicity, protect against organ damage, prevent inflammation and decrease growth of pathogens and parasites *in vitro* and *in vivo*.

Such activities should be studied in greater detail for the development of pharmaceutically-active extracts.

The majority of studies indicate that *M. oleifera* is safe for use at the recommended dosages, although warnings have been raised for potential hepatonephrotoxicity, aenemia and genotoxicity with chronic and supra-supplementary usage. *M. oleifera* is regarded as the poor-person's plant with the promise to benefit rural Africa. Leaves can help decrease developing countries dependence on imported goods, such as expensive vitamin and mineral complexes used for health properties. Due to the wide panel of pharmacological activities, its nutritional value and the absence of toxicity at physiological concentrations, *M. oleiefera* extracts could lead to an increase in health and quality of life. *M. oleifera* has the potential to become an economical, safe supplement to the diet which may contribute to protection against disease and promote well-being.

References

[1] World Health Organization. Traditional medicine - fact sheet no 134. 2008; Available from: http://www.who.int/mediacentre/factsheets/ fs134/en/

[2] Andarwulan N, Kurniasih D, Apriady RA, Rahmat H, Roto AV, Bolling BW. Polyphenols, carotenoids, and ascorbic acid in underutilized medicinal vegetables. *Journal of Functional Foods*. 2012;4:339–47.

[3] Asare GA, Gyan B, Bugyei K, Adjei S, Mahama R, Addo P, et al. Toxicity potentials of the nutraceutical *Moringa oleifera* at supra-supplementation levels. *Journal of Ethnopharmacology* 2012;139:265–72.

[4] Makkar HP, Becker K. Nutrients and antiquality factors in different morphological parts of the *Moringa oleifera* tree. *Journal of Agricultural Science* 1997;128:311–22.

[5] Bamishaiye EI, Olayemi FF, Awagu EF, Bamshaiye OM. Proximate and Phytochemical Composition of *Moringa oleifera* Leaves at Three Stages of Maturation. *Journal of Food Science* 2011;3:233–237.

[6] Awodele O, Oreagba IA, Odoma S, da Silva J a T, Osunkalu VO. Toxicological evaluation of the aqueous leaf extract of *Moringa oleifera* Lam. (Moringaceae). *Journal of Ethnopharmacology* 2012;139:330–336.

[7] Sinha M, Das DK, Datta S, Ghosh S, Dey S. Amelioration of ionizing radiation induced lipid peroxidation in mouse liver by *Moringa oleifera* Lam. leaf extract. *Indian Journal of Experimental Biology* 2012;50:209–215.

[8] Oluduro OA, Idowu TO, Aderiye BI, Famurewa O, Omoboye OO. Evaluation of antibacterial potential of crude extract of *Moringa oleifera* seed on orthopaedics wound isolates and characterization of phenylmethanamine and benzyl isothiocyanate derivatives. Research *Journal of Medicinal Plant* 2012;6:383–394.

[9] Paliwal R, Sharma V, Pracheta, Sharma S, Yadav S. Anti-nephrotoxic effect of administration of *Moringa oleifera* Lam in amelioration of DMBA-induced renal carcinogenesis in Swiss albino mice. *Biology and Medicine* 2011;3:27–35.

[10] Sharma N, Gupta PC, C.V. R. Nutrient content, mineral content and antioxidant activity of Amaranthus viridus and *Moringa oleifera* leaves. *Research Journal of Medicinal Plant* 2012;6:253–259.

[11] De Saint SA. Conclusion on the potential of Moringa on the food supplement market in Europe. Moringa News 2012; Available from: www.moringanews.org
[12] Fuglie LJ. The Miracle tree. The multiple attibutes of Moringa. 2001.
[13] National Research Council. Lost Crops of Africa: Volume II: Vegetables. Natl. Acad. Press, Washington, DC; 2006.
[14] Ramachandran C, Peter KV, Gopalakrishnan P. Drumstick (*Moringa oleifera*): A Multipurpose Indian Vegetable. *Economic Botany* 1980;34:276–283.
[15] Gadzirayi CT, Masamha B, Mupangwa JF, Washaya S, Polytechnic U, Province E. Performance of Broiler Chickens Fed on Mature *Moringa oleifera* Leaf Meal as a Protein Supplement to Soyabean Meal. *International Journal of Poultry Science* 2012;11:5–10.
[16] Peter K. Drumstick - A multi-purpose vegetable. *Indian Horticulture* 1979;23:17–19.
[17] Wadankar GD, Malode SN, Sarambekar SL. Traditionally Used Medicinal Plants for Wound Healing in the Washim District, Maharashtra (India). *International Journal of PharmTech Research* 2011;3:2080–2084.
[18] Wadankar GD, Malode SN, Sarambekar SL. Indigenous Medicine Used for Treatment of Gynecological and other related Problems in Washim District, Maharashtra. *International Journal of PharmTech Research* 2011;3:698–701.
[19] Semenya S, Potgieter M, Erasmus L. Ethnobotanical survey of medicinal plants used by Bapedi healers to treat diabetes mellitus in the Limpopo Province, South Africa. *Journal of Ethnopharmacology* 2012;141:440–445.
[20] Estrella MC., Mantaring JB., David GZ. A double blind randomised controlled trial on the use of malunggay (*Moringa oleifera*) for augmentation of the volume of breastmilk among non-nursing mothers of preterm infants. *Philippine Journal of Pediatrics* 2000;49:3–6.
[21] Chopra IC, Handa KL, Kapur LD. Chopra's Indigenous Drugs of India. Dhur, Calcutta; 1958.
[22] Kumbhare M, Guleha V, Sivakumar T. Estimation of total phenolic content, cytotoxicity and in–vitro antioxidant activity of stem bark of *Moringa oleifera. Asian Pacific Journal of Tropical Disease* 2012;2:144–50.
[23] Palada MC, Chang LC. Suggested Cultural Practices for Moringa. International Cooperators' Guide. AVRDC; 2003.
[24] Ayssiwede SB, Zanmenou JC, Issa Y, Hane MB, Dieng A, Chrysostome CAAM. Nutrient Composition of Some Unconventional and Local Feed Resources Available in Senegal and Recoverable in Indigenous Chickens or Animal Feeding. *Pakistan Journal of Nutrition* 2011;10:707–717.
[25] Fadiyimu AA, Fajemisin AN, Alokan JA. Chemical composition of selected browse plants and their acceptability by West African Dwarf sheep. *Livestock Research for Rural Development* 2011;23:Article #256.
[26] Moyo B, Masika PJ, Hugo A, Muchenje V. Nutritional characterization of Moringa (*Moringa oleifera* Lam) leaves. *Journal of Biotechnology* 2011;10:12925–12933.
[27] Mukunzi D, Nsor-Atindana J, Xiaoming Z, Gahungu A. Comparison of volatile profile of *Moringa oleifera* leaves from Rwanda and china using HS-SPME. *Pakistan Journal of Nutrition* 2011;10:602–608.
[28] Shih M-C, Chang C-M, Kang S-M, Tsai M-l. Effect of Different Parts (Leaf, Stem and Stalk) and Seasons (Summer and Winter) on the Chemical Compositions and

Antioxidant Activity of *Moringa oleifera. International Journal of Molecular Sciences* 2011;12:6077–6088.

[29] Tesfay SZ, Bertling I, Odindo AO, Workneh TS, Mathaba N. Levels of anti-oxidants in different parts of moringa (*Moringa oleifera*) seedling. *African Journal of Agricultural Research* 2011;6:5123–5132.

[30] Vishwakarma KL, Dubey V. Nutritional analysis of indigenous wild edible herbs used in eastern Chhattisgarh, India. *Chemical Analysis* 2011;23:554–560.

[31] Bhatta R, Saravanan M, Baruah L, Sampath KT. Nutrient content, in vitro ruminal fermentation characteristics and methane reduction potential of tropical tannin-containing leaves. *Journal of the Science of Food and Agricultural* 2012

[32] Melesse A, Steingass H, Boguhn J, Rodehutscord M. In vitro fermentation characteristics and effective utilisable crude protein in leaves and green pods of Moringa stenopetala and *Moringa oleifera* cultivated at low and mid-altitudes. *Journal of Animal Physiology and Animal Nutrition* 2012;1–10.

[33] Maurya A, Gupta S, Srivastava SK. Preparative isolation of bioactive nitrile glycoside "niazirin" from the fruits of *Moringa oleifera* using fast centrifugal partition chromatography. *Separation Science and Technology* 2011;46:1195–1199.

[34] Anwar F, Rashid U. Physico-chemical characteristics of *Moringa oleifera* seeds and seed oil from a wild provenance of Pakistan. *Pakistan Journal of Botany* 2007;39:1443–1453.

[35] Ayerza R. Seed yield components, oil content, and fatty acid composition of two cultivars of moringa (*Moringa oleifera* Lam.) growing in the Arid Chaco of Argentina. *Industrial Crops amd Products* 2011;33:389–394.

[36] Kleiman R, Ashley D a., Brown JH. Comparison of two seed oils used in cosmetics, moringa and marula. *Industrial Crops and Products* 2008;28:361–364.

[37] Andrade GDF, Márcia T, Melo S, Dumans C, Novack KM, Cardoso R, et al. Biological Evaluation of Crude and Degummed Oil from *Moringa oleifera* Seeds. *Brazilian Archives of Biology and Technology* 2011;54:1003–1006.

[38] Bukar A, Uba A, Oyeyi TI. Antimicrobial profile of *Moringa oleifera* Lam. extracts against some food-borne microorganisms. *Bayero Journal of Pure and Applied Sciences* 2010;3:43–48.

[39] Arun T, Rao CHP. Phytochemical screening and antibacterial activity of *Moringa oleifera* Lam. against Proteus mirabilis from urinary tract infected patients. *International Journal of PharmTech Research* 2011;3:2118–2123.

[40] Srinivasa U, N AJ, Reena K, Rajan MS. Preliminary phytochemical investigation and anthelmintic activity of *Moringa oleifera* leaves. *International Research Journal of Pharmacy* 2011;2:130–131.

[41] Adejumo OE, Kolapo AL, Folarin AO. *Moringa oleifera* Lam. (Moringaceae) grown in Nigeria: In vitro antisickling activity on deoxygenated erythrocyte cells. *Journal of Pharmacy and BioAllied Sciences* 2012;4:118–122.

[42] Dholvitayakhun A, Cushnie TPT, Trachoo N. Antibacterial activity of three medicinal Thai plants against Campylobacter jejuni and other foodborne pathogens. *Natural Product Research* 2012;26:356–363.

[43] Khatun S, Khan MMH, Ashraduzzaman M, Pervin F, Bari L, Absar N. Antibacterial activity and cytotoxicity of three lectins purified from Drumstick (*Moringa oleifera* Lam.) leaves. *Journal of Biosciences* 2009;17:89–94.

[44] Verma AR, Vijayakumar M, Mathela CS, Rao CV. In vitro and in vivo antioxidant properties of different fractions of *Moringa oleifera* leaves. *Food and Chemical Toxicology* 2009;47:2196–2201.

[45] Singh BN, Singh BR, Singh RL, Prakash D, Dhakarey R, Upadhyay G, et al. Oxidative DNA damage protective activity, antioxidant and anti-quorum sensing potentials of *Moringa oleifera*. *Food and Chemical Toxicology* 2009;47:1109–1116.

[46] Kurma SR, Mishra SH. Hepatoprotective principles from the stem bark of *Moringa pterygosperma*. *Pharmaceutical Biology* 1998;36:295–300.

[47] Pontual EV, Napoleão TH, Dias de Assis CR, de Souza Bezerra R, Xavier HS, Navarro DMDAF, et al. Effect of *Moringa oleifera* flower extract on larval trypsin and acethylcholinesterase activities in Aedes aegypti. *Archives of Insect Biochemistry and Physiology* 2012;79:135–152.

[48] Ndong M, Uehara M, Katsumata S-I, Suzuki K. Effects of oral administration of *Moringa oleifera* Lam on glucose tolerance in Goto-Kakizaki and Wistar rats. *Journal of Clinical Biochemistry and Nutrition* 2007;40:229–233.

[49] Mathur M, Kamal R. Studies on trigonelline from *Moringa oleifera* and its in vitro regulation by feeding precursor in cell cultures. *Brazilian Journal of Pharmacognosy* 2012;22:994–1001.

[50] Chen K-H, Chen Y-J, Yang C-H, Liu K-W, Chang J-L, Pan S-F, et al. Attenuation of the extract from *Moringa oleifera* on monocrotaline-induced pulmonary hypertension in rats. *Chinese Journal of Physiology* 2012;55:22–30.

[51] Sahakitpichan P, Mahidol C, Disadee W, Ruchirawat S, Kanchanapoom T. Unusual glycosides of pyrrole alkaloid and 4′-hydroxyphenylethanamide from leaves of *Moringa oleifera*. *Phytochemistry* 2011;72:791–795.

[52] Cheenpracha S, Park E-J, Yoshida WY, Barit C, Wall M, Pezzuto JM, et al. Potential anti-inflammatory phenolic glycosides from the medicinal plant *Moringa oleifera* fruits. *Bioorganic and Medicinal Chemistry* 2010;18:6598–6602.

[53] Jerri H a, Adolfsen KJ, McCullough LR, Velegol D, Velegol SB. Antimicrobial sand via adsorption of cationic *Moringa oleifera* protein. *Langmuir* 2012;28:2262–2268.

[54] Francis JA, Jayaprakasam B, Olson LK, Nair MG. Insulin Secretagogues from *Moringa oleifera* with Cyclooxygenase Enzyme and Lipid Peroxidation Inhibitory Activities. *Helvetica Chimica Acta* 2004;87:317–326.

[55] Iffiú-Soltész Z, Wanecq E, Lomba A, Portillo MP, Pellati F, Szöko E, et al. Chronic benzylamine administration in the drinking water improves glucose tolerance, reduces body weight gain and circulating cholesterol in high-fat diet-fed mice. *Pharmacological Research* 2010;61:355–363.

[56] Kashiwada Y, Ahmed FA, Kurimoto S, Kim S-Y, Shibata H, Fujioka T, et al. New α-glucosides of caffeoyl quinic acid from the leaves of *Moringa oleifera* Lam. *Natural Medicines* 2012;66:217–221.

[57] Sashidhara KV, Rosaiah JN, Tyagi E, Shukla R, Raghubir R, Rajendran SM. Rare dipeptide and urea derivatives from roots of *Moringa oleifera* as potential anti-inflammatory and antinociceptive agents. *European Journal of Medicinal Chemistry* 2009;44:432–436.

[58] Bijina B, Chellappan S, Krishna JG, Basheer SM, Elyas KK, Bahkali AH, et al. Protease inhibitor from *Moringa oleifera* with potential for use as therapeutic drug and as seafood preservative. *Saudi Journal of Biological Sciences* 2011;18:273–281.

[59] Bijina B, Chellappan S, Basheer SM, Elyas KK, Bahkali AH, Chandrasekaran M. Protease inhibitor from *Moringa oleifera* leaves: Isolation, purification, and characterization. *Process Biochemistry* 2011;46:2291–2300.

[60] Gaikwad M, Kale S, Bhandare S, Urunkar V, Rajmane A. Extraction, Characterization and Comparison of Fixed oil of *Moringa oleifera* L and Moringa concanensis Nimmo Fam.Moringaceae. *International Journal of PharmTech Research* 2011;3:1567–1575.

[61] Mazumder MEH, Rahman S. Pharmacological Evaluation of Bangladeshi Medicinal Plants for Antioxidant Activity. Phar*maceutical Biology* 2008;46:704–709.

[62] Jacob SJP, Shenbagaraman S. Evaluation of antioxidant and antimicrobial activities of the selected green leafy vegetables. *International Journal of PharmTech Research* 2011;3:148–152.

[63] Kalpna R, Mital K, Sumitra C. Vegetable and fruit peels as a novel source of antioxidants. *Journal of Medicinal Plants Research* 2011;5:63–71.

[64] Latif S, Anwar F, Hussain AI, Shahid M. Aqueous enzymatic process for oil and protein extraction from *Moringa oleifera* seed. *European Journal of Lipid Science and Technology* 2011;113:1012–1018.

[65] Khalafalla MM, Abdellatef E, Dafalla HM, Nassrallah AA, Aboul-enein KM, Lightfoot DA, et al. Active principle from *Moringa oleifera* Lam leaves effective against two leukemias and a hepatocarcinoma. *African Journal of Biotechnology* 2010;9:8467–8471.

[66] Moyo B, Oyedemi S, Masika PJ, Muchenje V. Polyphenolic content and antioxidant properties of *Moringa oleifera* leaf extracts and enzymatic activity of liver from goats supplemented with *Moringa oleifera* leaves/sunflower seed cake. *Meat Science* 2012;91:441–447.

[67] Arabshahi-delouee S, Aalami M, Urooj A. Drumstick (*Moringa oleifera* L.) leaves: A potential source of natural lipid antioxidants. *Journal of Food Process Engineering* 2009;34:947–59.

[68] Chivapat S, Sincharoenpokai P, Saktiyasuthorn N, Shuaprom A, Thongsrirak P, Sakpetch A, et al. Acute and chronic toxicity of *Moringa oleifera* Linn leaves extracts. *Thai Journal of Veterinary Medicine* 2011;41:417–424.

[69] Kunyanga CN, Imungi JK, Okoth MW, Biesalski HK, Vadivel V. Total phenolic content, antioxidant and antidiabetic properties of methanolic extract of raw and traditionally processed Kenyan indigenous food ingredients. *LWT - Food Science and Technology* 2012;45:269–76.

[70] Ogbunugafor HA, Eneh FU, Ozumba AN. Physico-chemical and Antioxidant Properties of *Moringa oleifera* Seed Oil. *Pakistan Journal of Nutrition* 2011;10:409–14.

[71] Luqman S, Kaushik S, Srivastava S, Kumar R, Bawankule DU, Pal A, et al. Protective effect of medicinal plant extracts on biomarkers of oxidative stress in erythrocytes. *Pharmaceutical Biology* 2009;47:483–490

[72] Sreelatha S, Padma PR. Modulatory effects of *Moringa oleifera* extracts against hydrogen peroxide-induced cytotoxicity and oxidative damage. *Human and Experimental Toxicology* 2010.

[73] Khongrum J, Wattanathorn J, Muchimapura S, Thukhum-mee W, Thipkaew C, Wannanon P, et al. *Moringa oleifera* Leaves Extract Attenuates Neuropathic Pain

Induced by Chronic Constriction Injury. *American Journal of Applied Sciences* 2012;9:1182–1187.

[74] Ouédraogo M, Lamien-Sanou A, Ramdé N, Ouédraogo AS, Ouédraogo M, Zongo SP, et al. Protective effect of *Moringa oleifera* leaves against gentamicin-induced nephrotoxicity in rabbits. Experimental *and Toxicologic Pathology* 2011.

[75] Gupta R, Mathur M, Bajaj VK, Katariya P, Yadav S, Kamal R, et al. Evaluation of antidiabetic and antioxidant activity of *Moringa oleifera* in experimental diabetes. *Journal of Diabetes* 2012;4:164–171.

[76] Paliwal R, Sharma V, Sharma SH. Hepatoprotective and antioxidant potential of *Moringa oleifera* pods against DMBA-induced hepatocarcinogenesis in male mice. *International Journal of Drug Development and Research* 2011;3:128–138.

[77] Sasikala V, Rooban BN, Priya SGS, Sahasranamam V, Abraham A. *Moringa oleifera* prevents selenite-induced cataractogenesis in rat pups. *Journal of Ocular Pharmacology and Therapeutics* 2010;26:441–447.

[78] Ranjan R, Swarup D, Patra RC, Chandra V. Tamarindus indica L. and *Moringa oleifera* M. extract administration ameliorates fluoride toxicity in rabbits. *Indian Journal of Experimental Biology* 2009;47:900–905.

[79] Gupta R, Kannan GM, Sharma M, S Flora SJ. Therapeutic effects of *Moringa oleifera* on arsenic-induced toxicity in rats. *Environmental Toxicology and Pharmacology* 2005;20:456–464.

[80] Gunjal MA, Shah AS, Wakade AS, Juvekar AR. Protective effect of aqueous extract of *Moringa oleifera* Lam. stem bark on serum lipids, marker enzymes and heart antioxidants parameters in isoproterenol-induced cardiotoxicity in Wistar rats. *Journal of Natural Products* 2010;1:485–492.

[81] Fakurazi S, Hairuszah I, Nanthini U. *Moringa oleifera* Lam prevents acetaminophen induced liver injury through restoration of glutathione level. *Food and Chemical Toxicology* 2008;46:2611–2615.

[82] Uma N, Fakurazi S, Hairuszah I. *Moringa oleifera* enhances liver antioxidant status via elevation of antioxidant enzymes activity and counteracts paracetamol-induced hepatotoxicity. *Malaysian Journal of Nutrition* 2010;16:293–307.

[83] Al-Said MS, Mothana R a, Al-Yahya M a, Al-Blowi AS, Al-Sohaibani M, Ahmed AF, et al. Edible Oils for Liver Protection: Hepatoprotective Potentiality of *Moringa oleifera* Seed Oil against Chemical-Induced Hepatitis in Rats. *Journal of food science* 2012;77:T124–30.

[84] Julius P. Antimicrobial activities of *Moringa oleifera* Lam leaf extracts. *African Journal of Biotechnology* 2012;11:2797–2802.

[85] Walter A, Samuel W, Peter A, Joseph O. Antibacterial activity of *Moringa oleifera* and Moringa stenopetala methanol and n-hexane seed extracts on bacteria implicated in water borne diseases. *Journal of Microbiology* 2011;5:153–157.

[86] Sagar K, Reethi B, Akshatha G, Prasad S. Medicinal efficacy of some natural cosmoceuticals, neutraceuticals and medicinal plant extracts and ultrastructural alterations in food borne pathogens. *International Journal of Pharmacy and Pharmaceutical Sciences* 2012;4:113–120.

[87] Devendra BN, Srinivas N, Talluri VSSLP, Latha PS. Antimicrobial activity of *Moringa oleifera* Lam., leaf extract, against selected bacterial and fungal strains. *International Journal of Pharma and Bio Sciences* 2011;2:13–18.

[88] Jabeen R, Shahid M, Jamil A, Ashraf M. Microscopic evaluation of the antimicrobial activity of seed extracts of *Moringa oleifera*. *Pakistan Journal of Botany* 2008;40:1349–1358.

[89] Patel JP, Gami B, Patel K, Solanki R. Antibacterial activity of methanolic and acetone extract of some medicinal plants used in indian folklore. *International Journal* 2011;3:261–269.

[90] Kumar V, Pandey N, Mohan N, Singh RP. Antibacterial and antioxidant activity of different extract of *Moringa oleifera* leaves - an in-vitro study. *International Journal of Pharmaceutical Sciences Review and Research* 2012;12:89–94.

[91] Valarmathy K, Gokulakrishnan M, Kausar MS, Paul K. A study of antimicrobial activity of ethanolic extracts of various plant leaves against selected microbial species. *International Journal of Pharma Sciences and Research* 2010;1:293–295.

[92] Rafael J, Peixoto O, Silva GC, Costa RA, Fontenelle LDS, Hitzschky G, et al. In vitro antibacterial effect of aqueous and ethanolic Moringa leaf extracts. *Asian Pacific Journal of Tropical Medicine* 2011;201–204.

[93] Dougnon JT, Edorh AP, Bankole HS, Kpodekon M, Gbenou J. Efficiency of ethanol extract of *Moringa oleifera* lam leaves for the treatment of Staphylococcus aureus infections in chicks. *Journal of Medicinal Plants Research* 2011;5:6704–6708.

[94] Saadabi AM, Zaid IEA. An In vitro Antimicrobial Activity of *Moringa oleifera* L. Seed Extracts Against Different Groups of Microorganisms. *Australian Journal of Basic and Applied Sciences* 2011;5:129–134.

[95] Ali GH, El-Taweel GE, Ali MA. The cytotoxicity and antimicrobial efficiency of *Moringa oleifera* seeds extracts. *International Journal of Environmental Studies* 2004;61:699–708.

[96] Ali H, Ko GM, Khalid SA, Wright AD, Kaminsky R. Evaluation of selected Sudanese medicinal plants for their in vitro activity against hemoflagellates, selected bacteria, HIV-1-RT and tyrosine kinase inhibitory, and for cytotoxicity. *Journal of Ethnopharmacology* 2002;83:219–228.

[97] Doughari JH, Pukuma MS, De N. Antibacterial effects of Balanites aegyptiaca L. Drel. and *Moringa oleifera* Lam. on Salmonella typhi. *African Journal of Biotechnology* 2007;6:2212–2215.

[98] Rahman MM, Sheikh MM., Sharmin SA, Islam MS, Rahman MA, Rahman M., et al. Antibacterial Activity of Leaf Juice and Extracts of *Moringa oleifera* Lam. against Some Human Pathogenic Bacteria. *Journal of Natural Sciences* 2009;8:219–227.

[99] Rahman MM, Akhter S, Jamal M, Haque MA, Alam MF. Control of coliform bacteria detected from diarrhea associated patients by extracts of *Moringa oleifera*. *Nepal Medical College Journal* 2010;12:12–19.

[100] Morais MMC. Coagulant and antibacterial activities of the water-soluble seed lectin from *Moringa oleifera*. *Letters in Applied Microbiology* 2011;53:186–92.

[101] Lipipun V, Kurokawa M, Suttisri R, Taweechotipatr P, Pramyothin P, Hattori M, et al. Efficacy of Thai medicinal plant extracts against herpes simplex virus type 1 infection in vitro and in vivo. *Antiviral Research* 2003;60:175–180.

[102] Virmani M, Garg SL. In vitro antiviral activity of plant extracts against equine herpes virus-1. *Indian Journal of Comparative Microbiology, Immunology and Infectious Diseases* 2005;26:89–91.

[103] Chollom SC, Agada GOA, Gotep JG, Mwankon SE, Dus PC, Bot YS, et al. Investigation of aqueous extract of *Moringa oleifera* lam seed for antiviral activity against newcastle disease virus in ovo. *Journal of Medicinal Plants Research* 2012;6:3870–3875.

[104] Sengupta ME, Keraita B, Olsen A, Boateng OK, Pa R, Dalsgaard A, et al. Use of *Moringa oleifera* seed extracts to reduce helminth egg numbers and turbidity in irrigation water. *Water Research* 2012;46:3646–3656.

[105] Prabhu K, Murugan K, Nareshkumar A, Ramasubramanian N, Bragadeeswaran S. Larvicidal and repellent potential of *Moringa oleifera* against malarial vector, Anopheles stephensi Liston (Insecta: Diptera: Culicidae). *Asian Pacific Journal of Tropical Biomedicine* 2011;1:124–129.

[106] Coelho JS, Santos NDL, Napoleão TH, Gomes FS, Ferreira RS, Zingali RB, et al. Effect of *Moringa oleifera* lectin on development and mortality of Aedes aegypti larvae. *Chemosphere* 2009;77:934–938.

[107] Adisakwattana S, Chanathong B. Alpha-glucosidase inhibitory activity and lipid-lowering mechanisms of *Moringa oleifera* leaf extract. *European Review for Medical and Pharmacological Sciences* 2011;15:803–808.

[108] Duangjai A, Ingkaninan K. Potential mechanisms of hypocholesterolaemic effect of Thai spices/dietary extracts. *Natural Product Research* 2011;25:341–352.

[109] Jaiswal D, Kumar Rai P, Kumar A, Mehta S, Watal G. Effect of *Moringa oleifera* Lam. leaves aqueous extract therapy on hyperglycemic rats. *Journal of Ethnopharmacology* 2009;123:392–396.

[110] Divi SM, Bellamkonda R, Dasireddy SK. Evaluation of antidiabetic and antihyperlipedemic potential of aqueous extract of *Moringa oleifera* in fructose fed insulin resistant and STZ induced diabetic Wistar rats: A comparative study. *Asian Journal of Pharmaceutical and Clinical Research* 2012;5:67–72.

[111] Tende JA, Ezekiel I, Dikko AAU, Goji ADT. Effect of Ethanolic Leaves Extract of *Moringa oleifera* on Blood Glucose Levels of Streptozocin-Induced Diabetics and Normoglycemic Wistar Rats. *British Journal of Pharmacology and Toxicology* 2011;2:1–4.

[112] Kar A, Choudhary BK, Bandyopadhyay NG. Comparative evaluation of hypoglycaemic activity of some Indian medicinal plants in alloxan diabetic rats. *Journal of Ethnopharmacology* 2003;84:105–108.

[113] Mahajan SG, Mehta AA. Suppression of ovalbumin-induced Th2-driven airway inflammation by β-sitosterol in a guinea pig model of asthma. *European Journal of Pharmacology* 2011;650:458–464.

[114] Mahajan SG, Mehta A a. Anti-Arthritic Activity of Hydroalcoholic Extract of Flowers of *Moringa oleifera* Lam. in Wistar Rats. *Journal of Herbs, Spices and Medicinal Plants* 2009;15:149–163.

[115] Kumbhare M, Sivakumar T. Anti-Inflammatory and Analgesic Activity of Stem Bark of *Moringa oleifera*. *Pharmacologyonline* 2011;3:641–650.

[116] Sudha P, Asdaq SMB, Dhamingi SS, Chandrakala GK. Immunomodulatory activity of methanolic leaf extract of *Moringa oleifera* in animals. *Indian Journal of Physiology and Pharmacology* 2010;54:133–140.

[117] Gupta A, Gautam MK, Singh RK, Kumar MV, Rao CV, Goel RK, et al. Immunomodulatory effect of *Moringa oleifera* Lam. extract on cyclophosphamide induced toxicity in mice. *Indian Journal of Experimental Biology* 2010;48:1157–1160.
[118] Sulaiman MR, Zakaria ZA, Bunjarimin AS, Somchit MN, Israf DA, Moin S. Evaluation of *Moringa oleifera* aqueous extract for antinociceptive and anti-inflammatory activities in animal models. *Pharmaceutical Biology* 2008;46:838–845.
[119] Medhi B, Khanikor HN, Lahon LC, Mohan P, Barua CC. Analgesic, anti-inflammatory and local anaesthetic activity of *Moringa pterygosperma* in laboratory animals. *Pharmaceutical Biology* 2003;41:248–252.
[120] Park E-jung, Cheenpracha S, Chang LC, Kondratyuk TP, Pezzuto M, Pezzuto JM. Inhibition of lipopolysaccharide-induced cyclooxygenase-2 and inducible nitric oxide synthase expression by 4- [(2′-O-acetyl-α-L-rhamnosyloxy)benzyl]isothiocyanate from *Moringa oleifera*. *Nutrition and Cancer* 2011;63:971–982.
[121] Manaheji H, Jafari S, Zaringhalam J, Rezazadeh S, Taghizadfarif R. Analgesic effects of methanolic extracts of the leaf or root of *Moringa oleifera* on complete Freund's adjuvant-induced arthritis in rats. *Journal of Chinese Integrative Medicine* 2011;9:216–222.
[122] Upadhye KP, Rangari VD, Mathur VB. Evaluation of antinociceptive activities of fresh leaf juice and ethanolic extract of *Moringa oleifera* Lam. *Asian Journal of Pharmaceutical and Clinical Research* 2011;4:114–116.
[123] Amrutia JN, Lala M, Srinivasa U, Shabaraya AR, Moses RS. Anticonvulsant activity of *Moringa oleifera* leaf. *International Research Journal of Pharmacy* 2011;2:160–162.
[124] Patel AB, Prasanna SK, Shastry CS. Antiulcerogenic activity of *Moringa oleifera* root extract against ethanol-induced gastric ulcer in rats. *International Journal of Pharmaceutical and Chemical Sciences* 2012;1:243–249.
[125] Devaraj VC, Asad M, Prasad S. Effect of Leaves and Fruits of *Moringa oleifera* on Gastric and Duodenal Ulcers. *Pharmaceutical Biology* 2007;45:332–338.
[126] Debnath S, Biswas D, Ray K, Guha D. Phytomedicine *Moringa oleifera* induced potentiation of serotonin release by 5-HT 3 receptors in experimental ulcer model. *Phytomedicine* 2011;18:91–95.
[127] Costa-Lotufo LV, Khan MTH, Ather A, Wilke DV, Jimenez PC, Pessoa C, et al. Studies of the anticancer potential of plants used in Bangladeshi folk medicine. *Journal of Ethnopharmacology* 2005;99:21–30.
[128] Nair S, Varalakshmi KN. Anticancer, cytotoxic potential of *Moringa oleifera* extracts on HeLa cell line. *Journal of Natural Pharmaceuticals* 2011;2:138–142.
[129] Sreelatha S, Jeyachitra A, Padma PR. Antiproliferation and induction of apoptosis by *Moringa oleifera* leaf extract on human cancer cells. *Food and Chemical Toxicology* 2011;49:1270–1275.
[130] Smit HF, Woerdenbag HJ, Singh RH, Meulenbeld GJ, Labadie RP, Zwaving JH. Ayurvedic herbal drugs with possible cytostatic activity. *Journal of Ethnopharmacology* 1995;47:75–84.
[131] Khalafalla MM, Dafalla HM, Nassrallah A, Aboul-Enein KM, El-Shemy HA, Abdellatef E. Dedifferentiation of leaf explants and antileukemia activity of an ethanolic extract of cell cultures of *Moringa oleifera*. *African Journal of Biotechnology* 2011;10:2746–2750.

[132] Guevara AP, Vargas C, Sakurai H, Fujiwara Y, Hashimoto K, Maoka T, et al. An antitumor promoter from *Moringa oleifera* Lam. *Mutation Research* 1999;440:181–188.

[133] Purwal L, Pathuk AK, Jain UK. In vivo anticancer activity of the leaves and fruits of *Moringa oleifera* on mouse melanoma. *Pharmacologyonline* 2010;1:655–665.

[134] Promkun C, Kupradinun P, Tuntipopipat S, Butryee C. Nutritive evaluation and effect of *Moringa oleifera* pod on clastogenic potential in the mouse. *Asian Pacific Journal of Cancer Prevention* 2010;11:627–632.

[135] Sisenando HA, Napole TH, Felzenszwalb I, Claudia AF, Coelho LCBB, Medeiros SRB, et al. Genotoxicity Evaluation of *Moringa oleifera* Seed Extract and Lectin. *Journal of Food Science* 2011;76:T53–58.

[136] Adedapo AA, Mogbojuri OM, Emikpe BO. Safety evaluations of the aqueous extract of the leaves of *Moringa oleifera* in rats. *Journal of Medicinal Plants Research* 2009;3:586–591.

[137] Luqman S, Srivastava S, Kumar R, Maurya AK, Chanda D. Experimental Assessment of *Moringa oleifera* Leaf and Fruit for Its Antistress, Antioxidant, and Scavenging Potential Using In Vitro and In Vivo Assays. *Evidence-based Complementary and Alternative Medicine* 2012.

[138] Paul CW, Didia BC. The effect of methanolic extract of *Moringa oleifera* Lam roots on histology of kidney and liver of guinea pigs. *Asian Journal of Medical Sciencesc*2012;4:55–60.

[139] Nath D, Sethi N, Singh RK, Jain AK. Commonly used Indian abortifacient plants with special reference to their teratologic effects in rats. *Journal of Ethnopharmacology* 1992;36:147–154.

[140] Bhattacharya J, Guhu G. Powder microscopy of bark-poison used for abortion: *Moringa pterygosperma* Gaertn. *Journal of Indian Forensic Sciences* 1978;17:47–50.

[141] Oluduro AO, Aderiye BI. Effect of *Moringa oleifera* seed extract on vital organs and tissue enzymes activities of male albino rats. *African Journal of Microbiology Research* 2009;3:537–40.

[142] Mazumder UK, Gupta M, Chakrabarti S, Pal D. Evaluation of hematological and hepatorenal functions of methanolic extract of *Moringa oleifera* Lam. root treated mice. *Indian Journal of Experimental Biology* 1999;37:612–614.

[143] Ajibade TO, Olayemi FO, Arowolo ROA. The haematological and biochemical effects of methanol extract of the seeds of *Moringa oleifera* in rats. *Journal Of Medicinal Plants* 2012;6:615–621.

In: Plant Extracts
Editors: Adriana Giordano and Alfieri Costs
ISBN: 978-1-62417-534-3

Chapter 4

Agave Species: Potential Health Benefits and Medical Applications

María Teresa Carrillo-Rayas and Alejandro Blanco-Labra*
Centro de Investigación y de Estudios Avanzados del Instituto Politécnico Nacional, Unidad Irapuato, Department of Biotechnology and Biochemistry, Irapuato, Guanajuato, Mexico

Abstract

An analysis and a summary of the reported uses of the genus *Agave,* mainly related to their medicinal applications, are briefly outlined. Based on years of experience from traditional medicine to the present time, different extracts, as well as pure isolated chemical compounds, have been obtained from the *Agave* genus. This chapter attempts to combine the existing information on uses, distribution, ethnobotany, and medicinal and social importance of the different *Agave* species. Over the last decade, ethnobotanical studies of medicinal plants of Yucatán, Mexico revealed that some *Agave* species were already used by the Mayans since the pre-hispanic times to treat illnesses with cancer-like symptoms (Arellano-Rodríguez et al., 2003).

Research on natural products with biological activity has led to a boom of commercially significant compounds with potential health benefits and applications in the pharmaceutical industry. Hauad-Marroquin (2010) referred to the Phytochemistry as a branch of science that has allowed for the detection and subsequent identification of active principles responsible for some property attributed to a plant. On those bases, several research groups have been focusing on the study of Agave extract applications in the treatment of different diseases such as kidney cancer, diabetes, gonorrhea and diseases of the excretory system. Agave has also been used as a diuretic, as poultices or agents to cure skin infections, and in the treatment of snake bites, among others (Gentry, 1982; Quílez et al., 2006). Recently, Ramachandran et al. (2011) pointed out that various parts of *Agave cantula* were used as a laxative, emmenagogue, in the treatment of scurvy,

* Corresponding autor: ablanco@ira.cinvestav.mx, alejandroblancolabra@gmail.com.

syphilis, scrutula, swelling, retention of urine, cancer, citotoxic or diuretic conditions, and even as an aphrodisiac and antisyphilitic agent. Similarly, several reports describe cytostatic activity in aqueous extracts of *Agave* species (Quílez, 1996; Popoca et al., 1998; Sáenz et.al., 2000; Calderón et al., 2006; Mena-Rejón et al., 2009; Háuad-Marroquin et al., 2010; Chetan et al., 2010; Alonso-Castro et al., 2011; Ramos et al., 2012). There are also reports describing antitumor activity from *Agave* species (Bianchi et al., 1969; Koduru et al., 2007; Madhuri and Pandey 2008, 2009; Gómez-Flores et al., 2009). Furthermore, the medical applications of steroidal saponins from Agave have also been described (Corbiere et al., 1993; Abdel-Khalik et al., 2002; da Silva et al., 2002; Raju et al., 2009; Man et al., 2010), including the treatment of cognitive dysfunction (Xia et al., 2004), as well as for anti-fertility treatment (Fazli, 1968; Crabbe, 1979; Basilio, 1989). Also, the antimicrobiological properties of Agave extracts have been known for a long time (Davidson and Ortíz, 1983; Verastegui et al., 1996; García et al., 1999; Verastegui et al., 2008).

In general, monocotyledonous families contain compounds with potential industrial applications as sources of alkaloids, non-protein amino acids, anthraquinones, cyanogens, mono- and sesquiterpenoids, and saponins. In this regard, the steroidal saponins are mainly found in monocotyledonous families such as *Agavaceae*, *Dioscoreaceae* and *Liliaceae*. This has been addressed by Bedour et al. (1979); Gupta (1995); Sparg et al. (2004) and Güçlü-Üstündag et al. (2007). Therefore, several *Agave* species can be a good source of saponins (mainly heconin and tigonin) and sapogenins, from which derivative compounds can be obtained, such as contraceptives, anti-inflammatories, sexual hormones, etc. (Gupta, 1995; Peana et al., 1997; García et al., 2000; da Silva et al., 2002). The cytotoxicity of Agave saponins has been well documented (Sati, 1985; Ohtsuki, 2004; Yokosuka and Mimaki, 2009). Furthermore, Uribe et al. (2009) describe one formulation with anticancer and antioxidant properties, based on Agave syrup, containing flavonoids, polycosanols and sapogenins with potential anticancer and antioxidant properties. Following this, Ramachandran et al. (2011) pointed out that *Agave cantula* contains chemical constituents such as flavone and sterol glycosides, neotigogenin, hecogenin and tigogenin.

Polyphenols are also bioactive compounds with potential health benefits present in *Agave* species, based mostly on their antioxidant properties.

In summary, there are many investigations directed to the use of Agave extracts, as well as to the isolation, purification, characterization and structural elucidation of bioactive compounds with potential health benefits (Méndez et al., 2003; Peña-Álvarez et al., 2004; Guerra de León et al., 2007; Gutiérrez et al., 2008; Almaraz-Abarca et al., 2009; Menyar et al., 2012; Rizwan et al., 2012).

On the other hand, the presence of fructans in *Agave* species and their health applications have been well documented (Ulloa et al., 2010; Morris and Morris, 2012; Charalampopoulos and Rastall, 2012). These properties stem from the fact that prebiotic food ingredients (e.g., inulin-type fructans) reach the colon undigested, and once there, they are fermented by colonic microflora to short chain fatty acids (e.g., lactate, acetate, propionic and butyric acid). It is now believed that the beneficial effect on the colonic crypts, together with the antiproliferative neoplastic properties and the prevention of colon cancer (via enhanced apoptosis of colon cancer cells), are mostly due to the production of butyric acid (Verghese et al., 2002). Additionally, Ulloa et al. (2010) list several diseases and pathological conditions that can be prevented or reduced by inulin-type fructans, including constipation and diarrhea, especially those associated with intestinal infections, reducing the risk of osteoporosis, and cardiovascular atherosclerosis associated with dyslipidemia, hypertriglyceridemia and insulin resistance, obesity, diabetes type 2 associated with insulin resistance, irritable bowel syndrome, and chemically induced colon cancer.

Introduction

Plants of the *Agave* genus, meaning "noble" or admirable, have been used for at least 9,000 years (Gentry, 1982), particularly in Mexico, where they have shared a millenary relationship with its history.

The *Agave* genus belongs to the order *Asparagales* and to the *Agavaceae* family. It ranges throughout the Southwest of the United States, Mexico, Central America and the Antilles. Gentry (1982) included 136 species, dividing the subgenera *Agave* into 12 groups with 83 species, and the subgenera *Littaea* into eight groups with 53 species. Moreover, Gentry organized the genus by floral and vegetative characters, dividing it into two subgenera, based on the type of inflorescence and the arrangement of flowers, leaving out the Caribbean and South American species.

The investigations of García-Mendoza (2007) described approximately 200 *Agave* species, 75% found in Mexico, plus 36 species belonging to infraspecific categories. She reported that the genus extends from the southern United States to Colombia and Venezuela, covering the Caribbean islands from the Bahamas to Aruba, Curacao and Trinidad and Tobago. Mexico, the United States, Cuba and Guatemala are the countries with the largest number of taxa, whereas the other countries represent only 3% of the total. Mexico appeared as the country with the highest diversity.

According to Gentry (1982), the *Agave* genus spreads by seeds, vegetative stems or propagules in the inflorescence. Smaller *Agave* species reach maturity after 4 to 5 years, but larger species need 10 to 25 years to do so.

Agaves are xerophytic plants with leaves spirally arranged in rosettes at the apex of a stem, which can be short or long and erect (García-Mendoza, 2007). Since their stomata open late to prevent excess of water loss, they are adapted to live in suboptimal weather conditions, with long periods of drought and high temperatures. Most *Agavaceae* family members exhibit the classic Crassulacean Acid Metabolism (CAM), since at least seventeen taxa of Agave utilize this type of metabolism, including some economically important species such as *A. americana, A. fourcroydes, A. Lechuguilla* and *A. tequilana,* whereas other *Agave* species show some degree of plasticity, being also able to open their stomata during the day, to absorb CO_2 following a C3 metabolism. An example of this plasticity is *Agave deserti*, in which C3-like carbon fixation occurs during a short period of time in the morning, while in *A. mapisaga* (Figure 1) it occurs both during the morning and during the afternoon (García-Mendoza, 2007).

The reported applications developed manyfold over time, including their use as a source of food, drink, fuel and fiber, in medicine; for shelter, textiles, and religious use; for compost, home construction and even as the basis of ornamental components. Long before the arrival of the Spaniards to America, Agaves were used for sugar production and fibers (Háuad-Marroquin et. al., 2010); however, its use declined as the cultivation of sugarcane was established. For the people of America, culturally, Agaves have been of great economic and social importance. *Agave fourcroydes* (henequén) and *Agave sisilana* are species of importance in the production of Agave fibers (Gentry, 1982). Nevertheless, these natural fibers have been losing ground, because of their high production costs, to synthetic materials and other natural fibers.

Figure 1. *Agave mapisaga.*

Colunga-García Marín and May-Pat (1993) described that in Mexico during the colonial period, the importance of henequén persisted as a source of hard fiber becoming important at international markets. However, they mention that after this booming period, the international market was dominated by sisal, a fiber of *Agave sisalana*, and at about 1836, sisal was transported from the port of Sisal, Yucatan, via Florida, U.S.A., to be cultivated in the United States of America and the British colonies in Africa, and later on to Brazil. *Agave sisalana* was found to be most suitable for high performance textile uses. Each leaf contains an average of around 1,000 fibers. FAO (2009) reported that the global production of sisal in 2008 amounted to 247.8 thousand tons of which the largest producing country was Brazil, followed by China, Tanzania and Kenya. Interestingly, *Agave tequilana* (Weber), is commercially harvested for the production of the distilled alcoholic beverage tequila (Cedeño, 1995), which was first commercially produced in 1621 near the town of Tequila, Jalisco (Nobel, 1987). The *Agave tequilana* Weber var. azul, (Figure 2 and Figure 3) is grown only at a very restricted area established as protected areas by the appellation of the origin of tequila. These territories comprise the entire Jalisco state and some municipalities in the states of Nayarit, Guanajuato, Michoacán and Tamaulipas (Iñiguez-Covarrubias et al., 2001).

Figure 2. *Agave tequilana* Weber var. azul.

Figure 3. *Agave tequilana* Weber var. azul.

In addition to *Agave tequilana* (Weber), other *Agave* species are also used to produce alcoholic beverages, such as Mezcal, the fermented must of several *Agave* species, (e.g., *A. angustifolia*; *A. potatorum* and *A.salmiana*), Sotol (e.g., *A. dasylirion*; *Dasylirion ssp.*), Bacanora (e.g., *A. angustifolia*; *A. potatorum* and *A. pacifica*), and Pulque (e.g., *A. salmiana*; *A. atrovirens* and *A. mapisaga*) being the most traditional, nondistilled beverage, that represents a Mexican icon (Cedeño and Álvarez-Jacobs, 1999). Commonly, it is obtained by the spontaneous fermentation of the sap or from the liquid obtained directly out of the core of the plant (aguamiel) (Lappe-Oliveras, 2008). The *Agave* cores, or "piñas," produce these natural juices (aguamiel), which fresh or cooked, are the source of polysaccharides, fructans, high fructose syrup, biofuel or Maillard compounds.

The production of the distilled alcoholic beverage from the *Agave* species engages a complex fermentation in which bacteria (lactic and acetic acid) and yeasts (non-*Saccharomyces* and *Saccharomyces)* are responsible for the production of various chemical and volatile compounds that provide special characteristics to the final product (Lappe-Oliveras, 2008). Some of these compounds have been reported as part of the aroma and flavor (e.g., alcohols, fatty acids, esters, aldehydes, terpenes, phenols, lactones and sulfur) (Nykänen and Nykänen, 1991). Also, some volatile compounds have been reported in relation to the attraction of the insect *Scyphophorus Acupunctatus* gyllenhal (Coleoptera: Curculionidae), which produces serious looses of *Agave fourcroydes* in the field (Valdés-Rodrígues et al., 2004).

On the other hand, Agave fibrous wastes can be used as a source of paper-making fibers including handicrafts, geotextiles, filters, packaging, composites and fertilizers (Iñiguez-Covarrubias et al., 2001). They have also been used as substrates for the production of commercially important enzymes (Huitron et al., 2008) and in the production of edible insects (García-Herrera et al., 2009).

An additional feature of *Agave* species is that some of them possess, on top of the previously mentioned anticancer properties, antimicotic, antiviral, antimicrobial and antituberculosis atributes and they are also active against pathogens. The most investigated species of the *Agave* genus is *A. americana* L. (Figure 4) that has been reported for its antibacterial and anti-inflammatory properties (Peana et al., 1997). Recently, Rizwan et al. (2012) working with *Agave attenuata* showed the presence of important phytochemicals with different biological activities, such as natural antioxidants and functional food nutraceutical applications.

Figure 4. *Agave Americana.*

As mentioned above, several species of *Agave* genus constitute an important source of steroidal sapogenins, whose identification and studies on their biological activity have greatly contributed to the medicinal value of Agave, and some of them are presently used in clinical practice. These compounds have shown high potential against several health disorders.

Secondary Metabolites of Pharmacological Importance

Agave as a Source of Saponins

Saponins are a complex group of compounds that occur naturally in plants. Based on the carbon skeletons, saponins are classified into triterpenes and steroids (steroid aglycone or steroidal alkaloids). According to Sparg et al. (2004), their glycone parts are mostly oligosaccharides, arranged either in a linear or branched fashions, attached to a hydroxyl group through an acetal linkage. Sapogenins are the aglycone nonsugar portions of the saponin molecule (Peana et al., 1997). Steroidal saponins (27C) (steroid precursors), with a 6-ring spirostane or a 5-ring furostane skeleton, are a steroid "building block" or a secondary metabolite, that constitute a scaffold for the production of new steroids.

Triterpenoid saponins (30C) are triterpenes belonging to a large group of compounds arranged in a four or five ring configurations of 30 carbons with several oxygens attached. Triterpenes are assembled from a C5 isoprene unit through the cytosolic mevalonate pathway to make a C30 compound and are steroidal in nature, with different sugar molecules attached to the triterpene unit (Figure 5).

(A)

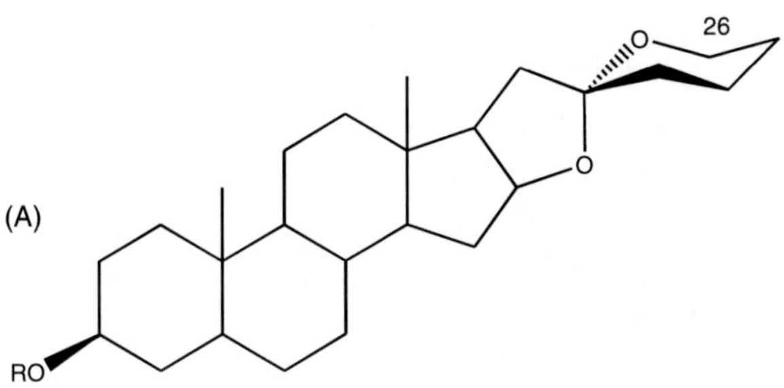

(B)

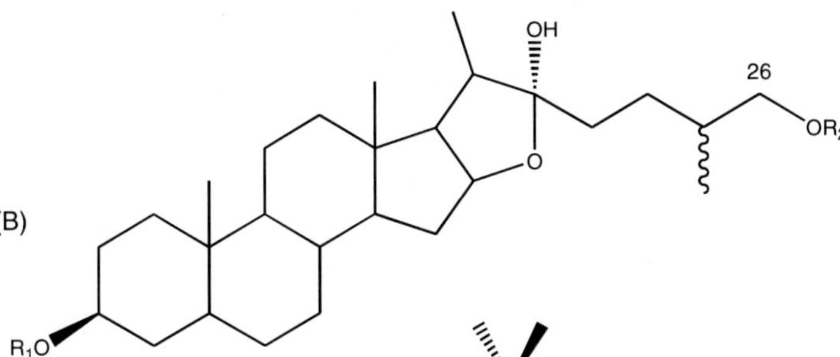

(C)

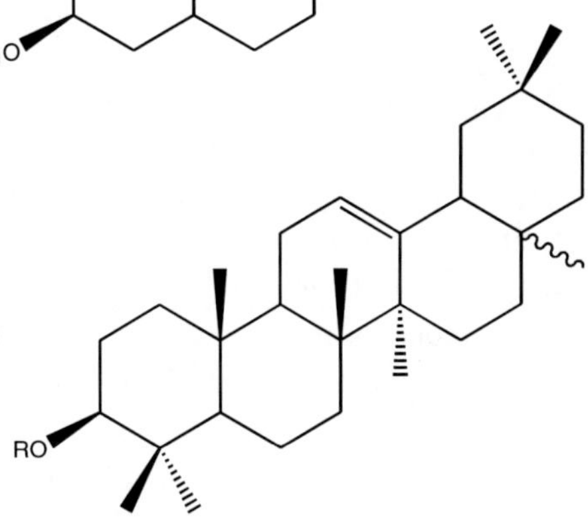

Figure 5. Aglycone skeletons of (A) steroidal spirostane, (B) steroidal furostane, and (C) triterpenoid saponins. R, sugar moiety.

In general, saponins exert a wide range of pharmacological activities including: antiinflammatory, hypocholesterolemic, immunomodulatory, hypoglycaemic, vasoprotective, expectorant, molluscicidal, antifungal and antiparasitic effects, among others (Sparg et al., 2004; Sahu et al., 2008).

As mentioned in the Introduction, the steroidal saponins are mainly found in *Agavaceae, Dioscoreaceae, Liliaceae, Solanaceae, Scrophulariaceae, Amaryllidaceae, Leguminosae* and *Rhamnaceae*; whereas triterpene saponins are predominantly present in *Acanthopanax, Leguminosae, Araliaceae, Scrophulariaceae, Campanulaceae* and *Caryophyllaceae* (Sparg et al., 2004; Güçlü-Üstündag et al., 2007). Therefore, the *Agave* genus presents sapogenin steroid-producing species that are important raw materials for the synthesis of steroidal drugs (Bedour et al., 1979; Gupta, 1995).

Sparg et al. (2004) list an *Agavaceae* family from where steroidal saponins have been isolated: *Agave americana* L., *A. attenuata* Salm-Dick, *A. decipiens* Baker, *Cordyline stricta* (Sims) Endl, *Furcraea foetida* (L.) Haw, *Yucca filamentosa* L., *Yucca schidigera* Roezl. ex Ortgies. In addition, several other *Agave* species were extensively investigated for the identification of steroidal saponins: *A. Lophantha*, *A. parasan* (Bedour, 1979), *A. Sisalana*

(Cripps and Blunden, 1978; Blunden, 1986; Ding et al., 1989; Pi-Yu et al., 2011), *Agave macroacantha* Zucc (Eskander et al., 2010), and *Agave cantala* (Uniyal, 1991). In the case of *A. lechuguilla* Torrey, *Agave aurea, A. avellanidens, A. cerulata, A. cerulata* ssp. *subcerulata, A. cock, A. goldmaniana,* and *A. shawii*, hecogenin and tigogenin were the major sapogenins isolated (Blunden et al., 1980). Four new spirostane saponins were obtained from the leaves of *Agave brittoniana* Trel. ssp *Brachypu* (Macías et al., 2007). In addition, the spherical saponin isolated from *Agave attenuata* leaves displays anti-inflammatory activity and hemolytic potential (da Silva et al, 2002, Paz-Mendes et al., 2004).

A bioactive steroidal saponin (tigogenin) extracted from *Agave shrevei* Gentry leaves, was described by Pereira da Silva et al. (2006) and also possess powerful hemolytic properties that have been related to the amphipathic characteristic of its structure (hydrophobic steroidal nucleus as well as hydrophilic carbohydrate moiety). Sparg et al. (2004) reported over 11 different classes of saponins, among them, dammaranes, tirucallanes, lupanes, hopanes, oleananes, taraxasteranes, ursanes, cycloartanes, lanostanes, cucurbitanes, and steroidals. Among these, only 5 classes of saponins have proved to have strong antitumor effects on different type of cancers (cycloartanes, dammaranes, oleananes, lupanes and steroids). Their biological effects also depend on the saponin structure including the type, position, and number of sugar moieties attached by a glycosidic bond at different positions of the rings. Thus, observations from numerous studies confirm that the biological activity of saponins is influenced by both the aglycone and the sugar moiety (Ohtsuki et al., 2004; Pereira da Silva et al., 2006). However, studies directed to the establishment of structure-activity relationships relevant to cytotoxicity are still incomplete.

The importance of the number of sugar moieties in a sugar chain and the structure of aglycone was confirmed by the studies on saponins isolated from *Agave fourcroydes*. Following this, chlorogenin, hecogenin and tigogenin hexasaccharides (with the same sequence of sugar chains), showed cytotoxic activity against HeLa cells (IC_{50} 13.1, 5.2, and 4.8 μg/mL respectively), whereas diglucosides were inactive (Ohtsuki et al., 2004). As an additional feature of the presence of a lipid-soluble aglycone and water soluble sugar chain(s) in their structure (amphiphilic nature), saponins are surface active compounds with detergent, wetting, emulsifying, and foaming properties.

Uribe et al. (2009) established a specific formulation based in Agave syrup containing phytochemicals (flavonoids, polycosanols and sapogenins) with anticancinogenic and antioxidant properties. An additional feature of *Agave* species is that anordin and dinordin prepared with steroids derived from sisal plants (*Agave sisilana*), whose anti-fertility properties have been confirmed, constitute a new family of contraceptives (Crabbe, 1979). Sterols, steroidal sapogenins, steroidal alkaloids and alkaloidal amines, all derived from *Agave sisilana*, help the pharmacological properties of the plant. Therefore, the juice obtained from *Agave sisalana* leaves stimulates the intestinal and uterine musculature, lowers the blood pressure and produces abortion in pregnant animals (Debnath et al., 2010).

Consequently Rizwan et al., (2012) reported an extract from the leaves of *Agave attenuata*, with various biological activities. They extracted 31 compounds as major components, among them: mono-2-ethylhexyl phthalate (11.37%), 1,2-benzenedicarboxylic acid (6.33%), *n*-docosane (6.30%) and eicosane (6.02%). The leaves included total phenolic contents (10.541–39.35 GAE, mg/100 g) and total flavonoid contents (43.35–304.8 CE, mg/100 g). They reported that the extract and some of its fractions showed moderate antimicrobial effects and a good antioxidant potencial. Rizwan suggest that this Agave may be used as a source of functional ingredients for the production of healthy foods due to its antioxidant properties.

Afterwards Madhuri and Pandey (2009) found that leaves from *Agave americana* L. contained steroidals saponins, alkaloids, coumarins, isoflavonoids, hecogenin and vitamins (A, B, C). Moreover, a freeze-dried extract of *Agave americana* administered via an intraperitoneal route at doses equivalent to 200 and 300 mg/kg of fresh plant, starting material, showed good anti-inflammatory activity (Peana, et al., 1997). In addition, Basilio et al. (1989), working with a crude extract of *A. americana,* reported that it contains two utero-active compounds, one of them presented pharmacological activity similar to that of acetylcholine; however, their electrophoretic mobility and chromatographic characteristics were different.

It should be noted that the most studied effects of hecogenin, diosgenin and tigogenin (Figure 6) are related to cancer treatment. These molecules were found to exhibit an antiproliferative activity on human osteosarcoma cells (Corbiere et al., 2003).

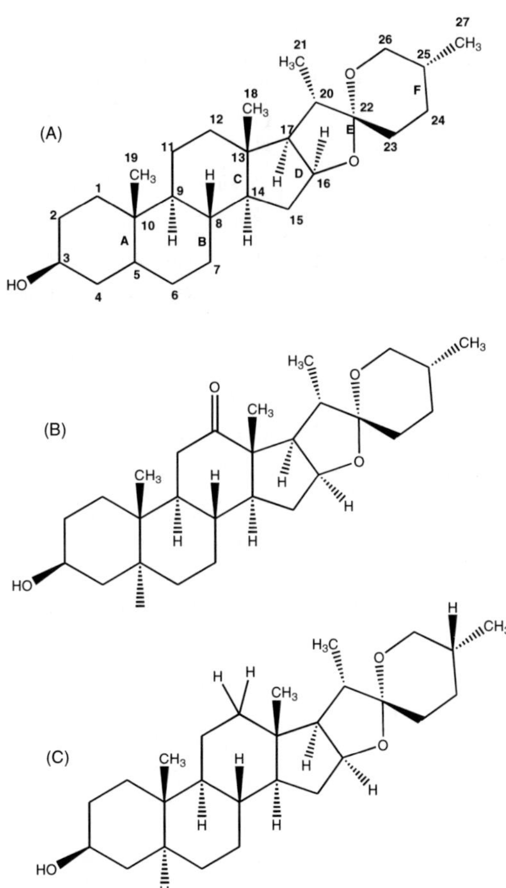

Figure 6. Chemical structures of (A) diosgenin, (B) hecogenin and (C) tigogenin.

Numerous studies were performed to understand the role of diosgenin as a chemopreventive/therapeutic agent against several types of cancer (Raju and Mehta, 2009). Yokosuka et al. (2000), working with *A. americana,* isolated a saponin bisdesmosidic spirostanol, along with three known saponins. Among the isolated saponins, the hecogenin tetraglycoside showed cytotoxic activity against human promyelocytic leukemia cells HL-60, with an IC_{50} of 4.3 μg/mL. In addition, Anajwala et al. (2010) reported a methanol extract of *A.americana* and aqueous extract of *Areca catechu* with potent cytotoxic effects against the MCP-7 cancer cell line with IC_{50} of 545.9 μg/mL for *A. americana* and 775.1 μg/mL for *Areca catechu.* These extracts were also tested against the Vero cell line, obtaining a IC_{50} of 1854 μg/mL for the *A. americana* methanol extract, indicating that both extracts have a high cytotoxicity on the MCF7 cell line and a good relation for the Inhibition vs. Log [dose] Proportion. In contrast, Singh (2011) tested the effect of leaf extracts of *A. americana* L. in the disease called visceral leishmaniasis (VL). Over 20,000 cases of VL are reported annually in India, representing a large public health problem in India and around the world. The effects of CSAs (crude soluble antigens) extracted from leaves of *A. americana* L. (0.5 mg/mL) killed 100% of promastigotes of *L. donovani* in a period of 24 hours ($p = 0.00098$).

A well-known toxic effect of saponin is its ability to lyse erythrocytes. This, in general, is due to its interaction with cholesterol leading to membrane destabilization in the erythrocyte (Pereira da Silva et al., 2006). Recently, Sharma et al. (2012) reported an analysis of saponins from extracts of *Agave americana* L. leaves using thin layer chromatography (TLC), where the plate was previously immersed in a suspension of sheep erythrocytes. The detection of 3 saponins was based on their hemolytic property. On the other hand, *Agave schottii* (*Amaryllidaceae*) (Bianchi, 1969) extracts, showed anti-tumor properties, and one saponin extracted from *A. fourcroydes* identified as chlorogenin had an IC_{50} of 7.5 to 10 μg/mL against HeLa cells (Ohtsuki et al., 2004). Ramos-Casillas et al. (2012) investigated the ethanol extract from leaves of *A. lechuguilla* Torr, finding cytotoxic activity against HeLa cells *in vitro*, with a (IC_{50}) = 89 μg/mL. They also isolated a 5β-steroidal saponin with a IC_{50} = 79 μg/mL agaist HeLa cells. Investigations by Yokosuka et al., on *A. utahensis* reported the presence of steroidal saponins (Yokosuka and Mimaki, 2009; Yokosuka et al., 2009). This group also evaluated the cytotoxic activity of steroidal saponins isolated from fresh whole plants of *Agave utahensis* against HL-60 human promyelocytic leukemia cells. In their early work they found that 4 out of the 15 steroidal saponins obtained showed moderate cytotoxicity, relative to the control, etoposide (IC_{50} 0.20 μg/mL), with IC_{50} values ranging from 5.5 -12.3 μg/mL, whereas the rest did not exhibit apparent cytotoxicity even at higher concentrations (20 μg/mL). Later on, this group isolated eight new spirostanol saponins (1-8) and three new furostanol saponins (9-11) from the same Agave. They referred to two smilagenin diglycosides and one trisaccharide with high cytotoxicity (IC_{50} 4.9–7 μg/mL). The additional C-2 hydroxyl group or C-12 carbonyl group in the 5b-spirostanol structure significantly reduced cytotoxic activity and induced apoptosis-activating caspase-3.

Recently, Háuad-Marroquin et al. (2010) observed that increasing the polarity of *Agave tequilana* Weber extracts resulted in a negative effect on the growth of murine B16F10 melanoma cells, in a dose-dependent manner, using a dose of 1.38 mg/mL.

Extracts from different parts of *Agave cantala*, a perennial, stout, and scapigerous herb, have been used a laxative; emmenagogue; anticancer treatment; citotoxic; diuretic; aphrodisiac and in the treatment of syphilis, swelling, retention of urine, scurvy, and scrutula (Ramachandran et al., 2011). This Agave has been shown to have chemical constituents such as flavone glycosides, sterol glycosides, neotigogenin, hecogenin and tigogenin (Ramachandran et al., 2011). The cytotoxic activity of this Agave was also tested using a *Brine Shrimp (Artemia salina)* lethality bioassay. It was reported that the aqueous and alcoholic leaf extracts showed potent lethality (LC_{50}) at concentrations of 15 and 12.5 mg, respectively (Ramachandran et al., 2011).

On the other hand, Magalhaes de Brito et al. (2011) evaluated the behavioral effects of hecogenin (10 mg/kg) obtained from *Agave sisilana*, on the oxidative stress parameters in the central nervous system of mice. This steroidal sapogenin caused increased lipid peroxidation, compared to the control, and showed a biphasic effect on stress, while the smaller dose (5 mg/kg) did not cause significant changes in the oxidative stress and has antidepressant and anxiolytic effects. On the other hand, Santos-Cerqueira et al. (2012) showed that the same compound, hecogenin, isolated from the same Agave (*A. sisilana*), presented a significant gastroprotective effect on experimental models of gastric ulcer in mice, and suggested that a possible mechanism of action seems to be mediated by the opening of K^+_{ATP} channels and by the COX-2/PG pathway. They also demostrated antioxidant and anti-inflammatory properties that may play a role in the gastroprotective drug effect. On the same line, hecogenin has been widely studied due to its pharmacological activities, including antifungal and hipotensive activities (Gondim, 2006). In sum, because of their biological activity, saponins became commercially important compounds with possible applications in food and cosmetics industry, where they are utilized as emulsifiers or sweeteners, and in pharmaceutical sectors (Güçlü-Üstündag et al., 2007), as well as chemopreventives in cancer control (Raju and Mehta, 2009).

Application of Agave Extracts in the Treatment of Different Diseases

A decoction of the roots of *A. angustifolia* has been used in the treatment of kidney cancer. Moreover, the decoction of its leaves was used to treat diabetes, gonorrhea, diseases of the excretory system and as a diuretic. It has also been used to cure skin infections, as well as in the treatment of snake bites (Gentry, 1982). In addition, antibacterial, antimicotic, and antiviral effects, among others, have been atributed to the *Agave* genus, and extracts from *Agave* species have been used as a cardioactive or larvicide, and in the treatment of different diseases (Achenbach et al., 1994). Moreover, Popoca et al. (1998) refers to tea made of *Agave salmiana* leaves (Figure 7) that has been used in inflammatory and respiratory diseases, wounds, ulcers, venereal diseases and in skin tumors.

Agave intermixta L. and *Cissus sicyoides* L. (*Vitaceae*) are two tropical plants from the Dominican Republic. Gupta (1995) reported some of the therapeutic uses of their aqueous extracts from both plants, in the treatment of inflammations, arthritis, tumors and infections. García et al. (2000) reported that the decoction extracts of this Caribbean species have shown potent *in vivo* anti-inflammatory activity after both oral (300 and 500 mg/kg (po)) and topical administrations (2 and 5 mg/mouse ear) where *A. intermixta* Trel presented a higher activity, especially after the oral administration. They suggested that these positive anti-inflammatory effects might be related to the improvement of the symptoms associated with cancerous processes treated with these species in Caribbean traditional medicine. He also reported a distinct decrease in the level of the enzyme myeloperoxidase in tissues from inflamed areas. Both species have been used as anticarcinogens in the Dominican Republic in the treatment of gall bladder cancer. This group studied the synergistic action of both species (*Agave intermixta* Trel. and *Cissus sicyoides* L.) to address different types of leukemia and a colangicarcinoma. They reported that the use of this mixture decreased toxicity, maintaining the anti-inflammatory and cytostatic effect in both species, also displaying significant antimicrobial actions. In addition, it was shown that a direct application of a poultice prepared with a mixed leaf decoction was benefical in the treatment of cancerous processes (Quilez et al., 2006).

Figura 7. *Agave salmiana* var. crassispina.

In this respect, ethanolic extracts of *Agave lechuguilla* Torr (Agavaceae), *Baccharis glutinosa* Pers. (Compositae) and *Larrea tridentata* (DC.) Cov. (Zygophyllaceae) were more active that other Agave extracts previously tested, on the growth of yeasts, molds and bacteria. They inhibited the growth C. *perfringens, S. dysenteriae* and all the tested fungi (*Cryptococcus neoformans*, *Microsporum gypseum*, *Trichophyton tonsurans*, and *Sporotrix schenckii*), except C. *krusei,* which was only partially inhibited. Yeasts (*C. neoformans*) were inhibited only by *A. lechuguilla*, as well as the molds (*M. gypseum, T. tonsurans, S. schenckii*) (Verastegui et al., 1996). Furthermore, these authors, using an agar diffusion method, examined *in vitro* antimicrobial activity of extracts of diferent *Agave* species: *A. lechuguilla, A. picta, A. scabra* and *A. lophanta* against the same pathogenic bacteria, yeast, and molds. They reported that the extracts of *A. picta* produced zones of inhibition of 9-13 mm for *E. coli*, *L. monocytogenes*, *S. aureus*, and *V. cholerae*, while *B. cereus* and *Y. enterocolitica* were not inhibited. The other *Agave* species showed no inhibitory activity against the bacteria tested. Nevertheless, the four *Agave* sp. tests were inhibitory against all yeasts and molds analyzed, with similar levels of inhibition in all cases (zones 9 to 20 mm of inhibition). The minimum microbicidal concentration (MMC) of the active extract ranged from 1.8 to 7.0 mg/mL for the sensitive bacteria, and 2.0–3.0 mg/mL for yeast. In the case of molds, the minimum inhibitory concentration (MIC) of the active extracts ranged from 3.0 to 6.0 mg/mL. They suggested that the *Agave sp.* analyzed were potential antimicrobial candidates with a wide range of activity (Verastegui et al., 2008).

On the other hand, *Aspergillus flavus* and *Aspergillus parasiticus,* two important contaminants in certain foods for their ability to produce aflatoxins (Farr et al., 1989), and *A. flavus* that also produces cyclopiazonic acid (CPA), another toxic secondary metabolite, (Gqaleni et al., 1996, Sánchez et al., 2005), were also inhibited. A study was done using different extracts (ethanolic, methanolic and aqueous) from flowers of *Agave asperrima*

Jacobi and *Agave striata* Gentry, against both strains (*Aspergillus flavus* and *Aspergillus parasiticus*), producing growth inhibition. They also noted the same effect (growth inhibition of these fungi) on corn storage. Moreover, the methanol extracts from both Agaves, exhibited the best antifungal effect compared with ethanolic and aqueous extracts with the lowest MIC (0.5-2.0 mg/mL) for the two species (*A. asperrima* and *A. Striata).*

Extracts from leaves of *Agave angustifolia* and *Agave potatorum* mixed with honey and sugar syrup were succesfully used to heal skin wounds or sores and, historically, the Aztecs in Mexico already used Agave extracts to heal all kinds stab wounds. Similarly, the Agave extract was successfully used in many diseases: cough, tuberculosis, rheumatism, kidney swelling, gonorrhea internal tumors, sprains and bruises (Davidson et al., 1983; Gómez de Lara, 2009). A laboratory analysis of one concentrated extract of maguey syrup (*Agave* spp.) indicates that its use as a remedy by the ancient and modern Mesoamericans could be attributed to the healing process of infected aerobic wounds. Both pyogenic and enteric bacteria appear to be susceptible to *Agave spp.* Salt addition seems to enhance the effectiveness associated with the material in the inhibition of growth of one of the major causes of infectious processes pyogenes, *Staphylococcus aureus.* Presently, the Agave extract mixed with honey or molasses, is still used in Mexico to relieve coughs, rheumatism and to heal wounds.

Agave as a Source of Fructans

The presence of fructans in many species of *Agave* genus in the last decade has boosted its economic importance (Roberfroid, 1999; Ritsema and Smeekens, 2003; Urías-Silvas and López, 2004; Robertfroid, 2007; Salazar-Leyva et al., 2010). Fructans are reserve polysaccharides of some plant species in nature, which are indigestible to vertebrates, and are considered as soluble dietary fiber. They are composed of non-reducing carbohydrates, mainly of fructosyl units with a complex and highly branched structure with β(2-1) and β(2-6) links with external glucose molecules (graminanes), as well as internal glucose (neofructans); the latter type of fructan has been called agavinas (López, 2003; Mancilla and López, 2006) presenting different degrees of polymerization, depending on the source (Roberfroid, 2005). The type of bonds present in the structure of fructan are resistant to enzymatic hydrolysis in the human stomach. The fructans may be classified as prebiotics, stimulating the growth of beneficial bacteria in the colon. The concept of prebiotics was introduced in 1995 by Gibson and Roberfoid as an alternative focus in the modulation of the gut microbiota. Gibson, et al. (2004) gives a more fresh definition of the term as ''a selectively fermented ingredient that allows specific changes, both in the composition and/or activity in the gastrointestinal microbiota that confers benefits upon host wellbeing and health''. All prebiotics, apart from inulin, are short-chain carbohydrates with low degrees of polymerisation (DP), often referred as oligosaccharides (Manning and Gibson, 2004). Monocotyledons (families *Agavaceae, Poaceae, Alliaceae, Asparagaceae Amaryllidaceae*, *Iridaceae* and *Haemodoraceae*) produce more complex fructans than dicotyledonous plants (Ulloa et al., 2010).

In the last decade, the presence of fructans in *Agave* genus has been well documented. Mancilla and López (2006) reported that agave fructans consist of a complex mixture containing mainly graminans and branched neo-fructans. This group reported the molecular structure of fructans from *A. tequilana* Weber var. azul and other *Agave* species. Their findings show structural differences within species, probably due to their growth in different environmental regions. Javier-Arrizon et al. (2010) report the composition of water-soluble carbohydrates from *Agave tequilana*, were the profile of oligosaccharides changes in function of the age. The richest concentration of lower-molecular weight fructans was observed in younger plants (2-year-old agave plants), suggesting that they could have potential as prebiotics. They also observed in 4-year-old agave plants, a polymerisation and depolymerisation process associated with an increase and decrease in the branching degree of fructans.

Furthermore, the stems of *Agave veracruz* store a complex mixture of fructooligosaccharides, inulin, inulin neoseries and branched neofructans (López et al., 2003; Mancilla and López, 2006)). In *Agave americana*, several fructan structures have been found where inulin was identified as the main carbohydrate reserve (Bathia and Nandra, 1979). Also, Sánchez-Marroquín and Hope (1953) reported inulin as the principal storage carbohydrate in *Agave tequilana*. Studies with *A. deserti* exhibited the presence of fructans (Wang and Nobel, 1998). More recently, Urías-Silvas and López (2004) extracted fructan from diferent *Agave* species, testing them on *Bifidobacterium breve*, fnding that short- chain fructans, presented in *A. angustifolia* Oax., *Dasyrilion spp.* y *A. tequilana*, favored the growth of *B. breve* and generated short-chained fatty acids namely, acetic, propionic and butyric, considered to be the main end products of carbohydrate metabolism via fermentation by enzymes of probiotic origin, which are then used by the host organism as an energy source.

On the other hand, there have been a number of papers and reviews on the topic of prebiotics and their health benefits (Macfarlane et al., 2006; López and Urías-Silvas, 2007; Morris and Morris, 2012; Charalampopoulos and Rastall 2012; Saad et al., 2012; Sharma et al., 2012). Those benefits include an increase in mineral absorption, especially calcium and magnesium (Abrams et al., 2007; Wang et al., 2010), and also an improvement of the immune response (Macfarlane, Steed and Macfarlane, 2007; Seifert and Watzl, 2007; Chung-Hsiung et al., 2012). There is mounting evidence that prebiotics play a role in colorectal cancer prevention and therapy (Reddy, 1999; Asad et al., 2008; Liong, 2008; Taper and Roberfroid, 2008; Stein et al., 2011). Also, modifications of inflamatory conditions, such as irritable bowel syndrome, (IBS), ulcerative colitis and inflammatory bowel disease (IBD) have been reported (Spiller, 2008; Hedin et al., 2007; Steed et al., 2008) as well as hipolipidemic effects (Srih et al., 2012), regulation of metabolic disorders related to obesity (Parnell and Reimer, 2009).

Sharma et al. (2012) described the mechanisms of action of prebiotics and probiotics. "They affect intestinal bacteria by increasing the numbers of beneficial anaerobic bacteria and decreasing the population of potentially pathogenic microorganisms". Probiotics affect the intestinal ecosystem by stimulating mucosal immune mechanisms and by stimulating nonimmune mechanisms through antagonism/competition with potential pathogens. However, Sharma et al., suggests that probiotics reduce the risk of colon cancer in animal models probably due to their role in suppressing the activity of certain bacterial enzymes that may increase the levels of procarcinogens.

Recent studies have been directed to the sinergistic effect of prebiotics and probiotics in colon cancer prevention (Reddy, 1999; Liong, 2008; Stein et al., 2011). In this sense, the consumption of synbiotics, a mixture of probiotics and indigestible food constituents such as dietary fiber, has been reported to reduce colon cancer risk. Research of Zhu et al. (2011) proved that symbiosis resulted in a synergism in preventing CRC (colon rectal cancer), with more activity than either a probiotic or a prebiotic alone. Hence, there is a growing number of data indicating that specific bacterial species, in combination with dietary additions, can modulate overall gut performance as well as their effect on inflammation. Rowland et al. (1998) showed that synbiotics may be particularly efficacious for reducing colonic preneoplastic lesions. These studies showed that the combination of inulin and *Bifidobacterium longum,* produced a decrease of 80% in ACF (a preneoplastic lesion for colon cancer), whereas inulin alone decreased ACF (aberrant crypt foci) by 41%, and *B. longum* alone decreased ACF by 26%. Similarly, several studies have found that adding a relative large amount of inulin (10%) to a diet, reduced ACF (Rowland et al., 1998; Rao et al., 1998). The exact mechanisms by which probiotics inhibit colon cancer may involve multiple pathways, among them are: cell cycle, reactive oxygen species (ROS), apoptosis, production of specific bacterial enzymes and effects on the host metabolome. Zhu et al. (2011) report that probiotics can inhibit the inflammatory process by enhancing host immune responses, altering the bacterial phylotypes in the colon and affecting the gut metabolome.

Park et al. (2007) provided evidence on the effect of prebiotics, probiotics or a combination, in the inhibition of ACF (aberrant crypt foci). They anayzed the effect of a high-fat and low-fiber diet supplemented on rats fed daily with the probiotic *Bacillus polyfermenticus* (3×10^8 cfu/1.3 g), which resulted in a 50% reduction in ACF formation, compared to rats fed with the control diet.

In addition, *in vitro* assays were performed with Agave fructans resulting in the stimulation of growth of at least 6 differents strain of *Bifidobacterium*, and 4 different strains of *Lactobacillus,* which was more effective than inulin-type fructans (Urias-Silvas and López, 2009).

On the other hand, Umang Munjal et al. (2009) prepared fermentation products of insulin-type fructans and studied their biological properties in two human colon cell lines, LT97 and HT29 (representing early and late stages of colon cancer). They reported growth-inhibiting and apoptosis-inducing effects of fermentation supernatants of inulin.

To evaluate the effect of *Bifidobacterium lactis*, ''resistant starch'' (RS) and a combination of them (synbiotic), on their ability to protect against CRC (Colon Rectal Cancer), rats were fed with RS in combination with *B. Lactis*. This study showed a significantly lowered incidence and multiplicity of colonic neoplasms compared with the control group. *B. lactis* utilizes the carbohydrate ''Resistant Starch'' (RS) as substrate and up-regulates the acute apoptotic response to a carcinogen in the colon (colonic neoplasms were induced by 2 weekly injections of azoxymethane 15 mg/kg body wt) (Le Leu et al., 2010).

Chung-Hsiung et al. (2012) investigated the effect of diosgenin (steroidal sapogenin) on the growth of enteric Lactic Acid Bacteria (LAB). The presence of diosgenin significantly enhanced the growth of *Lactobacillus murinus* and *Lactobacillus reuteri*. Therefore, the steroidal sapogenins may be a novel class of prebiotics for the enteric lactic acid bacteria.

Also, another type of fructan, the levan polysaccharide, has been shown to have industrial applications as a new industrial gum in the fields of cosmetics, foods like dietary fiber, and pharmaceutical products. Moreover, these compounds have also been shown to express biological activities such as immunostimulating, antitumoral and anti-inflammatory activities (Szwengiel et al., 2004). Following this, the study of Belghith (2012), revealed that treatment with levan polysaccharide in rats fed with a high-cholesterol diet with a daily dose of levan equivalent to 5% resulted in positive changes of plasma enzyme activities and lipid levels.

She concluded that Levan could protect against oxidative stress linked atherosclerosis and decrease the atherogenic index.

In addition, Abdel-Fattah et al. (2012) reported that a novel levansucrase produced from *Bacillus subtilis* NRC1aza yielded two types of levan. One of them, the sulfated *B. subtilis* NRC1aza levan showed high selective cytotoxicity against hepatocellular carcinoma HepG2 cells. It led to DNA damage and fragmentation associated with the induction of apoptosis. They concluded that sulfated *B. subtilis* levan was a promising anticancer agent that induced intrinsic apoptosis in liver cancer cells.

Agave s a Source of Policosanol

Policosanol, or Long-Chain Alcohols (LCA), is the generic term for a natural extract of plant waxes. They have a mixture of long-chain alcohols as their active principle (24, 26, 27, 28, 29, 30, 32 and 34 carbon atoms) that is used as a nutritional supplement mainly to lower LDL Cholesterol ("bad" cholesterol) and increase HDL cholesterol ("good" or "healthy" cholesterol) and to help to prevent atherosclerosis. Agave leaves are covered by a thin film of wax that prevents loss of water. This wax consists mainly of alcohols, esters, aldehydes, ketones, fatty acids and hydrocarbons (Mendez et al., 2003). This group isolated, purified and characterized chemically a mixture of eleven aliphatic, straight and high molecular weight alcohols from leaves of henequen (*Agave furcroydes L.*) wax. Among them, 1-octacosanol and 1-triacontanol were the main components.The mixture consists of the following alcohols: 1-hexacosanol, 1-heptacosanol, 1-octacosanol, 1-nonacosanol, 1-triacontanol, 1-hentriacontanol, 1-dotriacontanol, 1-tritriacontanol, 1-tetratriacontanol, 1-pentatriacontanol and 1-hexatriacontanol. The amount of wax in the leaves of henequen ranges between 0.1 to 0.3%, depending mainly on its age, soil composition, and climatic conditions. Agave wax composition has also been studied by Guerra de Leon et al., (2007) in leaves of *Agave brittoniana* T., indicating that esters consisted of a mixture of high molecular weight alcohols: hexacosanol, heptacosanol, octacosanol, nonacosanol, triacontanol, dotriacontanol and tetratriacontanol.

In plants of *A. salmiana*, *A. angustifolia* and *A. tequilana*, long-chain fatty acids were identified by Peña-Álvarez et al. (2004). They analyzed the terpenes present in these three plants, identifying a total of 32 terpenes, 21 of which were exclusively found in *A. tequilana*. The linalool, geraniol, *p*-cimene, limonene, β-*trans* ocimene and *trans*-nerolidol were found in the three *Agave* species analyzed.

Agave as a Source of Polyphenols

Polyphenols are a group of chemical substances found in plants and food products that may have antioxidant characteristics with bioactive functions and potential health benefits (Dueñas et al., 2005).

Phenolic compounds comprise flavonoids, phenolic acids, and tannins, among others. Phenolic acids can be divided in two subgroups according to their structure: the hydroxyl-benzoic and the hydroxycinnamic acids. The most commonly found hydroxybenzoic acids include gallic, hydroxybenzoic, protocatechuic, vanillic and syringic acids, whereas among the hydroxycinnamic acids they include, caffeic, ferulic, p-coumaric and sinapic acids (Bravo, 1988).

Different processes were established for the extraction of polyphenols from different plants such as microwave-assisted extraction, ultrasonic extraction and various extraction methods using different solvents and conditions. Ben Hamissa et al. (2012), working with *A. americana* leaves using air or a nitrogen atmosphere, increased the contact time (15 to 240 minutes) and temperature (25 to 150 °C), reporting a major impact on total polyphenol (TP) and total flavonoid (TF) contents, and consequently on their antioxidant activity. Furthermore, Almaraz-Abarca et al., (2009) studied the polyphenol composition of *Agave durangensis* tissues and found that the most abundant polyphenols were flavonol, kaempferol and quercetin. According to Duke (2000), the polyphenolic fraction of *A. americana* leaves is essentially composed of the flavonoid kaempferol.

Flavonoids constitute the most numerous group of plant phenolics, representing over half of the eight thousand naturally occurring phenolic compounds (Harborne and Baxter, 1999).

Variations in the substitution patterns to ring C in the structure of these compounds result in the main flavonoid classes (e.g., flavonols, flavones, flavanones, flavanols, isoflavones, and anthocyanidins).

Parmar et al. (1992) report the isolation and structural elucidation of a new flavanone (5,7-dihydroxy-6,5'-dimethoxy-3',4'-methylenedioxy flavanone) from *Agave amaricana* named Agamanone. This was the first report of the occurrence of a flavanone in the genus *Agave.*

Later on, three flavones and seven homoisoflavonoids were isolated from the methanolic extract of the leaves of *Agave sisilana* Perrine ex Engelm (Chen et al., 2009). More recently, Morales-Serna et al. (2010) isolated from the ethyl acetate and acetone extracts of *Agave tequilana* Weber three compounds that belong to a small homogeneous group of naturally occurring oxygen heterocycles called homoisoflavanones, whose basic structure consists of a 16 carbon skeleton, including a chromanone system with a benzyl group at position 3.

On the other hand, Uribe and Saldivar (2010) confirmed the anticancerogenic and antioxidant properties of the Agave syrup.

Conclusion

The *Agave* genus (Agavaceae), widely distributed throughout the Americas and Africa, is now known to be an excellent source for a large number of chemicals with a high potential use in the medical field. Among them, saponins, sapogenins, and steroidal sapogenins, mainly hecogenin and tigogenin, are widespread in this genus. Some species of the *Agave* genus have also been used in different industries and in folk medicine. There are presently many investigations focusing on the chemical constituents present in the *Agave* genus, such as flavone glycoside, sterol glycosides, neotigogenin, hecogenin and tigogenin, and everyday new important compounds, are reported in the literature. Among the recent reports, fructans have shown important industrial applications, in foods like dietary fiber and also in the pharmaceutical industry. Levan polysaccharides have shown relevant biological activities such as immunostimulating, antitumoral and anti-inflammatory activities. We can find applications of the chemicals present in Agave in the treatment of many of the most important diseases, such as cancer, diabetes, and others like gonorrhea, skin infections (as poultices), treatment of snake bites; and as anti-inflammatory, antifungal, antivirus, and antimicrobiological agents among others. The present chapter attempts to compile and summarize the studies made on the chemical components isolated from the *Agave* genus and compared them with their reported biological activities, including those reported in ancient traditional medicine. However, in order to maximize the use of this important genus, there is a need to continue searching for the presence of the different chemical compounds in the *Agave* genus, as well as on their potential uses, and to establish their mechanism of action to allow a more rational use of these resources.

Acknowledgments

We express our gratitud to Dr. Nayelli Marsh for her critical review of this manuscript. To Dr. June Simpson, Katia del Carmen Gil-Vega and María Jazmín Abraham Juárez, for authorizing us to use their Agave photography for this chapter. Thank you also to Dr. José Carlos Alvarado Monzón for his help in the preparation of this manuscript.

References

Abdel-Khalik SM, Miyase T, Melek FR, el-Shabraway OA, Mahmoud II, Mina SA (2002). New steroidal saponins from *Agave lophantha* Schiede and their pharmacological evaluation. *Pharmazie* 57(8):562-566.

Abdel-Fattah AM., Gamal-Eldeen AM, Helmy WA, Esawy MA (2012). Antitumor and antioxidant activities of levan and its derivative from the isolate *Bacillus subtilis* NRC1aza. *Carbohydrate Polymers* xxx: xxx–xxx. Article In press doi:10.1016/j.carbpol.2012.02.041.

Abraham-Juárez MJ, Ramírez-Malagón R, Gil-Vega KC and Simpson J (2009). AFLP Analysis of genetic variability in three reproductuive forms of *Agave tequilana*. *Rev. Fitotec. Mex*. 32 (3): 171–175.

Abrams SA, Hawthorne KM, Aliu O, Hicks PD, Chen Z, and Griffin IJ (2007). An inulin-type fructan enhances calcium absorption primarily via an effect on colonic absorption in humans. *The J. Nutr* 2208-2212.

Achenbach H, Hübner H, Brandt W, and Reiter M (1994), Cardioactive steroid saponins and other constituents from the aerial parts of *Tribulus cistoides. Phytochem*. 35, 1527Ð1543.

Almaraz-Abarca N, Delgado-Alvarado EA, Hernández-Vargas V, Ortega-Chávez M, Orea-Lara G, Cifuentes-Díaz de León A, Ávila-Reyes JA, Muníz-Martínez R (2009). Profiling of phenolic compounds of somatic and reproductive tissues of *Agave durangensis* Gentry (*Agavaceae*). *Am. J. Appl. Sci.* 6:1076–1085.

Alonso-Castro AJ, Villarreal ML, Salazar-Olivo LA, Gómez-Sánchez M, Domínguez F, García-Carranca A (2011). Mexican medicinal plants used for cancer treatment: Pharmacological, phytochemical and ethnobotanical studies. *J. Ethnopharmacol*. 133:945–972.

Anajwala Chetan C, Patel RM, Dakhara SL, Jariwala JK (2010). *In vitro* cytotoxicity study of *Agave americana, strychnos Nuxvomica* and *Areca catechu* extracts using MCF-7 cell line. *J. Adv. Pharm. Technol. and Res.* 1 (2):245-252.

Arellano-Rodríguez JA, Flores JS, Tun-Garrido J, Cruz-Bojórquez MM (2003). Nomenclatura, Forma de Vida, Uso, Manejo y Distribución de las especies Vegetales de la Península de Yucatán. Etnoflora Yucatanense, Fascículo 20:1-815. Universidad Autónoma de Yucatan Press, México, pp. 312–380.

Asad U., Emenaker NJ, Milner JA (2008). Colorectal cancer prevention: The role of prebiotics. In G.R. Gibson, and M.B. Roberfroid, *Handbook of Prebiotics* (pp. 285-294). Editor Taylor and Francis (2008), 485pp Publisher: (CRC Press, Boca Raton, Fla.). USA.

Ávila-Fernández A, Olvera-Carranza C, Rudiño-Piñera E, Cassab IG, Nieto-Sotelo J, López-Munguía A (2007). Molecular characterization of sucrose: sucrose 1-fructosyltransferase (1-SST) from *Agave tequilana* Weber var. azul. *Plant Sci.* 173: 478-486.

Basilio CM, Seyler L, Bernstein J, Castro de la Mata R (1989). Isolation and characterization of an utero-active compound from *Agave americana. P. R. Health Sci. J.* 8 (3):295-299.

Bathia IS and Nandra KS (1979). Studies on fructosyl transferase from *Agave americana. Phytochem.* 18:923-927.

Bedour MS, e lgamal MH, El Tawil BA (1979). Steroid sapogenins, part XV. The constituents of *Agave utahensis* var. nevadensis, *A. lophantha* and *A. parasana. Planta Med.*36 (2):180-181.

Belghith KS, Dahecha I, Hamden K, Feki A, Mejdoub H, Belghith H (2012). Hypolipidemic effect of diet supplementation with bacterial levan in cholesterol-fed rats. *Int. J. of Biol. Macromol.* 50:1070– 1074.

Ben Hamissa AM, Seffea M, Aliakbarian B, Casazza AA, Perego P, Converti A (2012). Phenolics extraction from *Agave americana* (L.) leaves using high-temperature, high-pressure reactor. *Food Bioprod Process* 90:17–21.

Bianchi E, Cole JR (1969). Antitumor agents from *Agave schottii* (*Amaryllidaceae*). *J. Pharm. Sci.* 58 (5):589-591.

Blunden G, Carabot A, Cripps AL, jewers K (1980). Ruizgenin, a new steroidal sapogenin diol from *Agave lecheguilla. Steroids*, 35(5):503-510.

Blunden G, Patel AV and Crabb TA (1986), Barbourgenin, a new steroidal sapogenin from *Agave sisalana* leaves. *J. Nat. Prod.* 49: 687-689.

Bravo Laura (1998). Polyphenols: Chemistry, Dietary Sources, Metabolism, and Nutritional Significance. *Nutr. Rev* 56 (11):317-333.

Calderón AI, Vázquez Y, Solís PN, Caballero-George C, Zacchino S, Giménez A, Pinzón R, Cáceres A, Tamayo G, Correa M, Gupta MP (2006). Screening of Latin American plants for cytotoxic activity. *Pharm. Biol.* 44:130–140.

Cedeño CM (1995). "Tequila production", en Crit. *Rev. Biotechnol.*, núm. 15 (1):1- 11.

Cedeño CM and Álvarez-Jacobs J (1999). Tequila production from agave: historical influences and contemporary processes. In *The Alcohol Textbook*; Jacques, K. A., Lyons, T. P., Kelsall, D. R., Eds.; Nottingham University Press: Nottingham, U.K., 2003, Charper 15: 223 246.

Charalampopoulos D and Rastall RA (2012). Prebiotics in foods. *Curr. Opin. Biotechnol.* 23:187–191.

Chen PY, Kuo YC, Chen CH, Kuo YH, Lee CK, (2009). Isolation and immunomodulatory effect of homoisoflavones and flavones from *Agave sisalana* Perrine ex Engelm. *Molecules* 14:1789–1795.

Chung-Hsiung H, Jin-Yi C, Ming-Chung D, Chung-Hsi C, Tong-Rong J (2012). Prebiotic effect of diosgenin, an immunoactive steroidal sapogenin of the Chinese yam. *Food Chem*. 132: 428–432.

Colunga-GarcíaMarín P and May-Pat F (1993). *Agave* Studies in Yucatan-Mexico. I. Past and Present Germplasm Diversity and Uses. *Econ. Botan* 47(3):312-327.

Colunga-GarciaMarin P, Coello-Coello J, Eguiarte LE, Pinero D (1999). Isozymatic variation and phylogenetic relationships between henequen (*Agave fourcroydes*) and its wild ancestor *A. angustifolia* (Agavaceae). *Amer. J. Bot*, 86, 115–123.

Corbiere C, Liagre B, Bianchi A, Bordji K, Dauca M, Netter P, Beneytout JL (2003). Different contribution of apoptosis to the antiproliferative effects of diosgenin and other plant steroids, hecogenin and tigogenin, on human 1547 osteosarcoma cells. *Int. J. Oncol.* 22:899–905.

Crabbe P (1979). Mexican plants and human fertility. *UNESCO Cour.* 7: 33-34.

Cripps AL, Blunden G (1978). A quantitative gas-liquid chromatographic method for the estimation of hecogenin and tigogenin in the leaves, juice and sapogenin concentrates of *Agave sisalana*. *Steroids* 31 (5), 661-669.

da Silva BP, de Sousa AC, Silva GM, Mendes TP, Parente JP (2002). A new bioactive steroidal saponin from *Agave attenuata*. *J. Biosci.* 57 (5-6):423 - 428.

Davidson JR, Ortiz de Montellano BR (1983). The antibacterial properties of an Aztec wound remedy. *J. Ethnopharmacol.* 8 (2):149-161.

Ding Y, Chen YY, Wang DZ, Yang CR (1989). Steroidal saponins from a cultivated form of *Agave sisilana*. *Phytochem.* 28:2787-2791.

Dueñas M, Fernández D, Hernández T, Estrella I, Muñoz R (2005). Bioactive phenolic compounds of cowpeas (Vigna sinensis L.). Modifications by fermentation with natural microflora and with *Lactobacillus plantarum* ATCC 14917. *J. Sci. Food. Agr* 85:297–304.

Duke JA (2000). *Handbook of Medicinal Herbs*. Boca Raton, FL: CRC Press.

Eskander J, Lavaud C, Harakat D (2010). Steroidal saponins from the leaves of *Agave macroacantha*. *Fitoterapia* 81(5):371-374.

Farr DF, Billis GF, Chamuris GP, Rossman AY (1989). Fungi on plants and plant products in the United States. *American Phytopathological.* Society Press, St. Paul, MN, pp. 1– 152.

Fazli FR (1968). Contraceptives and other steroid drugs: their production from steroidal sapogenins. *Pak. J. Sci.*20 (1 and 2):64-67.

García-Herrera EJ, Méndez-Gallegos SJ, Talavera-Magaña D (2009). El Género *Agave* spp. en México: Principales Usos de Importancia Socioeconómica y Agroecológica. VIII Simposium-Taller Nacional y 1er Internacional de "Producción y Aprovechamiento del Nopal". RESPYN *Revista Salud Pública y Nutrición*, Edición Especial 2010 No. 5:109-129. Fac. de Salud Pública y Nutrición Universidad Autónoma de Nuevo León. Escobedo; N.L., México.

García MD, Quílez AM, Sáenz MT, Martínez-Domínguez ME, de la Puerta R (2000). Anti-inflammatory activity of *Agave intermixta* Trel. and *Cissus sicyoides* L., species used in the Caribbean traditional medicine. *Journal of Ethnopharmacology* 71(3): 395–400.

García MD, Sáenz MT, Quílez A, Fernández MA (1999). Antibacterial activity of *Agave intermixta* and *Cissu sicyoides*. *Fitoterapia* 70 (1): 71-73.

García-Mendoza A (2002). "Distribution of the genus *Agave* (Agavaceae) and its endemic species in Mexico", in *Cactus Succulent J.* (U.S.) 74: 177-187.

García Mendoza AJ (2007). *Los Agaves de México*. CIENCIAS 87 Julio-Septiembre. pp. 14-23 Jardín Botánico, Instituto de Biología, Universidad Nacional Autónoma de México. Editorial UNAM, Facultad de Ciencias. Responsable: Patricia Magaña Rueda.

Gentry HS (1982). *Agaves of Continental North America*, Ed. The University of Arizona Press, 1[a.] edición, Tucson, Az. U.S.A. 670 p. Editor-in-Chief: Allyson Carter.

Gibson GR, and Roberfroid MR (1995). Dietary modulation of the human colonic microbiota: introducing the concept of prebiotics. *J. Nutr.* 125: 1401-12.

Gibson GR, Probert HM, Loo JV, Rastall RA, Roberfroid MB (2004). Dietary modulation of the human colonic microbiota: Updating the concept of prebiotics. *Nutr. Res. Rev.* 17:259-275.

Gil-Vega K, Diaz C, Nava-Cedillo A, Simpson J (2006). AFLP analysis of *Agave tequilana* varieties. *Plant Sci*, 170:904–909.

Gómez-Flores R, Verástegui-Rodríguez L, Quintanilla-Licea R, Támez-Guerra P, Monreal-Cuevas E, Támez-Guerra R, Rodríguez-Padilla C (2009). Antitumor properties of Gymnosperma glutinosum leaf extracts. *Cancer Investigation* 27: 149–155.

Gómez de Lara JL (2009). Tratamientos curativos naturales aplicados en el hospital de San Pedro, Puebla. *Revista Digital Universitaria* UNAM Volumen 10 Número 9. art.57.

Gondim KL (2006). Avaliação da atividade cardiovascular de fitoestrógenos e caracterização do mecanismo de ação de diosgenina em ratos. Tese de doutorado em Produtos Naturais e Sintéticos Bioativos, Universidade Federal da Paraíba, UFPB, Brasil.

Gqaleni N, Smith JE, Lacey J (1996). Co-production of aflatoxins and cyclopiazonic acid in isolates of *Aspergillus flavus*. *Food Addit Contam* 13:677–685.

Güçlü-Üstündag Ö, Mazza G (2007). Saponins: Properties, applications and processing. *Crit. Rev. Food Sci. Nutr* 47: 231-258.

Guerra de León JO, Nogueiras Lima C, Laguna Granja A, De Rojas Pérez M (2007). Estudio de las ceras obtenidas de hojas del *Agave brittoniana* T. *Revista Cubana de Química*. Vol. XIX 1:62-64.

Gutiérrez A, Rodríguez IM, del Río JC (2008). Chemical composition of lipophilic extractives from sisal (*Agave sisalana*) fibers. *Ind. Crops Prod* 28:81-87.

Gupta M (1995). 270 *Plantas medicinales Iberoamericanas*. Programa Iberoamericano de Ciencia y Tecnología para el Desarrollo, CYTED. Convenio Andrés Bello. Santa Fé de Bogotá, Colombia. pp 3-6. Editorial Presencia Ltda 1[a]. Edición Bogotá pp 250-254.

Háuad-Marroquín LA, Rodríguez Padilla C, Franco M, Castillo P, Vázquez Alvarado R, Rojas MA (2010). Propiedades terapéuticas de *Agave tequilana* Weber. *Rev. Int C Tecnol Biom* pp1-7.

Harborne JB, and Baxter H (1999). *The Handbook of Natural Flavonoids*. Chichester, UK: Publisher: Wiley. Chichester, 1999, 1838pp Edited by J.B. Harborne FRS and H. Baxter.

Hedin C, Whelani K, Lindsay JO (2007). Evidence for the use of probiotics and prebiotics in inflammatory bowel disease: a review of clinical trials. In *Proceedings of the Nutrition Society*, 66:307-315. Editor(s): K R. Westerterp, Maastricht University, The Netherlands. Editorial: Board.

Huitron C, Pérez R, Sánchez AE, Lappe P, Rocha-Zavaleta L (2008). Agricultural waste from the tequila industry as substrate for the production of commercially important enzymes. *J. Environ. Biol* 29 (1):37-41.

Iñiguez-Covarrubias G, Díaz-Teres R, Sanjuan-Dueñas R, Anzaldo-Hernández J, Rowell RM (2001). Utilization of by-products from the tequila industry. Part 2: potential value of *Agave tequilana* Weber azul leaves. *BioRes Technol.* 77:101-108.

Javier A, Morel S, Gschaedler A, Monsan P (2010). Comparison of the water-soluble carbohydrate composition and fructan structures of *Agave tequilana* plants of different ages. *Food Chem* 122:123-130.

Koduru S, Grierson DS, Afolayan AJ (2007). Ethnobotanical information of medicinal plants used for treatment of cancer in the Eastern Cape Province, South Africa. *Curr. Sc.* 92:906– 908.

Lappe-Oliveras P, Moreno-Terrazas R, Arrizón-Gaviño J, Herrera-Suárez T, García-Mendoza A, Gschaedler-Mathis A (2008). Yeasts associated with the production of Mexican alcoholic nondistilled and distilled Agave beverages. *FEMS Yeast Res* 8: 1037-1052.

Le Leu Richard K., Hu Y., Brown IL, Woodman RJ., Young GP. (2010). Synbiotic intervention of *Bifidobacterium lactis* and resistant starch protects against colorectal cancer development in rats. *Carcinogenesis* 31 (2):246–251.

López MG, Mancilla-Margalli NA, Mendoza-Diaz G (2003). Molecular structures of fructans from *Agave tequilana* Weber var azul. *J. Agric. Food Chem.* 51:7835–7840.

López MG and Urías-Silvas JE (2007). Agave fructans as prebiotics. In *Advances in Fructooligosaccharides Research* [Shiomi N.; N Benkeblia and S Onodera, editors]. Kerala, India: Ed. Research Signpost 37/661 (2), Fort P.O., Trivandrum-695 023. pp 1-14.

Liong MT (2008). Roles of probiotics and prebiotics in colon cancer prevention: postulated mechanisms and in vivo evidence. *Int. J. Mol. Sci.* 9:854-863.

Macías FA, Guerra JO, Simonet AM, Nogueiras CM (2007). Characterization of the fraction components using 1D TOCSY and 1D ROESY experiments. Four new spirostane saponins from *Agave brittoniana* Trel. spp. *Brachypus. Magn. Reson. Chem.* 45: 615–620.

Madhuri S and Pandey G (2008). Some dietary agricultural plants with anticancer properties. *Plant Arch.* 8:13–16.

Madhuri S and Pandey G (2009). Some anticancer medicinal plants of foreign origin. *Current Sci.*, 96 (6):779-783.

Macfarlane S, Macfarlane GT and Cummings JH (2006). Review article: prebiotics in gastrointestinal tract. *Aliment Pharmacol and Ther* 24:701-714.

Macfarlane GT, Steed H and Macfarlane S (2007). Bacterial metabolism and health-related effects of galacto-oligosaccharides and other prebiotics. *J. Appl. Microbiol.* 104 (2):305-344.

Magalhaes de Brito E, Cavalcante de Abreu D, Honorio Junior R, Barbosa-Filho, Florenco de Sousa, Chaves Silva C, Cortez Lima B, Menezes RG, Fonteles F, Mendes Vasconcelos (2011). Behavioural effects and parameters of oxidative stress with hecogenin in central nervous system of mice. Ed. P.4.d. *Anxiety disorders and anxiolytics – Anxiolytics* (basic) S543. P.4.d004.

Man Shuli, Gao W, Zhang Y, Huang L, Changxiao (2010). Chemical study of medical applications of saponins as anti-cancer agents. *Fitoterapia* 81:703-714.

Mancilla-Margalli NA and López MG (2006). Water soluble carbohydrates and fructan structure patterns from *Agave* and *Dasylirion* species. *J. Agric. Food Chem.* 54:7832–7839.

Manning TS, Gibson GR (2004). Prebiotics. *Best Pract Res Clin. Gastroenterol.* 18 (2): 287-298.

Martínez-Hernández A, Pastrana-Chávez J, Sánchez-Villarreal A, Lara-Reyna J, Herrera-Estrella L, Herrera-Estrella A, Martínez de la Vega O, Simpson-Williamson J. Genómica de *Agave tequilana*: Identificación de genes útiles para la industria tequilera y el desarrollo de usos alternativos del agave. http://www.ncbi.nlm.nih.gov

Martínez M (1989). *Las Plantas Medicinales de México*.Editorial: Ediciones Botas, Fondo de Cultura Económica. 6ª. Edición ilustrada. Editor: Martínez M., Fecha publicación 14 de Enero 2010, México, p. 656.

Mena-Rejón G, Camal-Fuentes E, Cantillo-Ciau Z, Cedillo-Rivera R, Flores-Guido J, Moo-Puc R (2009). In vitro cytotoxic activity of nine plants used in Mayan traditional medicine. *J. Ethnopharmacol.* 121:462–465.

Méndez E, Blanco M, Laguna A, García E (2003). Isolation and characterization of a mixture of higher primary aliphatic alcohols of high molecular weight from henequen (*Agave furcroydes* L.) wax. *Rev. CENIC Cie. Quim.* 34 (1):35-40.

Menyar Ben HA, Seffen M, Aliakbarian B, Casazza AA, Perego P, Converti A (2012). Phenolics extraction from *Agave americana* (L.) leaves using high-temperature, high-pressure reactor. *Food Bioprod Process* 90:17-21.

Morales-Serna JA, Jiménez A, Estrada-Reyes R, Márquez C, Cárdenas J, Salmón M (2010). Homoisoflavanones from *Agave tequilana* Weber. *Molecules* 15 (5):3295-3301.

Morris C and Morris GA (2012). The effect of inulin and fructo-oligosaccharide supplementation on the textural, rheological and sensory properties of bread and their role in weight management: A review. *Food Chem.* 133:237-248.

Mousumi D, Mukeshwar P, Rohi S, Thakur GS and Lal P (2010). Biotechnological intervention of *Agave sisilana*: A unique fiber yielding plant with medicinal property. *J. Med. Plant Res.* 4 (3): 177-187.

Nobel PS (1987). Environmental responses and productivity of the CAM Plant, *Agave tequilana*. *Agric. For Meteorol*, 39:319-334.

Nykänen L and Nykänen I (1991). *Volatile Compounds in Foods and Beverages,* Chapter 15, in: H. Maarse (Ed.), Marcel Dekker Inc., New York. Editor: Taylor and Francis, 1991. 784pp.

Ohtsuki T, Koyano T, Kowithayakorn T, Sakai S, Kawahara N, Goda Y, Yamaguchi N, Ishibashi M (2004). New chlorogenin hexasaccharide isolated from *Agave fourcroydes* with cytotoxic and cell cycle inhibitory activities. *Bioorg. Med. Chem.* 12 (14):3841-3845.

Park E, Jeon GI, Park JS, Paik HD (2007). A probiotic strain of *Bacillus polyfermenticus* reduces DMH-induced precancerous lesions in F344 male rat. *Biol. Pharm. Bull* 30:569–74.

Paz-Mendes T, Silva GM, da Silva BP, Parente JP (2004). A new steroidal saponin from *Agave attenuata*. *Nat. Prod. Res.*18 (2):183-188.

Parmar VS, Jha HN, Gupta AK, Prasad AK (1992). Agamanone, a Flavanone from *Agave americana*. *Phytochem*, 31 (7):2567 2568.

Parnell JA and Reimer RA (2009). Weight loss during oligofructose supplementation is associated with decreased ghrelin and increased peptide YY in overweight and obese adults. *Am. J. Clin. Nutr.*, 89 (6), 1751-1759.

Peana AT, Moretti MDL, Manconi V, Desole G, Pippia P (1997). Anti-inflammatory activity of aqueous extracts and steroidal sapogenins of *Agave americana*. *Planta Med.* 63 (3): 199–202.

Peña-Álvarez A, Díaz L, Medina A, Labastida C, Capella S, Vera LE (2004). Characterization of three *Agave* species by gas chromatography and solid-phase microextraction–gas chromatography–mass spectrometry. *J. Chromatogr A*, 1027:131-136.

Pereira da Silva B, Valente AP and Paz Parente J (2006). A new steroidal saponin from *Agave shrevei*. Nat Prod Res, 20 (4):385–390.

Pi-Yu C, Chin-Hui C, Ching-Chuan K, Tzong-Huei L, Yueh-Hsiung K, Ching-Kuo L (2011). Cytotoxic Steroidal Saponins from *Agave sisalana*. *Planta Med.* 77 (9):929-933.

Popoca J, Aguilar A, Alonso D, Villarreal ML (1998). Cytotoxic activity of selected plants used as antitumorals in Mexican traditional medicine. *J. Ethnopharmacol.* 59:173–177.

Poulain N, Dez I, Perrio C, Lasne MC, Prud'homme MP, Nakache E (2003). Microspheres based on inulin for the controlled release of serine protease inhibitors: preparation, characterization and in vitro release. *J. Contr. Release* 92:27–38.

Quílez Guerrero A, García MD, Sáenz MT (2006). Plantas utilizadas en procesos inflamatorios y cancerosos en el área del Caribe. *Rev Fitoterapia* 6 (1):59-63.

Quílez A (1996). Investigación fitoquímica y bioensayos de sustancias obtenidas de plantas empleadas en diferentes tipos de cáncer en República Dominicana. IX Jornadas Farmacéuticas de la Universidad Nacional Pedro Henríquez Ureña. Santo Domingo, República Dominicana.

Raju J, Mehta R (2009). Cancer chemopreventive and therapeutic effects of diosgenin, a food saponin. *Nutr. Cancer* 61:27–35.

Ramachandran SM, Vamsikrishna KV, Gowthami H, Dhanaraju MD (2011). Assessment of cytotoxic activity of *Agave cantula* using brine shrimp (Artemia salina) Lethality bioassay. *Asian J. Scient Res.* 4 (1): 90-94.

Ramos Casillas F, Oranday Cárdenas A, Rivas Morales C, Verde Star MJ, Cruz-Vega DE (2012). Cytotoxic activity of *Agave lechuguilla* Torr. *Afr. J. Biotechnol.* 11 (58):12229-12231.

Rao CV, Chou D, Simi B, Ku H, Reddy BS (1998). Prevention of colonic aberrant crypt foci and modulation of large bowel microbial activity by dietary coffee fiber, inulin and pectin. *Carcinogenesis.* 19:1815-1819.

Ritsema T, Smeekens S (2003). Fructans: beneficial for plants and humans. *Curr. Opinion. Plant Biol.* 6:223–230.

Rizwan K, Zubair M, Rasool N, Riaz M, Zia-Ul-Haq M, de Feo V (2012). Phytochemical and Biological Studies of *Agave attenuata. Int. J. Mol. Sci.* 13:6440-6451.

Roberfroid MB (1999). Concepts in functional foods: the case of inulin and oligofructose. *J. Nutr.* 129:1398S–1401S.

Roberfroid MB (2007). Inulin type fructans: Functional food ingredients. *J. Nutr.* 137:2493S-2502S.

Robert ML, Herrera JL, Contreras F, Scorer KN (1987). Invitro propagation of *Agave fourcroydes* Lem (Henequen). *Plant Cell Tissue Organ Cult*, 8:37–48.

Robert ML, Herrera-Herrera JL, Castillo E, Ojeda G, Herrera-Alamillo MA (2006). An efficient method for the micropropagation of Agave species. *Meth Mol Biol*, 318, 165–178.

Robert ML, Lim KY, Hanson L, Sanchez-Teyzr F, Bennett MD, Leitch AR, Leitch IJ (2008). Wild and agronomically important Agave species (Asparagaceae) show proportional increases in chromosome number, genome size, and genetic markers with increasing ploidy. *Bot. J. Linn Soc*, 158:215–222.

Rowland IR, Rumney CJ, Coutts JT, Lievense LC (1998). Effect of *Bifidobacterium longum* and inulin on gut bacterial metabolism and carcinogen-induced aberrant crypt foci in rats. *Carcinogenesis*. 19:281-285.

Saad N, Delattre C, Urdaci M, Schmitter JM, Bressollier P (2012). An overview of the last advances in probiotic and prebiotic field. LWT-Food *Sci Technol* xxx:1-16. Art in press.

Sáenz MT, García MD, Quílez A, Ahumada MC (2000). Cytotoxic Activity of *Agave intermixta* L. (*Agavaceae*) and *Cissus sicyoides* L. (*Vitaceae*) *Phytother. Res.* 14 (7):552–554.

Sahu NP, Banerjee S, Mondal NB, Mandal D (2008). Steroidal saponins. In: *Fortschritte der Chemie Organischer Naturstoffe/progress in the chemistry of organic natural products*, vol. 89. Springer, Vienna, 89: 45–141.

Salazar-Leyva JA, Zazueta-Patrón IE, Brito-Rojas HD, Osuma-Ruiz I, Rodríguez-Tirado VA, Regalado-Rivera R, Lizárraga Vidal VM (2010). Obtención de Fructanos a partir de *Agave tequilana* Weber cv. azul cultivado en el estado de Sinaloa. VII Congreso del Noroeste y III Nacional de Ciencias Alimentaria y Biotecnología Centro de las Arte, Universidad de Sonora. Hermosillo Son. BT-21:141-162.

Sánchez E, Heredia N, García S (2005). Inhibition of growth and mycotoxin production of *Aspergillus flavus* and *Aspergillus parasiticus* by extracts of *Agave* species. *Int. J. Food Microbiol*. 98:271–279.

Sánchez-Marroquín A and Hope PH (1953). Agave juice: Fermentation and Chemicals composition of some species. *J. Agric. Food Chem*., 1:246-249.

Santos-Cerqueira G, dos Santos e Silva G, Rios Vasconcelos E, Fragoso de Freitas AP, Arcanjo Moura B, Silveira Macedo D, Lopes Souto A, Barbosa Filho JM, de Almeida Leal LK, de Castro Brito GA, Caden Souccar, de Barros Viana GS (2012). Effects of hecogenin and its possible mechanism of action on experimental models of gastric ulcer in mice. *Eur. J. Pharmacol.* 683:260-269.

Sati OP, Pant G, Nohara T and Sato A (1985).Cytotoxic saponins from *Asparagus* and *Agave*. *Pharmazie* 40 (8):586.

Seifert Stephanie and Bernhard Watzl (2007). Inulin and Oligofructose: Review of Experimental Data on Immune Modulation. *The J. Nutr.* 2563S-2567S.

Sharma Sheel, Nidhi Agarwal, Preeti Verma (2012). Miraculous Health Benefits of Prebiotics. IJPSR (*International Journal of Pharmaceutical Sciences and Research).* 3 (6):1544-1553.

Silva GM, De Souza AM, Lara LS, Mendes TP, da Silva BP, Lopes AG (2005). A new steroidal saponin from *Agave brittoniana* and its biphasic effect on the Na+-ATPase activity. En: *Z Naturforsc*. C, 60 (1-2):121-127.

Simpson J, Martínez Hernández A, Abraham Juárez MJ, Delgado Sandoval S, Sanchéz Villareal A and Cortés Romero C (2011). Genomic resources and transcriptome mining in *Agave tequilana*. *GCB Bioeng* 3:25–36.

Singh SK, Bimal S, Narayan S, Jee C, Bimal D, Das P Bimal R (2011). Leishmania donovani: Assesment of leishmanicidal effects of herbal extracts obtained from plants in the visceral leishmaniasis endemic area of Bihar, India. *Exper Parasitol.* 127:552-558.

Sparg SG, Light ME, van Staden J (2004). Biological activities and distribution of plant saponins. *J. Ethnopharmacol.* 94:219–43.

Spiller R (2008). Review article: probiotics and prebiotics in irritable bowel syndrome. *Aliment Pharmacol Ther* 28:385-396.

Srih Belghith K, Dahech I, Hamden K, Feki A, Mejdoub H, Belghith H (2012). Hypolipidemic effect of diet supplementation with bacterial levan in cholesterol-fed rats. *Int. J. Biol. Macromol.* 50:1070– 1074.

Steed H, Macfarlane GT, Macfarlane S (2008). Prebiotics, synbiotics and inflammatory bowel disease. *Mol. Nutr. and Food Res.* 52:898-905.

Stevens CV, Meriggi A, Booten K (2001). Chemical modification of inulin, a valuable renewable resource, and its industrial applications. *Biomacromol.* 2:1–16.

Szwengiel A, Czarnecka M, Roszyk H, Czarnecki Z (2004). Levan Production by *Basillus subtilis* DSM 347 strain. Institute of Food Technology of Plant Origin. August Cieszkowski Agricultural University of Poznań Wojska Polskiego 31:60-624 Poznań, Poland.

Taper HS and Roberfroid MB (2008). Prebiotics and cancer therapy. In GR. Gibson and MB. Roberfroid (Eds.), *Handbook of prebiotics* (pp. 329–339). Editor Taylor and Francis (2008), 485pp Publisher: (CRC Press, Boca Raton, Fla.) USA.

Ulloa JA, Espinosa AH, Cruz Rodríguez GK, Rosas Ulloa P, Ulloa Rangel BE, Ramírez Ramírez JC (2010). Los fructanos y su papel en la promoción de la salud. *Revista Fuente* Año 2, 5:57-62.

Umang Munjal, Glei M, PoolZobel BL, Scharlau D (2009). Fermentation products of inulintype fructans reduce proliferation and induce apoptosis in human colon tumour cells of different stages of carcinogenesis. *Brit. J. Nutr.* 102:663-671.

Uniyal GC, Agrawal PK, Sati OP, Thakur R (1991). A spirostane hexaglycoside from *Agave cantala. Phytochem.* 30 (4):1336-1339.

Urías-Silvas JE, Cani PD, Delmeé E, Neyrinck AM, López MG, Delzenne NM (2008). Physiological effect of dietary fructans extracted from *Agave tequilana* Gto. and *Dasylirion* spp. *British J. Nutr.* 99:254-261.

Urías-Silvas JE and López MG (2004). Efecto Prebiótico de los fructanos de Agave. 1er encuentro participación de la Mujer en la Ciencia.

Uribe JG, Saldivar SS, Ozuna AV (2009). The present invention comprises an *Agave* syrup that contained flavonoids, polycosanols and sapogenins, which provide anticancerigen and antioxidant properties. *Patent* application number: 20090124685.

Uribe JG and Saldivar SS (2010). *Agave* syrup extract having anticancer activity. *Food Nutr Agricul*, 2:93-95.

Valdés-Rodríguez S, Ramírez-Choza JL, Reyes-López J, Blanco-Labra A (2004). Respuestas del insecto Max (Scyphophorus acupunctatus gyllenhal (Coleoptera: Curculionidae)) hacia algunos compuestos atrayentes del henequén. *Acta Zoológica Mexicana* (n.s.) 20 (3):157-166.

Vega Oliver B (1995). "El uso del maguey". México. México *Desconocido*. Disponible en www.méxicodesconocido. com.mx (fecha de consulta: 12 de marzo de 2009).

Verástegui MA, Sánchez CA, Heredia NL, García-Alvarado JS (1996). Antimicrobial activity of extracts of three major plants from the Chihuahuan desert. *J. Ethnopharmacol* 52:175-177.

Verástegui A, Verde J, García S, Heredia N, Oranday A, Rivas C (2008). Species of *Agave* with antimicrobial activity against selected pathogenic bacteria and fungi. *World J. Microbiol. Biotechnol.* 24:1249–1252.

Verghese MD, Rao R, Chawan CB, Williams LL, Shackelford L (2002). Dietary Inulin Supresses Azoxymethane-Induced, Preneoplastic Aberrant Crypt Foci in Mature Fischer 344 Rats. *Am. Soc. Nutr. Sci.* 2809-2813.

Wang N and Nobel P (1998). Phloem transport of fructans in the crassulacean acid metabolism species *Agave deserti.Plant Physiol.*, 116: 709-714.

Wang Yu MA, Tao Zeng MD, Shu-e Wang MA, Wei Wang MA, Qian Wang MA, Hong-Xia Yu MA (2010). Fructo-oligosaccharides enhace the mineral absorption and counteract the adverse effect of phytic acid in mice. *Nutrition* 26: 305-311.

Xia Z, Rubin I, Whittle B, Gunning P, Hu Y, Brostoff J, Wang W (2004). A specific formulations including saponin isolated form agave for the treatment of cognitive dysfunction. BN1034. *US PATENT* # 6,258,386 (2001). Copyrighted 2004 Raintree Nutrition, Inc. Carson City, Nevada 89701.

Yokosuka A, Mimaki Y, Kuroda M, Sashida Y (2000). A new steroidal saponin from the leaves of *Agave americana*. *Planta Med.* 66 (4):393-396.

Yokosuka A and Mimaki Y (2009). Steroidal saponins from the whole plants of *Agave utahensis* and their cytotoxic activity. *Phytochem* 70 (6):807–815.

Yokosuka A, Jitsu M, Yui S, Yamazaki M, and Mimaki Y (2009). Steroidal glycosides from *Agave utahensis* and their cytotoxic activity. *J. Nat. Prod.* 72:1399–1404.

Zou P, Fu J, Yu HS, Zhang J, Kang LP, Ma BP, Yan XZ (2006). The NMR studies on two new furostanol saponins from *Agave sisalana* leaves. *Magn. Reson. Chem.* 44:1090-1095.

Zhu Y, Michelle Luo T, Jobin C, Young HA (2011). Gut microbiota and probiotics in colon tumorigenesis. *Cancer Lett* 309: 119-127.

In: Plant Extracts
Editors: Adriana Giordano and Alfieri Costs

ISBN: 978-1-62417-534-3

Chapter 5

Herbal Medicinal Plants Used in the Treatment of Diabetes in Africa

Oluwatoyin A. Odeku *
Department of Pharmaceutics and Industrial Pharmacy,
University of Ibadan, Ibadan, Nigeria

Abstract

Diabetes mellitus, a metabolic disorder of the pancreas in which the blood glucose levels are abnormally high (hyperglycaemia), is a growing threat to public health in modern society. The number of people suffering from diabetes is increasing due to urbanization, population growth, aging and increasing prevalence of obesity and physical inactivity. While some Non Insulin Dependent Diabetic Mellitus (NIDDM) patients can be managed by diet alone, others require hypoglycaemic therapy and/or insulin. Although insulin therapy affords an effective glycaemic control, drawbacks such as oral ineffectiveness, short shelf life, requirement of constant refrigeration, parenteral therapy with its attendant abscesses and fatal hypoglycaemia in the event of excess dosage has limited its usage. On the other hand, pharmacotherapy with sulphonylureas, biguanides and thiazolidones is also associated with side effects. In recent years, the popularity of complementary medicine has increased considerably and the WHO has suggested the evaluation of the potentials of plants as effective therapeutic agents, especially in areas where the modern drugs are not readily available. Due to their relative safety and low costs, herbal medicinal plants are prescribed for the treatment and prevention of a variety of diseases. These medicinal plants are used in their crude forms as herbal teas, syrups, infusions, ointments, liniments and powders. There are many herbs available having antidiabetic activity and they have been shown to act by different mechanisms. Some have been shown to have stimulating or regenerating effects on β cell or possess extra pancreatic effect for hypoglycemic activity. This article examines some of the herbal medicinal plants that have been used in the folkloric treatment of diabetes in Africa and the experimental evidences to support the antidiabetic properties.

* Correspondig author: Oluwatoyin A. Odeku. E-mail: pejuodeku@yahoo.com; o.odeku@mail.ui.edu.ng, Tel.: +234 805 57320466.

Introduction

Diabetes mellitus is a metabolic disorder of the pancreas in which the blood glucose levels are abnormally high (hyperglycaemia), either because the body does not produce enough insulin or the insulin so produced cannot be utilized by the body (Mayfield, 1998). Diabetes is a growing threat to public health in modern society and continues to be one of the highest causes of morbidity and mortality in recent times. The number of people suffering from diabetes is currently estimated to over 30 million people worldwide (WHO, 1985). The number of people suffering from diabetes is increasing due to urbanization, population growth, aging and increasing prevalence of obesity and physical inactivity. While some Non Insulin Dependent Diabetic Mellitus (NIDDM) patients can be managed by diet alone, others require hypoglycaemic therapy and/or insulin. Although insulin therapy affords an effective glycaemic control, drawbacks such as oral ineffectiveness, short shelf life, requirement of constant refrigeration, parenteral therapy with its attendant abscesses and fatal hypoglycaemia in the event of excess dosage has limited its usage. Pharmacotherapy with sulphonylureas, biguanides and thiazolidones is also associated with many side effects.

The global increase in prevalence of diabetes mellitus has accelerated due to the aging population structure in the developed countries and due to increasing obesity, as well as stressing life style. Diabetes mellitus is the sixth leading cause of death globally (Nash et al., 2001). Considerably large numbers of anti-diabetic herbs are known through folklore, however pharmacological evaluation by scientific methods is required to establish the anti-diabetic activity. The study of such medicines might offer a natural key to unlock a diabectologist's pharmacy for the future (Venkateshwarlu et al., 2009)

In recent years, the popularity of complementary medicine has increased considerably and the WHO has suggested the evaluation of the potentials of plants as effective therapeutic agents, especially in areas where the modern drugs are not readily available (WHO, 1985). Due to their relative safety and low costs, herbal medicinal plants are prescribed for the treatment and prevention of a variety of diseases. A wide variety of plants have been used by herbal medicine practitioners for the treatment of non-insulin-dependent (type 2) diabetes but their benefit in type 1 (insulin- dependent) diabetics has not been well established (Dieye et al., 2008).

Plants have been very good source of drugs for years. Authors have suggested that there are about 800 plants that have been shown to possess antidiabetic properties which are considered good sources for a new drug or a lead compound for a new drug. Plant extract have been said to bring about hypoglycemic effect by increasing the pancreatic secretion of insulin from β-cells or by potentiating the effect of insulin or by insulin-like activities (Stanely et al., 2000; Ko et al., 2005). Some other researchers have attributed the antihyperglycemic effects of plants to be due to the ability of the plant to restore the function of pancreatic tissues by causing an increase in insulin output, the inhibition of the intestinal absorption of glucose and the facilitation of metabolites in insulin dependent processes. Most plant extracts used in the treatment of diabetes have been found to contain glycosides, alkaloids, terpenoids, carotenoids (Loew and Kaszkin, 2002).

The use of medicinal plants in the treatment of diabetes have been reported in several clinical studies in African regions and up to 80% usage has been reported in Morocco among the type II diabetic patients (Eddouks et al., 2002; Erato et al., 2005; Dieye et al., 2008).

Ethnobotanical studies of the plants used in the treatment of diabetes in parts of African have shown that about 49 plant species are used in Lagos, southwestern Nigeria (Gbolade, 2009), 34 plant species in North Western region of Nigerian plants (Etuk et al., 2010), 14 species in Enugu state of Nigeria (Aiyeloja and Bello, 2006), 28 plant species in Cote d'ivoire (N'guessan et al. 2009b), 32 species in South Africa (Deutschlander et al., 2009) and 54 plant species in North Central region of Morocco (Jouad et al., 2001). This paper reports scientific evidence to support the antidiabetic properties of the plants that are used in traditional medicine for the treatment of diabetes in Africa.

Some of the plants that have shown varying degree of hypoglycemic activity include:

Abelmoschus Esculentus (L.) Moench

Abelmoschus esculentus (L.) Moench., synonym of okra, known in many English-speaking countries as lady's fingers, is a flowering plant valued for its edible green seed pods. It is an important vegetable and widely distributed from Africa to Asia, Southern Europe and America. Okra has a wide range of medicinal values and has been used to control various diseases and disorders including weakness, exhaustion, depression, ulcers, lung inflammation, sore throat, asthma as well as irritable bowel (Sengkhamparn et al., 2009). The administration of okra peel and seed powder (200 mg/kg) to streptozotocin induced diabetic rats showed significant reduction ($p < 0.001$) in blood glucose level and increase in body weight than diabetic control rats. A significantly increased ($p < 0.001$) levels of hemoglobin and total protein, and decreased levels of glycosylated hemoglobin and serum glutamate-pyruvate transferase were observed in the treated rats after 28 days of treatment with okra peel and seed powders. The elevated lipid profile levels in diabetic rats returned to near normal after the administration of 100 and 200 mg/kg dose of the powders compared to diabetic control rats (Sabitha et al., 2011).

Ageratum Conyzoides L

Ageratum conyzoides L. (Asteracées) is a well-known medicinal plant that has been used in several countries including Cameroon for treating various diseases including diabetes mellitus (Tsabang et al. 2001). A study of the hypoglycaemic and antihyperglycaemic properties of the aqueous extracts of the leaves of *A. conyzoides* L. in normoglycemic and streptozotocin-induced diabetic rats has shown that the aqueous extracts of the plant at the doses of 100, 200 and 300mg/kg showed statistically significant hypoglycaemic and antihyperglycaemic activities while the 100mg/kg dose only attenuated significantly the rise of blood glucose in normal fasted rats (Nyunaï et al., 2009).

Allium Cepa L (Onion)

Oral administration of S-methyl cysteine sulphoxide (SMCS) obtained from *Allium cepa* L. (Onion) daily at a dose of 200 mg/kg body weight for a period of 45 days to alloxan induced diabetic rats controlled the blood glucose and lipids in serum and tissues, and altered the activities of liver hexokinase, glucose 6- phosphatase and HMG CoA reductase towards normal values. These effects of SMCS were comparable to those of glibenclamide and insulin (Kumari *et al.*, 1995). Oral administration of onion SMCS to diabetic rats for one month, ameliorated the diabetic condition similar to rats treated with glibenclamide and insulin. The effect of feeding a 15% capsaicin diet or 3% freeze-dried onion powder containing diet produced a significant reduction in the hyperglycaemic status of diabetic animals. This study revealed that onion feeding improves the metabolic status in diabetes probably because of its hypocholesterolemic as well as its hypoglycaemic effect (Babu and Srinivasan, 1997).

Allium Sativum

Allium sativum L., (Garlic) family Liliaceae, grows in all parts of the world and the bulb bas been used extensively in both traditional and popular medicine for more than 5000 years. Interest in the study of the medical properties of this plant has increased during the last decade. Garlic is reputed to offer protection against strokes, coronary thrombosis, atherosclerosis and platelet aggregation (Adamu et al., 1982). Gastric administration of garlic capsules to anaesthetized dogs induced statistically significant, dose-dependent diuretic and natriuretie responses which reached maximum 30- 40 min after garlic administration and decreased to basal levels after 100- 150 min (Pantoja et al., 1991) S-allyl cysteine sulphoxide (SACS), a sulphur-containing amino acid of *A. sativum* that is the precursor of allicin and garlic oil, has been found to show significant antidiabetic effects in alloxan diabetic rats. Administration of a dose of 200 mg/kg significantly decreased the concentration of serum lipids, blood glucose and activities of serum enzymes like alkaline phosphatase, acid phosphatase and lactate dehydrogenase and liver glucose-6-phosphatase. It significantly increased liver and intestinal HMG CoA reductase activity and liver hexokinase activity (Sheela and Augusti, 1992). Oral administration of SACS to diabetic rats for one month ameliorated the diabetic conditions of treated rats comparable with rats treated with glibenclamide and insulin and controlled lipid peroxidation better than the other two drugs (Sheela *et al.*, 1995). Furthermore, SACS significantly stimulated *in vitro* insulin secretion from beta cells isolated from healthy rats. Hence it can be surmised that the beneficial effects of SACS could be due to both its antioxidant and its secretagogue actions (Augusti and Sheela, 1996).

Aloe Barbadensis Mill (Barbados)

The oral administration of the exudates of *Aloe barbadensis* Mill. (Barbados) leaves (500 mg/kg) produced no reduction in blood glucose level whereas its bitter principle (5 mg/kg) administered intraperitoneally produced a significant hypoglycaemic effect that extended over a period of 24h. In chronic studies, *A. barbadensis* and its bitter principle produced a maximum effect after 5 days. The hypoglycaemic effect of the plant and its bitter principle may be mediated through stimulating synthesis and/or release of insulin from the beta-cells of the islets of Langerhans (Ajabnoor, 1990). Administration of the extract slightly decreased the area under the glucose tolerance curve compared to control (1.4 %) or tolbutamide (14.3 %) in healthy rabbits (Ramon-Ramos *et al.*, 1992).

Aloe Ferox Mill

A. ferox Mill also known as "Cape Aloe", is indigenous to South Africa and is widely distributed from the Western to Eastern Cape provinces (Deutschländer et al., 2009). Extracts from various parts of *A. ferox* have been used for arthritis, eczema, conjunctivitis, inflammation, and in the treatment of hypertension (Van Wyk, 2008; Deutschländer *et al.*, 2009). However the anti-diabetic properties are reported to be the most common application (Deutschländer *et al.*, 2009; Botes, 2009). A recent study by Botes, (2009) investigated the anti-diabetic properties of *A. ferox* and found that its leaf gel has anti-oxidant properties which may explain some of the health benefits in reducing symptoms associated with/or prevention of cardiovascular diseases, cancers, neurodegeneration, and diabetes. Furthermore, streptozotocin induced diabetic rats, treatment with *A. ferox* at 300 mg/kg/day resulted in a modest increase in the serum insulin levels and a slight improvement of the lipid profile (Botes, 2009).

Anacardium Occidentale Linn

Anacardium occidentale Linn (Anacardiaceae) is a plant used in South Cameroon as well as in other tropical countries, as a remedy for diabetes mellitus (Kamtchouing et al., 1998). The hypoglycemic effect of the aqueous leaf extract of *A. occidentale* in diabetic rats and its prophylactic diabetogenic action in streptozotocin induced diabetic rats have been reported (Kamtchouing et al., 1998; Sokeng et al, 2001). More recent studies have also shown the effects of the methanol leaf extract and the dichloromethane, ethyl acetate and n-hexane fractions of the extracts on streptozotocin-induced diabetic rats. Oral administration of methanol extract at doses of 35, 175 and 250 mg/kg bw significantly reduced blood glucose levels in diabetic rats 3 h after administration. The maximum of 37% reduction in blood glucose levels was observed which became more pronounced (48%) after repeated administration. The n-hexane fraction produced the maximum hypoglycemic effect (45%) and the same dose of the dichloromethane and ethyl acetate fractions reduced hyperglycemia by 21% and 41% respectively, at the end of the treatment. Significant decrease in urine glucose levels was also observed in diabetic rats after repeated administration of the methanol extract and the fractions. These results suggest the hypoglycemic effect of the methanol extract and fractions of *A. occidentale* in streptozotocin-induced diabetic rats (Sokeng,et al. 2007). Another study had shown that the administration of *A. occidentale* stem-bark extract significantly prevented changes in plasma glucose, triglyceride, total cholesterol/HDL-cholesterol ratio, malonyldialdehyde, urea, and creatinine induced by enriched fructose diet. The chronic oral administration of the methanol extract at a dose of 200.0 mg/kg body weight was found to be a safe alternative antihyperglycemic agent and has the beneficial effects of improving plasma glucose and lipids in fructose-induced diabetic rats, which is associated with a reduced lipid peroxidation (Olatunji et al. 2005).

Recent study has also shown that the aqueous leaf extract significantly lowered ($p>0.05$) the blood glucose concentration and the serum triglyceride and serum total protein of normoglycemic albino rats (Saidu et al., 2012). Treatment with the methanolic extract of *Anacardium occidentale* has been shown to result in reduction in hyperglycaemia and regeneration of beta cells in streptozotocin-induced diabetic rats (Ukwenya et al., 2012).

Azadirachta Indica A. Juss (Neem)

Azadirachta indica A. Juss belongs to the family Meliaceae and has a long history of use in folkmedicine as a treatment against various ailments (Sonia and Srinivasan, 1999). The hypoglycemic actions of its leaf, stem bark and seeds have been articulated in a review by Biswas et al. (2002), *A. indica* leaf extract was found to have no action on peripheral utilization of glucose or on hepatic glycogen in healthy and streptozocin-induced diabetic rabbits. The reduction in peripheral utilization of glucose and glycogenolytic effect due to epinephrine was blocked by *A. indica* leaf extract, almost completely in diabetic rabbits and to a certain extent in healthy animals (Chattopadhyay et al., 1987a, b) More recently, it has been demonstrated that in an *in vitro* rat pancreas preparation, *A. indica* leaf extract significantly blocked the inhibitory effect of serotonin on insulin secretion mediated by glucose (Chattopadhyay et al., 1996; 1999).

Boerhaavia Diffusa L

The root and aerial parts of Boerhaavia *diffusa* Linn. (Nyctaginaceae) were used in parts of Africa for the treatment of diabetes. The daily oral administration of aqueous solution of *Boerhaavia diffusa* L. leaf extract (200 mg/kg) for 4 weeks has been shown to cause a significant decrease in blood glucose and significant increase in plasma insulin levels in normal and alloxan induced diabetic rats. Treatment with the extract resulted in a significant reduction of glycosylated haemoglobin and an increase in total haemoglobin level and a significant improvement in glucose tolerance in rats. The effect of the extract was more prominent than those of glibenclamide (600μg/kg) (Pari and Satheesh, 2004). In another study, the chloroform extract of the leaf of *B. diffusa* produced a dose-dependent reduction in blood glucose level comparable to that of glibenclamide in streptozotocin-induced diabetic rats. The results indicate that the reduction in blood glucose produced by the extract is probably through rejuvenation of pancreatic b-cells or through extrapancreatic action (Nalamolu et al., 2004).

Caesalpinia Bonducella Fleming

The aqueous and ethanolic extracts of *Caesalpinia bonducella* Fleming seeds (100 mg/kg) exhibited hypoglycaemic activity four hours after administration but the hypoglycaemia produced by the aqueous extract was of more prolonged duration than the ethanolic extract. In diabetic rats, both extracts produced marked antihyperglycaemic effects after five days (Sharma *et al*, 1997).

Cassia Italic Mill.) Lam. Ex F.W. Ander

The hypoglyceamic effects of the whole plant extract of *Cassia italica* (Mill.) Lam. ex F.W. Ander family Caesalpiniaceae, in the alloxan induced diabetic rabbits has been compared with the hypoglycemic efficacy of glibenclamide. The oral dose administration of

100 mg/kg body weight (bw) of plant extract did not cause significant changes in blood glucose level but at the dose of 200 mg/kg, hypogyceamic effect similar to glibenclamide (2.5mg/kg body weight) was obtained.

The comparable effect of the extract (200mg/kg) with glibenclamide (2.5mg/kg) was attributed to the similarity in their mode of action. Glibenclamide acts by stimulating the release of insulin and the inhibition of glucagon secretion. The findings also suggest that the plant extracts may regenerate β cells and have protective effect on β cells from glucose toxicity. (Qamar et al., 2011).

Cassia Simea Lam

Cassia siamea Lam. (Ceasalpiniaceae) commonly called Kassia or Cassod tree is a tropical woody tree. The different plant parts are used in the treatment of intestinal worm infestation, malaria, typhoid, fever, convulsion in children, diabetes and scabies. The aqueous extract of *Cassia siamea* (1000-3000mg/kg bw) given orally showed significant hypoglyceamic effects 2-8 hours after administration, in normal and alloxan induced diabetic rats comparable with chlorpropamide (400mg/kg) (Odason and Kolawola, 2007).

Chrysophyllum Cainito L

Chrysophyllum cainito (Sapotaceae) is a plant recognized by the traditional healers in Côte-d'Ivoire as an antidiabetic agent. The study of the effect of an aqueous decoction of the plant's leaves on alloxan induced diabetic rabbits showed that at the doses of 10 g/l, *C. cainito* does not induce the hypoglycaemic effect but a graded decrease in hyperglycaemia was observed with increase in the dose administered. But after 6 weeks of treatment, the diabetic rabbits stopped eating and succumbed to death between the 8th and the 9th weeks of experimentation. It was concluded that *C. cainito* leaves have glucose lowering effect at doses > 10 g/l but appears to be toxic and lethal at 30 g/l. The study showed that the aqueous decoction of *C. cainito* leaves has hypoglycaemic activity, at a dose of 20 g/l but would exert a toxic effect from 30 g/l. Thus, its use as antidiabetic plant would require close supervision (N'Guessan et al., 2009a).

Combretum Micranthum G. Don

Combretum Micranthum G. Don, family Combretaceae, is a medicinal plant used in the treatment of diabetes in Northwestern Nigeria. The plant has also been shown to possess antioxidant, antimicrobial as well as anti-inflammatory properties (Olajide, et al., 2003; Karou et al., 2005). The aqueous leaf extract of *C. micranthum* dissolved in normal saline administered to the normal and diabetic rats at 100 mg/kg, produced a 24.6% reduction in blood glucose level while 200 mg and 400 mg/kg body weight doses, 21.9% and 18.9% reduction respectively. The aqueous leaf extract of *C. micranthum* was found to possess antidiabetic property (Chika, and Bello, 2010).

Crescentia Cujete L

Crescentia cujete L. (family Bignoniaceae) commonly known as the calabash tree, is species of flowering plant that is native to Central and South America that has been used as a medicinal plant for centuries, even though the fruit from this tree cannot be safely eaten by humans. Some of the most popular uses for this herb include the treatment of respiratory illnesses, stomach ailments, and intestinal disturbances. The aqueous extract of the leaves of *C. cujete* has been shown to lowers hyperglycaemia at 20 g/l (N'Guessan et al., 2008). The extract is widely believed to stimulate insulin production and could be useful in the treatment of diabetes.

Dorstenia Picta Bureau

Dorstenia picta Bureau (Family Moraceae) is an herbaceous plant used in south Cameroon for the treatment of diabetes and hypertension. Single oral administration of the methanolic extract of *D. picta* to normal and streptozotocin-induced diabetic rats at a dose of 75 mg/kg, significantly, reduced the blood glucose levels by 23.74 % and 29.79 %, in normal and diabetic rats, respectively after 5 hours. *D. picta* a significantly ($p < 0.01$) decreased in blood glucose at the dose of 75 mg/kg (53.88 %) and 150 mg/kg (81.96 %) and prevented body weight loss, polyphagia and polydipsia, after daily oral administration for 14days. The effect was similar to those of insulin and was attributed to insulin-like activity, and its ameliorate effects on the kidney and liver functions. *D. picta* twigs have antidiabetic effect and could be a potential source of new oral antidiabetic drug (Ngueguim et al., 2007).

Eugenia Jambolana Lam. (Gambol)

The oral administration of the extract from the pulp from the fruit of *Eugenia jambolana* Lam. (Gambol) resulted in the enhancement of insulinemia in normoglycaemic and diabetic rats (Achrekar *et al.*, 1991). The incubation of isolated pancreatic islet cells of normal and streptozosin induced diabetic rats with the plant extracts resulted in increased insulin secretion and an inhibition of insulinase activity from liver and kidney (Achrekar *et al.*, 1991).

Ficus Bengalensis L. (Banyan)

More than twenty-five species of Ficus exist within Africa. Two anti-diabetic glycosides (a dimethoxy derivative of perlargonidin 3-O-alpha- L rhamnoside have been isolated from *Ficus bengalensis* L. (banyan tree) bark. The glycosides were active at a dose of 10-25 mg/kg body weight in diabetic rabbits. Single dose glycoside treatment decreased fasting blood glucose by 19 % and improved glucose tolerance by 29 %. After one-month of treatment, the fasting blood glucose level went down to almost half of the pre-treatment levels in both the groups and their glucose tolerance improved by 41 % in the glibenclamide group and by 15 % in the glycoside treated group. They were found to be useful not only in mild diabetes but also severe diabetes in which pancreas is almost completely destroyed. The two glycosides were found effective even when given intermittently. The beneficial effects continued for 10 days after the treatment has been discontinued. There were no toxic effects during the period of treatment and the histopathological abnormalities seen in the pancreas of untreated rabbits were not observed in the treated rabbits. The mechanism of action seems to be both by increasing serum insulin levels as well as by directly acting on liver and muscle enzymes involved in glucose metabolism i.e. pancreatic and extra pancreatic. (Achrekar et al., 1991).

Glycoside of leucopelargonidin isolated from the bark of *F. bengalensis* has also demonstrated significant hypoglycaemic, hypolipidemic and serum insulin raising effects in moderately diabetic rats with close similarities to the effects of a minimal dose of glibenclamide (Cherian and Augusti, 1993). Dimethoxy ether of Leucopelargonidin-3- O-alpha-L rhamnoside isolated from the bark of *F. bengalensis* administered orally at a dose of 100 mg/kg showed significant hypoglycaemic effect and appears to stimulate insulin secretion in healthy and alloxan induced-diabetic dogs during a period of 2h (Augusti *et al.*, 1994). A leucodelphinidin derivative isolated from the bark of *F. bengalensis* L. also showed hypoglycaemic action at a dosage of 250 mg/kg when administered orally to both healthy and alloxan diabetic rats and the action was similar to that of glibenclamide (2 mg/kg) tested under the same conditions (Geetha *et al.*, 1994).

Ficus Carica L

A decoction of *Ficus carica* L. (Common Fig) leaves used as a supplement to breakfast in insulin-dependent diabetes mellitus (IDDM) patients has been found to help to control postprandial glycaemia (Serraclara et al., 1998). The study showed that postprandial glycaemia was lower during supplementation with *F. carica* [156.6 ± 75.9 mg/dl versus commercial tea 293.7 ± 45.0 mg/dl ($p<0.001$)] without pre-prandial differences (145.0 ± 41.5 and 196.6 ± 43.2 mg/dl, respectively). The organic phase of the aqueous decoction of *F. carica* leaves (after treatment with HCl, centrifuging, treatment with NaOH and extraction with chloroform) have been shown to produce a decrease in hyperglycaemia and a reduction in the level of total cholesterol and reduction in the total cholesterol/HDL cholesterol ratio in diabetic rats (Canal *et al.*, 2000).

Gongronema Latifolium Benth

Gongronema latifolium family Asclepiadaceace, an edible rainforest climbing plant native to South-Eastern Nigeria, is used widely as a staple vegetable and spice (Morebise et al., 2002). It is used in tradomedicine for healthy glycemic control and to support the pancreas (Morebise and Fafunso, 1998; Okafor and Ham, 1999; Ugochukwu et al., 2003). In the United State, it is used as a constituent of herbal tea blend for maintenance of healthy glycemic control. Scientific studies have shown the antihyperglycemic, antioxidant, antilipidemic and antihypercholesterolemic activities of the leaves of *G. latifolium* in both normal and STZ diabetic rats have been reported (Ugochukwu and Babady, 2003; Ugochukwu et al., 2003)

Loranthus Bengwensis L

The African mistletoe, *Loranthus bengwensis* L. (Loranthaceae), has been widely used in Nigerian folk medicine to treat diabetes mellitus. The aqueous extract or infusion (1.32 g/kg per day) of the leaves of African mistletoe parasitic on lemon, *Citrus limon (*L.) Brum f. (Rutaceae), guava, *Psidium guajava* L. (Myrtaceae) and jatropha, *Jatropha curcas* L. (Euphorbiaceae), admisitered to both normal and streptozotocin-induced diabetic rats for 28 days showed that the infusions of mistletoe parasitic on both lemon and guava trees significantly decreased serum glucose levels in normal ($P < 0.05$) and diabetic ($P < 0.001$) rats, whereas infusions of mistletoe parasitic on jatropha did not have any effect on the blood glucose level. This indicates that African mistletoe possesses significant anti-diabetic activity in streptozotocin-induced diabetic rats and the activity appears to be dependent on the host plant species (Obatomi et al., 1994).

Momordica Charantia L. (Karela); (Balsam Pear)

Momordica charantia L. (Family Cucurbitaceae) called bitter melon or bitter gourd is widely grown in Asia, Africa, and the Caribbean for its edible fruit, which is among the most bitter of all fruits. The oral administration of aqueous extract of *M. charantia* has been shown to attenuate the glycaemic response to both oral and intraperitoneal glucose, without altering the insulin response. This aqueous extract and the residue after alkaline chloroform extraction reduced hyperglycaemia in diabetic mice after 1 h and the effect was independent of intestinal glucose absorption and involved an extra pancreatic activity (Day *et al.*, 1990). The methanolic extract (30 mg/kg) caused a decrease in blood glucose level (BGL) 3 h after oral administration to streptozotocin-induced diabetic rats.

The n-butanol fraction of *M. charantia* inhibited the increase of BGL significantly after intraperitoneal glucose load in a similar manner to the sulfonylureas (Higashino *et al.*, 1992). *Momordica charantia* extract seems to act like insulin or via insulin secretion from the pancreas. The aqueous fruit extract decreased the fasting glucose level in normoglycaemic and cyproheptadine induced hyperglycaemic mice. This effect did not improve the tolerance to glucose in treated mice (Cakici, *et al.*, 1994). The alcoholic extract of the pulp (500 mg/kg), administered to healthy glucose-primed rats depressed plasma glucose levels 1 h in a similar manner to tolbutamide (100 mg/kg). This reduction in plasma glucose levels was not accompanied by increased insulin secretion.

In streptozotocin-induced diabetic rats, *M. charantia* extract improved oral glucose tolerance causing significant reduction in plasma glucose. The extract caused a 4-5 fold increase in the rate of glycogen synthesis from U-14C glucose in the livers of normally fed rats. These data suggest that the mechanism of action of this plant could partly be attributed to increased glucose utilization by the liver rather than an insulin secretion effect (Sarka *et al.*, 1996).

The activity of the juice *of M. charantia* fruit was tested on streptozotocin (STZ)-treated RIN cells and isolated islets *in vitro*. Feeding the juice of *M. charantia* fruit produced a reduction in hyperglycaemia in STZ-induced diabetic mice. It strongly reduced the STZ-induced lipid peroxidation in the pancreas of mice, RIN and islet cells. Moreover, it reduced the STZ-induced apoptosis in RIN cells (Sitasawad *et al.*, 2000).

Morinda Lucida Benth

Morinda lucida Benth (Rubiaceae), is a medicinal plant used in folklore African medicine in the treatment of uncontrolled diuresis not necessarily associated with diabetes but linked to general body weakness and rapid loss of weight. The oral administration of aqueous root extract of *Morinda lucida* exhibited potent hypoglycaemic effects in both normoglycemic and alloxan-induced diabetic mice. This effect was dose-dependent and more potent than chlorpropamide. The extract (140 mg/kg body weight) produced a 32% fall in blood sugar while a 51% decrease was observed in alloxan-induced diabetic mice after four hours (Kamanyi et al., 1994).

Toxicity studies have also shown that the extract was relatively safe and large doses can be administered without danger of hypoglycaemia, but the extract is not recommended in cases of diabetic coma due to the delay in the onset of action (Kerharo and Adam, 1974).

Moringa Oleifera Lam

Moringa oleifera Lam (family Moringacaea), a fast growing drought-resistant tree native to sub-Hamalayan tracts of northern India but now distributed worldwide in the tropic and sub tropics, is traditionally used for the treatment of a number of ailments including rheumatism, venomous bites, microbial infections, diarrhea, as diuretic and stimulant in paralytic affliction, epilepsy and hysteria (Quisumbing, 1978). *M. oleifera* (250 and 500 mg/kg i.p.) produced a dose-dependent, significant reduction ($p<0.05$) in blood glucose levels of streptozocin induced diabetic rats while for the fasted normal rats, there was significant increase in the blood glucose levels when compared to control (Tende et al., 2011). *Moringa oleifera* have been shown to have glucose lowering effect only in hyperglycemic diabetic rats and not in hypoglycemia by possibly stimulating the β-cells and or due to its insulin-like activity.

The plant extract has been shown to have ameliorating effect for glucose intolerance probably due to the quercetin-3-glucoside mediated effect and fiber contents of the leaf powder (Ndong et al., 2007). The progression of diabetes in streptozocin induced diabetic rats has been significantly reduced after treatment with methanol extract of *M. oleifera* with significant reduction in serum glucose and nitric oxide, with concomitant increases in serum insulin and protein levels (Gupta et al., 2012).

Mucuna Pruriens L

One of the known antidiabetic medicinal plants used worldwide is *Mucuna pruriens* L (Fam. Leguminoseae) also known as velvet beans or Cowitch in English. It is an annual climbing plant indigenous to tropical regions, especially Africa, India and the West Indies where it is found in bushes, bush paths and hedges.

The plant also produces clusters of pods, which contains seeds known as Mucuna beans and the seed pods are covered with reddish-orange hair-like needles that contain mucunain and serotonin and causes itching, blisters and dermatitis. The seed is a natural source of L-dopa (about 7-10%) which account for its use in the management of Parkinson's disease and neurological illnesses (Katzenschlager et al., 2004). Oral administration of the crude ethanolic extract of *M. pruriens* seeds (100mg/kg/day) to alloxan-induced diabetic rats (plasma glucose > 450mg/dL) resulted in a dose dependent reduction in blood glucose level of the diabetic rats comparable to the effects of glibenclamide (5mg/kg/day). The antidiabetic activity of *M. pruriens* seeds was shown to reside in the methanolic and ethanolic fractions of the extract (Majekodunmi et al. 2010).

Musa Sapientum Kuntze

Musa sapientum Kuntze (Banana) is among the plants most widely used in the treatment of diabetes mellitus. Oral administration of 1.5, 0.2 and 0.25 g/kg body weight of the chloroform extract of the flowers of *M. sapientum* during a 30-day period significantly decreased the hyperglycaemic peak and the area under the glucose tolerance curve in hyperglycaemic rabbits. The extract showed antihyperglycaemic action and was found to be more effective than glibenclamide (Pari and Umamaheswari, 2000). In another study, The methanolic extract of mature green fruit of *M. sapientum* (100-800mg/kg p.o.) was found to induce significant dose-related ($p<0.05$) reduction in the blood gluocose concentrations of both normal and streptozocin induced diabetic mice and the effect was found to be comparable with chlorpropamide although the precise mechanism of the hypoglycemic actiob of the extract was not clear (Ogewole and Adewunmi, 2003).

Persea Americana Mill

Extracts of *Persea americana* Mill (Lauraceae) ("Avocado") have been traditionally used to treat hypertension and diabetes mellitus. The ethanolic extracts of *P. americana* leaf has been shown to induce a dose-dependent hypoglycaemic responses in STZ-induced diabetic rats while subchronic treatment with the extract additionally increased hepatic glycogen concentrations.

Acute administration of the infusion decreased urine flow and electrolyte excretion rates, whilst subchronic treatment reduced plasma creatinine and urea concentrations. The results indicate the basis of the ethnomedicinal use of *P. americana* leaf extract in diabetes management (Gondwe et al., 2008).

Prunus Africana Hook F

Prunus africana Hook f. known as "African cherry", is a medium to large canopy tree reaching 30-40 m in height widely distributed throughout Africa. The well known medicinal value of *P. africana* is its ability to reduce benign prostatic hyperplasia symptoms; however its use for the treatment of fever, gonorrhea, stomach pain, chest pain, intercostals pain, inflammatory disorders, kidney disease, urinary tract complications, wound healing, malaria and as an appetite stimulant or purgative has been reported (Stewart, 2003). Furthermore Karachi (2009) investigated its anti-diabetic properties using *in vitro* and *in vivo* animal models. The study reported that *P. africana* bark extract has strong anti-oxidant properties, and in the cell cultures it improved glucose utilization.

Moreover, a dramatic increase in the serum insulin levels was observed in rats after treatment with *P. africana* extract for four weeks, suggesting the extract may function as an insulin secretagogue (Karachi, 2009).

Ricinus Communis L

Ricinus communis L. family Euphorbiaceae is used in traditional medicine for the management of diabetes mellitus. It is cultivated for its seed oil known as castor oil. The leaf, root and seed oil of this plant have been used for the treatment of inflammation and liver disorders. Fifty percent of ethanolic extract of the root, stem and leaves of *R. communis* has been shown to possess hypoglycemic activity in normal animals and antihyperglycemic activity in diabetic animals (Dhar et al., 1968).

The administration of 50% ethanolic root extract for prolonged period led to a significant dose dependent reduction in blood glucose level of diabetic rats up to a dose of 500 mg/kg body weight and a maximum decrease in blood glucose was observed after 8 hours.

But, higher doses up to 2000 mg/kg body weight did not show any dose-dependent effect. Out of the twenty column-purified fractions of the 50% ethanolic root extracts tested for antihyperglycemic activity in diabetic rats, only one fraction effected a significant decrease (60.6% decrease after 8 hours) in blood glucose of the diabetic rats (Shokeen et al., 2008). The study further revealed that administration of this ethanolic extract to the diabetic rats for 20 days, not only significantly lowered the blood glucose of the diabetic animals to almost normal level, but also increased the insulin levels and caused improvement in lipid profile and body weight of the diabetic animals (Shokeen et al., 2008).

Sclerocarya Birrea (A. Rich.) Hochst

Sclerocarya birrea (A. Rich.) Hochst. (Anacardiaceae) is one of the plant species widely used in Africa against various diseases, including diabetes (Dieye et al. 2008). *S. birrea* stem-bark extracts have been shown to exert hypoglycaemic effects in animal models. In rats with streptozotocin-induced diabetes, acute administration of *S. birrea* stem-bark extract reduces blood glucose levels (Ojewole 2003, Dimo et al. 2007).

Administrations of the single dose of aqueous extract (800 mg/kg p.o.) significantly reduced ($p < 0.001$) the blood glucose levels of both fasted normal (normoglycemic) and fasted STZ-treated, diabetic rats. Chronically, the efficacy of the extract administered for 5 weeks was shown to be similar to metformin treatment with regards to lowering of glycaemia (Gondwe et al. 2008).

The insulin-secreting INS-1E cells and isolated rat islets following 24 h of treatment with 5 μg/ml of the aqueous extracts of *S. birrea* has been shown to markedly potentiate glucose-stimulated insulin secretion without affecting the basal insulin release nor non-nutrient stimulation.

The extracts had no acute effects at the effective concentration and could be regarded safe. The mechanism of action of the extract was related to glucose metabolism as both ATP generation and glucose oxidation were enhanced following the 24-h treatment. In streptozotocin-induced diabetic rats, administration of the extract corrected glycaemia and restored plasma insulin levels after 2 weeks of treatment.

These authors showed the direct action of *S. birrea* on insulin-secreting cells and favour further delineation for use of the plant in the management of diabetes (Ndifossap et al., 2010).

Sphenostylis Stenocarpa Hochst Ex A. Rich

The leaves of *Sphenostylis stenocarpa* Hochst ex A. Rich is widely used in Nigerian for the treatment of diabetes mellitus. Studies have shown that the oral administration of water extracts of the leaves of *S. stenocarpa* demonstrate hypoglycaemic action in normal rats but did not produce a decrease in the blood glucose level after treatment for 6 weeks in diabetic rats, although a reduction in the blood cholesterol concentration (13% after five weeks) was observed in diabetic rats (Onoagbe et al, 1999; 2010). The lack of effect of *S. stenocarpa* extract in the diabetic rats has cast some doubt on the relevance and usefulness of this plant in the treatment of diabetes mellitus.

Sutherlandia Frutescens L

Sutherlandia frutescens L. also known as "Cancer Bush" is one of the most valuable South African indigenous medicinal plants. The medicinal uses include fever reduction, cancer tonic, stress, tuberculosis, colds, influenza, chicken pox, diabetes, chronic fatigue syndromes, inflammation, kidney and liver conditions, back ache, rheumatism, peptic ulcers, anxiety, osteoarthritis, menopausal symptoms, gastritis, poor appetite and HIV infection (Van Wyk, 2008). *S. frutescens* has been used widely by diabetic patients, especially along the western coastal region of South Africa and has been marketed as potential hypoglycemic agent for the treatment of the type II diabetes (Van Wyk and Albrecht, 2008; Chadwick *et al.,* 2007).

Its hypoglycemic properties have been identified in several animal studies (Van Wyk and Albrecht, 2008). Chadwick *et al.* (2007) observed statistically significant increase in glucose uptake in both muscle and adipose tissues, a reduction in intestinal glucose uptake with no weight gain in pre-diabetic rats treated with *S. frutescens* extracts via their drinking water for eight weeks. These authors concluded that *S. frutescens* showed promise as a therapy for type II diabetics.

More recently, MacKenzie *et al.* (2009) showed that a commercial *S. frutescens* extract prevented the onset of high fat diet-induced insulin resistance by attenuating plasma free fatty acid levels.

Telfairia Occidentalis Hook

Telfairia occidentalis Hook is a tropical vine grown in West Africa as a leaf vegetable and for its edible seeds. Aqueous extract of this plant is reported to increase haematological parameters and has been shown to be hepatoprotective against garlic induced oxidative stress (Olorunfemi et al, 2005).

The ethanolic extract has demonstrated hypoglycaemic properties both in normoglycaemic and alloxan-induced diabetic rat (Nwozo et al., 2004).

Recent study has also shown that rats fed on a diet of *Telfairia occidentalis* leaves for 30days showed statistical significant increase ($p<0.05$) in the plasma glucose level compared to those fed on normal diet, and would not be fit for consumption by diabetic patients (Adisa et al., 2012).

Terminalia Catappa L

Terminalia catappa L. (Combretaceae) is used in traditional medicine for the treatment of diabetes in Africa. Hyperglycaemic rabbits treated with a decoction of *T. catappa* leaves showed a significant dose dependent reduction in the blood glucose level. The decoction has a normoglycaemic activity at lower doses ($\leq$ 2.5 mg/ml) and a hypoglycaemic activity at higher doses (> 2.5 mg/ml). The dose of 40 mg/ml of the decoction reduced hyperglycaemia by about 30% and the effects was comparable to that of glibenclamide (0.25 mg/ml) (N'guessan et al., 2011).

Trigonella Foenum Graecum L. (Fenugreek)

Fenugreek is one of the most widely used herbs in the treatment of diabetes (Al-Rowais, 2002). Fenugreek diet significantly reduced fasting blood glucose and improved the glucose tolerance test in insulin-dependent diabetic patients with a 54% reduction in the 24 h urinary glucose excretion (Sharma *et al.*, 1990). Oral administration of *T. foenum graecum* (2 and 8 g/kg) produced a significant dose dependent reduction in blood glucose level in healthy and alloxan induced diabetic rats (Khosla *et al.*, 1995). When steroid saponins extracted from fenugreek seed was mixed with food (12.5 mg/day per 300 g body weight) and administered to healthy and streptozotocin-induced diabetic rats, food intake and the motivation to eat in healthy rats were significantly decreased and the food consumption in diabetic rats was also stabilized. In both healthy and diabetic rats, steroid saponins decreased total plasma cholesterol without any change in triglycerides (Petit *et al.*, 1995). More recently, it has been shown that the disrupted free radical metabolism in diabetic animals may be normalized by fenugreek seed supplementation in the diet (Ravikurma and Anuradha, 1999). Moreover, fenugreek significantly decreased the hyperglycaemic peak and the area under the glucose tolerance curve in hyperglycaemic rabbits (Alarcon-Aguilar *et al.*, 1998).

Triplochiton Scleroxylon K. Schum

Triplochiton scleroxylon K. Schum is one of the herbs used in the treatment of diabetes mellitus widely cultivated in Nigeria for their medicinal values. The aqueous extract of *T. scleroxylon* administered to alloxan - induced diabetic rabbits for 28 days led to a significant decrease in the plasma glucose level. The crude extract of this herb has hypoglycemic properties but did not affect the body weight and some hematological properties of rabbits (Prohp et al., 2006).

Urena Lobata L

Oral administration of aqueous extracts of *Urena lobata* (roots or leaves) to streptozotocin induced diabetic rats significantly decreased the blood glucose level within the 6 weeks. The roots of *U. lobata* were more efficacious than the leaves in reducing hyperglyceamia in the diabetic animals. The ability of *U. lobata* to effectively reduce the elevated levels of blood glucose and cholesterol, as well as liver triglyceride in the diabetic rats renders valid the claimed anti-diabetic activities of the medicinal plant (Onoagbe et al., 2010)

Viscum Album Stapf

The aqueous leaf extract of *Viscum album* Stapf (family Loranthaceae) administered intraperitoneally at 200 mg/kg and 400 mg/kg body weight to fasted normal white albino rats and alloxan induced diabetic rabbits respectively, produced a significant dose-dependent blood sugar lowering effects comparable with those of tolbutamide. The mechanism underlying the observed hypoglycemic effect of the extract has not been fully understood but was related to enhancement in peripheral utilization of glucose and increase in insulin secretion (Ohiri et al. 2003).

Vernonia Amygdalina Del

Vernonia amygdalina (African bitter leaf) family Asteraceace (Compositae) is a small tree (1-3 m high) that grows throughout tropical Africa and has been domesticated in some parts of West Africa including Nigeria, where it is locally known as bitter leaf (Igile et al., 1995). Besides its use as a vegetable, all parts of the plant has found usefulness in folkmedicine (Igile et al., 1994; Igile et al., 1995; Babalola et al., 2001; Ojiako and Nwanjo, 2006; Abosi and Raseroka, 2003; Izevbigie et al., 2004). Scientific and pharmacologic studies have revealed antihyperglycemic action of the roots (Nimenibo-Uadia, 2003) and leaves (Akah and Okafor, 1992; Akah et al., 2004) and hypoglycemic effect of the leaves (Gyang et al., 2004) of this plant. Extracts from the leaves have been reported to possess hypolipidemic and antihyperlipidemic properties (Atangwho et al., 2007a) and to protect the kidneys (Atangwho et al., 2007b) and the liver (Atangwho et al., 2007c) of alloxan diabetic rats against complications.

Zingiber Officinale Roscoe

Zingiber officinale Roscoe (family, Zingiberaceae) is a medicinal plant that has been widely used in herbal medicines all over the world, for a wide array of unrelated ailments that include arthritis, rheumatism, sprains, muscular aches, pains, sore throats, cramps, constipation, indigestion, vomiting, hypertension, dementia, fever, infectious diseases and helminthiasis. The ethanolic extract of *Zingiber officinale* (200 mg/kg) administered orally for 20 days to streptozotocin -induced diabetic rats produced significant antihyperglycaemic effect ($P < 0.01$) comparable to gliclazide (25 mg/kg, orally). Treatment with the extract also led to lower lipid, total serum cholesterol and triglycerides and increase in the HDL-cholesterol levels when compared with pathogenic diabetic rats ($p < 0.01$). *Zingiber officinale* extract treatment also lowered the liver and pancreas thiobarbituric acid reactive substances values ($p < 0.01$) as compared to pathogenic diabetic rats (Bhandari et al., 2005).

References

Abosi, A. O. and Raseroka, B. H. (2003) In vitro antimalarial activity of *Vernonia amygdalina*. *Brit. J. Biomed. Sci.* 60(2): 89-91.

Achrekar, S., Kaklij, G. S., Pote, M. S., and Kelkar, S. M. (1991) Hypoglycemic activity of *Eugenia jambolana* and *Ficus bengalensis*: mechanism of action. *In Vivo.* 5(2): 143-147.

Adamu, I., Joseph, P. K. and Augusti, K. T. (1982) Hypolipidemic action of onion and garlic unsatured oils in sucrose-fed rats over a two-month period. *Expetientia* 38.899-902.

Adisa, W. A., Otamere, H. O., Osifo, C. U., Ediangbe, A. P., Ogarah, P. A., and Aigbogun, T. O. (2012) Plasma glucose in *Telfairia occidentalis* treated rats. *Advan. Appl. Sci. Res.* 3 (3): 1551-1553.

Aiyeloja, A. A., Bello, O. A. (2006). Ethnobotanical potentials of common herbs in Nigeria: A case study of Enugu state. *Acad. J. Educ. Res. Rev. 1* (1): 16-22.

Ajabnoor, M. A. (1990). Effect of aloes on blood glucose levels in normal and alloxan diabetic mice. *J. Ethnopharmacol.* 28:215–220.

Akah, P., Okafor, C. L. (1992) Blood lowering effects of *Vernonia amygdalina* Del. in an experimental rabbit model. *Phytother. Res.* 6: 111-114.

Akah, P., Njoku, O., Nwanguma, A., Akunyili, D. (2004). Effects of aqueous leaf extracts of *Vernonia amygdalina* on blood glucose and triglyceride level of alloxan-induced diabetic rats (*Rattus rattus*). *Ani. Res. Int.* 1(2): 90-94.

Alarcon-Aguilar, F. J., Roman-Ramos, R., Perez-Gutierrez, S., Aguilar-Contreras, A., Contreras-Weber, C. C., Flores-Saenz, J. L. (1998) Study of the anti-hyperglycemic effect of plants used as antidiabetics. *J. Ethnopharmacol.* 61: 101–110.

Al-Rowais, N. A. (2002). Herbal medicine in the treatment of diabetes mellitus. *Saudi Med. J.*, 23: 1327-1331.

Atangwho, I. J., Ebong, P. E., Egbung, G. E., Eteng, M. U., Eyong, E. U. (2007a). Effect of *Vernonia amygdalina* Del. on liver funtion in alloxan-induced hyper glycemic rats. *J. Pharm. Bioresour.* 4(1): 1-7.

Atangwho, I. J., Ebong, P. E., Eteng, M. U., Eyong, E. U., Obi, A. U. (2007b). Effects of *Vernonia amygdalina* Del. leaf on kidney function of diabetic rats. *Intl. J. Pharm.* 3(2): 143-148.

Atangwho, I. J., Ebong, P. O., Eyong, M. U., Eteng, M. U., Uboh, F. E. (2007c). *Vernonia amygdalina* Del.: A potential prophylactic antidiabetic agent in lipids complication. *Glob. J. Pur. Appl. Sci.* 13(1): 103-106.

Augusti, K. T., Daniel, R. S., Cherian, S., Sheela, C. G., Nair, C. R. (1994) Effect of Leucoperalgonin derivative from *Ficus bengalensis* Linn. on diabetic dogs. *Indian J. Med. Res.* 99:82–86.

Augusti, K. T., Shella, C. G. (1996). Antiperoxide effect of S-allyl cysteine sulfoxide, an insulin secretagogue in diabetic rats. *Experientia*. 52:115–120.

Babalola, O. O., Anetor, J. I., Adeniyi, F. A. A. (2001). Amelioration of carbon tetrachloride induced hepatotoxicity by terpenrid extract from leaves of Vernonia amygdalina. *Afr. J. Med. Med. Sci.* 30: 91-93.

Babu, P. S., Srinivasan, K. (1997). Hypolipidemic action of curcumin, the active principle of turmeric (*Curcuma longa*) in streptozotocin induced diabetic rats. *Mol. Cell Biochem.* 166: 169–175.

Babu, P. S., Prabuseenivasan, S., Ignacimuthu, S. (2007). Cinnamaldehyde-A potential antidiabetic agent. *Phytomedicine*, 14 (1): 15–22.

Bhandari, U., Kanojia, R., Pillai, K. K. (2005). Effect of ethanolic extract of *Zingiber officinale* on dyslipidaemia in diabetic rats. *J. Ethnopharmacol.* 97(2): 227-230.

Biswas, K., Chattepadhya, I., Banergee, R. K., Bandyopadhyayi, U. (2002). Biological activities and medicinal properties of neem (*Azadirachta indica*). *Curr. Sci.* 82(11): 1336-1346.

Botes, L. (2009). The phytochemical content and anti-diabetic properties of Aloe ferox and Aloe greatheadii var. davyana. *Unpublished PhD thesis*. North-West University South Africa.

Cakici, I., Hurmoglo, C., Tunctan, B., Abacioglu, N., Kanzik, I., Sener, B., (1994). Hypoglycaemic effect of *Momordica charantia* extracts in normoglycaemic or cyproheptadine induced hyperglycaemic mice. *J. Ethnopharmacol.* 44: 117-121.

Canal, J. R., Torres, M. D., Romero, A., Perez, C. (2000) A chloroform extract obtained from a decoction of *Ficus carica* leaves improves the cholesterolaemic status of rats with streptozotocin-induced diabetes. *Acta Physiologica Hungarica*, 87: 71–76.

Chadwick, W. A., Roux, S., Van de Venter, M., Louw, J., Oelofsen, W. (2007). Anti-diabetic effect of Sutherlandia frutescens in Wistar rats fed a diabetogenic diet. *J. Ethnopharmacol.* 109: 121-127.

Chattopadhyay, R. R., Chattopadhyay, R. N., Nandy, A. K., Poddar, G., Maitra, S. K. (1987a). Preliminary report on antihyperglycemic effect of fraction of fresh leaves of *Azadiracta indica* (Beng neem) *Bull. Calcutta. Sch. Trop. Med.* 35: 29–33.

Chattopadhyay, R. R., Chattopadhyay, R. N., Nandy, A. K., Poddar, G., Maitra, S. K. (1987b). The effect of fresh leaves of *Azadiracta indica* on glucose uptake and glycogen content in the isolated rat hemidiaphragm. *Bull. Calcutta. Sch. Trop. Med.* 35: 8–12.

Chattopadhyay, R. R. (1996). Possible mechanism of antihyperglycaemic effect of *Azadirachta indica* leaf extract, part IV. *Gen. Pharmacol.*, 27: 431-434.

Chattopadhyay, R. R. (1999). Possible mechanism of antihyperglycaemic effect of *Azadirachta indica* leaf extract, part V. *J. Ethnopharmacol.*, 67: 373-376.

Cherian, S. and Augusti, K. T. (1993). Antidiabetic effects of glycoside of Leucopelargonidin isolated from *Ficus bengalensis* Linn. *Indian J. Exp. Biol.* 31: 26-29.

Chika, A., Bello, S. O. (2010). Antihyperglycaemic activity of aqueous leaf extract of Combretum micranthum (Combretaceae) in normal and alloxan-induced diabetic rats. *J. Ethnopharmacol.* 129: 34–37.

Day, C., Cartwright, T., Provost, J., Bailey, C. J. (1990). Hypoglycaemic effect of *Momordica charantia* extracts. *Planta Med.* 56: 426-429.

Deutschländer, M. S., Lall, N., van de Venter, M. (2009). Plants species used in the treatment of diabetes by South African traditional healers: An inventory. *Pharm. Biol.* 47(4): 348-365.

Dhar, M. L., Dhar, M. M., Dhawan, B. N., Mehrotra, B. N., Ray, C. (1968). Screening of Indian plants for biological activity, Part I. *Indian J. Exp. Biol.* 6, 232–247.

Dieye, A. M., Sarr, A., Diop, S. N., Ndiaye, M., Sy, G. Y., Diarra, M., Rajraji Gaffary, I., Ndiaye Sy, A., Faye, B. (2008). Medicinal plants and the treatment of diabetes in Senegal: survey with patients. *Fund. Clin. Pharmacol.* 22: 211–216.

Dimo, T., Rakotonirina, S. V., Tan, P. V., Azay, J., Dongo, E., Kamtchouing, P., Cros, G. (2007) Effect of *Sclerocarya birrea* (Anacardiaceae) stem bark methylene chloride/methanol extract on streptozotocin-diabetic rats. *J. Ethnopharmacol.* 110: 434–438.

Diè ye, A. M., Sarr, A., Diop, S. A., Ndiaye, M., Sy, G. Y., Diarra, M., Rajraji-Gaffary, I., Ndiaye-Sy, A., Faye, B. 2008. Medicinal plants and the treatment of diabetes in Senegal: survey with patients. *Fund. Clin. Pharmacol.* 22: 211-216.

Eddouks, M., Maghrani, M., Lemhadri, A., Ouahidi, M. L., Jouad, H. (2002). Ethnopharmacological survey of medicinal plants used for the treatment of diabetes mellitus, hypertension and cardiac diseases in the south-east region of Morocco (Tafilalet). *J. Ethnopharmacol.* 82: 97-103.

Erato, P., Adebola, P. O., Grierson, D. S., Afolayan, A. J. (2005). An ethnobotanical study of plants used for the treatment of diabetes in the Eastern Cape Province, South Africa. *Afr. J. Biotech.* 4: 1458-1460.

Etuk, E. U., Bello, S. O., Isezuo, S. A., Mohammed, B. J. (2010) Ethnobotanical Survey of Medicinal Plants used for the treatment of diabetes mellitus in the North Western region of Nigeria. *Asian J. Exp. Biol. Sci.* 1 (1): 55-59.

Gbolade, A. A. (2009). Inventory of antidiabetic plants in selected districts of Lagos State, Nigeria. *J. Ethnopharmacol.* 121: 135–139.

Geetha, B. S., Mathew, B. C., Augusti, K. T. (1994). Hypoglycaemic effects of leucodelphinidin derivative isolated from *Ficus bengalensis* (Linn.). *Indian J. Physiol. Pharmacol.,* 38: 220-222.

Gondwe, M., Kamadyaapa, D. R., Tufts, M., Chuturgoon, A. A., Musabayane, C. T. (2008). *Sclerocarya birrea* [(A. Rich.) Hochst.] [Anacardiaceae] stem-bark ethanolic extract (SBE) modulates blood glucose, glomerular filtration rate (GFR) and mean arterial blood pressure (MAP) of STZ-induced diabetic rats. *Phytomed.* 15: 699–709.

Gondwe, M., Kamadyaapa, D. R., Tufts, M. A., Chuturgoon, A. A., Ojewole, J. A., Musabayane, C. T. (2008). Effects of *Persea americana* Mill (Lauraceae) "Avocado" ethanolic leaf extract on blood glucose and kidney function in streptozotocin-induced diabetic rats and on kidney cell lines of the proximal (LLCPK1) and distal tubules (MDBK). *Methods Find Exp. Clin. Pharmacol.* 30 (1): 25-35.

Gupta, R., Mathur, M., Bajaj, V. K., Katariya, P., Yadav, S., Kamal, R., Gupta, R. S. (2012) Evaluation of antidiabetic and antioxidant activity of *Moringa oleifera* in experimental diabetes. *J. Diabetes.* 4 (2): 164–171.

Gyang, S. S., Nyam, D. D. and Sokomba, E. N. (2004). Hypoglycaemic activity of *Vernonia amygdala* (chloroform extract) in normoglycaemic and alloxan induced hypergycaemic rats. *J. Pharm. Biores.* 1: 61 -66.

Higashino, H., Suzuki, A., Tanaka Pootakham, K. (1992). Hypoglycaemic effects of Siamese *Momordica Charantia* and *Phyllantus urinaria* extracts instreptozotocin-induced diabetic rats. *Nippon Yakurigaku Zasshi*, 100: 515-521.

Igile, G. O., Oleszek, W., Jurzysta, M., Burda S., Fafunso M., Fasanmade. A. A. (1994) Flavonoids from *Vernonia amygdalina* and their antioxidant activities. *J. Agric. Food Chem.* 42: 2445-2448.

Igile, G. O., Oleszek, W., Jurzysta, M., Burda, S., Fafunso, M., Fasanmade, A. A. (1995). Nutritional assesment of Vernonia amygdalina leaves in growing mice. *J. Agric. Food Chem.* 43: 2162-2166.

Izevbigie, E. A., Bryant, J. L., Walker, A. (2004). A novel natural inhibitor of extracellular signal-regulated kinases and human breast cancer cell growth. *Exptal. Biol. Med.* (Maywood). 229(2): 163-169.

Jouad, H., Haloui, M., Rhiouani, H., EL Hilaly, J., Eddouks, M. (2001). Ethnobotanical survey of medicinal plants used for the treatment of diabetes, cardiac and renal diseases in North centre region of Morocco (Fez-Boulemane). *J. Ethnopharmacol.* 77 (2-3): 175-182.

Kamanyi, A., Njamen, D., Nkeh, B. (1994). Hypoglycaemic properties of the aqueous root extract of *Morinda lucida* (Benth) (Rubiaceae). Studies in the mouse. *Phytotherapy Research*, 8, 369-371.

Kamtchouing, P., Sokeng, S. D., Moundipa, P., Watcho, P., Jatsa, B. H., Lontsi, D. (1998). Protective role of *Anacardium occidentale* extract against streptozotocin-induced diabetes in rats. *J. Ethnopharmacol.* 62: 95-99.

Karachi, J. (2009). Efficacy of selected Kenyan medicinal plants used in the treatment and management of type II diabetes. M.Sc. dissertation. Nelson Mandela Metropolitan University, South Africa.

Karou, D., Dicko, M. H., Simpore, J., Traore, A. S. (2005). Antioxidant and antibacterial activities of polyphenols from ethnomedicinal plants of Burkina Faso. *Afr. J. Biotech.* 4: 823–828.

Katzenschlager, R., Evans, A., Manson, A., Patsalos, P.N., Ratnaraj, N., Watt, H., Timmermann, L., Van der Goessel, R., Lees, A. J. (2004). *Mucuna pruriens* in Parkinson's disease: a double blind clinical and pharmacological study. *J. Neurosurg. Psych.* 75: 1672-1677.

Kerharo, T., Adam, J. G (1974) la pharmacopia senegalaise traditionanelle. In: *Plantes medicates et toxiques* Editions Vigot Frerez, Paris.

Khosla, P., Gupta, D. D., Nagpal, R. K. (1995). Effect of *Trigonella foenum graecum* (Fenugreek) on blood glucose in normal and diabetic rats. *Indian J. Physiol. Pharmacol.*, 39: 173-174.

Ko, B. S., Choi, S. B., Park, S. K., Jang, J. S., Kim, Y. E., and Park, S. (2005) Insulin sensitizing and insulinotropic action of berberine from *Cortidisrhizoma. Biol. Pharm. Bull.* 28 (8): 1431-1437

Kumari, K., Mathew, B. C., Augusti, K. T. (1995) Antidiabetic and hypolipidaemic effects of S-methyl cysteinesulfoxide, isolated from *Allium cepa* Linn. *Ind. J. Biochem. Biophys.* 32: 49-54.

Loew, D. and Kaszkin, M. (2002) Approaching the problem of bioequivalence of Herbal Medicinal Products. *Phytother. Res.*; 16: 705-711.

MacKenzie, J., Koekemoer, T., van de Venter, M., Dealtry, G., Roux, S. (2009) *Sutherlandia frutescens* limits the development of insulin resistance by decreasing plasma free fatty acid levels. *Phytother. Res.* 23 (11): 1609-1614.

Majekodunmi, S. O., Oyagbemi, A. A., Umukoro, S., Odeku, O. A. (2011). Evaluation of the anti–diabetic properties of *Mucuna pruriens* seed extract. *Asian Pacific J. Trop. Med.* 4 (8): 632-636.

Mayfield, J. (1998). Diagnosis and classification of Diabetes mellitus- New Criteria. *J. Amer Acad. Fam. Physic.* 58: 11-15.

Morebise, O., Fafunso, M. A. (1998). Antimicrobial and phytotoxic activities of saponin extracts from two Nigerian edible medicinal plants. *Biochem.* 8(2): 69-77.

Morebise, O., Fafunso, M. A., Makinde, J. M., Olajide, O. A, Awe, E. O. (2002). Antiinflammatory property of the leaves of *Gongronema latifolium*. *Phytother. Res. Suppl.* 1: 75-77.

Nalamolu, R. K., Boini, K. M., Nammi, S. (2004). Effect of chronic administration of *Boerhaavia diffusa* Linn. leaf extract on experimental diabetes in rats. *Trop. J. Pharm. Res.* 3 (1): 305-309.

Nash, D., Koenig, J., Novielli, K., Liberoni, R., and Reisman, M. (2001). The importance of the individualized pharmaceutical therapy in the treatment of diabetes mellitus. *Dis. Manag.* 4 (1): 5-23.

Ndifossap, I. G. M., Frigerio, F., Casimir, M., Tsofack, F. N., Dongo, E., Kamtchouing, P., Dimo, T., Maechler, P. (2010). *Sclerocarya birrea* (Anacardiaceae) stem-bark extract corrects glycaemia in diabetic rats and acts on β-cells by enhancing glucose-stimulated insulin secretion. *J. Endocrinol.* 205: 79-86.

Ndong, M., Uehara, M., Katsumata, S., Suzuki, K. (2007). Effects of oral administration of *Moringa oleifera* Lam on glucose tolerance in Goto-Kakizaki and wistar rats. *J. Clin. Biochem. Nutr.* 40 (3): 229–233.

Ngueguim, T. F., Dimo, T., Dzeufiet, D. P. D., Vouffo, B., Dongo, E., Renaud, B., Asongalem, A. E., Zapfack, L., Kamtchouing, P. (2007). Antidiabetic activities of methanol-derived extract of Dorstenia picta twigs in normal and streptozotocin-induced diabetic rats. *Asian J. Trad. Med.* 2 (4): 140-148.

N'Guessan, K,, Aké-Assi, E., Adou, Y. C., Traoré, D. (2008). Effet de l'extrait aqueux des feuilles de *Crescentia cujete* sur la glycémie de lapins diabétiques. *Revue Méd. Pharm. Afr.* 21: 89-96.

N'guessan, K., Amoikon K. E., Tiébré, M., Kadja, B., Zirihi, G. N. (2009a). Effect of aqueous extract of *Chrysophyllum cainito* leaves on the glycaemia of diabetic rabbits. *African J. Pharm. Pharmacol.* 3(10): 501-506.

N'guessan, K., Kouassi, K. É., Kouadio K. (2009b). Ethnobotanical study of plants used to treat diabetes, in traditional medicine, by Abbey and Krobou people of Agboville (Côte-d'Ivoire). *Amer. J. Sci. Res.* 4, 45-58.

N'Guessan, K., Fofie, N. B. Y., Kouassi, K. H. (2011) Effect of aqueous extract of *Boerhavia diffusa* leaves on the glycaemia of rabbits. *Int. J. Appl. Biol. Pharm. Technol.* 2 (3): 330-338.

Nimenibo-Uadia, R. (2003). Effect of *Vernonia amygdalina* in alloxan induced diabetic albino rats. *J. Med. Lab. Sci.* 12(1): 25-31.

Nwozo, S. O., Adaramoye, O. A., Ajaiyoba, E. O. (2004). Antidiabetic and hypolipidemic studies of *Telfairia occidentalis* on alloxan – induced diabetic rats. *Nig. J. Nat. Prod. Med.* 8: 45-47.

Nyunaï, N., Njikam, N., Abdennebi, E.H., Mbafor, J. T. and Lamnaouer, D. (2009). Hypoglycaemic and antihyperglycaemic Activity of *Ageratum Conyzoides* L. in Rats. *Afr J Tradit. Complement. Altern. Med.* 6 (2): 123–130.

Obatomi, D.K., Bikomo, E.O., Temple, V.J. (1994). Anti-diabetic properties of the African mistletoe in streptozocin-induced diabetic rats. *J. Ethnopharmacol.* 43: 13-17.

Odason, E. E., Kolawola, J. A. (2007). Antidiabetic properties and brine shrimp toxicity of the aqueous extract of the root of *Cassia siamea* Lam (Ceasalpiniaceae). *Nig. J. Pharm. Res.* 6 (1): 66-69.

Ohiri, F. C., Esimone, C. O, Nwafor, S. V., Okoli, C. O., Ndu, O. O. (2003). Hypoglycemic Properties of *Viscum album* (Mistletoe) in alloxan-induced diabetic animals. 41 (3): 184-187.

Ojewole, J. A. (2003). Hypoglycemic effect of *Sclerocarya birrea* [(A. Rich.) Hochst.] [Anacardiaceae] stem-bark aqueous extract in rats. *Phytomedicine.* 10: 675–68.

Ojewole, J.A., Adewunmi, C.O. (2003). Hypoglycemic effect of methanolic extract of *Musa paradisiaca* (Musaceae) green fruits in normal and diabetic mice. *Methods Find Exp. Clin. Pharmacol.* 25 (6): 453.

Ojiako, O. A. and Nwanjo, H. U. (2006). Is *Vernonia amygdalina* hepatotoxic or hepatoprotective? Response from biochemical and toxicity studies in rats. *Afri. J. Biotechnol.* 5(18): 1648-1651.

Okafor, J., Ham, R. (1999). Identification, utilization and cultivation of medicinal plants in South-eastern Nigeria. *Iss. Afr. Biochem.* (3): 1-8.

Olajide, O. A., Makinde, J. M., Okpako, D. T. (2003). Evaluation of the anti-inflammatory property of the extract of *Combretum micranthum* G. Don (Combretaceae). *Inflammopharmacol.* 11, 293–298.

Olatunji, L. A., Okwusidi, J. I., Soladoye, A. O. (2005). Antidiabetic effect of *Anacardium occidentale* stem-bark in fructose-diabetic rats. *Pharm. Biol.* 4 (7): 589–593.

Olorunfemi, A. E., Arnold, C. I., Emmanuel, O., Nkaima, N., Akeem, A., (2005) *J. Pharm. Biores.*, 2: 36-42.

Onoagbe, I. O., Negbenebor, E. O., Ogbeide, V. O., Dawha, I. H., Attah, V., Lau, H. U., Omonkhua, A. A. (2010). A Study of the anti-diabetic effects of *Urena lobata* and *Sphenostylis stenocarpa* in streptozotocin-induced diabetic rats. *European J. Sci. Res.* 43 (1): 6-14.

Onoagbe, I. O., Ebhota, A. O., Udegbe, H. C., Omondia, M., Edeni, D., Ebengho, S. O. (1999). Assessment of some medicinal plants for hypoglycemic activities in rats and rabbits. *Biosci. Res. Commun.* 11: 159-163.

Pantoja, C. V., Chiang, L. Ch., Norris, B. C., Concha, J. B. (1991). Diuretic, natriuretic and hypotensive effects produced by *Allium sativum* (garlic) in anaesthetized dogs. *J. Ethnopharmacol.* 31: 325-331.

Pari, L., Satheesh, M. A. (2004). Antidiabetic activity of *Boerhaavia diffusa* L.: effect on hepatic key enzymes in experimental diabetes. *J. Ethnopharmacol.* 91 (1): 109–113.

Pari, L. and Umamaheswari, J. (2000): Antihyperglycaemic activity of *Musa sapientum* flowers: effect on lipid peroxidation in alloxan diabetic rats. *Phytother. Res.*, 14:136-138.

Petit, P. R., Sauvaire, Y. D., Hillaire-Buys, D. M., Leconte, O. M., Baissac, Y. G., Ponsin, G. R., Ribes, G. R. (1995). Steroid saponins from fenugreek seeds: extraction, purification, and pharmacological investigation on feeding behavior and plasma cholesterol. *Steroids.* 60: 674-680.

Prohp, T. P., Onoagbe, I. O., Onyebuagu, P. C., Omeni, A. A., Okoli, R. I., Obeto, N. P. (2006). Effects of aqueous extract of *Triplochiton scleroxylon* on red blood cells and associated parameters in alloxan - induced diabetic rabbits. *Pakistan J. Nutr.* 5 (5): 425-428,

Qamar, F., Afroz, S., Feroz, Z., Siddiqui, S., Ara, A. (2011). Evaluation of hypoglycemic effect of *Cassia italica. J. Basic and Applied Sci.* 7 (1): 61-64.

Quisumbing, E. (1978). *Moringa oleifera* Lam, medicinal plants of the Philippines. Katha Publication Company, Inc., pp: 346-349.

Ravikumar, P., Anuradha, C. V. (1999). Effect of fenugreek seeds on blood lipid peroxidation and antioxidants in diabetic rats. *Phytother. Res.* 13:197–201.

Roman-Ramos, R., Alarcon-Aguilar, F., Lara-Lemus, A., Flores-Saenz, J. L. (1992). Hypoglycemic effect of plants used in Mexico as antidiabetics. *Archiv. Med. Res.* 23: 59–64.

Sabitha, V., Ramachandran, S., Naveen, K. R., Panneerselvam, K. (2011). Antidiabetic and antihyperlipidemic potential of *Abelmoschus esculentus* (L.) Moench. in streptozotocin-induced diabetic rats. *J. Pharm. Bioallied Sci.* 3(3): 397–402.

Saidu, A. N., Mann, A., Balogun, S. (2012). The hypoglycemic effect of aqueous extract of the *Anacardium occidentale* Linn leaves grown in Nigeria on normoglycemic albino rats. *J. Emerging Trends in Engin. Appl. Sci.* 3 (2): 302-308.

Sarkar, S., Pranava, M., Marita, A. (1996). Demonstration of the hypoglycemic action of *Momordica charantia* in a validated animal model of diabetes. *Pharmacol. Res.* 33: 1-4.

Sengkhamparn, N., Verhoef, R., Schols, H. A., Sajjaanantakul, T., Voragen, A. G. (2009). Characterisation of cell wall polysaccharides from okra (*Abelmoschus esculentus* (L.) Moench) *Carbohydr. Res.* 344:1824–32.

Serraclara, A., Hawkins, F., Perez, C., Dominguez, E., Campillo, J. E., Torres, M. D. (1998). Hypoglycemic action of an oral fig-leaf decoction in type-I diabetic patients. *Diabetes Res. Clin. Pract.* 39: 19–22.

Sharma, R. D., Raghuram, T. C., Rao, N. S. (1990). Effect of fenugreek seeds on blood glucose and serum lipids in type I diabetes. *Euro. J. Clin. Nutr.* 44: 301–306.

Sharma, S. R., Dwivedi, S. K., Swarup, D. (1997). Hypoglycemic, antihyperglycemic and hypolipidemic activities of *Caesalpinia bonducella* seeds in rats. *J. Ethnopharmacol.*58: 39–44.

Sheela, C. G., Augusti, K. T. (1992). Antidiabetic effects of *S*-allyl cysteine sulphoxide isolated from garlic *Allium sativum* Linn. *Indian J. Exp. Biol.* 30: 523–526.

Sheela, C. G., Kumud, K., Augusti, K.T. (1995). Anti-diabetic effects of onion and garlic sulfoxide amino acids in rats. *Planta Med.* 61: 356-357.

Shokeen, P., Anand, P., Krishna Murali, Y., Tandon, V. (2008). Antidiabetic activity of 50% ethanolic extracts of *Ricinus communis* and its purified fractions. *Food Chem. Toxicol.* 46: 3458–3466.

Sitasawad, S. L., Shewade, Y., Bhonde, R. (2000). Role of bittergourd fruit juice in STZ-induced diabetic state *in vivo* and *in vitro*. *J. Ethnopharmacol.* 73: 71-79.

Sokeng, D. S., Kamtchouing, P., Watcho, P., Jatsa, B. H., Moundipa, F. P., Lontsi, D. (2001). Hypoglycemic activity of *Anacardium occidentale* L. in streptozotocin-induced diabetic rats. *Diabetes Res.* 36: 1-9.

Sokeng, S.D., Lontsi D., Moundipa P.F., Jatsa H.B., Watcho P., Kamtchouing P. (2007). Hypoglycemic effect of *Anacardium occidentale* L. methanol extract and fractions on streptozotocin-induced diabetic rats. *Global J. Pharmacol.* 1 (1): 01-05.

Sonia, B., Srinivasan, B. P. (1999). Investigations into the anti-diabetic activity of *Azadirachta indica. Ind. J. Pharm.* 31: 138-141.

Stanely, P., Prince, M., Menon, V. P. (2000). Hypoglycaemic and other related actions of *Tinospora cordifolia* roots in alloxan-induced diabetic rats. *J. Ethnopharmacol.* 70: 9–15.

Stanley, P., Prince, M., Menon, V. P. (2000). Hypoglycaemic and other related actions of *Tinospora cordifolia* roots in alloxan-induced diabetic rats. *J. Ethnopharmacol.* 70:9-15.

Stewart, K. M. (2003) The African cherry (*Prunus africana*): can lesson be learned from an over-exploited medicinal tree? *J. Ethnopharmacol.* 89: 3-13.

Tende, J. A., Ezekiel, I., Dikko, A. A. U., and Goji, A. D. T. (2011) Effect of ethanolic leaves extract of *Moringa oleifera* on blood glucose levels of streptozocin-induced diabetics and normoglycemic wistar rats. *British J. Pharmacol. Toxicol.* 2(1): 1-4.

Tsabang, N., Nkongmeneck, B. A., Zapfack, L., Dongmo, Z., Nguenang, G. M., Lando, G., Carlson, T. J., Lowry, P. P., (2001) Inventaire des plantes à vertus antidiabétiques dans la région de Yaoundé au Cameroun. *Revue de médecines et pharmacopées africaines.* 15:87–94.

Ugochukwu, N. H., Babady, N. E. (2003). Antihyperglycemic effect of aqueous and ethanolic extracts of *Gongronema latifolium* leaves on glucose and glycogen metabolism in livers of normal and streptozotocin-induced diabetic rats. *Life Sci.* 73(15): 1925-1938.

Ugochukwu, N. H., Babady, N. E., Cobourne, M., Gasset, S. R. (2003). The effect of*Gongronema latifolium* extracts on serum lipid profile and oxidative stress in hepatocytes of diabetic rats. *J. Biosci.* 28 (1): 1-5.

Ukwuenya, V. O., Ashaolu, J. O., Adeyemi, A. O., Akinola, O. A., Caxton-Martins, E. A. (2012). Antihyperglycemic activities of methanolic leaf extract of *Anacardium occidentale* (Linn.) on the pancreas of streptozotocin-induced diabetic rats. *J. Cell Animal Biol.* 6 (11): 169-174.

Van Wyk, B. E. (2008). A board review of commercially important Southern African medicinal plants. *J. Ethnopharmacol.* 119: 342-355.

Van Wyk, B. E., Albrecht, A. (2008). A review of the taxonomy, ethnobotany, chemistry and pharmacology of *Sutherlandia frutescens* (Fabaceae). *J. Ethnopharmacol.* 119: 620-629.

Venkateshwarlu, E., Reddy, A. N., Sunder, A. S., Kiran, G., Rao, V. J., and Madhusudhan, S. (2009). Anti-hyperglycemic activity of methanolic extract of *Salacia fruticosa* leaves in alloxan induced diabetic rats. *Drug Invention Today.* 2: 95-97.

World Health Organization (1985) Diabetes mellitus report of a WHO study group on diabetes mellitus. *WHO Technical Report*, Geneva. pp. 727.

In: Plant Extracts
Editors: Adriana Giordano and Alfieri Costs

ISBN: 978-1-62417-534-3

Chapter 6

Herbal Medicine: Dry Extracts Production and Applications

Loreana C. Gallo**[*]**, *Verónica Bucalá, María V. Ramírez Rigo and Juliana Piña***

Department of Chemical Engineering, PLAPIQUI,
Universidad Nacional del Sur, CONICET, Bahía Blanca, Argentina

Abstract

Medicinal plants have provided therapeutic agents for thousands of years and continue being a great source of novel pharmaceutical products. Two-thirds of the world population still relies to some extent on natural medicines for primary health care. In addition, the demand for medicinal plant products is now increasing in both developing and developed countries as a consequence of a growing aging population and increasing consumer awareness about general health. Nevertheless, there is a widespread çoncern regarding safety, efficacy and quality of herbal medicines. Currently, many countries are conscious about the need of regulatory guidelines for medicinal herbs formulation, production and use. Phytomedicines represent a wide range of product types: from raw herbs (dried or fresh) to others with different processing degrees (such as tinctures and extracts). The production of herbal medicines involves adequate cultivation and post-harvest methodologies as well as botanical identification of the vegetable drugs. For herbal medicine end-products with added value, appropriate extraction procedures (i.e., in terms of bioactivity), standardization of extracts, selection and tuning of the drying processes operating conditions and design of pharmaceutical formulations are also necessary.

In this chapter the current global scenario of phytomedicine is first presented, evaluating its impact on public health and markets. The challenges associated to safety, efficacy and quality of herbal medicines due to the extracts complexity are then introduced. Understanding the great importance of providing accurate dosages of herbal therapeutic agents, the production of dry plant extracts is next discussed focusing on: the

[*] To whom correspondence should be addressed. Tel: 54-291-486-1700, Fax: 54-291-486-1600, E-mail: lgallo@plapiqui.edu.ar.

drying step including the formulation of the solution to be dried (which comprises the herbal extract) and the dried extract quality. Regarding the available technologies for preparing dry medicinal plants extracts, spray drying is particularly described. This technique is the most commonly used in the herbal processing industry due to its ability to generate particulate products with precise specifications in continuous operations. Several key studies about the effect of the operating conditions and feed composition on the yield of the spray drying process and the particles properties are subsequently summarized. For the production of tablets by direct compression, stable herbal dried powders with good flow properties are required. Consequently, the effects of different processing variables on the stability and flowability of dried extracts are also presented. The formulation of solid dosage forms, based on spray-dried plant extracts, for conventional and modulated delivery is then reviewed. Besides, existing strategies to modulate the drug delivery of plant extracts are also described. Since modified-release phytomedicines are still in an exploratory stage, the prospective for the development of novel products is finally discussed.

1. Introduction

Medicinal plants came into use in the modern medicine through the utilization of herbal material as indigenous cure in traditional medicine systems. Today, the world is moving towards phytomedicines for the prevention and treatment of various health ailments (Pandey *et al.*, 2011). About 5-15% of the total 250.000 recorded plants species were systematically investigated but, to date, the potential of many plants as a source of novel bioactive compounds is scarcely tapped (Sen *et al.*, 2011).

Nowadays, herbal medicines are used in almost all cultures worldwide. Moreover, around two-thirds of the world population with sufficient access to modern drugs still relies on natural medicines (Efferth, 2011). The World Health Organization (WHO) estimated that 4 billion people (80% of the world population) of developing countries, especially those in Asia, Africa, Latin America and the Middle East, use herbal medicines for some aspect of primary health care (Robinson and Zhang, 2011). As an example, Figure 1 shows the usage of herbal medicines in some selected countries of Africa. In the last years, the use of herbal medicines in some developed countries also increased rapidly. Canada, France, Germany and Italy reported that about 70-90% of their populations use herbal medicines under the title of complementary, alternative, or non-conventional medicines (Robinson and Zhang, 2011). Figure 2 shows the percentage of people in some developed countries that used herbal medicines at least one time.

The total sale of herbal medicines reflects its popularity and use by people all over the world. According to the Secretariat of the Convention on Biological Diversity, the global sales of herbal products were estimated to be US $60 billion in 2000. In 2008 the worldwide market was about US $83 billion, with a steady growth rate ranging between 3% and 12% per year (Zhang *et al.*, 2012). Moreover, the global herbal products market is forecast to reach US $107 billion by the year 2017, driven by a growing aging population and increasing consumer awareness regarding general health. Despite the recent economic recession, this market still exhibited robust growth with little or no significant decline (Global Industry Analysts, Inc., 2012). There is a growing trend of people moving from synthetic allopathic drugs to herbal medicines. This situation could be attributed to, among other reasons, the people interest in taking an active part of their health, the belief that herbal medicines are free from side effects because humans have been consuming it from immemorial time, the prevalence of chronic illnesses that cannot be cured by conventional drugs and the high cost of synthetic drugs (Pandey *et al.*, 2011).

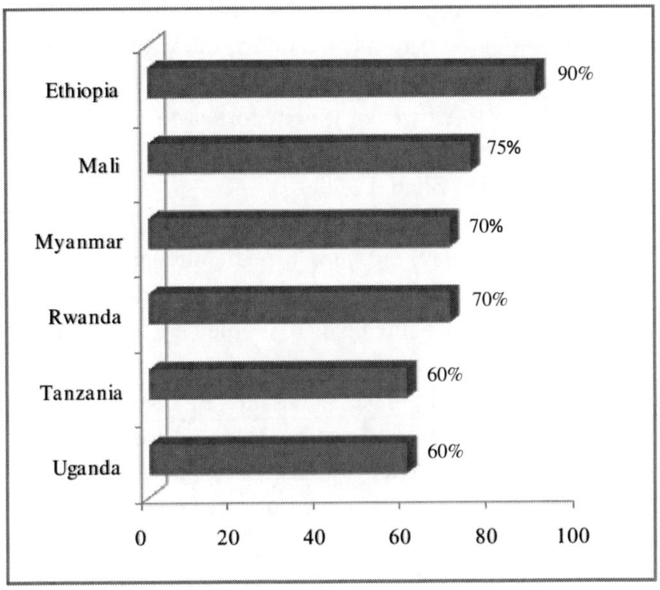

Figure 1. Percent use of herbal medicines in some developing countries (Robinson and Zhang, 2011).

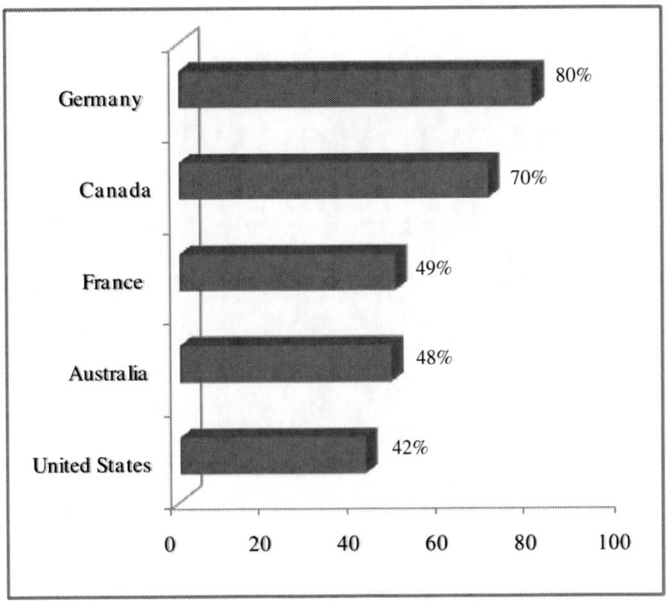

Figure 2. Percentage of people that used herbal medicines at least one time in some developed countries (Robinson and Zhang, 2011).

A recent publication by WHO on traditional medicine revealed that, in China, herbal medicinal preparations constituted between 30% and 50% of the total consumption of medicines, equating to a sales value of US $ 14 billion in 2005 (28% more than the previous year). In Japan, spending on traditional medicines in 2006 came to just over US $1 billion while during 2004-2005 Australia spent a total of US $ 1.86 billion on herbal medicines. The expenditure in UK and Brazil in 2007 was relatively low, only US $ 230 million and US $ 160 million, respectively. Among the European nations, the value of the known market sales of herbal medicines in 2003 was estimated to be US $ 5 billion. Figure 3 shows the distribution of sales within Europe. However, it is important to note that all these data are not representative of the actual cost to consumers because they were based on the manufacturers' prices to wholesalers (Robinson and Zhang, 2011).

Nearly 1 of 5 adults in the United Stated (USA) reported taking a herbal product at least one time (Bent, 2008). In that country, the herbal medicines market climbed after the 1994 Dietary Supplements and Health Education Act. This law permitted labeling herbal medicines as dietary supplements (i.e. a product intended to supplement the diet that contains one or more of the following ingredients: vitamin, mineral, amino acid, herb or other botanical). Dietary supplements are not allowed to make any claims to "treat, prevent, cure, mitigate, or diagnose a specific disease". These claims are reserved to drugs, and drug manufacturers must submit proof of safety and efficacy to receive approval from the Food and Drug Administration (FDA) before sale and marketing. For dietary supplements, the FDA requires good manufacturing practices that imply processing in a consistent manner and meeting quality standards. Since herbal medicines are not regulated as drugs, they are quite available to people (Bent and Ko, 2004).

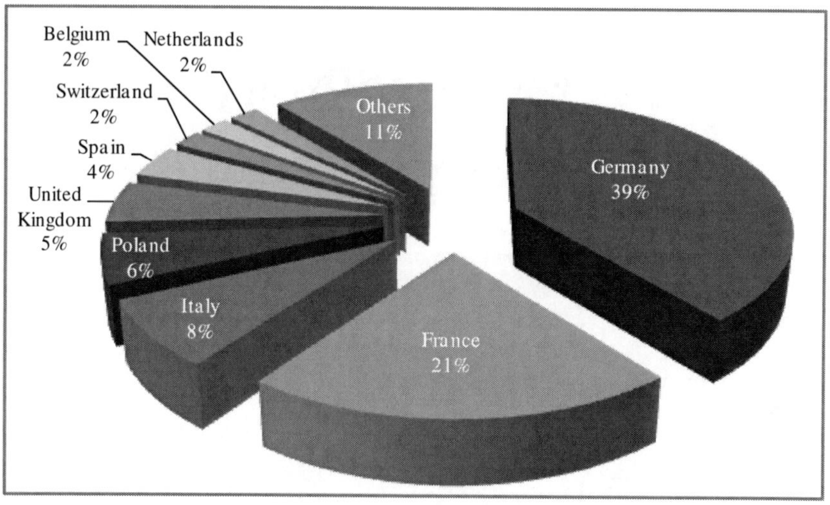

Figure 3. Market sales of herbal medicines in Europe, 2003 (Robinson and Zhang, 2011).

The USA contribution to the world herbal medicines market in 1996 was about US $ 4 million and its herbal medicines sales showed an increase of 101% between 1996 and 1998 (Sen *et al.*, 2011) as a consequence of the law signed during the 1994 Dietary Supplements and Health Education Act. At the present time, the global market of herbal medicines is increasing exponentially. Nonetheless, important aspects as efficacy, safety, quality control and regulation of herbal products are still great challenges worldwide.

Currently, herbal products are marketed in the form of liquid preparations (i.e., teas, tinctures) and powders. Powders are produced by drying and comminuting of plants or by drying liquid plant extracts. Dried extracts are available as capsules and tablets. The advantages of dried extracts over liquid preparations are higher concentrations, more stability of their active components and lower storage costs (Banker and Rhodes, 2002; Souza *et al.*, 2008). Tablets, the most common used oral solid forms, are preferably prepared by direct compression (DC) (Sarraguça *et al.* 2010). However, as dried extracts are hygroscopic and poor flow materials, they cannot be directly processed by DC (Tong *et al.*, 2008). For this reason, the development of new technologies and/or formulations to obtain dried plant extracts with appropriate characteristics for DC is a main issue for the herbal processing industry. Besides, strategies to modulate the delivery of plant extracts constitute an important challenge in the phyto-pharmaceutical research. In this context, this chapter initiates with a description of the actual worldwide situation concerns about efficacy, safety, quality control and regulation aspects of herbal medicines. Then, the production by spray drying of plant extracts with desirable properties for DC is described. Besides, other applications of spray drying related to the improvement of the pharmacological activity of herbal extracts and to their nutritional use are discussed. Finally, the production of microparticles by spray drying to modulate plant extracts release and the prospective for the development of novel herbal products is presented.

2. Herbal Medicines: Efficacy, Safety, Quality Control and Regulation Aspects

All plants produce chemical compounds as part of their normal metabolic activities. These include primary metabolites (sugars and fats) and secondary metabolites (alkaloids, flavonoids, glycosides, saponins, tannins, terpenes and phenolics) that have therapeutic effects and can be isolated for production of synthetic drugs (Barnes *et al.*, 2007). The assessment of efficacy and safety of herbal products is more difficult than for synthetic drugs because they contain a variety of chemical compounds. Moreover, it is well known that the efficacy and safety of whole plants or extracts may be quite different than those of their isolated ingredients (Jordan *et al.*, 2010).

Figure 4 summarizes the main issues for the development of phytomedicines. The chemical profile of medicinal plants can be affected by conditions of cultivation, storage, handling and processing (Sahoo *et al.*, 2010). Considering that the last three steps can be standardized, the domestication and planting of medicinal plants instead of the use of wild harvested herbs will offer great advantages, since it will be possible to provide uniform and high quality raw materials (Calixto, 2000).

Herbal medicines and/or their products require an adequate chemical characterization to attain reproducibility in efficacy studies and to guarantee the product safety. This is usually attempted through standardization, which is a process that defines the required amount of active constituent (or if the active ingredient is not known, a suitable marker constituent) in a herbal medicine.

Standardization usually takes into account one or two compounds; thus variation may still remain in other ingredients and influence both the efficacy and safety of the product (Banker and Rhodes, 2002; Ernst, 2005).

For appropriate standardization and quality control of herbs, the development of adequate analytical techniques becomes an important matter. Several markers (taxonomic, chemical, genomic, proteomic) together with macroscopic, microscopic and chemical analysis are employed for identification purposes. Particularly, electrophoretic and chromatographic techniques as thin-layer chromatography (TLC), gas chromatography (GC), high-performance liquid chromatography (HPLC), mass spectrometry (MS), infrared-spectrometry and ultraviolet/visible spectrometry used alone or in combination (hyphenation procedures, e.g. HPLC/MS, GC/MS) are frequently used to identify and quantify the active ingredients or chemical markers (Calixto *et al.*, 2000; Liang *et al.*, 2004; Sahoo *et al.*, 2010).

Clinical trials are viable to study herbal medicines efficacy; however, specified literature reveals that few well-controlled clinical trials were carried out. Various topics contributed to this situation, among others, lack of standardization and quality control of herbal drugs, different dosages use, non suitable selection and number of patients to achieve statistical significance and the usage of different treatments times (Calixto *et al.*, 2000). Against this background, few herbal medicines were adequately studied. In USA, only five of the most commonly used 10 herbs (Echinacea, Ginseng, Garlic, Ginkgo, St. John's wort, Soy, Peppermint, Ginger, Chamomile and Kava kava) have considerable scientific evidence about their herbal pharmacological effect (Table 1). In general, insufficient evidence does not indicate absence of herbal benefit in people health, but shows a lack of conclusive information about the efficacy of herbal product (Bent, 2008).

Generally, most of the population believes that all natural treatments are inherently safe. Nevertheless, plants are not completely free from the possibility of toxicity or other adverse effects (Robinson and Zhang, 2011).

Among others reported cases, serious adverse reactions associated with plants as Ephedra and Aristolochia have shown that herbs can be toxic. The inherent toxicity of certain herbs is well recognized, however many adverse effects from the use of herbal medicines are unknown. In addition, adverse effects could be the consequence of different external situations, such as herbal contamination with toxic metals, pesticides and microbes and adulterations with undeclared chemical substances (Sahoo *et al.*, 2010).

Therefore, quality control of phytomedicines is essential for the acceptance of herbal products worldwide. Besides and as above mentioned, the number of clinical trials in humans is usually low; thus, the reported data related to herbal medicines safety are limited. Additionally, people do not voluntary report the incidence of adverse effects from phyto-medicines.

Consequently, the necessity of pharmacovigilance is of important concern for the phytomedicine area (Banker and Rhodes, 2002; Jordan *et al.*, 2010). Besides, the same herbal product is sold around the world with different titles, such as prescription or over-the-counter medication, self-medication or self-care, home remedy or dietary supplement, health food, functional food or phytoprotectant, with a poor correspondence between descriptions. Moreover, the legislative and regulatory framework of herbal medicines and products changes between countries over the world mainly due to different cultural aspects (Sahoo *et al.*, 2010; Robinson and Zhang, 2011).

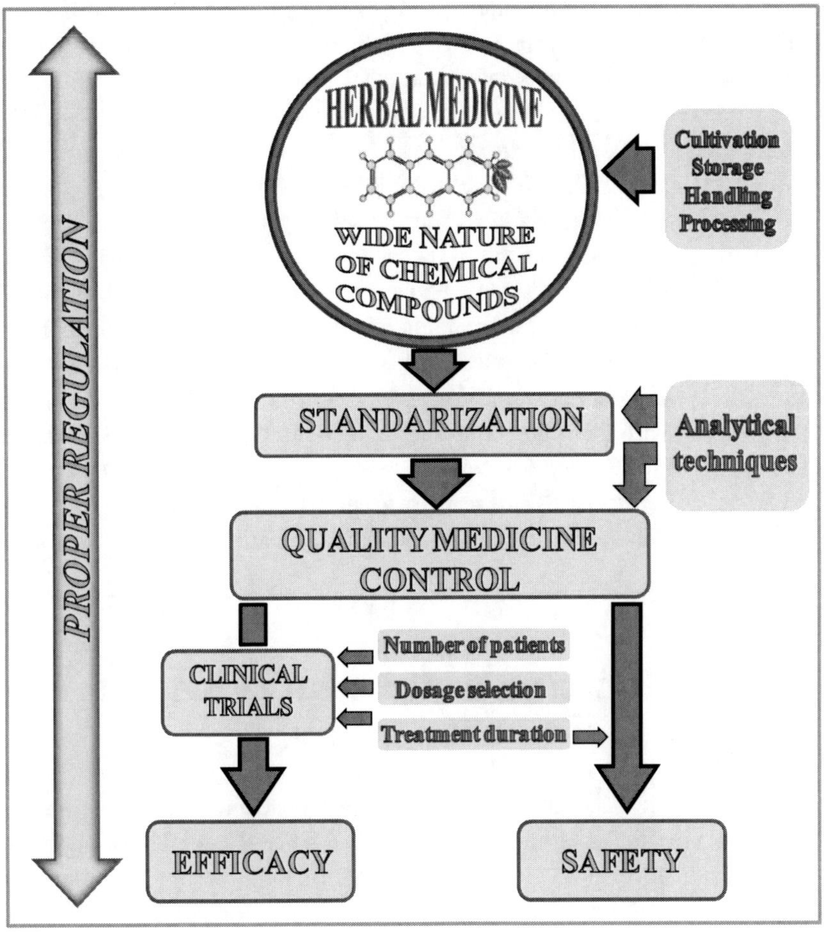

Figure 4. Main issues for the development of phytomedicines.

Table 1. The five most commonly used herbal medicines in USA with scientific evidence of efficacy (Bent, 2008)

Herb	Indication	Scientific evidence for efficacy
Garlic	Hypercholesterolaemia	Likely effective
Ginkgo	Dementia	Likely effective
St. John's wort	Depression	Likely effective for mild-moderate depression
Soy	Menopausal symptoms	Not effective
Kava kava	Anxiety	Likely effective

Despite these various challenges, considerable progress was achieved during the last years on building institutional and regulatory frameworks for registration, research and production of herbal medicine products. Nevertheless, many countries still need technical assistance and/or the development of technological knowledge for producing, distributing, and performing quality, efficacy and safety studies of herbal products. For this reason, WHO reported guidelines to define criteria for the evaluation of efficacy, safety and quality of herbal medicines, aiming to assist regulatory authorities and manufacturers. To ensure that the raw materials selected for medicinal products are of assured quality, the Good Agricultural and Collection Practice (GACP) guidelines are also available. Besides, the Good Manufacturing Practices (GMP) are established to provide assistance to convert herbal medicines into assured quality medicines. Furthermore, WHO also prepared pharmacopeic monographs of herbal medicines and the basis for the assessment of herbal drugs (Calixto *et al.*, 2000; Robinson and Zhang, 2011). Today, advances in analytical technologies are allowing a great comprehension of herbal chemical compounds to better understand the therapeutic and toxic effects of herbal medicines. In addition, nations worldwide are making efforts to guarantee their quality and improve their regulatory aspects. The current circumstances, together with the high economic impact and the widespread population use of herbal products, suggest the need of further research in the herbal medicine area. The investigation of suitable processes to transform herbal raw materials into high quality pharmaceutical dosage forms indeed constitutes a major challenge for the science of herbal medicines.

3. Herbal Medicine Solid Oral Dosage Forms: Spray Drying

Tablets are still the most commonly consumed pharmaceutical dosage form because of their manufacturing simplicity, administration convenience, accurate dosing and stability compared to oral liquid forms (Wu and Seville, 2009). DC is the most efficient method for tablet manufacturing. This process produces tablets by direct compression of a powder blend constituted by the active ingredients and suitable excipients. DC only involves three manufacturing steps: the powder mixing, lubrication, and compression. No pre-treatment of the powder blends, as wet or dry granulation, is needed. For these reasons, DC is a simple and economical process that requires low production times and energy consumption. However, to apply this technique, drugs and excipients have to meet good physico-mechanical properties as flowability and compactability. Adequate flowability is essential to ensure a powder fast flow for uniform die filling in high-speed rotary tablet machines. Another important requirement is the powder compactability, i.e., the powder blend must be compactable and remain in that form once the compression force is released (Swarbrick and Boylan, 2007). In the production of herbal medicine tablets, dried plant extracts are usually used as the therapeutic active ingredients (Newall, 1996). Generally, dried plant extracts tend to be hygroscopic due to their high affinity for the surrounding water vapour (Tong *et al.*, 2008). The powder hygroscopicity plays an important role in particle-particle interactions and contributes to poor flow properties (Jin *et al.*, 2007). Besides, herbal medicine tablets frequently contain a high dose of dry plant extract; thus, the amount of excipients that can be incorporated becomes a critical subject in the manufacture of tablets with reasonable size (Rocksloh *et al.*, 1999; De Souza *et al.*, 2001). Consequently, DC of dry plant extracts constitutes a complex task. Appropriate excipients can be added to herbal extracts, and preferably co-processed by means of a proper technology, to overcome the above-mentioned problems. However, dry extracts standardization is also difficult because different processing conditions can lead to products with different active compounds' qualities. Thus, the development of new technologies to transform dry plant extracts with poor physico-mechanical properties into high quality powders adequate for DC is probably the main challenge for the herbal medicine industry.

Numerous drying techniques (as freeze, spray and fluid bed drying) are reported in the literature to produce high quality solid materials. Among these methods, spray drying is the preferred route due to its capacity to convert a liquid (i.e. aqueous or organic solution, emulsion, dispersion or suspension) into a dry powder with precise specifications (moisture content, size, solubility, bulk density and physico-chemical stability) through continuous and relatively economic operations (León-Martínez *et al.*, 2010; Oliveira and Petrovick, 2010; Souza and Oliveira, 2006). Generally, spray dryers are open systems, i.e. the drying gas (usually air) is discharge after used. A schematic representation of a spray drier is shown in Figure 5. In the spray-drying process, the liquid feed is pumped from the feed recipient to an atomizer usually located at the top of the drying chamber. As shown in Figure 6, there are various types of atomizers; among others, two-fluid nozzles, rotary discs and ultrasonic nebulizers. The selection of the atomizer depends on the nature and flowrate of the feed and on the characteristic desired for the dried powder. Two-fluid nozzles operate with the liquid feedstock and compressed air (or steam), which enter to the atomizers through different channels and exit together at the nozzle tip. This design utilizes the energy of the compressed gas to atomize the liquid into very fine droplets (between 10-1000 μm) and is suitable to atomize highly viscous feeds, but is relatively expensive to operate because of the high cost of compressed air. Rotary atomizers use the energy of a high speed rotating wheel to divide bulk liquid into droplets (> 200 μm) fed at a relatively low pressure. For this type of atomizer, the particle size can be easily varied by changing the wheel speed. However, rotary atomizers can produce large quantities of fine particles, which may cause dust control problems. Unlike conventional atomizing nozzles that rely on pressure and high-velocity motion to shear the fluid into small drops, ultrasonic atomizers only use ultrasonic vibrational energy to generate the spray. For this design, the droplet size is mainly governed by the frequency at which the nozzle vibrates, usually between 1-10 μm (Masters, 1979).

For the cases that use air as drying medium, the gas is taken from the atmosphere by vacuum or pressure systems. The air that enters the unit is filtered and then heated. The droplets produced by the atomizer meet the hot air and evaporation takes place. There are two types of dryer designs: countercurrent and co-current flow. For counter-current dryers and atomizers located at the top of the drying chamber, the hot air is introduced at the end of the dryer. The counter-current design offers faster evaporation and higher energy efficiency than the co-current one, basically because the driest particles are contacted with the hottest air. The counter-current design is not suitable for heat-sensitive products. In co-current dryers, the spray is directed into the hot air that enters at the top of the chamber. Therefore, co-current dryers are the preferred design for heat sensitive products since the hottest air contacts the droplets at their maximum moisture content. The dried droplets are mainly collected downstream in a cyclone, where the powder is separated from the air. Fine particles are usually entrained by the air; for this reason, the outlet air is filtered (Mujumdar, 2006). The residence time of a droplet in the dryer varies according to its design; however, only few seconds (5-30 s) are recommended in order to process complex herbal extracts and other labile materials without significant loss of potency or flavor.

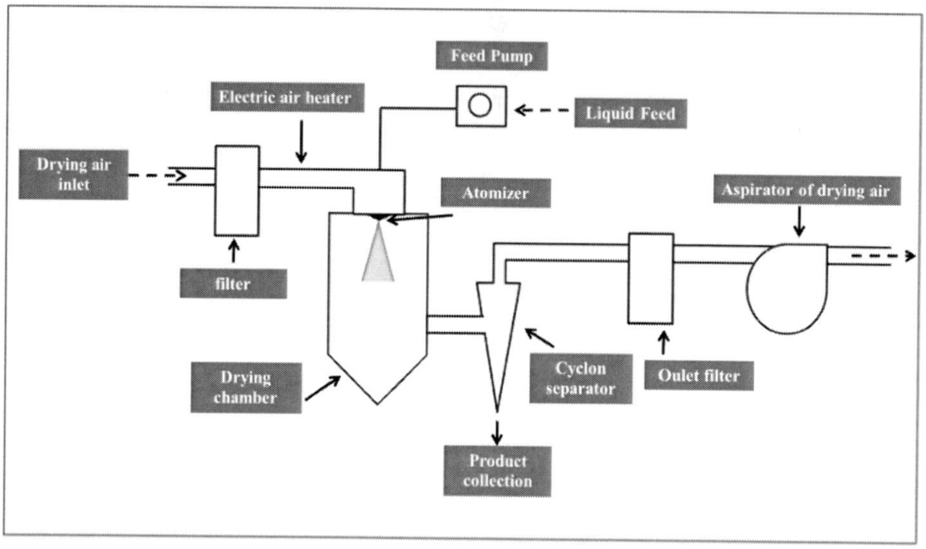

Figure 5. Spray-dryer scheme.

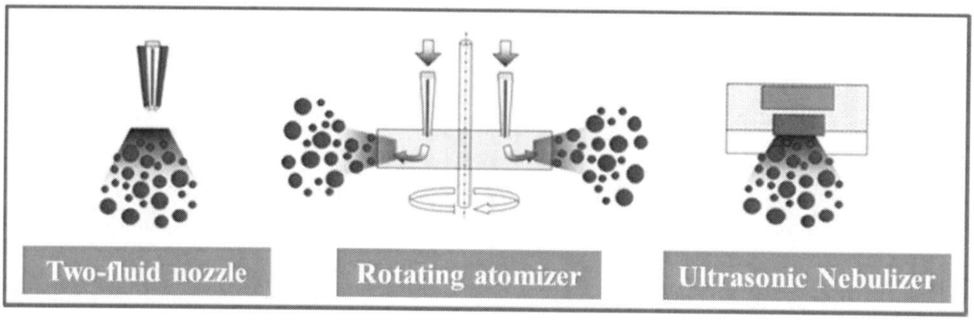

Figure 6. Main types of spray-drying atomizers.

The operating variables inlet air (or other drying gas) temperature, atomization air flowrate (for two-fluid nozzles), pump speed (i.e., liquid flowrate) and drying air flowrate, together with the composition of the liquid feed, strongly affect the properties of the spray dried powders (as particle size and final moisture content) as well as the process yield. The optimization of these parameters is typically achieved by using trial and error procedures and/or experimental factorial designs (Swarbrick and Boylan, 2007; Tajber *et al.*, 2009; Tonon *et al.*, 2008). Some guidelines for process optimization include: high drying gas flowrate to enhance separation in the cyclone and thus, to attain high process yields, low pump speed to have high outlet temperatures, small temperature differences between the inlet and outlet with high drying gas flowrates (i.e. short particles residence time in the drying chamber) to obtain powders with low moisture content. Atomization is the most important operation in the drying process because it defines the particle size distribution. Low atomization air flowrates tend to produce relatively big particles. In addition, high liquid feed concentrations or viscosities may also increase the particles size (Training Paper Mini Spray Drying, 2002; Ré, 2006).

The particles' shape and structure depend strongly on the nature of the sprayed liquids and the process operating conditions (Nandiyanto and Okuyama, 2011). In the pharmaceutical industry, spray drying is employed to produce microparticles for dry oral solid dosage forms, or for parenteral, nasal, or pulmonary administration routes. These particles require appropriate flowability, and for respiratory delivery also adequate aerodynamic properties. Besides, particles must be capable to stabilize the active component and to supply physical stability to the dosage form during storage (Vehring *et al.*, 2007). According to the desired purpose, different particle shapes can be obtained: dense, hollow, encapsulated, hairy, with doughnut forms, etc. (Figure 7). The particles generally have a spherical shape, which is the most stable form for a droplet.

The use of drying adjuvants in the spray drying process is a common practice. The selection of the adjuvant is a critical factor in the spray drying of fluid plant extracts because it determines the stability and quality of the extract powders. Among others, the most frequently used drying adjuvants are: starch, ciclodextrins, maltrodextrin, lactose, arabic gum and colloidal silicon dioxide (SiO_2) (Oliveira and Petrovick, 2010).

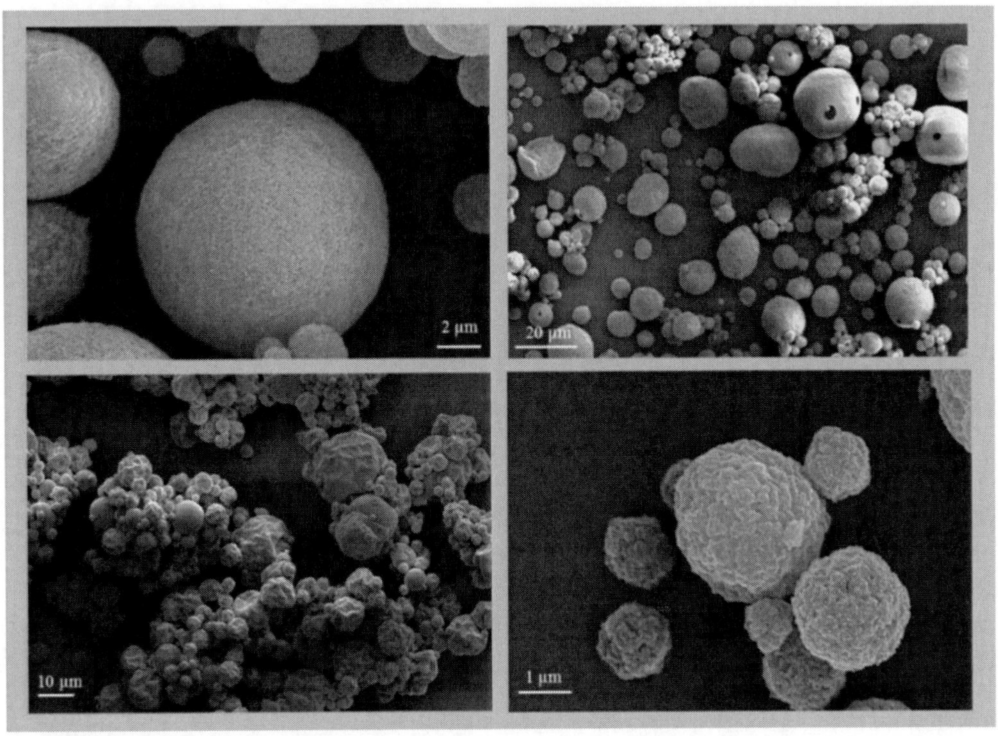

Figure 7. Shape of spray-dried particles from different herbal medicine extracts.

Several authors investigated the influence of spray-drying operating conditions on dried medicinal herbal extracts and the use of different drying adjuvants, in order to improve the extracts powder physical properties (i.e. moisture content, hygroscopicity, flowability and compactability) and the process yields. Soares *et al.* (2005), Souza *et al.* (2007) and Petrovick *et al.* (2010) spray dried different medicinal extracts at fixed operating conditions. In all these cases, the obtained dried extracts showed poor flow and compaction properties. Since the operating conditions were not optimized, the poor powders flowability cannot be attributed to the use of the spray-drying technology. In fact, there are opportunities to improve the powder quality by fitting the operating variables of the dryers.

Table 2 summarizes various additional studies about spray drying of plant extracts; with particular detail on the therapeutic effect, employed drying adjuvants and operating conditions as well as measured variables. Vasconcelos *et al.* (2005) studied the effect of the inlet drying temperature and adjuvant concentration (SiO_2) on the process yield and powder moisture content and hygroscopicity for the spray drying of a *Schinus terebinthifolius* Raddi extract.

The authors found that an inlet air temperature of about 140 ºC and the highest tested adjuvant concentration led to the powders with lowest moisture content and hygroscopic tendency. In addition, high process yields (around 80%) were achieved. Souza and Oliveira (2006) studied the effect of the spray- drying process variables on the product properties and dryer performance during the production of *Bauhinia forficata* dried extract using a fixed concentration of SiO_2.

The highest tested temperature and drying air mass flowrate led to the dried extract with lowest moisture content. Even though these were the best conditions reported by the authors, the process yield was low (49%) and the powders flowability was poor. Souza *et al.* (2008) studied the spray drying of *Rosmarinus officinalis* Linné. The selected drying adjuvant was a given mixture of SiO_2 and maltodextrin. The highest tested temperature caused the highest loss of extract active compounds. For all the operating conditions, the obtained dried extracts showed poor flow properties. Gallo *et al.* (2011a) used a factorial statistical design to find the optimal spray-drying operating conditions and appropriate SiO_2:SR ratio to produce a Cascara sagrada dried extract with good rheological and compactability properties for DC, low moisture content and hygroscopicity, and high process yield. Five factors at two levels (low and high) were considered.

The factorial experimental design was an interesting tool to determine optimal sets of spray-drying operating conditions and the feed design in order to produce dry extracts with desirable characteristics. The adjuvant concentration and the atomization air flowrate were significant variables on the flow properties, particle size, moisture content and hygroscopicity of the dried extracts. The use of relatively high concentrations of adjuvant together with low atomization air flowrates allowed obtaining a powder with good flow and compactability for DC, relatively big particle size, low moisture content and hygroscopicity. Within the studied range, the inlet temperature was not found as a significant variable for the above-mentioned properties. The use of high aspiration flowrates led to high process yields. Figure 8 shows the Cascara sagrada spray dried particles obtained by using high concentrations of adjuvant together with low atomization air flowrates. As it can be seen, the particles were spherical and showed smooth surfaces. On the other hand, Figure 9 presents the particles obtained by setting low adjuvant concentrations together with high atomization air flow rates. In this case, the Cascara sagrada spray-dried particles were agglomerates with different surface patterns, some smooth and others shriveled (like the morphology of the dried plant extract without carrier), constituting the powder with worst flowability. Gallo *et al.* (2011a) concluded that by modifying the operating conditions of the spray-drying process, it is possible to optimize the powder properties of dried plant extracts.

Tabla 2. Studies about spray drying of herb extracts

Authors	Herb	Therapeutic Effect	Drying adjuvant	Variables for parametric studies[†]	Measured variables[¥]
Vasconcelos *et al*. (2005)	*Schinus terebinthifolius* Raddi	Antimicrobial, cicatrizant, antiinflamatory	SiO_2 (Aerosil®)	T_0= [120; 130; 140; 150;160] R= [0.25:1; 0.33:1; 0.43:1]	MC, H and PY
Souza and Oliveira (2006)	*Bauhinia forficata*	Hypoglycemic, depurative, diuretic	SiO_2 (Tixosil 333®) R= [0.8:1]	T_0= [80; 150] $\dot{m}_{da}$= [0.0118, 0.0168; 0.0227 kg/s]	MC, PY and F
Souza *et al*. (2008)	*Rosmarinus officinalis* Linné	Antispasmodic,, alleviate respiratory disorders, stimulant hair growth	SiO_2 (Tixosil 333®) Maltodextrin DE 14 (Mor Rex®) R= [(1:0.5):1]	T_0= [80; 150] $\dot{m}_{da}$= [0.0340; 0118 kg/s]	Active components degradation and F
Gallo *et al*. (2011 a)	Cascara sagrada (*Rhamnus purshianus* D.C.)	Laxant	SiO_2 (Aerosil®)	T_0= [130; 170] $\dot{m}_{da}$= [30; 35 m^3/h] $\dot{m}_f$= [1, 3 ml/min] $\dot{m}_{at}$= [400; 800 l/h] R= [0.5:1; 1:1]	F, MC, H, PY and PS
Zétola *et al*. (2002)	*Lippia alba* (Mill.) N. E. Br.	Sedative, anticonvulsant, myorelaxant,	SiO_2 (Aerosil®) Polysorbate 80 β-cyclodextrin	R= [(0.53 SiO_2: 0.13 Polysorbate 80):1; (0.26 SiO_2: 0.13 Polysorbate 80:0.26 β-cyclodextrin):1]	TE

Table 2. (Continued)

Authors	Herb	Therapeutic Effect	Drying adjuvant	Variables for parametric studies[†]	Measured variables[¥]
Da Silva (2007)	*Ilex paraguariensis*	Central nervous system stimulant, thymoleptic, diuretic, antirheumatic, analgesic, antioxidant	SiO_2 (Tixosil 333®) Maltodextrin (Glucidex®) Microcrystatalline cellulose (Avicel®)	R= [0.43 SiO_2:1; (0.21 SiO_2:0.21 Maltodextrin):1; 0.43 Maltodextrin:1; (0.21 SiO_2:0.21 Microcrystatalline cellulose):1]	MC, PY, photostability and antioxidant activity
Georgetti *et al.* (2008)	Soybean	Antioxidant	SiO_2 (Tixosil 333®) Maltodextrin Starch	R= [0.1 SiO_2:1; 0.1 Maltodextrin :1; 0.1 Starch:1; 0:1] T_0= [80; 115;150]	F and antioxidant activity
Quek *et al.* (2007)	Watermelon *(Citrulus lanatus)*	Antioxidant	Maltodextrin (Dridex 9®)	T_0= [145;155; 165; 175] R= [0.03:1; 0.05:1]	PY, MC and active components degradation
Tonon *et al.* (2008)	*Açai* (*Euterpe oleraceae* Mart.)	Antioxidant	Maltodextrin (MOR-REX®)	T_0= [138; 150; 170; 190; 202] R= [0.1:1; 0.14:1; 0.2:1; 0.26:1; 0.3:1] $\dot{m}_f$= [5; 9; 15; 21; 25 g/min]	PY, H, MC, PS and active components degradation

† T_0 = Inlet air temperature (°C), R= (Adjuvant):SR(extract solid residue) ratio (w/w), $\dot{m}_{da}$=drying air flow rate, $\dot{m}_f$= feed liquid flowrate, $\dot{m}_{at}$= atomization air flowrate.

¥ F=flowabilty, H= hygroscopicity, MC= moisture content, PS=particle size, PY=process yield, TE= therapeutic effect.

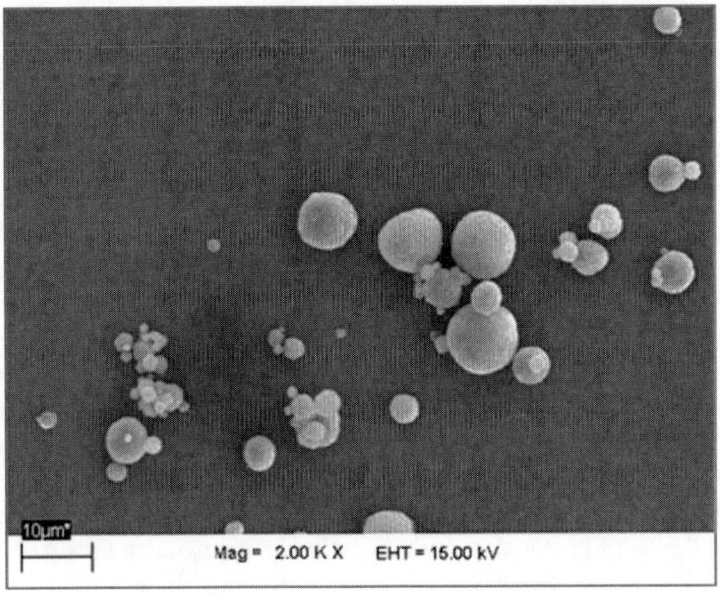

Figure 8. Spray-dried Cascara sagrada particles with good flow properties obtained by Gallo *et al.* (2011a).

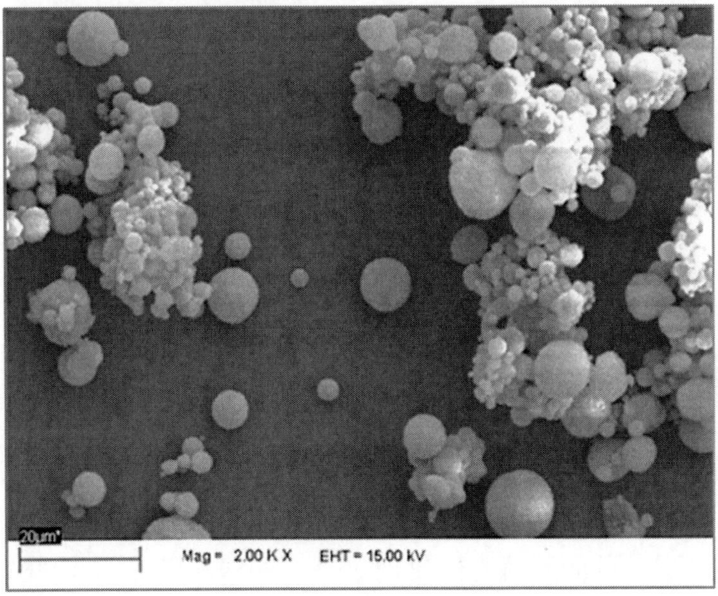

Figure 9. Spray-dried Cascara sagrada particles with poor flowability properties obtained by Gallo *et al* (2011a).

Other authors investigated the spray drying effect on the biological and chemical properties of different plant extracts. Among others, Zétola *et al.* (2002) studied the activity of *Lippia alba* (Mill.) N. E. Br. Spray- dried extracts on the central nervous system. The authors demonstrated the technological feasibility of spray drying the liquid extract using a mixture of SiO_2, Polysorbate 80 and β-cyclodextrin as adjuvants. For a given set of operating conditions, it was found that the drying process did not affect the herbal pharmacological activity. However, it was reported that the use of β-cyclodextrin had influence on the therapeutic effect. Da Silva (2007) spray dried an *Ilex paraguariensis* extract. The author employed SiO_2, maltodextrin and microcrystatalline cellulose, alone or combined, as drying excipients. The dried extract containing just SiO_2 showed the best physical and photostability properties. Georgetti *et al.* (2008) evaluated the effects of different process and formulation variables on the antioxidant activity and flowability of spray-dried soybean extracts. By using a given set of selected operating conditions, the authors also studied the influence of different drying adjuvants (SiO_2, maltodextrin and starch). The powder obtained using SiO_2 had the best flow properties and the lowest agglomeration tendency. For this reason, the authors selected SiO_2 as drying adjuvant and subsequently performed spray drying experiments to analyze the effect of different inlet temperatures on the antioxidant activity. According to the reported results, the higher the inlet air temperature the lower the antioxidant activity. However, the dried extract with SiO_2 exhibited higher antioxidant activity than the powder obtained without adjuvants, suggesting a possible protective effect of SiO_2 against degradation of antioxidants compounds. Considering the limited number of published studies regarding spray drying of herbal medicines, the open literature about drying of similar natural compounds as other plant extracts or fruit pulps was also reviewed to gain understanding. Quek *et al.* (2007) spray dried watermelon juice, which is widely used due to its antioxidant properties. Two different maltodextrin concentrations and four different inlet temperatures were evaluated. Maltodextrin was an effective drying adjuvant; the highest its concentration, the lowest the product stickiness on the dryer walls and the best the powder physicochemical properties. Besides, the highest tested inlet air temperature led to the powders with lowest moisture content. Nevertheless, higher thermal levels caused more degradation of the antioxidant components. Tonon *et al.* (2008) studied the influence of the spray drying operating conditions and of the feed maltodextrin concentration on the physicochemical properties of Açai dried extracts. The authors reported that increases in the inlet air temperature decreased the powder moisture content, increased the process yield, powder hygroscopicity and particle size and led to the loss of the antioxidant components. On the other hand, the use of maltodextrin in high concentrations decreased the process yield and powder hygroscopicity and increased the particle size probably due to an increase in the feed viscosity. Besides, high liquid flowrates were found to decrease the process yield and powder hygroscopicity and to increase the powder moisture content.

From the above-mentioned studies, it can be concluded that SiO_2 is a widely used adjuvant to process herbal medicine extracts by spray drying. Generally, the use of high SiO_2:SR ratios leads to powders with good flow properties, low moisture content and hygroscopicity and high process yields. In addition, it seems that the presence of SiO_2 does not affect the chemical composition of dried extracts. Moreover, for high inlet air temperatures, the SiO_2 appears to be a protective agent against the degradation of active components. The manipulation of the spray-drying operating variables allows improving powders properties as moisture content, flowability and hygroscopicity as well as the process yield.

Summarizing, spray drying becomes an adequate technique to produce dried herbal medicines with good rheological properties to be used in the production of phytomedicine tablets by DC. Furthermore, the proper selection of the process operating conditions and type and concentration of the drying adjuvant constitutes a major challenge to both formulators and manufacturers in order to efficiently produce high quality dry herbal extracts.

4. Spray Drying and Herbal Medicine Modified-Release Oral Dosage Forms

The design of pharmaceutical dosage forms has experienced an important evolution in the last decades, being modified-release dosage forms extensively investigated. Modified release drug delivery systems are widely studied to improve drug efficiency and patient adherence to treatments as well as to reduce drug toxicity. These systems are particularly useful in the treatment of chronic pathologies because they provide an effective drug blood level. Typically and as it is shown in Figure 10, modified release drug delivery systems provide the drug gradually to maintain the therapeutic effect over a predetermined period (Ansel, 2009). There are diverse modified-release dosage forms, including oral, ocular, parenteral, subdermal and vaginal products. The oral route of administration is the most used for both conventional and modified-release systems. For oral dosage forms, modified drug action is achieved by affecting the drug release rate and/or by slowing the transit time of the dosage form through the gastrointestinal tract (Banker and Rhodes, 2002). Table 3 shows the different orally modified-release systems described by Aulton (2004). Among others drug transport mechanisms, diffusion is a common drug delivery phenomenon. Unlike reservoirs and osmotic pumps, the drug release rate from matrix systems is not affected by coating defects. This benefit, together with the simple and low cost production technique, the high level of reproducibility, ease of scale-up and process validation make matrices the most popular oral modified-release system (Varma *et al.*, 2004).

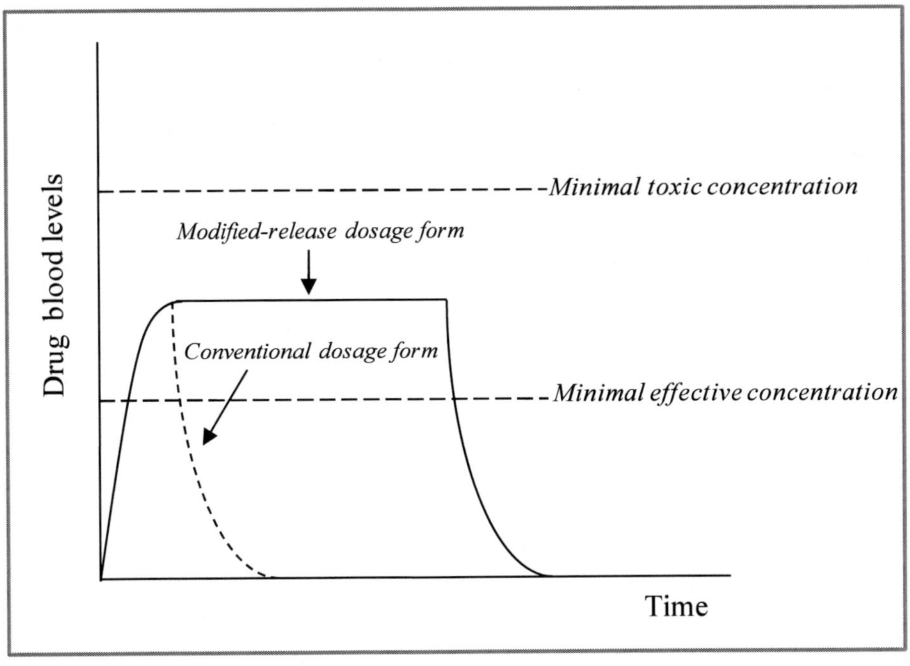

Figure 10. Typical drug blood level profiles for conventional and modified-release forms.

Table 3. Modified release systems (Aulton, 2004)

Modified release system	System description	Variables that modulate the release rate
Reservoir	Drug reservoir (tablet or granules) coated with a membrane	Coating thickness and type
Matrix	*Lipidic*: drug and lipid *Insoluble*: drug and insoluble polymer *Hydrophilic*: drug and hydrophilic polymer	Polymers and lipids properties (type, concentration, viscosity, etc.)
Osmotic pump	Core tablet (drug and polymeric osmotic agent) coated by a semi-permeable membrane with a delivery orifice.	Membrane thickness and composition and/or orifice diameter

As reported in Table 3, different matrix systems are available. Lipidic matrices are constituted by a mixture of the drug with a hydrophobic substance, which is then compacted. The hydrophobic material (e.g. wax) remains intact during the drug release. The release is subjected to the addition of a channelling agent that allows the formation of matrix pores, through which body fluids enter and dissolve the drug. Nowadays, these matrices are out of use (Aulton, 2004). Insoluble matrices are comprised by the drug and a polymer insoluble in gastrointestinal fluids. The release rate depends on the drug diffusion through the capillaries formed between the compacted polymer particles. These matrices also remain intact during gastrointestinal transit (Ansel, 2009). On the other hand, hydrophilic matrices, more known as swellable soluble matrices, are obtained by compression of the drug and a water-swellable hydrophilic polymer. These systems undergo swelling, followed by erosion of the formed gel layer and dissolution of the hydrophilic components in the body fluids; thus, there are not matrix ghosts (Aulton, 2004). Besides this advantage, hydrophilic matrices are suitable to deliver high molecular weight drugs and have a good in *vitro-in vivo* correlation (Maderuelo *et al.*, 2011).

Modified release concepts applied to plant medicinal extracts are still in an exploratory stage (Saraf, 2010). Since several dried extracts are used in the treatment of chronic pathologies (Bent and Ko, 2004; Bustamante Delgado *et al.*, 2009), hydrophilic matrix systems based on medicinal plant extracts offer an appropriate therapeutic strategy. As it was afore-mentioned, dried extracts have poor flow properties for DC and are hygroscopic powders. For this reason, the plant fluid extracts are usually co-processed with different drying adjuvants by means of suitable drying technologies. To produce hydrophilic matrices based on herbal medicine extracts, the co-processed materials (plant extract and drying excipients) need to be mixed with hydrophilic polymers. This production method would require two processing steps. Instead, the development of a spray dried co-processed powder constituted by the medicinal plant extract, a drying adjuvant and the hydrophilic polymer can be an overwhelming approach. Though, the obtained powder should have good flow properties for DC and be useful to modify the extract release rate.

The drug release from hydrophilic matrices constitutes a complex phenomenon because of the interaction between different physical events. When a hydrophilic matrix is exposed to an aqueous medium, the up taken water decreases the polymer glass transition temperature (T_g) to values close to 37 °C. This phenomenon causes a physical transformation in the polymer from its glassy into its rubbery state. Macroscopically and as it can be seen in Figure 11, this behavior is verified by the formation of a gel layer on the surface of the matrix, which is responsible of the drug release rate control (Colombo *et al.*, 2000). The gel layer acts as a diffusional barrier that retards the tablet water uptake and the release of the dissolved drug. Firstly, and as it was above-mentioned water enters in the matrix and the gel layer is formed. The thickness of this layer increases as more and more water access into the system. Simultaneously and at the matrix surface, the earlier hydrated polymer chains relax allowing the drug dissolution until the erosion of the system begins (Colombo et al., 1999). As a result, different fronts can be observed during this process. Figure 12 shows the swelling front (polymer glassy - rubbery transition boundary), the diffusion front (solid drug - dissolved drug solution boundary) and the erosion front (swollen matrix - solvent boundary). The erosion front moves inwards when the matrix dissolves and outwards when the matrix swells. The swelling front moves inwards after water penetration. The diffusion front moves inwards and near to the swelling front when the matrix contains soluble drugs because of the drug dissolution. Conversely, the diffusion front moves outwards when insoluble drugs are present, as non-dissolved drug particles can be translocated by the polymer swelling (Colombo *et al.*, 2000).

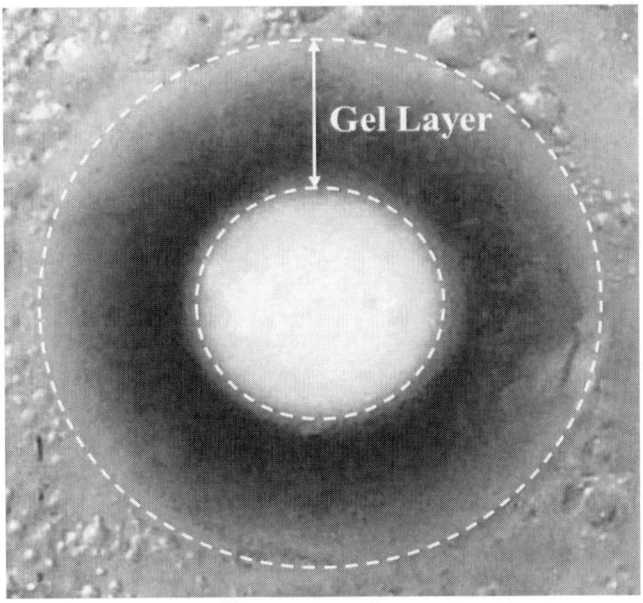

Figure 11. Hydrophilic matrix system in water. Gel layer formation.

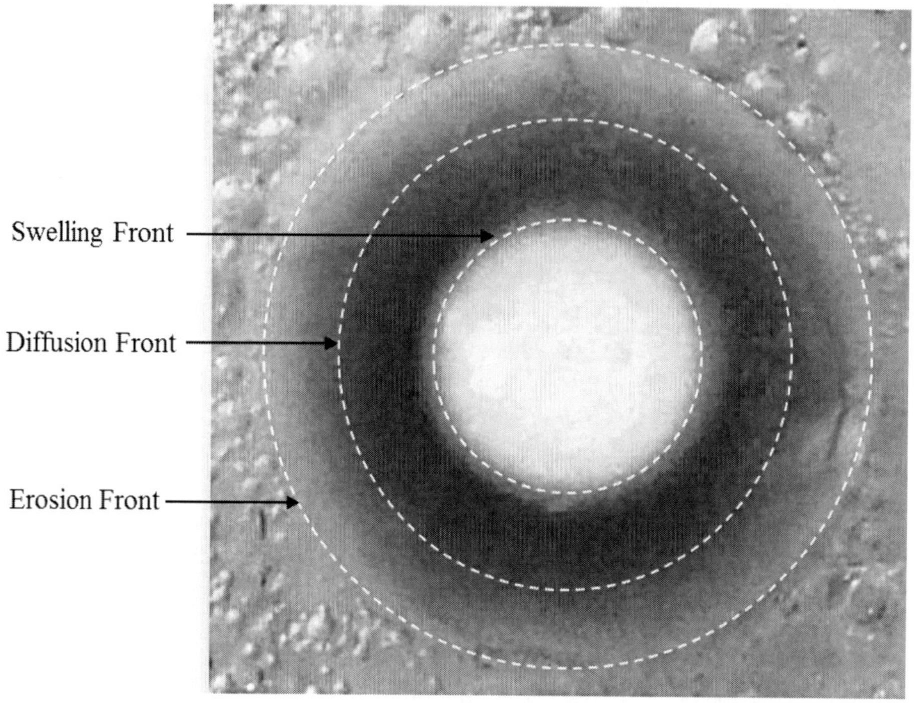

Figure 12. Hydrophilic matrix system in water. Swelling, diffusion and erosion fronts.

The thickness of the gel layer, drug dose and polymer concentration and viscosity affect the drug release rate. In hydrophilic matrixes, the drug diffusion through the gel layer and the polymer relaxation/erosion are the main drug transport mechanisms (Varma *et al.*, 2004). Due to the high complexity of the physical and chemical phenomena involved in the drug release from the matrices, different kinetics models were used to analyze the drug dissolution profiles (Sandoval *et al.*, 2008).

Fick's first and second laws describe diffusion of a solute across a particular medium and are the basis for several of the semi-empirical models proposed to describe the drug release from matrix systems. One of them is given by the Higuchi equation (Eq. 1), which states that the amount of drug released is proportional to the square root of time (i.e., a typical mathematical function found for systems where the drug release rate is controlled by pure diffusion):

$$\frac{M_t}{M_\infty} = kt^{1/2} \tag{1}$$

where M_t/M_∞ is the drug released fraction at time t and k is a kinetic constant that has to be fitted for each drug/polymer system. Since swellable systems can suffer erosion that enhances the drug release, the Higuchi model considering pure drug diffusion is not suitable for this type of matrices. Besides, the contribution of polymer chains relaxation has to be considered. Therefore, a model capable of taking into account all the probable drug transport mechanisms is preferred. The simple semi-empirical equation proposed by Korsmeyer *et al.* (1983) (Eq. 2), is frequently used for the analysis of release data from swellable matrices.

$$\frac{M_t}{M_\infty} = kt^n \tag{2}$$

The drug release is represented by a time power law, where M_t/M_∞ is the fraction of drug released at time t, k is the kinetic constant which accounts for the structural and geometrical properties of the matrix, and n is the diffusional exponent indicative of the drug release mechanism. This exponent characterizes the drug release mechanism, which is also dependent on the shape of the matrix as described in Table 4. The Fickian diffusion implies that the release mechanism is Case-I, i.e. diffusion is the process that controls the drug release. The non-Fickian or anomalous transport represents the case for which drug diffusion and polymer relaxation-erosion contribute to the overall release, and it can be represented by first-order kinetics. In contrast, when the drug release is controlled by relaxation-erosion of the polymer, the extract transport is characterized by zero-order kinetics (Case-II) (Peppas and Shalin, 1989).

Other drug release model which considers both the drug diffusion and polymer relaxation-erosion mechanisms was proposed by Peppas and Sahlin (1989) for pharmaceutical matrices (Eq. 3):

$$\frac{M_t}{M_\infty} = k_d t^m + k_r t^{2m} \tag{3}$$

where k_d and k_r are the diffusional and relaxational constants, respectively and m is the diffusional exponent for a matrix of any geometrical shape. According to Peppas and Sahlin (1989) and aiming to identify the main drug transport mechanism, Equation 4 can be used to approximately calculate the drug release ratio between relaxational (R) and diffusional (F) contributions:

$$\frac{R}{F} = \frac{k_r}{k_d} t^m \tag{4}$$

Table 4. Diffusion exponent and drug release mechanism

	Exponent, n		Drug release mechanism
Thin film	Cylinder	Sphere	
0.5	0.45	0.43	Fickian diffusion/Case-I transport
0.5 < n < 1.0	0.45 < n < 0.89	0.43 < n < .85	Anomalous transport
1.0	0.89	0.85	Case-II transport

All the above-presented equations (Eq. 1, 2, 3 and 4) are semi-empirical models that can be useful for the understanding and comparison of different drug release profiles. However, special attention is required for results interpretation, bearing in mind the model assumptions. The use of these equations all together may help in the identification of the drug release mechanisms from hydrophilic matrix systems (Ferrero *et al.*, 2000; Miranda *et al.*, 2006; Escudero *et al.*, 2008; Escudero *et al.*, 2010).

To the best of our knowledge, matrix systems based on spray-dried plant extracts are not extensively studied or reported in the open literature. In particular and as nutraceutical product, Gavini *et al.* (2005) produced a matrix system based on a *Cynara scolymus* (antioxidant agent) spray-dried extract. The fluid extract was spray dried at given operating conditions using lactose and hypromellose as polymers. The reported results show that the microparticles containing lactose quickly release the extract active component. On the other hand, the microspheres based on hypromellose release only 20% of the loaded extract after 8 h. In spite of this, the matrix system prepared by compression of extract-lactose and extract-hypromellose microparticles (1:1 ratio, w/w) showed a slow delivery rate, which lasted 24 h for the complete extract release, being adequate for a one-a-day nutraceutical controlled release matrix formulation.

Specifically related to matrix systems based on spray-dried plant extracts, Gallo *et al.* (2011b and c) developed a dried extract with good flow properties for DC and useful as a modified release matrix system by spray drying a *Hamamelis virginiana* aqueous extract with sodium carboximetilcellulose medium viscosity (Na-CMC MV) and SiO_2. The selection of this polymer was based on its wide use in oral pharmaceutical formulations, its gel forming capacity (Rowe *et al*, 2009) and a detailed polymers screening procedure previously performed.

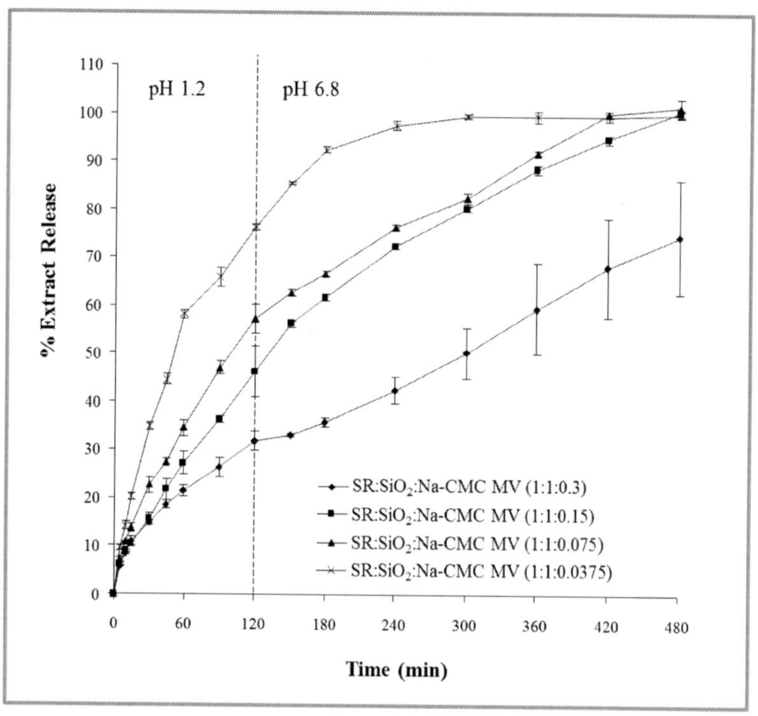

Figure 13. Extract release from matrix systems with different Na-CMC MV concentrations.

The drying adjuvant was added to improve the powder flowability and to reduce the adhesion of the product onto the spray dryer chamber wall (Albertini *et al.*, 2004). *H. virginiana* was selected as a model plant extract due to its widespread use for chronic varicose vein disease treatments (WHO, 2002). The selected spray-drying operating conditions were: drying air inlet temperature 150 ºC, atomization air volumetric flowrate 400 l/h, feed volumetric flowrate 10 % (2 ml/min) and drying air volumetric flowrate 100 % (about 35-38 m^3/h). The feed solids composition as SR:SiO_2:Na-CMC MV was 1:1:X, the polymer content was varied from X=0.0375 to X=0.3. The influence of the feed Na-CMC MV concentration on the powder flow properties, moisture content, process yield and extract release from compacted powders was evaluated. The results showed that all the obtained powders exhibited good flow properties with low moisture content (lower than 2.80 %) at high process yields (around 80%). Figure 13 shows the extract release from the matrix systems with different Na-CMC MV concentrations in simulated gastric fluid (SGF) for the first 2 hours and in simulated intestinal fluid (SIF) for the remaining 6 hours (in order to simulate physiological conditions). The 1:1:0.3 SR:SiO_2:Na-CMC MV matrix did not allowed the complete extract release within the studied time. On the other hand, the 1:1:0.0375 SR:SiO_2:Na-CMC MV system rapidly release the extract, being the polymer concentration insufficient to modify the extract release rate. Both matrices, the 1:1:0.075 and 1:1:0.15 SR:SiO_2:Na-CMC MV showed an adequate drug release profile. Owing to its lower Na-CMC MV content, the 1:1:0.075 SR:SiO_2:Na-CMC MV powder was selected as the preferred material for further characterization and extract release mechanism investigation. Based on the obtained experimental data and the models previously described, it was found that the extract transport was controlled by both the diffusion and erosion mechanisms.

Summarizing, spray drying is a promising technology to formulate co-processed modified drug release microparticles, particularly for those based on medicinal dried plant extracts that usually have poor flow properties.

Conclusion

Currently, the great herbal medicine market worldwide reflects its growing use in both developing and developed countries. New analytical technologies are contributing to understand the complex chemical composition of herbal products. In addition, more regulation aspects about herbal medicine are taken into account worldwide. These advances are directed to improve the efficacy, safety and quality of herbal medicines. In this context, the spray drying technology constitutes the most employed process to transform herbal raw material into high quality dried extracts to be used in pharmaceutical solid dosage forms. The optimal selection of the process operating conditions and type and proportion of the drying adjuvant are critical factors to obtain products with improved properties. Besides, the spray drying technology allows the production of microparticles useful in the design of herbal modified release systems, products that constitute a new challenge in the phytomedicines research area.

References

Albertini, B.; Passerini, N.; González-Rodríguez, M.L.; Perissutti, B.; Rodriguez, L. Effect of Aerosil® on the properties of lipid controlled release microparticles. *Journal of Control Release* 2004, *100*, 233-246.

Aulton, M.E. Farmacia. La ciencia del diseño de las formas farmacéuticas. Second Edition. España: Elsevier; 2004.

Ansel, H.C.; Popovich, N.G.; Allen, L.V. Pharmaceutical Dosage Forms and Drug Delivery Systems. Ninth Edition. Philadelphia: Lippincott Williams and Wilkins; 2009.

Banker, G.S.; Rhodes, C.T. Modern Pharmaceutics. Fourth Edition. New York: Marcel Dekker; 2002.

Barnes, J.; Anderson, L.A.; Phillipson, J.D. Herbal Medicines. Third Edition. London: Pharmaceutical Press; 2007.

Bent, S.M.D.; Ko, R. Commonly Used Herbal Medicines in the United States: A Review. *The American Journal of Medicine* 2004, *116*, 478-485.

Bent, S. Herbal Medicine in the United States: Review of Efficacy, Safety, and Regulation. *Journal of General Internal Medicine* 2008, *23*, 854-859.

Bustamante Delgado, S.; Torres Castro, R.; Morales Segura, M.A. Fitofármacos en el tratamiento de la insuficiencia venosa crónica. *Revista de Fitoterapia* 2009, *9*, 35-51.

Calixto, J.B. Efficacy, safety, quality control, marketing and regulatory guidelines for herbal medicines (phytotherapeutic agents). *Brazilian Journal of Medical and Biological Research* 2000, *33*, 179-189.

Colombo, P.; Bettini, R.; Peppas, N.A. Observation of swelling process and diffusion front position during swelling in hydroxypropyl methyl cellulose (HPMC) matrices containing a soluble drug. *Journal of Control Release* 1999, *61*, 83-91.

Colombo, P., Bettini, R.; Santi, P.; Peppas, N.A. Swellable matrices for controlled drug delivery: gel-layer behaviour, mechanisms and optimal performance. *PSTT* 2000, *3*, 198-204.

Da Silva, F.A. Avaliação tecnológica e atividade antioxidante de productos secos por spray-drying de *Ilex paraguariensis* A. St. Hill. - Aquifoliaceae (erva-mate). Porto Alegre, 243p. Tese de doutorado, Programa de Pós-graduação em Ciências Farmacêuticas, 2007.

De Souza, T.P.; Bassani, V.L.; González Ortega, G.; Dalla Costa, T.C.T.; Petrovick, P.R. Influence of adjuvants on the dissolution profile of tablets containing high doses of spray-dried extract of *Maytenus ilicifolia*. *Pharmazie* 2001, *56*, 730-733.

Efferth, T. Perspectives for Globalized Natural Medicines. *Chinese Journal of Natural Medicines* 2011, *1*, 0001-0006.

Ernst, E. The efficacy of herbal medicine-an overview. *Fundamental and Clinical Pharmacology* 2005, *19*, 405-409.

Escudero, J.J; Ferrero, C.; Jiménez-Castellanos, M.R. Compaction properties, drug release kinetics and fronts movement studies from matrices combining mixtures of swellable and inert polymers: Effect of HPMC of different viscosity grades. *International Journal of Pharmaceutics* 2008, *351*, 61-73.

Escudero, J.J.; Ferrero, C.; Jiménez-Castellanos, M.R. Compaction properties, drug release kinetics and fronts movement studies from matrices combining mixtures of swellable and inert polymers. II. Effect of HPMC with different degrees of methoxy/hydroxypropyl substitution. *International Journal of Pharmaceutics* 2010, *387*, 56-64.

Ferrero, C.; Muñoz-Ruiz, A.; Jiménez-Castellanos, M.R. Fronts movement as a useful tool for hydrophilic matrix release mechanism elucidation. *International Journal of Pharmaceutics* 2000, *1202*, 21-28.

Gallo, L.; Llabot, J.; Allemandi, D.; Bucalá, V.; Piña, J. Influence of spray-drying operating conditions on *Rhamnuspurshiana* (Cáscara sagrada) extract powder physical properties, *Powder Technology* 2011a, *208*, 205-214.

Gallo, L.; Piña, J.; Bucalá, V.; Allemandi, D.; Ramirez Rigo, M. Use of carboxymethylcellulose as release modifier of a medicinal plant extract. IX Simposio Argentino de Polímeros. 2011. Bahía Blanca. Ediuns, 2011b. CD. 27-30.

Gallo, L.; Piña, J.; Palma, S., Bucalá, V.; Ramírez Rigo M. Desarrollo de sistemas matriciales de liberación modificada de *Hamamelis virginiana* constituidos por un polvo co-procesado obtenido por secado por atomización. Exposición y Congreso Internacional de Farmacia y Bioquímica, 2011c.

Gavini, E.; Alamanni, M.C.; Cossu, M.; Giunchedi, P. Tabletted microspheres containing *Cynara scolymus* (var. Spinososardo) extract for the preparation of controlled release nutraceutical matrices. *Journal of Microencapsulation* 2005, *22*, 487-499.

Georgetti, S.A; Casagrande, R.; Fernandes Souza, C.R; Oliveira, W.P.; Vieira Fonseca, M.J. Spray drying of the soybean extract: Effects on chemical properties and antioxidant activity, *LWT-Food Science and Technology* 2008, *41*, 1521-1527.

Global Industry Analysts, Inc. Herbal Supplements and Remedies - A Global Strategic Business Report. 2012 [cited 2012 Aug 7]. Available from: http://www.strategyr.-com/Herbal_ Supplements_and_Remedies_Market_Report.asp.

Jin, P.; Madieh, S.; Augsburger, L.L. Selectec Physical and Chemical Properties of Feverfew (*Tanacetum parthenium*) Extracts Important for Formulated Product Quality and Performance. *AAPS PharmSciTech* 2007, *9*, 22-30.

Jordan, S.A.; Cunningham, D.G.; Marles, R.J. Assessment of herbal medicinal products: Challenges, and opportunities to increase the knowledge base for safety assessment. *Toxicology and Applied Pharmacology* 2010, *243*, 198-216.

Korsmeyer, R.W.; Gurny, R.; Doelker, E.; Buri, P.; Peppas, N.A. Mechanisms of solute release from porous hydrophilic polymers. *International Journal of Pharmaceutics*1983, *15*, 25-35.

León-Martínez, F.M.; Méndez-Lagunas, L.L.; Rodríguez-Ramírez, J. Spray drying of nopal mucilage (Opuntia ficus-indica): Effects on powder properties and characterization. *Carbohydrate Polymers* 2010, *81*, 864-870.

Liang, Y.Z.; Xie, P.; Chan, K. Quality control of herbal medicines. *Journal of Chromatography B* 2004, *812*, 53-70.

Maderuelo, C.; Zarzuelo, A.; Lanao, J.M. Critical factors in the release of drugs from sustained release hydrophilic matrices. *Journal of Control Release* 2011, *154*, 2-19.

Masters, K. Spray Drying Handbook. Third Edition. London: George Godwin Limited; 1979.

Miranda, A.; Millán, M.; Caraballo I. Study of the critical points of HPMC hydrophilic matrices for controlled drug delivery. *International Journal of Pharmaceutics* 2006, *311*, 75-81.

Mujumdar, A.S. Handbook of Industrial Drying. Third Edition. Singapore: CRC Press; 2006.

Nandiyanto, A.B.D.; Okuyama K. Progress in developing spray-drying methods for the production of controlled morphology particles: From the nanometer to submicrometer size ranges. *Advanced Powder Technolog*y 2011, *22*, 1-19.

Newall, C.A.; Anderson, L.A.; Phillipson, J.D., Herbal Medicines: A Guide for Health-Care Professionals. First Edition. London: The Pharmaceutical Press; 1996.

Oliveira, O.W.; Ros Petrovick, P. Secagem por aspersao (spray drying) de extractos vegetais: bases e aplicaçoes. *Revista Brasilera de Farmacognosia* 2010, *20*, 641-650.

Pandey, M.; Debnath, M.; Gupta, S.; Chikara, S.K. Phytomedicine: An ancient approach turning into future potential source of therapeutics. *Journal of Pharmacognosy and Phytotherapy* 2011, *3*, 27-37.

Peppas, N.A.; Sahlin, J.J. A simple equation for the description of solute release. III. Coupling of diffusion and relaxation. *International Journal of Pharmaceutics* 1989, *57*, 169-172.

Petrovick G.F.; Ros Petrovick, P.; Bassani, V.L. *Achyrocline satureioides* (Lam.) DC., Asteraceae: development of granules from spray dried powder. *Brazilian Journal of Pharmacognosy* 2010, *20*, 803-811.

Quek, S.Y.; Chok, N.K.; Swedlund, P. The physicochemical properties of spray-dried watermelon powders. *Chemical Engineering. Process* 2007, *46*, 386-392.

Ré M.-I. Formulating Drug Delivery Systems by Spray Drying. *Drying Technology* 2006, *24*, 433-446.

Robinson, M.M.; Zhang, X. The world medicine situation 2011 (Traditional medicine: global situation, issues, and challenges). Third Edition. Geneva: World Health Organization; 2011.

Rocksloh, K.; Rapp, F.R.; Abu Abed, S.; Müller, W.; Reher, M.; Gauglitz, G.; Schmidt, P.C. Optimization of crushing strength and disintegration time of a high-dose plant extract tablet by neural networks. *Drug Development and Industrial Pharmacy* 1999, *25*, 1015-1025.

Rowe, R.C.; Sheskey, P.J.; Quinn, M.E. Handbook of Pharmaceutical Excipients. Sixth Edition. London: Pharmaceutical Press; 2009.

Sahoo, N.; Manchikanti, P.; Dey, S. Herbal drugs: Standards and regulation. *Fitoterapia* 2010, *81*, 462-471.

Sandoval, P.A.; Baena, Y.; Aragón, M.; Rosas, J.E.; Ponce D'León, L.F. Mecanismos generales de cesión de principios activos a partir de matrices monolíticas hidrofílicas preparadas con éteres de celulosa. *Revista Colombiana de Ciencias Químicas Farmacéuticas* 2008, *37*, 105-121.

Saraf, A. Applications of novel drug delivery system for herbal formulations. *Fitoterapia* 2010, *81*, 680-689.

Sarraguça, M.C.; Cruza, A.V.; Soares, S.O.; Amaral, H.R.; Costa, P.C.; Lopes, J.A. Determination of flow properties of pharmaceutical powders by near infrared spectroscopy. *Journal of Pharmaceutical and Biomedical Analysis* 2010, *52*, 484-492.

Sen, S.; Chakraborty, R.; De, B. Challenges and opportunities in the advancement of herbal medicine: India's position and role in a global context. *Journal of Herbal Medicine* 2011, *1*, 67-75.

Soares, L.A.L.; González Ortega, G.; Ros Petrovick, P.; Schmidt, P.C. Dry granulation and compression of spray-dried plant extracts. *AAPS PharmSciTech* 2005, *6*, 359-366.

Souza, C.R.F.; Oliveira, W. P. Powder Properties and System Behavior during Spray Drying of *Bauhinia forficata* Link Extract. *Drying Technology* 2006, *24*, 735-749.

Souza, C.R.F.; Schiavetto, I.A.; Thomazini, F.C.F.; Oliveira, W.P. Processing of *Rosmarinus officinalis* Linne extract on spray, and spouted bed dryers. *Brazilian Journal of Chemical Engineering* 2008, *25*, 59-69.

Souza, T.P.; Martínez-Pacheco, R.; Gómez-Amoza, J.L.; Ros Petrovick, P. Eudragit E as excipient for production of granules and tablets from *Phyllanthus niruri* L spray dried extract, *AAPS PharmSciTech* 2007, *8*, E1–E7.

Swarbrick, J.; Boylan, JC. Encyclopedia of Pharmaceutical Technology. Second Edition. New York: Marcel Dekker; 2007.

Tajber, L.; Corrigan, O.I.; Healy, A.M. Spray drying of budesonide, formoterol fumarate and their composites - II. Statistical factorial design and in vitro deposition properties. *International Journal Pharmaceutics* 2009, *367*, 86-96.

Tong, H.H.; Wong, S.Y.; Law, M.W.; Chu, K.K.; Chow, A.H. Anti-hygroscopic effect of dextrans in herbal formulations. *International Journal of Pharmaceutics* 2008, *363*, 99-105.

Tonon, R.V.; Brabet, C.; Hubinger, M.D. Influence of process conditions on the physicochemical properties of açai (*Euterpe oleraceae* Mart.) powder produced by spray drying, *Journal of Food Engineering* 2008, *88*, 411-418.

Training Papers Spray Drying BÜCHI Labortechnik AG, 2002.

Varma, M.V.S.; Kaushal, A.M.; Garg, A.; Garg S. Factors Affecting Mechanism and Kinetics of Drug Release from Matrix-Based Oral Controlled Drug Delivery Systems. *American Journal of Drug Delivery* 2004, *2*, 43-57.

Vasconcelos, E.A.F.; Medeiros, M.G.F.; Raffin, F.N.; Moura, T.F.A.L. Influência da temperatura de secagem e da concentração de Aerosil® 200 nas características dos extratos secos por aspersão da *Schinus terebinthifolius* Raddi (Anacardiaceae), *Brazilian Journal of Pharmacognosy* 2005, *15*, 243-249.

Vehring, R.; Foss W.R.; Lechuga-Ballesteros D. Particle formation in spray drying. *Aerosol Science* 2007, *38*, 728-746.

WHO, Monographs on Selected Medicinal Plants. Geneva: World Health Organization 2, 2002.

Wu, C.Y.; Seville, J.P.K.A comparative study of compaction properties of binary and bilayer tablets. *Powder Technology* 2009,*189*, 285-294.

Zétola, M.; De Lima, T.C.M.; Sonaglio, D.; González-Ortega, G.; Limberger, R.P.; Petrovick, P.R.; Bassani, V.L. CNS activities of liquid and spray-dried extracts from *Lippia alba* Verbenaceae (Brazilian false melissa). *Journal of Ethnopharmacology* 2002, *82*, 207-215.

Zhang, J.; Wider, B.; Shang, H.; Li, X.; Ernst, E. Quality of herbal medicines: Challenges and solutions. *Complementary Therapies in Medicine* 2012, *20*, 100-106.

In: Plant Extracts
Editors: Adriana Giordano and Alfieri Costs
ISBN: 978-1-62417-534-3

Chapter 7

Phenolic Compounds Extracted from Medicinal Plants: An Alternative for Cancer Prevention and Treatment

Marily González-Castillo[1], Raúl Rodríguez-Herrera[1], Yesenia Silva-Belmares[1], Cristóbal N. Aguilar-González[1], Ana María Rivas-Estilla[2] and Jesús A. Morlett-Chávez[3,*]

[1]Department of Food Research, School of Chemistry, Universidad Autónoma de Coahuila

[2]Molecular Infectious Disease Lab., Department of Biochemistry and Molecular Medicine, Universidad Autónoma de Nuevo León

[3]Laboratory of Clinical Analysis and Molecular Diagnostics, School of Chemistry, Universidad Autónoma de Coahuila, Saltillo, Coah., México

Abstract

Cancer is a chronic degenerative disease that involves multiple environmental, chemical, physical and metabolic factors as well as genetic predisposition to this condition. It is the principal cause of mortality in the world and worldwide, 7.6 million of deaths in 2008 were attributed to cancer. However, recent epidemiological evidence indicates that a diet based on fruits and vegetables rich in antioxidants significantly reduces risk of developing several types of cancer. This suggests that some antioxidants such as polyphenols, which are found in different foods, may be effective for cancer prevention and treatment. There are reports that this type of phytochemicals are involved in prevention mechanisms against development of tumors through inhibition of genotoxic effects, increasing anti-inflammatory activity, inhibition of proteases and cell proliferation, and protection of transduction pathway signals that modulate apoptosis process. In addition, some studies claim that many natural chemo-preventive agents induce cell cycle arrest or apoptosis by activating tumor suppressors such as p53 and its target genes. To elucidate the mechanisms involved in the action mode of these phyto-chemicals, extracts

[*] Corresponding author: morlett17@gmail.com; morlett17@yahoo.com.mx. Ph. +52 (844) 4-15-95-34, 4-15-57-52.

of some plants like strawberry, raspberry, blueberry, and pomegranate had been evaluated. Rich-polyphenol extracts were tested on different cancer cells including, MCF-7 (breast cancer), HT-29, HCT-116 (colon cancer), LNCaP and DU-145 (prostate cancer) cell lines causing tumor growth inhibition. Similarly, it has been demonstrated that pecan shell extracts reduce growth of kidney cancer cells (A-498, and 769-P). Moreover, recent studies using semi-desert plants (*Turnera diffusa* and *Origanum vulgare*) extracts reported high toxicity against leukemia cells (K562) and different cervical cancer cell lines (HeLa and SiHa), increasing levels of cell apoptosis. However, it is necessary to know the mechanism of action of these compounds against tumor cells and how affect their functionality. Hence, this chapter is intended to inform about cytotoxic effect of phenolic extracts obtained from plants and fruits used in traditional medicine, as well as the mechanisms involved in such activity, in addition, information about how natural compounds generally decrease the adverse effects caused by current therapies in treating cancer is presented too.

Introduction

Use of natural sources of medicinal compounds has increased by the necessity of find new treatments for chronic and degenerative diseases control such as cancer. Herbal or botanical medicines have been used traditionally by ancient civilizations as China and Egypt for control of several disease conditions [1,2]. However, in recent years *in vitro* and *in vivo* studies have demonstrated the therapeutic potential of a large number of plants extracts [3, 4, 5]. In this type of extracts can be found some compounds such as polyphenols, which are the most important phytochemicals for their biological properties. Polyphenols are the most abundant antioxidants in human diet as are present in plant foods including fruits, vegetables and cereal as well as in beverages as juices, wine, coffee and tea [6]. Despite of phenolic compounds abundance, their applications in clinical studies have not been possible, because the diversity and complexity of their chemical structures and of their behavior in the organism [7].

An important group of phenolic compounds present in vegetables are denominated tannins, which are secondary plant metabolites, and can be found in the condensed or hydrolyzed form. These molecules are of great nutritional and medicinal interest due to their biological properties [8]. One of these properties is their antioxidant capacity [9,10], from where are derived their anti-cancer and anti-mutagenic effect [11]. The antioxidant compounds give cellular protection against free radicals, which that are produced as result of several physiological and biochemical processes in the human body. Overproduction of free radicals may cause oxidative damage to biomolecules, triggering many degenerative diseases, such as diabetes, atherosclerosis and cancer [1].

Cancer is a systemic disease characterized by the uncontrolled multiplication of cells and metastases [12]. This disease is a major health problem in the USA and other developed countries besides being the second leading cause of death with 565,650 projected cancer deaths and 1,437,180 new cases of cancer in 2008 [13]. According this information is necessary to implement new alternatives for prevention and control of different types of cancer also known the action mechanisms of compounds used in the treatment of this condition.

Phenolic Compound Characteristics

Phenolic compounds are metabolites of high molecular weight and are widely distributed in the vegetal kingdom including fruits and vegetables [14]. Mainly involved in protection, structure, photosynthesis and plant nutrients absorption [15]. Chemically are formed by one aromatic ring with one or more hydroxyl groups [6]. As polyphenols are called to approximately 8,000 natural compounds, which are divided into several classes according to the number of phenol rings that they contain and to the structural elements that bind these rings to one another [16]. The main groups of polyphenols are: flavonoids, phenolic acids, phenolic alcohols, stilbenes and lignans.

The most abundant polyphenols are the flavonoids, which are low-molecular-weight compounds consisting of 15 carbon atoms, arranged in C6-C3-C6 configuration. Essentially, their structure consists of 2 aromatic rings joined by a 3-carbon bond, usually in the form of a heterocyclic ring [17]. Flavonoids are themselves divided into six subgroups: flavones, flavonols, flavanols, flavanones, isoflavones, and anthocyanins, according to the oxidation state of the central ring [18]. Phenolic acids could be divided in two classes: benzoic and cinnamic acid derivatives denominated hydroxybenzoic and hydroxycinnamic acids respectively. The hydroxybenzoic acids, such as gallic, phydroxybenzoic, protocatechuic, vanillic and syringic acids, are found in very few plants eaten by humans, such as red fruits (raspberryes and blackberries), tea and olive oil per example [16]. Hydroxycinnamic acids consist chiefly of coumaric, caffeic and ferulic acids which are rarely found in the free form. Caffeic acid is the most abundant phenolic acid in many fruits and vegetables, most often esterified with quinic acid as in chlorogenic acid, which is the major phenolic compound in coffee [19]. Another common phenolic acid is ferulic acid, which is present in cereals and is esterified to hemicelluloses in cell walls [20]. Phenolic alcohols are found at lower concentrations in food. Tyrosol and hydroxytyrosol are the main phenolic alcohols reported; these compounds contributing to the stability of those oils contained in the sample, per example olive oil [21]. Tyrosol and hydroxytyrosol are present in white wines, while hydroxytyrosol is additionally produced *in vivo* after wine ingestion [22]. Stilbenes can be found in oligomeric and polymeric forms in wine, and the main representative is resveratrol, and its derivatives which exhibit a wide range of pharmacological and biological properties [23]. Lignans are produced by oxidative dimerization of two phenylpropane units; they are present in nature mainly as free form, while their glycoside derivatives are only a minor form [24].

Tannin is a term commonly used to refer to plant polyphenols, these are divided into 2 chemically and biologically distinctive groups: condensed tannins or proanthocyanidins (found in tea, grapes, cranberries, nuts and so on) and hydrolysable tannins or ETs (found in raspberries, strawberries, pomegranates and so on) [25]. Hydrolysable tannins are compounds containing a central core of glucose or another polyol esterified with gallic acid, also called gallotannins, or with hexahydroxydiphenic acid, in this case called ellagitannins [26]. Phenols are associated with color, sensory (flavor, astringency and hardness), nutritional characteristics and antioxidants properties of vegetal foods. These compounds have gained popularity for their numerous biological activities and health benefits [27]. Antioxidant capacity is the most important property of these compounds; this property is due to reactivity of the phenol group present in their structure [28].

Antioxidant Capacity of Polyphenols

Antioxidants are compounds that can delay, inhibit, or prevent oxidation of oxidizable materials by scavenging free radicals and diminishing oxidative stress. The oxidizable substrate may be any molecule that is found in foods or biological materials, including carbohydrates, DNA, lipids, and proteins. Also, is commonly used a classification which categorizes antioxidants into primary, oxygen scavenging, and secondary, enzymatic and chelating/sequestering antioxidants [29].

The antioxidant activity of polyphenols should easily to reduce production of free radicals, for the inhibition of enzyme involves in these processes like dismutase, superoxide, catalase, reductase and peroxidase [22] or by chelation of transition metals which are responsible for the generation of free radicals [30]. All polyphenols have ability to sequester simple oxygen and free radicals O_2, OH, NO, through electrons donation generating a relatively stable phenoxyl radical. Particularly, flavonoids are able to reduce reactive oxygen species (ROS) highly oxidized due to their low redox potential [31]. Phenolic compounds have been considered powerful antioxidants *in vitro* and proved to be more potent antioxidants than C and E Vitamins and carotenoids [32]. Numerous studies have reported the relationship between antioxidant activity and total phenolic content using many plants extracts and different analytical techniques.

The most common and reliable analytical methods for determination of antioxidant capacity are 2, 2-azinobis (3-ethylbenzothiazoline-6-sulfonic) acid radical (ABTS. +) and 1,1-diphenyl-2-picrylhydrazyl radical (DPPH) which involve estimation of disappearance of free radicals using a spectrophotometer, [1]. Generation of more effective ABTS [2, 29-azinobis-(3-ethylbenzothiazoline-6-sulfonic acid)] radical cation involves direct production of blue/green $ABTS^{•+}$ chromophore through reaction between ABTS and potassium persulfate. Addition of antioxidants to the pre-formed radical cation reduces it ABTS, to an extent and on a time-scale depending on antioxidant activity, concentration of antioxidant and reaction time. Thus, the extent of decolorization as percentage inhibition of $ABTS^{•+}$ radical cation is determined as a function of concentration and time and calculated relative to reactivity of Trolox as a standard, under the same conditions [33]. 1,1-Diphenyl-2-pycrylhydrazyl (DPPH) is a stable nitrogen-centered free radical whose color changes from violet to yellow upon reduction by either the process of hydrogen or electron donation. Substances which are able to perform this reaction can be considered as antioxidants and therefore, radical scavengers [34].

Another technique for antioxidant capacity determination is Folin-Ciocalteu assay. Its basic mechanism is an oxidation/reduction reaction, although, it has been used as a measurement of total phenolics content for many years [6]. This measurement is influenced by the chemical nature of the analyte, as well as assay method, selection of standards and presence of interfering substances. Methods used for analysis of phenolics are usually classified as either those measuring total phenolics content, or those quantifying a specific group or class of phenolic compounds. [35].

Hydrolysable tannins can be quantified by a number of approaches including the potassium iodate method [36], rhodanine method [37] and sodium nitrite method [38]. The proanthocyanidin or HCL-Buthanol assay and vanillin assay are the principal methods used for quantification of condensed tannins [6]. However, HPLC technique currently represents the most popular and reliable method for analysis of phenolic compounds. Various supports and mobile phases are available for analysis of phenolics including anthocyanins, proanthocyanidins, hydrolysable tannins, flavonols, flavan-3-ols, flavanones, flavones, and phenolic acids in different plant extracts and food samples [39,40,41]. Through the HPLC technique can be analyzed simultaneously all components of interest together with their possible derivatives or degradation products in a single sample [42].

Application of Phenolic Compounds in Cancer Treatment

The cancer term is applied to several groups of diseases which can be affecting any part of human body. Cancer has been characterized by the fast and uncontrolled multiplication of abnormal cells [43]. When tumor cells go beyond the limits of the initial target organ and spread to other organs, was triggered the metastasis process. Many causes have been associates to cancer such as multiple environmental, chemical, physical and metabolic factors as well as genetic predisposition to contracting this disease [44]. There are several types of cancer. Carcinoma is a cancer that begins in the skin or in tissues that line or cover internal organs. Sarcoma is a cancer that begins in bone, cartilage, fat, muscle, blood vessels, or other connective or supportive tissue. Leukemia is a cancer that starts in blood-forming tissue such as bone marrow, and causes large numbers of abnormal blood cells to be produced, which may enter to the blood. Lymphoma and multiple myeloma are cancers that begin in the cells of the immune system. Central nervous system cancers are cancers that begin in the tissues of the brain and spinal cord [45].

The cancer treatment usually is performed through of clinical therapies in which are included radiation, chemotherapy, immuno-modulation and surgery [46]. However, high rate of mortality and morbidity associated to this disease, suggest that is necessary to implement news alternatives for its control [11]. In the last two decades molecular techniques have been implemented as part of therapies for cancer detection, however, these have not been effective for all types of tumors. This is probably to the complexity degree that implies the damage of several genes controlling cellular proliferation, apoptosis and metastasis processes [47,48].

There is recent epidemiological evidence which indicates that a diet composed of fruits and vegetables with a high content of antioxidants reduces significantly the risk of contract several cancer types. This information suggest that dietetic antioxidants as polyphenols may be effective in prevention and treatment of this disease [49]. In this respect, much research has been oriented to evaluation of extract plant as prophylactic agents, which demonstrated have potential for inhibition of the carcinogenesis process [50]. These types of substances have been used in cancer prevention as well as to stop or return the cellular and molecular process of this condition according to multiplex intervention strategies [51]. Phenolic compounds have been associated with reduction of oxidative stress and consequently classified as an important agent in prevention or treatment of chronic degenerative diseases including coronary heart disease, atherosclerosis, diabetes, rheumatoid arthritis, and cancer [43]. Polyphenol effects on cancer cells have been evaluated especially in cellular lines of different cancer types, also, have been performed laboratory studies on animals such as rats, mice and rabbits because the advantages in management and maintenance of these animals compared to other species. However, have been reported few clinical assays due to the stability problems of this compounds and the secondary effects that can generate in humans.

In Vitro Assays with Phenolic Compounds

The *in vitro* screening has contributed to discoveries and developments of new clinically useful anticancer drugs. There are a large number of fruits and plants that contain different types of phenolic compounds with cyto-toxic, anti-tumoral or anticancer activity, for this reason, has been evaluated on several cellular lines. According with The US National Cancer Institute (NCI) a cytotoxic agent is toxic to tumor cells *in vitro* and if this toxicity transfers through to tumor cells *in vivo*, the agent has anti-tumor activity. The term anticancer is reserved for materials that are toxic to tumor cells in clinical trials on humans [52].

Grape (*Vitis vinifera*) is one of the fruits with the highest concentration of polyphenols. It was found that grape skin and seed extracts induce apoptosis of prostate tumor cells [53,54] while, the residual pulp extract of the wine production process presented an important anti-proliferative effect on cellular line of colon adenocarcinoma (Caco-2)[55, 56]. Also, It has been observed that grape juice and proanthocyanidins extracted from grapes inhibit DNA synthesis and decreased cellular viability and proliferation of breast cancer, respectively [57,58]. The major concentration of antioxidants can be found in pomegranates (*Punica granatum*), which is considered in the botanical dietary supplements to provide a convenient alternative form for consuming bioactive polyphenols [59]. In addition, pomegranate is used in several medical types and is considered safe for consumers [60]. Pomegranate has valuable compounds in different fruit parts, such as peel, seeds, and arils. Another important product obtained from pomegranate fruit is the juice that can be obtained from arils or from the whole fruit [61].

It has been reported some pomegranate effects on several cancer types like breast, prostate, lung, colon and skin cancer [62], where it was demonstrated that whole pomegranate juice is more active than individual and purified compounds [63]. Some specific pomegranate properties for prevention of breast cancer as well as the involved mechanisms have been reported. Antocyanins extracted from pomegranate inhibited prostaglandin E2 (PGE2) synthesis, which decreased the aromatase enzyme responsible for estrogen synthesis which is potentially carcinogenic. On the other hand, ellagic acid induces cellular apoptosis mechanisms and inhibits activation of inflammatory pathways and angiogenesis process [64]. Ellagic acid is present in pomegranate juice and seed oils and is reported acting against skin, pancreas, breast, prostate, colon, intestine, o esophagus, bladder, oral, leukemia, liver and neuroblastoma cancer [65]. In addition, pomegranate juice, seed oil, and peel extracts have been tested on DU-145, LNCaP, and PC-3 cancer cell lines, resulting in decreasing of invasiveness and proliferation by causing cell cycle disruption, inducing apoptosis, and inhibiting prostate cancer growth [66]. HT-29 colon cancer cells have been treated with pomegranate juice, total pomegranate tannins and concentrated pomegranate punicalagin, in this study, a significant decrease in COX-2 expression was inducing. COX-2 intervenes in inflammatory cell signaling process, which may be a cause of cancer initiation and progression [67].

On the other hand, there are in literature reports on some fruits where only therapeutic effect of the extracted juice is mentioned, per example *prickly pear cactus fruit* (*Opuntia spp*).

Authors reported the juice activity of different *prickly pear cactus fruit* types on mama (MCF-7), prostate (PC39), colon (Caco2) and liver (HepG2) line cancer cells, results indicated that viability of colon and prostate cells was the most affected trait especially by *Opuntia violaceae* juice. *O. rastrera* juice was effective against the four cell lines used in this study, presenting the highest antioxidant capacity. *O. robusta* juice of the Gavia variety was the most effective against colon cancer and *O. robusta* and *O. rastrera* juices were the only ones that decreased hepatic cancer cells viability [68]. This information suggested that fruits of different species but into the same generous can have distinctive chemical properties which trigger also different biological properties.

Regarding the nut tree species, nuts have flavonoids as well as compounds as reservatrol, which are associated with reduction of cancer risk [69]. Carvalho *et al.*, (2010) determined the effect of walnut (*Juglans regia* L) shells for inhibition of colon cancer (Caco-2) and kidney cancer (A-498 and 769-P) growth. They reported that the phytochemical composition of this nut shells extracts can inhibit maligns cell growth in a range of 17% to 79% [70]. In other study was observed that hazelnut skin extract significantly inhibited the growth of colorectal adenocarcinoma cell line (HT-29) by 96% and 89% of the total cancer cells [71].

Medicinal plants are one of the most amply used treatments for cancer control due to the ancient knowledge about them. In Mexico, biological activities of some plants of the semiarid region have been studied. *Lippia berladieri* Shauer, *Larrea tridenata*, *Fluorensia cernua* DC and *Carya illinoensis* k. are some of the most studied plants for their abundance, chemical composition and sub-utilization [72]. Oregano (*Lippia berladieri* Shauer) has multiple properties mainly attributed to its essential oils such as timol and carvacrol. Timol presented an important anticancer activity. Also, the crude extract of *Fluorensia cernua* and its fractions are cytotoxic against human breast cancer cell lines [73]. According to Medi *et al.* (2011), the gradual addition of carvacrol inhibits growth of HeLa and SiHa cervical cancer cells and increases cellular apoptosis percent [74]. Also, *L. tridentata* extract and nordihydroguaiaretic acid (NDGA), its main compound, have been described as potential agents for cancer treatment. Different extracts from this plant obtained using organic solvent have been evaluated for their cytotoxic and anti-neoplastic effects on Vero and Hep 2 cell lines [75]. NDGA has the capacity of inhibit some receptors such as tirosin-kinasa (RTKs) and lipoxygenases important in the apoptosis process such as the cytochrome c and caspases [50]. Also, there are reports on induction of caspase CK18 activity and promotion of the process of programmed death cell on colon cancer cells (HCT-116) by different Egyptian plants extract [48].

Sometimes, the cytotoxic damage is caused by several chemical reagents as observed during performance of *in vitro* assay with cellular lines. Torres-Gonzalez *et al.* (2011) evaluated *Centaurea americana, Krameria ramosissima, Juglans mollis* and *Turnera diffusa* extracts on Huh7 human hepatoma cell line affected by CCl_4, showing decreased cellular damage after chemical exposition [76]. Itharat et al. (2004) studied cytotoxicity of diverse traditional Thai medicinal plants against cancer cell lines. Extracts were obtained using ethanol and water and were tested against human large cell lung carcinoma (COR-L23), human breast adenocarcinoma (MCF-7), human colon adenocarcinoma (LS-174T), and one normal human keratinocyte cell line (SVK-14). Results showed that three extract plants, *Dioscorea membranacea* Pierre, *Dioscorea birmanica* Burkill (Dioscoreaceae) and *Siphonodon celastrineus* Griff. (Celastraceae), exhibited high cytotoxic activity, showing a certain degree of selectivity against the different cell types [77]. Thus, the ethnobotanical data of the investigated plants appear to corroborate the efficacy of the cancer treatment in Thai traditional medicine. In addition, it has demonstrated that *Rubus idaeus* L., *Juniperus communis* L. and *Artemisia californica* extracts decreased significantly viability and invasion capacity of breast cancer cells (MCF-7/AZ). Invasion capacity is attributed to the ERK mediator activity. This mediator is important in the malign phenotype acquisition and involved in several specific signaling pathways for cancer development [50].

The most important advantage of *in vitro* screening panel is the potential to identify tumor-type-selective compounds. The enhanced sensitivity of this screening will also enable discovery of active constituents present at low concentrations in plant extracts. However, active extracts and purified compounds have to be further tested *in vivo* to confirm anti-tumor activity because some of the active compounds, which have been tested *in vitro*, may be metabolized as non-active compounds during the *in vivo* test.

In Vivo Assays with Phenolic Compounds

In vivo assays are performed when the compounds evaluated *in vitro* assays showed differential cytotoxicity for a particular tumor type using the same sensitive cell lines [78]. Plant extracts or its pure compounds have been tested *in vivo*. Pomegranate fruit extracts were tested on mice implanted with the prostate cancer PC-3 cell line, observing that the extract inhibits cell growth and induces apoptosis modulation [79,80]. Other investigation on mice showed that pomegranate fruit extract inhibits lung and skin cancer growth. During the lung cancer study, mice were supplied with a daily oral dosage of this extract comparable to that humans could reasonably consume, treated mice exhibited significantly less lung tumor growth than mice not treated. In a study treating mice with skin-cancer-inducing 12-O-tetra-decanoylphorbol-13-acetate (TPA), animals treated topically with pomegranate extract had significantly reduced incidence of skin tumors. This result was attributed to inflammation suppression (COX-2, MAPKs, NFκB) and tumor proliferation markers (ornithine decarboxylase) [81]. Campos *et al.*, (2010) evaluated the ability of *P. granatum* ethanolic leaf extract and *P. granatum* ethanolic fruit extract to modulate genotoxic effects induced by the alkylating agent cyclophosphamide on Swiss male mice. They found that important reductions in cellular damage by ethanolic fruit extract in doses of 50 and 75 mg/kg. These results demonstrated *P. granatum* offers a dose-dependent protective effect against CP-induced oxidative DNA damage [82]. Cyanidin-3-glucoside (C3G), the major anthocyanin in blackberry was investigated for the potential ability to inhibit 7,12-dimethylbenz [a] anthracene(DMBA)-12-O-tetradecanolyphorbol-13-acetate (TPA), which induced skin papillomas in animal skin model. Results showed that treatment of animals with C3G decreased significantly. Besides, C3G reduced the size of human lung carcinoma A549 tumor xenograft growth and significantly inhibited metastasis in athymic male nude mice. Suggesting that, C3G exhibits chemoprevention and chemotherapeutic potential activities by inhibiting tumor promoter-induced carcinogenesis and metastasis [83].

Some plant and plant extracts had showed positive effects on cancer treatment. Willow leaves aqueous extracts delayed death and inhibited tumor growth of transplanted animals with Ehrlich Ascites Carcinoma Cells (EACC), compared with untreated animals which died within two weeks [3]. Positive effects on colon cancer control were observed on rats feeding with almond [84]. This experimental chemical carcinogenesis-based approach prompted a follow-up experiment using a genetic cancer animal model, the Min mouse, which develops gastrointestinal tumors as a result of genetic mutations. However, the study apparently did not confirm a positive effect but this information helped to understanding pathways of growth control in gastrointestinal cancers. In the Min mice study, p27 kip1 was induced by almond feeding. Subsequent, studies have identified p27 kip1, a cyclin-dependent kinase inhibitor, as a tumor suppressor, which is induced by chemopreventive compounds [85, 86]. There is evidence of walnuts ability to reduce endothelin, potent vasoconstrictor associated with cellular growth [87]. Spinella et al. reported that this reduction is associated with less tumor growth in an ovarian cancer xenograft animal [88]. Additionally, a recent study showed that green tea polyphenolics reduce endothelin and accentuate the positive effects of a Cox-2 inhibitor in prostate cancer [62].

Curcumin is a natural polyphenolic compound extracted from rhizome of *Curcuma longa* Linn. Curcumin suppresses angiogenesis, abrogates FGF-2-induced angiogenic response, matrix metalloprotease expression and cyclooxygenase-2 *in vivo* assays [89]. The varied biologic properties of curcumin and absence of toxicity make curcumin an ideal and very attractive compound in attempt of therapeutic agents against various tumors. Oral administration of curcumin has shown to prevent cancer in colon, skin, stomach, duodenum, soft palate, and breasts of rodents [90,91]. Epigallocatechin-3-gallate (EGCG) is the major bioactive polyphenol present in green tea and displays *in vitro* and *in vivo* antiproliferative, antiangiogenic, antimetastatic, proapoptotic, and cell cycle perturbation activities in various tumor models [92,93,94]. Ahn *et al.* (2003) demonstrated that the antiproliferative effect of EGC on HPV-16 associated cervical cancer cell line CaSki depends by the arrest of cell cycle in G1 phase that precedes in time the consequent programmed cell death by apoptosis. Antitumor effects of EGCG were also observed in nude mice challenged with CaSki cells and then treated orally with EGCG [95].

Clinical Assays with Phenolic Compounds

Although most of the evidence about antioxidant activity of polyphenols is based on *in vitro* studies, increasing evidence indicates they may act *in vivo* in ways beyond antioxidant functions [96], for this reason is important to test these compounds in cancer patients to analyze bioavailability in human. In a clinical study phase II testing 46 men with recurrent prostate cancer, 16 patients (35%) showed a significant decrease in serum prostate specific antigen (PSA) levels (average=27%) during treatment with eight ounces of pomegranate juice. Also, corresponding *in vitro* assays using patient plasma and serum demonstrated significant decreases in prostate cancer cell line proliferation and increased apoptosis [97]. Cheng *et al.* in 2001 piloted a prospective phase-I study to evaluate anticancer properties and also toxicology, pharmacokinetics and the biologically effective dose of curcumin compound in patients with uterine CIN [98]. The study demonstrated that curcumin is not toxic to humans up to 8 mg/d when taken by mouth for 3 months and suggested a biological effect of curcumin on chemoprevention of cancer lesions at least in 25% of the cases under treatment. A curcumin testing started in 2008 in order to follow up its safety and efficacy against HPV infection, in a Phase II multicentric clinical trial on 280 women enrolled on the basis of clinical examination, cytology, histo-pathology and presence of HPV [99].

On the other hand, in a 6-month chemopreventive pilot study on patients with Barrett's esophagus (BE) which were treated with 32 or 45 g (female and male, respectively) of freeze-dried black raspberries (FBRs) [100]. BE is a premalignant esophageal condition in which the normal stratified squamous epithelium changes to a metaplastic columnar-lined epithelium and is underscored by the fact that it increases the risk for development of esophageal adenocarcinoma, a rapidly increasing and extremely deadly malignancy by 30- to 40-fold [101]. Their results suggested that daily consumption of FBRs reduced urinary excretion of 8-Iso-PGF2 and 8-OHdG, among patients with BE, indicating reduced oxidative stress.

However, toxic or lethal effects associated with moderate concentrations of active phenolic compounds may limit their application in cancer patients, although these compounds have showed positive results *in vitro* and/or *in vivo* assays. Indicine N-oxide from *Heliotropium indicum* L. showed substantial activity against acute leukemia but hepatotoxicity in patients was more severe [52]. *Brucea antidysenterica* other Thai plant, showed high antileukemic activity at a wide range of doses, so this plant passed to the next level of assessment in clinical assays but all evaluated patients presented secondary effects (nausea, vomiting, anorexia, fever, chills, and weakness) and although showed low hepatic damage was decided to conclude the treatment because bruceantin had only limited activity against malignant melanoma and was unlikely to contribute for systemic therapy of this disease [102].

Polyphenol Action Mechanisms during Cancer Treatment

Cancer is a disease that involves many dynamic changes in the genome. The central attention is on discovery of mutations which produce oncogenes with dominant gain of function and tumor suppressor genes with recessive loss of function [103]. Modulation of cell signaling pathways by polyphenols may help significantly to explain the mechanisms of actions of polyphenol-rich diets. This signaling is characterized by numerous and different agents that promote changes at cell level. Moreover, the same compound activates different signaling pathways depending on the cell types. The main signaling pathways activated by dietary chemopreventive agents are described below [13].

p53 Family Members

p53 tumor suppressor acts as transcriptional activator, controlling expression of a variety of genes important in cell cycle regulation and apoptosis, also DNA repair in response to various genotoxic stresses. Approximately half of all cases of human cancer may be attributed to a defective p53 protein [104]. Many natural chemopreventive agents induce cell cycle arrest or apoptosis by activating p53 and its target genes. Epigallocatechin-3-gallate (EGCG) induced the expression of p53 and its target p21 and BAX in prostate cancer cells with wild-type p53, but not with inactive p53 [105]. Luteolin also induced cell cycle arrest and apoptosis and increased chemo-sensitization by activating p53 and its targets p21 and BAX [106].

Nuclear Factor-Kappa B

Nuclear factor kappa B (NF-KB) is an ubiquitous rapid response transcription factor in cells involved in immune and inflammatory reactions, and exerts its effect by expressing cytokines, chemokines, cell adhesion molecules, growth factors, and immunoreceptors [107]. Most of the natural chemopreventive agents including curcumin, resveratrol, EGCG, lycopene, genistein, and luteolin act as potent inhibitors of NF-KB pathways. These compounds may block one or more steps in the NF-KB signaling pathway such as inhibition of the most upstream growth factor receptors that activate the NF-"B signaling cascade, translocation of NF-"B to the nucleus, DNA binding of the dimmers, or interactions with the basal transcriptional machinery [108-112].

Activator Protein 1

The AP-1 signal transduction pathway is known to be an important molecular target for chemopreventive strategies. AP-1, a ubiquitous transcriptional activator, is composed of members of the *Jun* and *Fos* families that form homodimers or heterodimers and bind to a distinct DNA response element. Green tea polyphenols inhibit the transcriptional activity of AP-1 in multiple cell types, which is essential for their growth inhibitory effects and pretreatment with resveratrol inhibits TPA-induced AP-1 DNA binding by inhibiting the nuclear expression of c-Jun and c-Fos [113].

The inhibitory effect of natural phenolics in developed carcinogenesis and tumor growth may be through two main mechanisms: modifying the redox status and interfering with basic cellular functions. Due to their ability to scavenge and reduce the production of free radicals and act as transition metal chelators, natural phenolic compounds can exert a major chemo-preventive activity. Indeed, it has been shown that natural polyphenols can inhibit carcinogen/toxin-induced cellular oxidative damage [114].

Regarding tumor cell interference, natural phenolics can affect basic cell functions related with cancer development by many different mechanisms. This start with the antioxidant, and antimutagenesis processes and proteases inhibition; during the promotion phase of cancer, when growth cells is uncontrolled, phenolic compounds act on the anti-inflammation mechanism and induction of apoptosis; finally, to occur the metastasis process, polyphenols are involved in the inhibition of cell proliferation [6]. In addition, phenol compounds protect intracellular communications and signal transduction [115]. The complete process is explained as follow: During initiation stage, phenolics may inhibit activation of procarcinogens by inhibiting phase I metabolizing enzymes, such as cytochrome P450 [116] and also facilitate detoxifying and elimination of the carcinogens by induction of phase II, metabolizing enzymes such as glutathione S-transferase (GST), NAD(P)H quinine oxidoreductase (NQO), and UDP glucuronyltransferase (UGT) [117].

Phenolics may inhibit tumors formation and growth by induction of cell cycle arrest and apoptosis. Any perturbation of cell cycle specific proteins by phenolics can potentially affect and/or block continuous proliferation of these tumorigenic cells. Induction of apoptosis and/or inhibition of proliferation/survival by polyphenols have been reported to result from a number of mechanisms including inducing cell cycle arrest; blocking extracellular regulated kinase (ERK), c-Jun N-terminal kinase (JNK), and P38 mitogen-activated protein kinase (MAPK) pathway; inhibition of transcription factors activation, NF-κB and activator protein-1 (AP1); suppression of protein kinase C (PKC); suppression of growth factor-mediated pathways [118]. Concerning previous information, polyphenols have been reported inducing cell cycle arrest at different cell phases: G1, S, S-G2, and G2 by directly down-regulating cyclins and cyclins-dependent kinases (CDKs) or indirectly inducing expression of p21, p27 and p53 genes [49]. Regarding the inflammation process, natural phenolics have been reported to inhibit transcription factors closely linked to inflammation such as NF-κB factor [119,120] pro inflammatory cytokines release [121] and enzymes such as COX-2 [122] lipoxygenases (LOX) [123] inducible nitric oxide synthase (iNOS) [124], that mediate inflammatory processes [125]. In many cases, polyphenols exhibit anti-inflammatory properties through blocking MAPK-mediated pathway. Polyphenols were found to suppress malignant cell migration, invasion and metastasis process. The inhibition effect has been shown to be related to their ability to down-regulate the matrix metalloproteases (MMPs), namely, MMP-2 and MMP-9, as well as urokinase-plasminogen activator (uPA) and uPA receptor (uPAR) expression. In addition, these compounds possess antiangiogenesis effects which are an important aspect in the tumor elimination [126].

Conclusion

Many fruits and vegetables that usually are consumed a daily basis synthesize active compounds with beneficial properties for human. Similar properties may be found in plants used in traditional medicine for cancer treatment but without knowledge about their action mechanisms. These therapeutic effects have been attributed to phenolic compounds present in vegetal sources. In many studies are described biological activities of this phytochemical group and its application in several areas of medicine, technology, food, pharmacology and others. Antioxidant capacity is the most important characteristic of these compounds which is associated to its anti-cancer property. Around the world, cancer is one of the main causes of death, its high incidence has generating search of therapeutic alternatives more efficient than conventional methods. Regarding this, use of phenolic compounds extracted from natural plants is considered a type of effective treatment for prevention and control of cancer. Multiplex *in vitro* and *in vivo* assays have been performed for study of grade and therapeutic levels of different phenolic extracts and pure compounds, results showed that in most of cases, polyphenolics reduce or stop tumor growth. This affirmation has been corroborated in some clinical studies with patient affected by several cancer types with different levels of progression. However, it is necessary carried out a more extensive analysis in order to know the biochemical mechanisms implicated in the polyphenols biological activity and determine the best way to apply these compounds.

References

[1] Cai, Y., Luob, Q., Sunc, M., Corkea, H. (2004). Antioxidant activity and phenolic compounds of 112 traditional Chinese medicinal plants associated with anticancer. *Life Sciences, 74,* 2157–2184.

[2] Nassr-Allah, A. A., Aboul-Enein, A. M., Aboul-Enein, K. M., Lightfoot, D. A., Cocchetto, A., El-Shemy, H. A. (2009). Anti-cancer and anti-oxidant activity of some Egyptian medicinal plants. *Journal of Medicinal Plants Research, 3 (10),* 799-808.

[3] El-Shemy, H. A., Aboul-Enein, A. M., Aboul-Enein, K. M., Fujita, K. (2007). Willow leaves' extracts contain anti-tumor agents effective against three cell types. *Plos One, 178 (1),* 1-5.

[4] Azadmehr, A., Hajiaghaee, R., Afshari, A., Amirghofran, Z., Refieian-Kopaei, M., Darani, H., Shirzad, H. (2011). Evaluation of in vivo immune response activity and in vitro anti-cancer effect by *Scrophularia megalantha. Journal of Medicinal Plants Research, 5(11),* 2365-2368.

[5] Aweng, E. R., Nur, H., Mohd, M. A., Nurhanan, Y., Shamsul, M. (2012). Antioxidant activity and phenolic compounds of *Vitex trifolia* var, simplicifolia associated with anticancer. *ISCA Journal of Biological Sciences, Vol. 1(3),* 65-68.

[6] Dai, J. & Mumper, R. (2010). Plant Phenolics: Extraction, analysis and their antioxidant and anticancer properties. *Molecules, 15,* 7313-7352.

[7] Scalbert, A., Johnson, I. T., Saltmarsh, M. (2005). Polyphenols: antioxidants and beyond. *Am J Clin Nutr, 81,* 215S–7S.

[8] Santos-Buelga, C.E., Scalbert, A. (2000). Proantocyanidins and tannin-like compounds-nature, occurrence dietary intake and effects on nutrition and health. *Journal of the Science of Food and Agriculture, 80,* 1094 - 1117.

[9] Wojdyło, A., Oszmian´ski, J., Czemerys, R. (2007). Antioxidant activity and phenolic compounds in 32 selected herbs. *Food Chemistry, 105,* 940–949.

[10] Pinheiro do Prado, A. C., Aragão, A., Fett, R., Block, J. M. (2009). Antioxidant properties of pecan nut [*Carya illinoinensis* (Wangenh.) C. Koch] Shell infusion. *Grasas y Aceites, 60 (4),* 330-335.

[11] Nawab, A., Yunus, M., Mahdi, A., Gupta, S. (2011). Evaluation of anticancer properties of medicinal plants from the Indian Sub-Continent. *Mol Cell Pharmacol, 3(1),* 21-29.

[12] Leung, P-C. & Fong, H. (2007). *Alternative Treatment for Cancer.* World Scientific Publishing, Vol. 3, USA.

[13] Amin, A.R.M., Kucuk, O., Khuri, F.R., Shin, D.M. (2009). Perspectives for cancer prevention with natural compounds. *Journal of Clinical Oncology, 12 (16),* 2712-2725.

[14] Isaza, J. (2007). Taninos o polifenoles vegetales. *Scientia et Technica, 13 (33),* 13-18.

[15] Lattanzio, V., Lattanzio, V.M.T., Cardinali, A. (2006). Role of phenolics in the resistance mechanisms of plants against fungal pathogens and insects. *Phytochemistry: Advances in Research,* 23-67.

[16] D'Archivio, M., Filesi, C., Di Benedetto, R., Gargiulo, R., Giovannini, C., Masella, R. (2007). Polyphenols, dietary sources and bioavailability. *Ann Ist Super Sanità, 43 (4),* 348-361.

[17] Balasundram, N., Sundram, K., Samman, S. (2006). Phenolic compounds in plants and agri-industrial by-products: antioxidant activity, occurrence, and potential uses. *Food Chem, 99,* 191–203.

[18] Martins, S., Mussatto, S., Martínez-Avila, G., Montañez-Saenz, J., Aguilar, C.N., Teixeira, J.A. (2011). Bioactive phenolic compounds: Production and extraction by solid-state fermentation. A review. *Biotechnology Advances, 29,* 365–373.

[19] Monteiro, M., Farah, A., Perrone, D., Trugo, L.C., Donangelo, C. (2007). Chlorogenic acid compounds from coffee are differentially absorbed and metabolized in humans. *The Journal of Nutrition, 137,* 2196–2201.

[20] Sun, R., Tomkinsona, J., Sun, X.F., Wang, N.J. (2000). Fractional isolation and physico-chemical characterization of alkali-soluble lignins from fast-growing poplar wood. *Polymer, 41,* 8409–8417.

[21] Fernández-Bolaños, J., Felizón, B., Brenes, M., Guillén, R., Heredia, A. (1998). Hydroxytyrosol and Tyrosol as the main compounds found in the phenolic fraction of steam-exploded olive stones. *JAOCS, 75 (11),* 1643–1649.

[22] Gresele, P., Cerletti, C., Guglielmini, G., Pignatelli, C., Gaetano, G., Violic, F. (2011). Effects of resveratrol and other wine polyphenols on vascular function: an update. *Journal of Nutritional Biochemistry, 22*, 201–211.

[23] Basli, A., Soulet, S., Chaher, N., Merillon, J-M., Chibane, M., Monti, J-P., Richard, T. (2012). Wine Polyphenols: Potential agents in neuroprotection. *Oxidative Medicine and Cellular Longevity, 012*, 1-14.

[24] Slanina, J. & Glatz, Z. (2004). Separation procedures applicable to lignan analysis. *J Chromatogr B Analyt Technol Biomed Life Sci,* 812(1-2), 215-29.

[25] Seeram, N., Lee R., Hardy M., Heber D. (2005). Rapid large scale purification of ellagitannins from pomegranate husk, a by-product of the commercial juice industry. *Separation and Purification Technology,* 41, 49–55.

[26] Khanbabaee, K. & van Ree, T. (2001). Tannins: classification and definition. *Nat. Prod. Rep., 18,* 641-649.

[27] Peteros, N. & Uy, M. (2010). Antioxidant and cytotoxic activities and phytochemical screening of four Philippine medicinal plants. *Journal of Medicinal Plants Research, 4(5),* 407-414.

[28] Guendez, R., Kallithraka, S., Makris D., Kefalas, P. (2005). Determination of low molecular weight polyphenolic constituents in grape (*Vitis vinifera* sp.) seed extracts: Correlation with antiradical activity. *Food Chemistry, 89*, 1-9.

[29] Wanasundara, P. K. J. P. D. & Shahidi, F. (2005). *Antioxidants: Science, Technology, and Applications*. Bailey's Industrial Oil and Fat Products, (Sixth Edition). Canada, Fereidoon Shahidi.

[30] Wong, S.P., Leong, L.P., Koh, J.H.W. (2005). Antioxidant activities of aqueous extracts of selected plants. *foodchem, 07(58),* 777-783.

[31] Budzyn, M., Iskra, M., Wielkoszynski, T., Malecka, M., Gryszczynska, B. (2009). The influence of black currant (*Ribes nigrum*) seed extract on effectiveness of human ceruloplasmin in Fe (II) ions elimination. *J. Elementol, 14(2)*, 217-227.

[32] Rice-Evans, C.A., Miller, N.J., Paganga, G. (1996). Structure-antioxidant activity relationships of flavonoids and phenolic acids. *Free Radic. Biol. Med. 20*, 93 3-956.

[33] Re, R., Pellegrini, N., Proteggente, A., Pannala, A., Yang, M., Rice-Evans, C. (1999). Antioxidant activity applying an improved ABTS radical cation decolorization assay. *Free Radical Biology & Medicine, 26 (9/10)*, 1231–1237.

[34] He, Z., Fu, M., Mao, L. (2011). Total phenolic, condensed tannin and antioxidant activity of four *Carya* species from China. *African Journal of Biotechnology, 10(51),* 10472-10477.

[35] Naczk, M., Shahidi, F. (2006). Phenolics in cereals, fruits and vegetables: occurrence, extraction and analysis. J. *Pharm. Biomed. Anal., 41,* 1523-1542.

[36] Hartzfeld, P.W., Forkner, R., Hunter, M.D., Hagerman, A.E. (2002). Determination of hydrolysable tannins (gallotannins and ellagitannins) after reaction with potassium iodate. *J. Agric. Food Chem., 50,* 1785-1790.

[37] Inoue, K.H., Hagerman, A.E. (1988). Determination of gallotannin with rhodanine. *Anal. Biochem., 169*, 363-369.

[38] Wilson, T.C., Hagerman, A.E. (1990). Quantitative determination of ellagic acid. *J. Agric. Food Chem., 38,* 1678-1683.

[39] Harris, C.S., Burt, A.J., Saleem, A., Le, P.M., Martineau, L.C., Haddad, P.S., Bennett, S.A., Arnason, J.T. (2007). A single HPLC-PAD-APCI/MS method for the quantitative comparison of phenolic compounds found in leaf, stem, root and fruit extracts of *Vaccinium angustifolium. Phytochem. Anal., 18,* 161-169.

[40] Rodriguez, A., Lopez, J., Ferraces, P., Lage, M.A. (2007). Analysis of non-anthocyanin phenolic compounds in wine samples using high performance liquid chromatography with ultraviolet and fluorescence detection. *J. Sep. Sci., 30,* 1262-1266.

[41] Mertz, C., Cheynier, V., Gunata, Z., Brat, P. (2007). Analysis of phenolic compounds in two blackberry species (*Rubus glaucus* and *Rubus adenotrichus*) by high-performance liquid chromatography with diode array detection and electrospray ion trap mass spectrometry. *J. Agric. Food Chem., 55,* 8616-8624.

[42] Downey, M.O., Rochfort, S. (2008). Simultaneous separation by reversed-phase high-performance liquid chromatography and mass spectral identification of anthocyanins and flavonols in Shiraz grape skin. *J. Chromatography, 1201,* 43-47.

[43] Uttara, B., Singh, A.V., Zamboni, P., Mahajan, R.T. (2009). Oxidative stress and neurodegenerative diseases: A review of upstream and downstream antioxidant therapeutic options. *Current Neuropharmacology, 7*, 65-74.

[44] World Health Organization (OMS). (2012). Cancer, descriptive note N° 247. Available from: http://www.who.int/mediacentre/factsheets/fs297/es/index.html.

[45] National Cancer Institute (U.S.A). (2012). NCI Dictionary of Cancer Terms. Available from: http://www.cancer.gov/dictionary?cdrid=45333.

[46] Mehta RG., Murillo G., Naithani R., Peng X. (2010). Cancer chemoprevention by natural products: how far have we come? *Pharm Res., 27*, 950-61.

[47] Hanahan, D., Weinberg, R. A. (2000). The hallmarks of cancer. *Cell, 100,* 57-70.

[48] El-Menshawi, B., Fayad, W., Mahmoud, K., El-Hallouty, S., El-Manawaty, M., Olofsson, M., Linder, S. (2009). Screening of natural products for therapeutic activity against solid tumors. *Indian Journal of Experimental Biology, 48,* 258-264.

[49] Fresco, P., Borges, F., Diniz, C., Marques, M.P. (2006). New insights on the anticancer properties of dietary polyphenols. *Med. Res. Rev. 26,* 747-766.

[50] Van Slambrouck S., Daniels A., Hooten C., Brock S., Jenkins A., Ogasawara M., Baker J., Adkins G., Elias E., Agustin V., Constantine S., Pullin M., Shors S., Kornienko A., Steelant W. (2007). Effects of crude aqueous medicinal plant extracts on growth and invasion of breast cancer cells. *Oncology Reports, 17,* 1487-1492.

[51] Mehdi S., Ahmad A., Irshad M., Manzoor N., Rizvi M. (2011). Cytotoxic effect of Carvacrol on human cervical cancer cells. *Biology and Medicine, 3 (2),* 307-312.

[52] Itharat, A. & Ooraikul, B. (2007). Research on Thai medicinal plants for cancer treatment. *Advances in Medicinal Plant Research*, 287-317.

[53] Hudson, T.S., Hartle, D.K., Hursting, S.D., Nunez, N.P., Wang, T.T.Y., Young, H.A., Arany, P., Green, J.E. (2007). Inhibition of prostate cancer growth by muscadine grape skin extract and resveratrol through distinct mechanisms. *Cancer Res., 67,* 8396–8405.

[54] Veluri, R., Singh, R.P., Liu, Z., Thompson, J.A., Agarwal, R., Agarwal, C. (2006). Fractionation of grape seed extract and identification of gallic acid as one of the major active constituents causing growth inhibition and apoptotic death of DU145 human prostate carcinoma cells. *Carcinogenesis, 27,* 1445-1453.

[55] God, J.M., Tate, P., Larcom, L.L. (2007). Anticancer effects of four varieties of muscadine grape. *J. Med. Food., 10,* 54–59.

[56] Lazze, M.C., Pizzala, R., Pecharroman, F.J.G., Garnica, P.G., Rodriguez, J.M.A., Fabris, N., Bianchi, L. (2009). Grape waste extract obtained by supercritical fluid extraction contains bioactive antioxidant molecules and induces antiproliferative effects in human colon adenocarcinoma cells. *J. Med. Food., 12,* 561–568.

[57] Singletary, K.W., Stansbury, M.J., Giusti, M., Breemen, R.B.V., Wallig, M., Rimando, A. (2003). Inhibition of rat mammary tumorigenesis by Concord grape juice constituents. *J. Agric. Food Chem., 51,* 7280–7286.

[58] Faria, A., Calhau, C., Freitas, V., Mateus, N. (2006). Procyanidins as antioxidants and tumor cell growth modulators. *Journal of Agricultural and Food Chemistry, 52,* 2392-2397.

[59] Heber, D., Seeram, N.P., Wyatt, H., Henning, S.M., Zhang, Y., Ogden, L.G., Dreher, M., Hill, J.O. (2007). Safety and antioxidant activity of a pomegranate ellagitannin-enriched polyphenol dietary supplement in overweight individuals with increased waist size. *J Agric Food Chem, 55(24),* 10050–4.

[60] Jurenka, J. (2008) Therapeutic Applications of Pomegranate (*Punica granatum* L.): A Review. *Alternative Medicine Review, 13(2),* 128-144.

[61] Viuda-Martos, M., Fernández-López, J., Pérez-Álvarez, J.A. (2010). Pomegranate and its many functional components as related to human health: A review. Comprehensive Reviews in Food Science and Food Safety, 9, 635-654.

[62] Adhami, V.M., Malik, A., Zaman, N., Sarfaraz, S., Siddiqui, I.A., Syed, D.N., Afaq, F., Pasha, F.S., Saleem, M. (2007). Combined inhibitory effects of green tea polyphenols and selective cyclooxygenase-2 inhibitors on the growth of human prostate cancer cells both *in vitro* and *in vivo*. *Clin Cancer Res., 13*, 1611–9.

[63] Adhami, V.Q., Khan, N., Mukhtar, H. (2009). Cancer chemoprevention by pomegranate: Laboratory and clinical evidence. *Nutr. Cancer, 6*, 811- 815.

[64] Sturgeon, S.R. & Ronnenberg, A.G. (2010). Pomegranate and breast cancer: Possible mechanisms of prevention. *Nutr. Rev., 68*, 122-128.

[65] Miguel, M., Neves, M., Antunes, M. (2010). Pomegranate (*Punica granatum* L.): A medicinal plant with myriad biological properties - A short review. *Journal of Medicinal Plants Research, 4(25),* 2836-2847.

[66] Albrecht, M., Jiang, W., Kumi-Diaka, J. (2004). Pomegranate extracts potently suppress proliferation, xenograft growth, and invasion of human prostate cancer cells. *J Med Food, 7,* 274-283.

[67] Adams, L.S., Seeram, N.P., Aggarwal, B.B. (2006). Pomegranate juice, total pomegranate ellagitannins, and punicalagin suppress inflammatory cell signaling in colon cancer cells. *J Agric Food Chem 2006, 54*, 980-985.

[68] Chavez-Santoscoy, R., Gutierrez-Uribe, J., Serna-Saldívar, S. (2009). Phenolic composition, antioxidant capacity and *in vitro* cancer cell cytotoxicity of nine prickly pear (*Opuntia spp.*) Juices. *Plant Foods Hum Nutr., 64*, 146–152.

[69] Azadmard-Damirchi, S., Emami, Sh., Hesari, J., Peighambardoust, S.H., Nemati, M. (2011). Nuts composition and their health benefits. *Engineering and Technology, 57,* 508-512.

[70] Carvalho, M., Ferreira, P., Mendes, V., Silva, R., Pereira, J., Jerónimo, C., Silva, B. (2010). Human cancer cell anti-proliferative and antioxidant activities of *Juglans regia* L. *Food and Chemical Toxicology, 48,* (441–447).

[71] Li, H., Parry, J.W. (2011). Phytochemical compositions, antioxidant properties, and colon cancer antiproliferation effects of Turkish and Oregon hazelnut. *Food and Nutrition Sciences, 2,* 1142-1149.

[72] Ruiz, J., Ascacio, J., Rodríguez, R., Morales, D., Aguilar, C. (2011). Phytochemical screening of extracts from some Mexican plants used in traditional medicine. *Journal of Medicinal Plants Research, 5(13)*, 2791-2797.

[73] Laux, M.T., Aregullin, M., Flanders, J., Ketzis, J., Rodriguez, E. (2003). Comparison of three different cytotoxicity assays of *Flourensia cernua*, asteraceae in human breast cancer *in vitro*. *Rev Latinam Quim, 31,* 111–124.

[74] Mehdi, S., Ahmad, A., Irshad, M., Manzoor, N., Rizvi, M. (2011). Cytotoxic effect of Carvacrol on human cervical cancer cells. *Biology and Medicine, 3 (2),* 307-312.

[75] Zamora, J.M. (1988). Cytotoxic, antimicrobial and pytochemical properties of *Larrea tridentate* Cav. Doctoral dissertation. Auburn University, Auburn, Alabama, USA, 82.

[76] Torres-González, L., Muñoz-Espinosa, L., Rivas-Estilla, A.M., Trujillo-Murillo, K., Salazar-Aranda, R., Waksman, N., Cordero-Pérez, P. (2011). Protective effect of four Mexican plants against CCl4-induced damage on the Huh7 human hepatoma cell line. *Annals of hepatology, 10 (1),* 73-79.

[77] Itharat, A., Houghton, P.J., Eno-Ammguaye, E., Burke, P.J., Sampson, J.H. and Raman, A. (2004). *In vitro* cytotoxic activity of Thai medicinal plants used traditionally to treat cancer. *J. Ethnopharmacol, 90*, 33-38.

[78] Tsao, R. (2010). Chemistry and Biochemistry of Dietary Polyphenols. *Nutrients, 2,* 1231-1246.

[79] Malik, A., Mukhtar, H. (2006). Prostate cancer prevention through pomegranate fruit. *Cell Cycle, 5,* 371-373.

[80] Malik, A., Afaq, F., Sarfaraz, S. (2005). Pomegranate fruit juice for chemoprevention and chemotherapy of prostate cancer. *Proc Natl Acad Sci, 102,* 14813-14818.

[81] Afaq, F., Saleem, M., Krueger, C.G. (2005). Anthocyaninand hydrolyzable tannin-rich pomegranate fruit extract modulates MAPK and NF-kappaB pathways and inhibits skin tumorigenesis in CD-1 mice. *Int J Cancer, 113,* 423-433.

[82] Campos, M., Tavares, E.R., López, P., Realino, J. (2010). Assessment of mutagenic and antimutagenic effects of *Punica granatum* in mice. Brazilian Journal of Pharmaceutical Sciences, 46 (1), 121-127.

[83] Ding, M., Feng, R., Wang, S.Y., Bowman, L., Lu, Y., Qian, Y., Castranova, V., Jiang, B.H., Shi, X. (2006). Cyanidin-3-glucoside, a natural product derived from blackberry, exhibits chemopreventive and chemotherapeutic activity. *J. Biol. Chem., 281*, 17359-17368.

[84] Yamada, Y., Mori, H. (2007). Multistep carcinogenesis of the colon in Apc (Min/1) mouse. *Cancer Sci., 98*, 6–10.

[85] Eto, I. (2006). Nutritional and chemo-preventive anti-cancer agents up-regulate expression of p27Kip1, a cyclin-dependent kinase inhibitor, in mouse JB6 epidermal and human MCF7, MDA-MB-321 and AU565 breast cancer cells. *Cancer Cell Int., 6*, 20.

[86] Aggarwal, B.B., Banerjee, S., Bharadwaj, U., Sung, B., Shishodia, S., Sethi, G. (2007). Curcumin induces the degradation of cyclin E expression through ubiquitin-dependent pathway and up-regulates cyclin-dependent kinase inhibitors p21 and p27 in multiple human tumor cell lines. *Biochem Pharmacol., 73,* 1024–32.

[87] Davis, P., Jenab, M., Vanden, J.P., Furlong, T., Taylor, S. (2007). Tree nut and peanut consumption in relation to chronic and metabolic diseases including allergy. *The Journal of Nutrition, 138,* 1757S–1762S.

[88] Spinella, F., Rosano, L., Di Castro, V., Decandia, S., Albini, A., Nicotra, M.R., Natali, P.G., Bagnato, A. (2006). Green tea polyphenol epigallocatechin-3-gallate inhibits the endothelin axis and downstream signaling pathways in ovarian carcinoma. *Mol Cancer Ther., 5,* 1483–92.

[89] Di Domenico, F., Foppoli, C., Coccia, R., Perluigi, M. (2011). Antioxidants in cervical cancer: Chemopreventive and chemotherapeutic effects of polyphenols. *Biochimica et Biophysica Acta,* 4(5), 1-11.

[90] Rao, C.V., Rivenson, A., Simi, B., Reddy, B.S. (1995). Chemoprevention of colon carcinogenesis by dietary curcumin, a naturally occurring plant phenolic compound. *Cancer Res., 55*, 259–266.

[91] Kawamori, T., Lubet, R., Steele, V.E., Kelloff, G.J., Kaskey, R.B., Rao, C.V., Reddy, B.S. (1999). Chemopreventive effect of curcumin, a naturally occurring anti-inflammatory agent, during the promotion/progression stages of colon cancer. *Cancer Res., 59*, 597–601.

[92] Baliga, M.S., Meleth, S., Katiyar, S.K. (2005). Growth inhibitory and antimetastatic effect of green tea polyphenols on metastasis-specific mouse mammary carcinoma 4T1 cells *in vitro* and *in vivo* systems, *Clin. Cancer Res., 11*, 1918–1927.

[93] Kondo, T., Ohta, T., Igura, K., Hara, Y., Kaji, K. (2002). Tea catechins inhibit angiogenesis *in vitro*, measured by human endothelial cell growth, migration and tube formation, through inhibition of VEGF receptor binding. *Cancer Lett., 180*, 139–144.

[94] Fassina, G., Vene, R., Morini, M., Minghelli, S., Benelli, R., Noonan, D.M., Albini, A. (2004). Mechanisms of inhibition of tumor angiogenesis and vascular tumor growth by epigallocatechin-3-gallate. *Clin. Cancer Res., 10*, 4865–4873.

[95] Ahn, W.S., Huh, S.W., Bae, S.M., Lee, I.P., Lee, J.M., Namkoong, S.E., Kim, C.K., Sin, J.I. (2003). A major constituent of green tea, EGCG, inhibits the growth of a human cervical cancer cell line, CaSki cells, through apoptosis, G(1) arrest, and regulation of gene expression. *DNA Cell Biol., 22,* 217–224.

[96] Williams, R.J., Spencer, J.P., Rice-Evans, C. (2004). Flavonoids: Antioxidants or signaling molecules? *Free Radic. Biol. Med., 36*, 838-849.

[97] Pantuck, A.J., Leppert, J.T., Zomorodian, N. (2006). Phase II study of pomegranate juice for men with rising prostate-specific antigen following surgery or radiation for prostate cancer. *Clin Cancer Res, 12,* 4018-4026.

[98] Cheng, A.L., Hsu, C.H., Lin, J.K., Hsu, M.M., Ho, Y.F., Shen, T.S., Ko, J.Y., Lin, J.T., Lin, B.R., Ming-Shiang, W., Yu, H.S., Jee, S.H., Chen, G.S., Chen, T.M., Chen, C.A., Lai, M.K., Pu, Y.S., Pan, M.H., Wang, Y.J., Tsai, C.C., Hsieh, C.Y. (2001). Phase I clinical trial of curcumin, a chemopreventive agent, in patients with high-risk or pre-malignant lesions. *Anticancer. Res., 21,* 2895–2900.

[99] Talwar, G.P. (2008). A clinical study on the clearance of human papilloma virus (HPV) infection in uterine cervix by Basant (a polyherbal cream) and curcumin soft gelatin capsule in females infected with HPV Clinical Trial Registry — India. National Institute of Medical Statistics, ICMR, India.

[100] Kresty, L.A., Frankel, W.L., Hammond, C.D., Baird, M.E., Mele, J.M., Stoner, G.D., Fromkes, J.J. (2006). Transitioning from preclinical to clinical chemopreventive assessments of lyophilized black raspberries: interim results show berries modulate markers of oxidative stress in Barrett's esophagus patients. *Nutr. Cancer., 54*, 148-156.

[101] Spechler, S.J. (2002). Clinical practice. Barrett's Esophagus. *N. Engl. J. Med., 346*, 836-842.

[102] Dewick, P.M. (2002). *Tumor inhibitors from plants.* In Evans, W.C. (ed.) Trease and Evans Pharmacognosy, London, W.B. Saunders, 394-407.

[103] Hanahan, D. & Weinberg, R.A. (2000). The Hallmarks of Cancer. *Cell, 100*, 57–70.

[104] Goodsell, D. (1999). The molecular perspective: p53 tumor suppressor. *The Oncologist, 4,* 138-139.

[105] Hastak, K., Agarwal, M.K., Mukhtar, H. (2005). Ablation of either p21 or Bax prevent p53-dependent apoptosis induced by green tea polyphenol epigallocatechin-3-gallate. *FASEB J, 19*,789-791.

[106] Wu, B., Zhang, Q., Shen, W. (2008). Antiproliferative and chemosensitizing effects of luteolin on human gastric cancer AGS cell line. *Mol Cell Biochem, 313*, 125-132.

[107] Lee, J., Burckart, G. (1998). Nuclear factor kappa B: Important transcription factor and therapeutic target. *J Clin Pharmacol*, 38 (11), 981-993.

[108] Kasinski, A.L., Du, Y., Thomas, S.L. (2008). Inhibition of IkappaB kinase-nuclear factor-kappaB signaling pathway by 3,5-bis(2-flurobenzylidene)piperidin-4-one (EF24), a novel monoketone analog of curcumin. *Mol Pharmacol, 74*, 654-661.

[109] Manna, S.K., Mukhopadhyay, A., Aggarwal, B.B. (2000). Resveratrol suppresses TNF-induced activation of nuclear transcription factors NF-kappa B, activator protein-1, and apoptosis: Potential role of reactive oxygen intermediates and lipid peroxidation. *J Immunol, 164*, 6509-6519.

[110] Hastak, K., Gupta, S., Ahmad, N. (2003). Role of p53 and NF-kappaB in epigallocatechin-3-gallateinduced apoptosis of LNCaP cells. *Oncogene, 22*, 4851-4859.

[111] Kim, G.Y., Kim, J.H., Ahn, S.C. (2004). Lycopene suppresses the lipopolysaccharide-induced phenotypic and functional maturation of murine dendritic cells through inhibition of mitogen-activated protein kinases and nuclear factor-kappaB. *Immunology, 113*, 203-211.

[112] Li, Y., Kucuk, O., Hussain, M. (2006). Antitumor and antimetastatic activities of docetaxel are enhanced by genistein through regulation of osteoprotegerin/ receptor activator of nuclear factor-"B (RANK)/RANK ligand/MMP-9 signaling in prostate cancer. *Cancer Res, 66*, 4816-4825.

[113] Kundu, J.K., Shin, Y.K., Surh, Y.J. (2006). Resveratrol modulates phorbol ester-induced pro-inflammatory signal transduction pathways in mouse skin in vivo: NF-kappaB and AP-1 as prime targets. *Biochem Pharmacol, 72,* 1506-1515.

[114] Kampa, M., Nifli, A.P., Notas, G., Castanas, E. (2007). Polyphenols and cancer cell growth. *Rev. Physiol. Biochem. Pharmacol., 159*, 79-113.

[115] Soobrattee, M.A., Bahorun, T., Aruoma, O.I. (2006). Chemopreventive actions of polyphenolic compounds in cancer. *Biofactors, 27*, 19-35.

[116] Hodek, P., Trefil, P., Stiborova, M. (2002). Flavonoids-potent and versatile biologically active compounds interacting with cytochromes P450. *Chem. Biol. Interact., 139*, 1-21.

[117] Galati, G., Teng, S., Moridani, M.Y., Chan, T.S., O'Brien, P.J. (2000). Cancer chemoprevention and apoptosis mechanisms induced by dietary polyphenolics. *Drug Metabol. Drug Interact., 17,* 311-349.

[118] Ramos, S. (2008). Cancer chemoprevention and chemotherapy: dietary polyphenols and signallin pathways. *Mol. Nutr. Food Res., 52*, 507-526.

[119] Karlsen, A., Retterstol, L., Laake, P., Paur, I., Kjolsrud-Bohn, S., Sandvik, L., Blomhoff, R. (2007). Anthocyanins inhibit nuclear factor-kappaB activation in monocytes and reduce plasma concentrations of pro-inflammatory mediators in healthy adults. *J. Nutr., 137*, 1951-1954.

[120] Tsoyi, K., Park, H.B., Kim, Y.M., Chung, J.I., Shin, S.C., Lee, W.S., Seo, H.G., Lee, J.H., Chang, K.C., Kim, H.J. (2008). Anthocyanins from black soybean seed coats inhibit UVB-induced inflammatory cylooxygenase-2 gene expression and PGE2 production through regulation of the nuclear factor-kappaB and phosphatidylinositol 3-kinase/Akt pathway. *J. Agric. Food Chem., 56,* 8969-8974.

[121] Gauliard, B., Grieve, D., Wilson, R., Crozier, A., Jenkins, C., Mullen, W.D., Lean, M. (2008). The effects of dietary phenolic compounds on cytokine and antioxidant production by A549 cells. *J. Med. Food, 11,* 382-384.

[122] Hou, D.X., Masuzaki, S., Hashimoto, F., Uto, T., Tanigawa, S., Fujii, M., Sakata, Y. (2007). Green tea proanthocyanidins inhibit cyclooxygenase-2 expression in LPS-activated mouse macrophages: molecular mechanisms and structure-activity relationship. *Arch. Biochem. Biophys., 460*, 67-74.

[123] Hong, J., Smith, T.J., Ho, C.T., August, D.A., Yang, C.S. (2001). Effects of purified green and black tea polyphenols on cyclooxygenase and lipoxygenase-dependent metabolism of arachidonic acid in human colon mucosa and colon tumor tissues. *Biochem. Pharmacol., 62*, 1175-1183.

[124] Pergola, C., Rossi, A., Dugo, P., Cuzzocrea, S., Sautebin, L. (2006). Inhibition of nitric oxide biosynthesis by anthocyanin fraction of blackberry extract. *Nitric Oxide, 15,* 30-39.

[125] Jin, X.H., Ohgami, K., Shiratori, K., Suzuki, Y., Koyama, Y., Yoshida, K., Ilieva, I., Tanaka, T., Onoe, K., Ohno, S. (2006). Effects of blue honeysuckle (*Lonicera caerulea* L.) extract on lipopolysaccharide-induced inflammation *in vitro* and *in vivo*. Exp. *Eye Res., 82*, 860-867.

[126] Mojzis, J., Varinska, L., Mojzisova, G., Kostova, I., Mirossay, L. (2008). Antiangiogenic effects of flavonoids and chalcones. *Pharmacol. Res., 57*, 259-265.

In: Plant Extracts
Editors: Adriana Giordano and Alfieri Costs
ISBN: 978-1-62417-534-3

Chapter 8

New Possibilities of Using the Antifeedant and Repellent Effects of Plant Extracts in the Protection of Plants and Human Health

Roman Pavela*
Crop Research Institute, Prague, Czech Republic

Abstract

Mutual plant-insect interactions have developed in the course of many million years. Thanks to the high ability of plants to produce numerous various secondary metabolites that may lead to significant reduction of pest population dynamics, new and efficient extracts can be obtained today, which can be used in the production of botanical insecticides. Botanical insecticides (BI) play a significant role not only in plant protection against pests, but also in the protection of humans and domestic animals against significant vectors or against troubling insects.

Besides substances causing acute or chronic insect toxicity, a large group of compounds can be found in the plant kingdom, which is classified as allomones, thus substances exerting repellent, antiovipositional and antifeeding effects. The effects of some of these substances are so good that they can be used for the development of new BIs safe for the environment and for the health.

This chapter provides a critical summary of the most recent knowledge on plant allomones and their possible use in plant and human protection against pests and important vectors.

Keywords: Botanical insecticides, repellent, antifeedant, allelochemicals, plant extracts

* E-mail: pavela@vurv.cz; Phone: +420 233 022 111.

1. Introduction

Plants have developed together with insects since the very beginning of their existence. Thanks to their common history, very fragile mutual relationships formed between individual plant and insect species, which we are trying to sort out and clarify today using a scientific approach. In principle, plants utilize insects primarily for pollination, whereas insects use plants as their source of food. Therefore, for the mutual interactions between plants and insects to be balanced, and in order for the ecosystem to maintain its equilibrium with the maximum possible biodiversity, communication, as well as mutual influence or limitation, must exist to prevent the uncontrolled and excessive reproduction of any species. Plants have thus created many strategies in the course of their co-evolution to effective protect themselves against insect pests, while insects try to circumvent such strategies. Understanding their mutual relationships not only helps us to understand the world around us, but it also provides us with a lot of useful information which may have practical value as well.

As mentioned above, there are many plant-insect interactions; however, the interaction between plants – as a source of food for phytophagous insects – and their pests is one of the most interesting and important ones with regard to practical applications in agriculture. Since in this case the relationship is between "food" on the one hand and "consumer" on the other, the plants were forced to develop numerous defensive mechanisms to prevent uncontrolled attacks from destructive pests.

The ability of plants to develop active or passive defences against pests is called plant resistance. In general, plant resistance is classified based on the mechanism of its effect. There are three mechanisms of host plant resistance: 1) antixenosis - plants act as poor hosts to the insect, 2) antibiosis - plants have an adverse effect on the biology of the feeding insect, and 3) Tolerance - plants have an inherent ability to withstand the attack of the insect [85].

The potential of plants to achieve resistance is closely related to the genetically conditioned ability of morphological and/or biochemical intraspecies modifications. The ability of plants to synthesize biologically active substances of a defensive nature is one of the most significant mechanisms of plant resistance against pests. Plants produce chemicals for defensive purposes in two different ways; first, as constitutive substances to repel herbivores through direct toxicity or by reducing the digestibility of plant tissues, and second, as inducible substances synthesized in response to tissue damage by herbivores. These strategies are able to prevent most attacks from herbivores, although there are a small number of insects that are able to adapt to specific plant species [58].

In terms of practice, secondary metabolites can be classified into two large groups based on the way they affect insects: First, such substances may create direct insecticidal effects causing the mortality of phytophagous insects, and second, they may exert indirect insecticidal effects, merely influencing insect behaviour in some manner.

While the first group of compounds, i.e., those with direct insecticidal properties, has been used by man for millennia as extracts, both in fighting phytophagous pests as well as against parasites or storage pests, the other group of compounds generally called allelochemicals has become the subject-matter of more profound interest only in the past few decades [37, 66, 67].

The term allelochemicals is used to describe the chemicals involved in interspecific interactions. It is defined as a chemical significant to organisms of a species different from its source. Allelochemicals are divided into kairomones (which mediate interactions which are adaptively favorable to the receiving organism but not to the emitting organism), allomones (which mediate interactions which are adaptively favourable to the emitting organism but not to the receiving organism), synomones (which mediate interactions which are adaptively favourable to both receiving and emitting species), and apneumones (which are substances emitted by a non-living material that evoke behavioral or physiological responses that are adaptively favourable to the receiving organism, but detrimental to a second species found on or in the non-living material). In general, it can be said that the entire group of allelochemicals finds some application in plant protection [37, 43]. Nevertheless, metabolites falling in the allomone group represent the most studied and potential valuable group at present [26].

Many substances having repellent and/or antifeedant effects are found in the allomone group. Since these effects are closely related only to insect behaviour, plant extracts that contain these substances are considered to be selective and thus friendly to non-target organisms [26, 27, 37]. This is of great importance for the development of new botanical insecticides which are safe for the environment.

Although all known information on repellent and antifeedant plant metabolites cannot be presented in a single chapter, we shall attempt to provide at least a brief outline of the potential for plant and human health protection offered by the use of antifeedant and repellent plant extracts.

2. Repellents

According to their definition, repellents are substances that prevent an insect from making direct contact with a surface containing such substances, by causing an active change in the insect's behaviour. Aromatic plants are usually a natural source of repellent substances; such plants release volatile aromatic hydrocarbons, thereby repelling some insect species. However, compounds which may not be aromatic but still exhibit repellent effects when in direct contact with insect chemoreceptors can also be isolated from plants using a suitable method. [37, 85].

The present use of repellent substances falls under the domain of human and animal health protection against hazardous vectors, rather than the domain of plant protection itself. This is mainly due to the fact that most typical plant repellent substances are volatile, which means that their efficacy persists only for several hours after application. It is important for plant protection that the repellent effects persist for a sufficient duration. This can be accomplished only through the design of the production formulation itself.

2.1. Repellents for Human and Animal Health Protection

Repellent research was highly motivated by the loss of effectiveness in formulations used by military personnel during the Second World War. Consequently, thousands of repellent candidates were screened, and some synthetic repellents were developed [14]. Today, worldwide consumption of synthetic repellents has increased, due not only to the ever-greater resistance of mosquito populations and other vectors against insecticides (leading to an increase in population densities in the affected regions of occurrence), but also to changes in global temperatures. In fact, global warming has moved the mosquitoes that transmit malaria, yellow fever and dengue into some temperate and higher altitudes, affecting people who are vulnerable to these diseases. It has been predicted that the proportion of the world's population living within the potential zone for malaria transmission will increase by approximately 45–60% [16].

2.2.1. Synthetic Repellents

Several dozen synthetic repellents which have found a global application are currently in use (for example, N,N-Diethyl-2-phenyl-acetamide; Dimethyl benzene-1,2-dicarboxylate; Dodecanoic acid; Butyl 6,6-dimethyl-4-oxo-5Hpyran-2-carboxylate; 1-Piperidine carboxylic acid; 3-[N-butyl-N-acetyl]-aminopropionic acid ethyl ester; (1R,2R,5R)-2-(2-Hydroxy-propan-2-yl)-5-methyl-cyclohexan-1-ol; 2-Ethylhexane-1,3-diol; etc.). However, the purpose of this chapter is not to present them all, but rather just the most widely used active substances.

It is beyond doubt that the synthetic repellents most widely spread and used are based on N,Ndiethyl-m-toluamide (DEET), developed by the US Department of Agriculture in 1946 for use by the military. Since its promotion began in 1956, DEET has remained the most widely used and, to many, the most effective repellent [71].

DEET was historically believed to work by blocking insect olfactory receptors for 1-octen-3-ol, a volatile substance that is contained in human sweat and breath. The prevailing theory was that DEET effectively "blinds" the insect's senses so that the biting/feeding instinct is not triggered by humans or other animals which produce these chemicals. DEET does not appear to affect the insect's ability to smell carbon dioxide, as had been suspected earlier [41].

However, more recent evidence shows that DEET serves as a true repellent, in that mosquitoes intensely dislike the smell of the chemical repellent.[71]. A type of olfactory receptor neuron in the special antennal sensilla of mosquitoes that is activated by DEET, as well as other known insect repellents such as eucalyptol, linalool, and thujone, has been identified. Moreover, in a behavioral test, DEET had a strong repellent effect in the absence of body odor attractants such as 1-octen-3-ol, lactic acid, or carbon dioxide. Female and male mosquitoes showed the same response [87, 92].

2.2.2. Botanical Repellents

Historically, older repellent products were based on aromatic plants or essential oils obtained from them, as the case may be. For example, citronella oil was the most widely used repellent before the 1940s, and is still used today in many formulations even though its effectiveness is inferior to that of newer products [46].

Cymbopogon plants were traditionally used to repel mosquitoes in jungle regions such as the Bolivian Amazon [46, 47]. This genus produces the most often used natural repellents in the world [91]. Many extracts and essential oils isolated from these plants have been tested against different kinds of arthropods. When it was evaluated in the laboratory against Anopheles arabiensis, *Cymbopogon excavatus* provided 100% repellency for 2 h; its repellency decreased to 59.3% after 4 h [23].

Cymbopogon winterianus oil, mixed with 5% vanillin, provided 100% protection for 6 h against Aedes aegypti, Culex quinquefasciatus and Anopheles dirus, results comparable to those observed with 25% DEET (N,N-diethyl-3-methylbenzanmide) [88]. The oil of *Cymbopogon martinii* provided 100% repellency for 12 h against *Anopheles* mosquitoes, in a field test which was carried out using pairs of volunteers sitting together, one of whom was treated with the oil while the other was not [1].

Essential oils are complex mixtures of volatile organic compounds produced as secondary metabolites in plants; they are made up of hydrocarbons (terpenes and sesquiterpenes) and oxygenated compounds (alcohols, aldehydes, esters, ethers, ketones, lactones, phenols and phenol ethers) [24]. Some monoterpenes such as a-pinene, cineole, eugenol, limonene, terpinolene, citronellol, citronellal, camphor, and thymol are common constituents of a number of essential oils described in the literature as exhibiting mosquito repellent activity [18, 21, 25, 29, 31, 32]. Among sesquiterpenes, b-caryophyllene is most often cited as a strong repellent against *A. aegypti* [22].

Although the repellent properties of several essential oils appear to be regularly associated with the presence of monoterpenoids and sesquiterpenes [31, 35], other authors [53] have found that phytol, a linear diterpene alcohol, exhibits high repellent activity against *Anopheles gambiae*. Some essential oils also exhibit a promising repellent effect against flies (*Musca domestica*). Examples of some essential oils that exhibit repellent effects against mosquitoes and flies are included in Table 1.

Plant-derived terpenoids are repellent against several species of ticks (Table 2). For example, Dautel et al. [11] found that *Ixodes ricinus* nymphs spent significantly less time on filter paper treated with 1 mg/cm^2 of myrtenal, a bicyclic terpene that is a constituent of the essential oil of a number of plants, including *Cymbopogon nardus*, *Mentha piperita*, and *Melissa officinalis* [47], than they did on untreated controls.

The oxygenated compounds phenylethyl alcohol, b-citronellol, cinnamyl alcohol, geraniol, and a-pinene, isolated from the essential oil of *Dianthus caryophyllum*, showed strong repellent activities against *I. ricinus* [93]. Eight hours after treatment, the monocyclic terpene eugenol, isolated from both plants, provided >90% repellency, while the acyclic terpene alcohol b-citronellol, isolated from carnation flower oil, provided 84.1% repellency. Similarly, oil of citronella, containing citronellol and geraniol, repelled 83% of *I. ricinus* nymphs after 8 h, and lily of the valley essential oil, which also contains citronellol, provided 67% repellency 8 h after application to filter paper [90]. Eugenol isolated from fractioned sweet basil, *Ocimum basilicum* (L.), provided repellency equivalent to deet against *I. ricinus* in Petri dish bioassays [13].

Thorsell et al. [90] found that 10% clove oil, which contains high amounts of eugenol, provided 78% repellency, while 10% deet provided 71% repellency against *I. ricinus* nymphs for 8 h. Pâllson et al. [57] tested constituents in the essential oil from the flowers of aromatic tansy, *Tanacetum vulgare* L., against nymphal *I. ricinus*.

The potential of using repellent substances for plant and agricultural commodity protection against phytophagous insects has also been tested. In practice, however, one problem encountered is the short persistence of the effects. Botanical insecticides based on the repellent effect of essential oils are particularly promising when used in closed areas, where the fumigation effect can be used. They can thus be used in granaries and households for protection against storage pests [28].

Table 1. Plant essential oils that have shown high repellency to mosquito species

Scientific name	Insect species	References
Aloysia citriodora	Aedes aegypti	[22]
Baccharis spartioides	Aedes aegypti	[22]
Cinnamomum zeylanicum	Anopheles stephensi	[74]
Cinnamomum zeylanicum	Aedes aegypti	[74]
Cinnamomum zeylanicum	Culex quinquefasciatus	[74]
Conyza newii	Anopheles gambiae	[54]
Croton pseudopulchellus	Anopheles gambiae	[53]
Curcuma amada	Musca domestica	[84]
Curcuma longa	Anopheles dirus	[88]
Cymbopogon citratus	Aedes aegypti	[55]
Cymbopogon citratus	Culex. quinquefasciatus	[75]
Cymbopogon citratus	Musca domestica	[78]
Cymbopogon nardus	Aedes aegypti	[91])
Cymbopogon winterianus	Culex quinquefasciatus	[88]
Dianthus caryophyllus	Aedes aegypti	[93]
Endostemon tereticaulis	Anopheles gambiae	[53]
Eucalyptus globulus	Aedes albopictus	[97]
Eucalyptus camaldulensis	Culex pipiens	[17]
Illicium verum	Musca domestica	[84]
Lippia javanica	Anopheles gambiae	[54]
Lippia ukambensis	Anopheles gambiae	[54]
Mentha piperita	Anopheles annularis	[2]
Mentha piperita	Anopheles culicifacies	[2]
Mentha piperita	Culex quinquefasciatus	[2]
Mkilua fragrans	Anopheles gambiae	[53]
Moschosma polystachyum	Culex quinquefasciatus	[77]
Myristica fragrans	Musca domestica	[84]
Ocimum americanum	Culex quinquefasciatus	[88]
Ocimum basilicum	Culex pipiens	[17]
Ocimum basilicum	Anopheles stephensi	[74]
Ocimum basilicum	Aedes aegypti	[74]

Scientific name	Insect species	References
Ocimum basilicum	Culex quinquefasciatus	[74]
Ocimum fischeri	Anopheles gambiae	[53]
Ocimum forskolei	Anopheles gambiae	[53]
Ocimum gratissimum	Musca domestica	[84]
Ocimum selloi	Anopheles braziliensis	[56]
Pimpinella anisum	Culex pipiens	[17]
Plectranthus longipes	Anopheles gambiae	[53]
Plectranthus marrubioides	Anopheles gambiae	[54]
Pogostemon cablin	Aedes aegypti	[91]
Pogostemon cablin	Culex quinquefasciatus	[91]
Pogostemon cablin	Anopheles dirus	[91]
Rosmarinus officinalis	Anopheles stephensi	[74]
Rosmarinus officinalis	Aedes aegypti	[74]
Rosmarinus officinalis	Culex quinquefasciatus	[74]
Schinus molle	Musca domestica	[95]
Solanum xanthocarpum	Culex quinquefasciatus	[77]
Syzygium aromaticum	Aedes aegypti	[91]
Syzygium aromaticum	Culex quinquefasciatus	[91]
Syzygium aromaticum	Anopheles dirus	[91]
Tarchonanthus camphoratus	Anopheles gambiae	[54]
Tetradenia riparia	Anopheles gambiae	[54]
Thymus serpyllum	Musca domestica	[84]
Zanthoxylum limonella	Culex quinquefasciatus	[91]
Zanthoxylum limonella	Anopheles dirus	[91]
Zanthoxylum limonella	Aedes aegypti	[91]
Zanthoxylum piperitum	Aedes aegypti	[9]
Zingiber officinalis	Culex quinquefasciatus	[76]

For crops grown in fields, the use of such insecticides is suitable only where their timely application may help to prevent the airborne attack of pests for a short time, and thus significantly reduce any subsequent damage to the plants. In fast-growing plants susceptible to damage by pests, such an approach can be applied only during a certain phenophase of the plants. Winter oilseed rape, whose buds are attacked by *Meligethes aeneus* adults, is a typical example.

Table 2. Plants that exhibit repellency against ticks

Scientific name	Tick species	References
Achillea millefolium	Ixodes ricinus	[90]
Andropogon gayanus	Rhipicephalus microplus	[10], [89]
Artemisia abrotanum	Ixodes ricinus	[93]
Azadirachta indica	Ixodes ricinus	[20]
Callicarpa americana	Amblyomma americanum, Ixodes scapularis	[7]
Callicarpa japonica	Amblyomma americanum, Ixodes scapularis	[7]
Chamaecyparis nootkatensis	Ixodes scapularis	[15], [59], [60]
Cleome gynandra	Rhipicephalus appendiculatus	[39], [40]
Cleome monophylla	Rhipicephalus appendiculatus	[52]
Commiphora erythraea	Amblyomma americanum, Dermacentor variabilis, Ixodes scapularis	[6]
Commiphora holtziana	Rhipicephalus microplus	[5]
Commiphora swynnertonii	Rhipicephalus appendiculatus	[33]
Convallaria majalis	Ixodes ricinus	[90]
Corymbia citriodora	Ixodes ricinus	[32]
Cupressus funebris	Amblyomma americanum, Ixodes scapularis	[8]
Cymbopogon spp	Ixodes ricinus	[90]
Dianthus caryophyllum	Ixodes ricinus	[93]
Humiria balsamifera	Amblyomma americanum, Ixodes scapularis	[101]
Juniperus communis	Amblyomma americanum, Ixodes scapularis	[8]
Juniperus chinensis	Amblyomma americanum, Ixodes scapularis	[8]
Lavandula angustifolia	Ixodes ricinus	[32]
Lycopersicon hirsutum f. glabratum	Amblyomma americanum, Dermacentor variabilis Ixodes scapularis	[96]
Melinis minutiflora	Rhipicephalus appendiculatus	[51], [89]
Ocimum basilicum	Ixodes ricinus	[13]

Scientific name	Tick species	References
Ocimum suave	Rhipicephalus appendiculatus	[50]
Pelargonium graveolens	Ixodes ricinus	[32]
Rhododendron tomentosum	Ixodes ricinus	[30]
Stylosanthes hamata	Rhipicephalus microplus	[49]
Stylosanthes humilis	Rhipicephalus microplus	[49]
Syzygium aromaticum	Ixodes ricinus	[90]
Tanacetum vulgare	Ixodes ricinus	[57]

It was found based on laboratory tests that, for example, the essential oil obtained from *Thymus vulgaris* or *Carum carvi* significantly repels *M. aeneus* adults for a long enough time that the first open flowers can emerge from the phenological phase of oilseed rape – green to yellow buds, preferred by the pest [70]. The phenological phase of oilseed rape – green to yellow buds – is most prone to damage by the pollen beetle. However, under suitable temperature and rainfall conditions, oilseed rape quickly bursts into blossom, and therefore it is likely that damage caused by *M. aeneus* could be significantly reduced by using the repellent and antifeedant effect. Pollen beetles' natural behaviour is to fly to their host plants [80], so the odour of non-host plants may impact their behaviour. Nevertheless, field tests must be carried out in order to provide evidence for this theory.

In plant protection, the potential offered by botanical repellents is probably higher; these repellents are based on compounds that do not fumigate as quickly in the surrounding environment and are more stable on the plants. Such substances include numerous polyphenols and higher terpenes. Very often, the repellent effect is also associated with an antifeedant or antiovipositional effect. For example, a relatively long-term repellent effect was observed for oil made of *Pongamia pinnata* seeds. The repellent activities, including host deterrence and anti-oviposition, of pongam oil against the adults of the common greenhouse whitefly *Trialeurodes vaporariorum* (Westwood) were tested in greenhouses [64, 65]. Chrysanthemum plants treated with different concentrations (0.5% - 2.0%) of water-suspended pongam oil showed relatively long-lasting host deterrent and anti-oviposition effects on the adults of the greenhouse whitefly.

Although the repellent effect declined over time and with lower concentration, strong effects on the reduction of oviposition were found, with this effect lasting at least 12 days after application, depending on concentration.

2.3. Potential of Products Based on Repellent Substances

The potential of botanical insecticides based on repellent substances is particularly great in the field of protecting humans and animals against vectors and tiresome insects [28, 46, 47]. Evidence of this potential is provided by the relatively large increase in the number of patents in recent decades. A Chemical Abstracts search for mosquito-repellent inventions containing plant-derived essential oils revealed 144 active patents, mostly from Asia. Chinese-, Japanese- and Korean-language patents as well as those from India (in English) accounted for roughly 3/4 of all patents. Since 1998, patents for EO-containing mosquito-repellent inventions have nearly doubled every 4 years or so. Over 1/2 of all patents named just one EO. *Cymbopogon nardus*, *Cymbopogon winterianus* and *Eucalyptus* spp. EOs were each cited in approximately 1/3 of all patents. *Cinnamomum camphora*, *Cinnamomum zeylanicum*, *Syzygium aromaticum*, *Pelargonium graveolens*, *Lavandula angustifolia*, *Citrus* × *limon*, *Cymbopogon citratus,* and *Mentha* × *piperita* EOs were each cited in > 10 % of patents. Repellent chemicals present in EO compositions or added as pure natural ingredients, such as geraniol, limonene, p-menthane-3, 8-diol, nepetalactone, and vanillin were described in approximately 40 % of all patents [72].

In agriculture, such a potential depends particularly on the possibility of developing botanical repellents whose effectiveness will persist for a sufficient duration, and whose price will be convenient for growers. However, research in the field of interest should be intensified immediately.

3. Antifeedants

The group of substances called antifeedants deters phytophagous insects from consuming food. The history of the research of substances showing antifeedant effects dates back to the 1930s. As early as 1932, tested Metzger and Grant [42] about 500 plant extracts against *Popilla japonica*, though the results were not substantially encouraging. Later, Pradhan *et al.* [73] evaluated extracts of the Indian neem tree, *Azadirachta indica*, which prevented feeding by desert locusts. Although terrestrial plants produce a diverse array of secondary metabolites, likely more than 100,000 unique compounds, today only about 900 compounds have been identified as possessing feeding deterrence against insects [36, 37].

Food selection among insect herbivores is a highly specialized phenomenon. While olfactory and physical aspects of plants or their organs can be important in insect host-finding and acceptance [44], the choice of food is primarily based upon contact chemoreception of various allelochemicals [19, 86]. In particular, dietary experience has influenced the ability of insects to taste plant chemicals that can serve as signals of suitability or unsuitability. Certain dietary constituents appear to suppress the development of taste sensitivity to deterrents in an insect [79]. Avoidance of allelochemicals, when looked at from a behavioral point of view, is the outcome of interactions with chemoreceptors, characterized by an often broad sensitivity spectrum of deterrents [48].

According to Schoonhoven et al. [83], there are four basic reasons why the chemosensory perception of feeding deterrents by phytophagous insects warrants special attention:

- First, they are apparently more important in host plant recognition than phagostimulants are;
- Second, a huge number exist with variable molecular structures which add to their diversity;

- Third, there are fewer deterrent receptors; and
- Fourth, different deterrents may elicit different behavioral reactions, indicating the presence of a differential sensory coding system.

Research into the chemosensoric insect system is only at its beginning point, and knowledge of the mode of action of the substances is superficial and should be studied further. Such a lack of knowledge is due especially to the fact that only in the past several decades has a sufficient technical level been achieved, making it possible to perform reliable measurement of substance response at insect chemoreceptors.

3.1. Current Practical Use of Antifeedant Substances

There is still a long road ahead before we arrive at the full-range of application potentials for plant protection offered by plant allomones. No commercial products based purely on an antifeedant effectiveness basis are presently known. Nevertheless, some products which contain primarily substances having an insecticidal, as well as an antifeedant and anti-oviposition effect, are currently being used.

Perhaps the most widely known application of antifeedant effects relates to extracts made from the Indian plant *Azadirachta indica. A. indica* has been well known in India and its neighbouring countries for more than 2,000 years as one of the most versatile medicinal plants, with a wide spectrum of biological activity.

Extracts made from the seeds of this plant contain numerous biologically active substances having insecticidal, fungicidal, and bactericidal effects, and are used in many industries, ranging from medicine to agriculture. Many works have been published concerning the plant itself as well as its effective substances and biological effectiveness, including antifeedant effects [3, 81].

In the case of *A. indica*, substances falling in the group of limonoids (azadirachtin, salanin, nimbin, etc.) are responsible for antifeedant activity. Both primary and secondary antifeedant effects have been observed in the case of azadirachtin [3, 45]. Primary effects include the process of chemoreception by the organism, (e.g., sensory organs on mouthparts which stimulate the organism to begin feeding) whereas secondary processes are effects such as gut motility disorders due to topical application only [3, 81]. Inhibition of feeding behaviour by azadirachtin results from the blockage of input receptors for phagostimulants, stimulation of deterrent receptor cells, or both [45]. In a recent study by Yoshida and Toscano [100], the relative consumption rate of *Heliothis virescens* larvae treated with azadirachtin was 25% of the control, the lowest assimilation effectiveness of all the natural insecticides tested. In another study, larvae of *Heliothis virescens* consumed less food, gained less weight, and were less efficient at converting ingested and digested food into biomass [4]. Differences between species in sensitivity to the antifeedant effects of azadirachtin is profound. The Lepidoptera order appear most sensitive to azadirachtin's antifeedant effects, with Coleoptera, Hemiptera, and Homoptera being less sensitive [45].

A whole range of commercial products based on azadirachtin are currently being sold; however, they mainly utilize another significant effect of azadirachtin-type tetranotriter-penoids, namely the growth inhibition effect [81].

Extracts from the Indian tree of genus *Pongamia* provide another example of the start of practical commercial applications of products based on antifeedant and anti-oviposition effects. This genus has only one species, namely *Pongamia pinnata*, which belongs to the family Leguminosae; subfamily Papilionaceae [38].

P. pinnata is a rich source of flavonoids, and the B-ring is linked to either the furan or pyran ring. Some of these flavonoids are known to possess biological activity. The anti-feeding activities of various extracts of *P. pinnata* were observed against many insect pests of different crops. Under laboratory conditions, 0.1% water emulsion of pongam oil (so-called karanj oil) showed antifeeding activity against *Amsacta moorei* (Butler) [94] and *Spodoptera litura* [38]. Pongam oil is known to possess strong repellent activity against the egg laying of many insect pests [34, 38]. The aqueous extracts of seeds and plants of this species are known to possess ovicidal action against *Phthorimaea operculella* and *Helopeltis theivora* [12].

Another example of commercially produced extracts with antifeedant effects is represented by aromatic plants. It has been determined that many monoterpenes from plant sources are feeding deterrents against insects [36, 37]. For example: capillin, capillarin, methyl eugenol, and ar-curcumene isolated from *Artemisia capillaris* hold promise as anti-feedant compounds against cabbage butterfly larvae, *Pieris rapae* crucivora. The relative strong antifeedant activity of capillin and capillarin suggest that the C = O carbonyl group, instead of the CH_2 methylene group, a C-C in a side chain, and a lactone ring are some of the many factors that contribute to the biological activity [18, 98]. Various derivatives of these base compounds, such as methyl eugenol, reveal that the 3,4-dimethyl group and 1-substituent of 3,4-dimethoxy-1-substituted benzenes contribute to the antifeedant activity [99]. Aromatic hydrocarbons show significant direct insecticidal effectiveness, thanks to which they have been studied most of all [62, 63, 68, 69]. Nevertheless, antifeedant effectiveness is connected not only with aromatic hydrocarbons in aromatic plants, but also with other polyphenolic substances synthesized by such plants as part of their defence strategy against diseases and pests. These polyphenolic substances are currently being studied. Given their huge biochemical variety, it is highly probable that an antifeedant effectiveness which could lead to practical applications is going to be discovered.

Recent research results concerning the biological activity of substances obtained from Eurasian-region plants also provide evidence of the fact that primary research is necessary, and could lead to commercial application. Research on this aspect has been performed both in our laboratory and at other departments around the world, and is thus providing its first results. In this research, it was found that seed extracts and root extracts of *Leuzea carthamoides* plants have a very strong, long-lasting antifeedant effectiveness against the Colorado potato beetle (*Leptinotarsa decemlineata*) [61, 62]. This antifeedant effectiveness, determined only against the above-mentioned pest, led (with respect to the pest's nature) to formulation of a new product based on antifeedant effectiveness, and its production is currently being prepared.

3.2. Perspective of Products Based on Antifeedant Substances

Many feeding inhibitors from plant sources have provided excellent results in laboratory conditions so far. In field situations, only a few of them are satisfactory alternatives to traditional pest management. Chemical control is usually done with broad-spectrum insecticides, which have to be broad-spectrum by necessity. They have to sell in amounts large enough to accommodate financial development, research, and marketing.

Nevertheless, the use of antifeedants in pest-management programmes has enormous intuitive appeal. They satisfy the need to protect specific crops while avoiding damage to nontarget organisms, so their potential value is very great. In fact, insect damage to plants results from feeding or from transmission of pathogens during feeding, and therefore the chemicals that reduce pest injury by rendering plants unattractive or unpalatable can be considered potential substitutes for conventional insecticides. The host choice of generalists and to some extent specialists may be modified when feeding inhibitors are used.

The range of insect species targeted may be chosen by either the chemical structure of the inhibitor or the composition of a mixture of inhibitors, if different inhibitors are active against different species within the range. Therefore, the multicomponent defence strategy of plants themselves could be used, as shown in a number of recent studies on the non-azadirachtin type of limonoid inhibitors [37, 82], in which the potentiation among non-azadirachtin limonoids having two explicitly different modes of action, such as feeding deterrence and physiological toxicity, plays a significant role in the potentiation effect. Similarly, some other plant extracts have a strong potential for commercial application [67].

For the research to be successful, further assays must be performed and suitable, new substances with high antifeedant effectiveness must be sought. Determining the synergistic effects of biologically active substances will also be an important area of research, since this synergy seems to be very significant in increasing the biological activity of substances with insecticidal effects [69]. Most feeding inhibitors are less stable chemicals than traditional insecticides and act with lower residual activity and environmental impact. Natural predators and parasitoids remain unharmed by feeding deterrents targeting herbaceous host insects. Since the target sites of antifeedants are different, pesticide-resistant insect populations will still be affected by feeding inhibitors.

Multicomponent tactics will also slow down the development of resistance to these new compounds. In fact, lack of resistance is very useful for the practical application of antifeedants. It is unlikely that oligophagous insects could develop general resistance to such deterrents, because this would result in a rapid change in their host-plant range, which is determined mainly by the occurrence of such chemicals in non-host plants.

The different molecular structures of possible antifeedant compounds could be another advantage. The blend of active constituents might diffuse the selection process, mitigating the development of resistance compared to that expected with a single active ingredient. This also supports the previously-mentioned contention that combination mixtures of antifeedants could be more effective than individual compounds.

Conclusion

This group of substances has been the most studied in recent times, since the use of such substances for protecting plants against pests has a great potential. This group of chemicals again includes the group of substances from simple aromatic terpenes, phenols or alkaloids, to molecules falling in the group of polyphenols or limonoids. At present, high hopes are being placed on this group of substances, due to their practical use in plant protection [37, 67].

There are several reasons why antifeedants and repellents have the potential for practical use. Allomones with antifeedant activity include a high percentage of substances which also meet the requirements for developing new botanical insecticides:

1. They are natural plant metabolites, safe for human health and the environment;
2. Such substances have specific effects on individual pest species – they thus show high selectivity;
3. Mixtures of substances with synergistic effects can be obtained, which by their nature prevent the development of pest resistance;
4. Antifeedant substances usually exhibit other biological effects (for example, growth inhibition, mortality, and reduced fertility of surviving individuals) which increase the effectiveness of the products themselves [26, 27, 37].

All these characteristics indicate that the use of antifeedant and repellent substances for protection has great potential. Nevertheless, it must be said that although several hundred biologically active substances and extracts from plants with antifeedant and repellent effects are known, research and development of plant protection products has just begun.

The huge variety of defensive mechanisms of plants, including allelochemical synthesis, means that present research, which seeks new environmentally safe products that will provide protection for plants against insects, has great potential. However, such research should be intensified, and cooperation between basic or applied research, manufacturers, and end users should be increased. Such cooperation could lead subsequently to an important reduction in the basic chemization of the agriculture industry, as well as to an increased quantity of safe foodstuffs and, as a final consequence, the improved health of this planet's inhabitants. And that, in essence, is one of the most important goals ahead of us, one which must be worked toward at the present time.

Acknowledgment

This study was supported by grants from the Czech Republic Ministry of Education, Youth and Sports (LH11133).

References

[1] Ansari, M.A. and Razdan, R.K. (1994). Repellent action of *Cymbopogon martini martini* Stapf. var. Sofia against mosquitoes. *Indian Journal of Malariology*, *31*, 95–102.

[2] Ansari, M.A., Vasudevan, P., Tandon, M. and Razdan, R.K. (2000). Larvicidal and mosquito repellent action of peppermint (*Mentha piperita*) oil. *Bioresource Technology*, *71*, 267–271.

[3] Ascher, K.R.S. (1993). Nonconventional insecticidal effects of pesticides available from the Neem tree, *Azadirachta indica*. *Archives of Insect Biochemistry and Physiology*, *22*, 433-449.

[4] Barnby, M.A. and Klovme, J.A. (1987). Effects of azadirachtin on the nutrition and development of the Tobacco Budworm, *Heliothis virescens* (Fabr) (Lepidoptera: Noctuidae). *Journal of Insect Physiology*, *33*, 69-75.

[5] Birkett, M.A., Al Abassi, S., Kröber, T., Chamberlain, K., Hooper, A.M., Guerin, P.M., Pettersson, J.A., Slade, R. and Wadhams, L.J. (2008). Antiectoparasitic activity of the gum resin, gum haggar, from the East African plant, *Commiphora holtziana*. *Phytochemistry*, *69*, 1710–1715.

[6] Carroll, J.F., Maradufu, A. and Warthen Jr., J.D. (1989). An extract of *Commiphora erythraea*: a repellent and toxicant against ficka. *Entomologia Experimentalis et Applicata*, *53*, 111–116.

[7] Carroll, J.F., Cantrell, C.L., Klun, J.A. and Kramer, M. (2007). Repellency of two compounds isolated from *Callicarpa americana* (Lamiaceae) against *Ixodes scapularis* and *Amblyomma americanum* ficka. *Experimental and Applied Acarology*, *41*, 215–224.

[8] Carroll, J.F., Tabanca, N., Kramer, M., Elejalde, N.M., Wedge, D.E., Bernier, U.R., Coy, M., Becnel, J.J., Demirci, B., Başer, K.H.C., Zhang, J. and Zhang, S. (2011). Essential oils of *Cupressus funebris*, *Juniperus communis*, and *J. chinensis* (Cupressaceae) as repellents against ticks (Acari: Ixodidae) and mosquitoes (Diptera: Culicidae) and as toxicants against mosquitoes. *Journal of Vector Ecology*, 36, 258-268.

[9] Choochote, W., Chaithong, U., Kamsuk, K., Jitpakdi, A., Tippawangkosol, P., Tuetun, B., Champakaew, D. and Pitasawat, B. (2007). Repellent activity of selected essential oils against *Aedes aegypti*. *Fitoterapia*, *78*, 359–364.

[10] Cruz-Vazquez, C. and Fernández-Ruvalcaba, M. (2000). Anti-tick repellent effect of *Andropogon gayanus* grass on plots of different ages experimentally infested with *Boophilus microplus* larvae, *Parasitologia al Día*, *24*, 88–91.

[11] Dautel, H., Kahl, O., Siems, K., Oppenrieder, M., Müller-Kuhrt, L. and Hilker, M. (1999). A novel test system for detection of tick repellents. *Entomologia Experimentalis et Applicata*, *91*, 431–441.

[12] Deka, M.K., Singly, K. and Handique, R. (1998). Antifeedant and repellent effect of pongam (*Pongamia pinnata*) and wild sage (*Lantana camarata*) on tea mosquito bug (*Helopeltis theivora*). *Indian Journal of Agricultural Sciences*, *68*, 274-276.

[13] Del Fabbro and S. Nazzi F. (2008). Repellent effect of sweet basil compounds on *Ixodes ricinus* ticks, *Experimental and Applied Acarology*, *45*, 219–228.

[14] Dethier, V.G. (1956). Repellents. *Annual Review of Entomology*, *1*, 181–202.

[15] Dietrich, G., Dolan, M.C., Peralta-Cruz, J., Schmidt, J., Piesman, J., Eisen, R.J. and Karchesy J. (2006). Repellent activity of fractioned compounds from *Chamaecyparis nootkatensis* essential oil against nymphal *Ixodes scapularis* (Acari: Ixodidae). *Journal of Medical Entomology*, *43*, 957–961.

[16] EPA (1997). Climate Change and Public Health.

[17] Erler, F., Ulug, I. and Yalcinkaya, B. (2006). Repellent activity of five essential oils against *Culex pipiens*. *Fitoterapia*, *77*, 491–494.

[18] Frazier, J.L. (1985). Nervous system: Sensory system. In: M.S. Blum, (Eds.), *Fundamentals of Insect Physiology* (John Wiley and Sons, pp. 287-356), New York, U.S.A.

[19] Frazier, J.L. (1986). The perception of plant allelochemicals that inhibit feeding. In: L.B Brattsten. and S. Ahmad. (Eds.), *Molecular Aspects of Insect Plant Associations* (Plenum Press, pp. 1-42). New York, U.S.A..

[20] Garboui, S.S., Jaenson, T.G.T. and Pâlsson, K., (2006). Repellency of MyggA Natural spray (para-menthane-3,8-diol) and RB86 (neem oil) against the tick *Ixodes ricinus* (Acari: Ixodidae) in the field of east-central Sweden, *Experimental and Applied Acarology*, *40*, 271–277.

[21] Garboui, S.S., Jaenson, T.G.T., Borg-Karlson, A.-K. and Pâlsson K. (2007). Repellency of methyl jasmonate to *Ixodes ricinus* nymphs (Acari: Ixodidae), *Experimental and Applied Acarology*, *42*, 209–215.

[22] Gillij, Y.G., Gleiser, R.M. and Zygadlo, J.A. (2008). Mosquito repellent activity of essential oils of aromatic plants growing in Argentina. *Bioresource Technology*, *99*, 2507–2515.

[23] Govere, J., Durrheim, D.N., Du, T.N., Hunt, R.H. and Coetzee, M. (2000). Local plants as repellents against *Anopheles arabiensis*, in Mpumalanga Province, South Africa. *Central African Journal of Medicine*, *46*, 213–216.

[24] Guenther, E. (1972). *The Essential Oils*. (Krieger Publishing Copany) Florida, USA.

[25] Ibrahim, J. and Zaki, Z.M. (1998). Development of environment-friendly insect repellents from the leaf oils of selected Malaysian plants, ASEA. *Review of Biodiversity, Environmental and Conservation*, *6*, 1–7.

[26] Isman, M.B. (1994). Botanical insecticides and antifeedants: New sources and perspectives. *Pesticide Research Journal*, *6*, 11-19.

[27] Isman, M. B. (2002). Insect antifeedants. *Pesticide Outlook*, *13*, 152-157.

[28] Isman, M.B. (2006). Botanical insecticides, deterrents, and repellents in modern agriculture and an increasingly regulated world. *Annual Review of Entomology*, *51*, 45-66.

[29] Jacobson, M. (1966). Chemical insects attractants and repellents. *Annual Review of Entomology*, *11*, 403–422.

[30] Jaenson, T.G.T., Pâlsson, K. and Borg-Karlson, A.-K. (2005). Evaluation of extracts and oils of tick-repellent plants from Sweden. *Medical and Veterinary Entomology*, *19*, 345–352.

[31] Jaenson, T.G.T., Pâlsson, K. and Borg-Karlson, A.K. (2006). Evaluation of extracts and oils of mosquito (Diptera: Culicidae) repellent plants from Sweden and Guinea-Bissau. *Journal of Medical Entomology*, *43*, 113–119.

[32] Jaenson, T.G.T., Garboui, S. and Pâlsson, K. (2006). Repellency of oils of lemon eucalyptus, geranium and lavender and the mosquito repellent MyggA Natural to *Ixodes ricinus* (Acari: Ixodidae) in the laboratory and field. *Journal of Medical Entomology*, *43*, 731–736.

[33] Kaoneka, B., Mollel, M. and Lyatuu, F. (2007). Leaf essential oil composition and tick repellency activity of *Commiphora swynnertonii*, Burtt. *Journal of Biological Research-Thessaloniki*, *8*, 213–216.

[34] Khaire, V.M., Kachare, B.V. and Mote, U.N. (1993). Effect of vegetable oils on mortality of pulse beetle in pingeon pea seeds. *Seed Research*, *21*, 78-81.

[35] Kiran, S., Devi, P. (2007). Evaluation of mosquitocidal activity of essential oil and sesquiterpenes from leaves of *Chloroxylon swietenia* DC. *Parasitology Research, 101*, 413–418.

[36] Koul, O. (1982). Insect feeding deterrents in plants. *Indian Review Life Science, 2*, 97-125.

[37] Koul, O. (2005). *Insect Antifeedants.* (CRC Press). Bota Racon, FL.: U.S.A.

[38] Kumar, S.M.B. and Kalidhar, S.B. (2003). A review of the chemistry and biological activity of *Pongamia pinnata. Journal of Medicinal and Aromatical Plant Science, 25*, 441-465.

[39] Lwande, W., Neskala, A.J., Hassanali, A., Moreka, L., Nyandat, N., Ndungu, M., Amiani, H., Gitu, P.M., Malonza, M.M. and Punyua, D.K. (1999). *Gynandropsis gynandra* essential oil and its constituents as tick (*Rhipicephalus appendiculatus*) repellents. *Phytochemistry, 50*, 401–405.

[40] Malonza, M.M., Dipeolu, O.O., Amoo, A.O. and Hassan, S.M. (1992). Laboratory and field observations on anti-tick properties of the plant *Gynandropsis gynandra* (L.) Brig. *Veterinary Parasitology, 42*, 123–136.

[41] Mathias Ditzen, Pellegrino, M. and Vosshall L.B. (2008). "Insect Odorant Receptors Are Molecular Targets of the Insect Repellent DEET". *Sciencexpress, 319* (5871), 1838–1842.

[42] Metzger, F.W. and Grant, D. H. (1932). Repellency of the Japanese beetle of extracts made from plants immune to attack. *USDA Technical Bulletins, 299*, 21.

[43] Miller, J.R. and Miller, T.A. (1986). *Insect-Plant interaction.* (Springer-Verlag) New York.

[44] Miller, J.R. and Strickler, K.L. (1984). Finding and accepting host plants. In: W. J Bell. and R.Carde (Eds.) *Chemical Ecology of Insects.* (Sinauer Associates, pp. 127-157). Sunderland, MA:U.S.A.

[45] Mordue, A.J. and Blackwell, A. (1993). Azadirachtin: an update. *Journal of Insect Physiology, 39*, 903-924.

[46] Moore, S.J., Hill, N., Ruiz, C. and Cameron, M.M. (2007). Field evaluation of traditionally used plant-based insect repellents and fumigants against the malaria vector *Anopheles darlingi* in Riberalta, Bolivian Amazon. *Journal of Medical Entomology, 44*, 624–630.

[47] Moore, S.J., Lenglet, A. and Hill, N. (2007). Plant-based insect repellents. In: M. Debboun, S. Frances and D. Strickman (Eds.), *Insect Repellents: Principles, Methods, and Uses* (CRC Press, pp. 275–303). Boca Raton, FL, U.S.A.

[48] Mullin, C.A., Chyb, S., Eichenseer, H., Hollister, B. and Frazier, J.L. (1994). Neuroreceptor mechanisms in insect gustation: A pharmacological approach. *Journal of Insect Physiology, 40*, 913-931.

[49] Muro Castrjón, F., Cruz-Vázquez, C., Fernández-Ruvalcaba, M., Molina-Torres, J., Soria Cruz, J. and Ramos Parra, M. (2003). Repellence of *Boophilus microplus* larvae in *Stylosanthes humilis* and *Stylosanthes hamata* plants. *Parasitología latinoamericana, 58* 118–121.

[50] Mwangi, E.N., Hassanali, A., Essuman, S., Myandat, E., Moreka, L. and Kimondo, M. (1995). Repellent and acaricidal properties of *Ocimum suave* against *Rhipicephalus appendiculatus* ficka. *Experimental and Applied Acarology*, *19*, 11–18.

[51] Mwangi, E.N., Essuman, S., Kasta, G.P., Nandat, E., Munyinyi, D. and Kimondo, M., (1995). Repellence of the tick *Rhipicephalus appendiculatus* by the grass *Melinis minutiflora*. *Tropical Animal Health and Production*, *27*, 211–216.

[52] Ndungu, M., Lwande, W., Hassanali, A., Moreka, L. and Chhabra, S.C. (1995). *Cleome monophylla* essential oil and its constituents as tick (*Rhipicephalus appendiculatus*) and maize weevil (*Sitophilus zeamais*) repellents. *Entomologia Experimentalis et Applicata*, *76*, 217–222.

[53] Odalo, J.O., Omolo, M.O., Malebo, H., Angira, J., Njeru, P.M., Ndiege, I.O. and Hassanali, A. (2005). Repellency of essential oils of some plants from the Kenyan coast against *Anopheles gambiae*. *Acta Tropica*, *95*, 210–218.

[54] Omolo, M.O., Okinyo, D., Ndiege, I.O., Lwande, W. and Hassanali, A. (2004). Repellency of essential oils of some Kenyan plants against *Anopheles gambiae*. *Phytochemistry*, *65*, 2797–2802.

[55] Oyedele, A.O., Gbolade, A.A., Sosan, M.B., Adewoyin, F.B., Soyelu, O.L. and Orafidiya, O.O. (2002). Formulation of an effective mosquito-repellent topical product from Lemongrass oil. *Phytomedicine*, *9*, 259-262.

[56] Padilha de Paula, J., Gomes-Carneiro, M.R. and Paumgartten, F.J.R. (2003). Chemical composition, toxicity and mosquito repellency of *Ocimum selloi* oil. *Journal of Ethnopharmacology*, *88*, 253–260.

[57] Pâllson, K., Jaenson, T.G.T., Bæckström, P. and Borg-Karlson, A.-K. (2008). Tick repellent substances in the essential oil of *Tanacetum Bulhare*. *Journal of Medical Entomology*, *45*, 88–93.

[58] Panda, N. and Khush, G.S. (1995). *Host plant resistance to insects*. (CAB International). Wallingford, CT: U.S.A.

[59] Panella, N.A., Barchesy, J., Maupin, G.O., Malan, J.C.S. and Piesman, J. (1997). Susceptibility of immature *Ixodes scapularis* (Acari: Ixodidae) to plant-derived acaricides. *Journal of Medical Entomology*, *34*, 340–345.

[60] Panella, N.A., Dolan, M.C., Barchesy, J.J., Xiong, Y., Peralta-Cruz, J., Khasawneh, M., Montenieri, J.A. and Maupin, G.O. (2005). Use of novel compounds for pest control: insecticidal and acaricidal activity of essential oil components from heartwood of Alaska yellow cedar. *Journal of Medical Entomology*, *42*, 352–358.

[61] Pavela, R. (2004). The effect of ethanol extracts from plants of the family Lamiaceae on Colorado Potato Beetle adults (*Leptinotarsa decemlineata* SAY). *National Academy Science Letters – India*, *27*, 195-203.

[62] Pavela, R. (2006). The antifeedant effect of extracts from *Leuzea carthamoides* (Willd.) DC. on *Leptinotarsa decemlineata* Say, In: J.N. Govil, V.K Singh and C. Arunachalam (Eds.), *Recent Progress in Medicinal Plants*: *Vol. 14 – Biopharmaceuticals* (Studium Press, LLC, pp. 305-314) Nashua, NH, U.S.A..

[63] Pavela, R. (2006). Insecticidal activity of essential oils against cabbage aphid *Brevicoryne brassicae*. *Journal of Essential Oil-Bearing Plants*, *9*, 99-106.

[64] Pavela, R. and Herda, G. (2007). Repellent effects of pongam oil on settlement and oviposition of the common greenhouse whitefly *Trialeurodes vaporariorum* on chrysanthemum. *Insect Science*, *14*, 219-224.

[65] Pavela, R. and Herda, G. (2007). Effect of pongam oil on adults of the greenhouse whitefly *Trialeurodes vaporariorum* (Homoptera: Trialeurodidae). *Entomolia Generalis*, *30*, 193-201.

[66] Pavela, R. (2007). The feeding effect of Polyphenolic compounds on the Colorado Potato Beetle (*Leptinotarsa decemlineata* Say). *Pest Technology*, *1*, 81-84.

[67] Pavela, R. (2007). Possibilities of botanical insecticide exploitation in plant protection. *Pest Technology, 1*, 47-52.

[68] Pavela, R. (2008). Insecticidal properties of several essential oils on the house fly (Musca domestica L.). *Phytotherapy Research, 22*, 274-278.

[69] Pavela, R. (2008). Acute and synergistic effects of some monoterpenoid essential oil compounds on the House Fly (*Musca domestica* L.). *Journal of Essential Oil-Bearing Plants*, *11*, 451 – 459.

[70] Pavela, R. (2011). Insecticidal and repellent activity of selected essential oils against of the pollen beetle, *Meligethes aeneus* (Fabricius) adults. *Industrial Crops and Products*, *34*, 888– 892.

[71] Peterson, C. and Coats, J. (2001). Insect repellentsepast, present, and future. *Pesticide Outlook*,*12*, 154-158.

[72] Pohlit, A.M, Lopes, N.P., Gama, R.A., Tadei, W.P. and De Andrade Neto, V.F. (2011). Patent literature on mosquito repellent inventions which contain plant essential oils - A review (Review). *Planta Medica*, *77*, 598-617.

[73] Pradhan, S., Jotwani, M.S. and Rai, B.K. (1962). The neem seed deterrent to locusts. *Indian Farming*, *12*, 70-71.

[74] Prajapati, V., Tripathi, A.K., Aggarwal, K.K. and Khanuja, S.P.S. (2005). Insecticidal, repellent and oviposition-deterrent activity of selected essential oils against Anopheles stephensi, *Aedes aegypti* and *Culex quinquefasciatus*. *Bioresource Technology*, *96*, 1749–1757.

[75] Pushpanathan, T., Jebanesan, A. and Govindarajan, M. (2006). Larvicidal, ovicidal and repellent activities of *Cymbopogan citratus* Stapf. (Graminae) essential oil against the filarial mosquito *Culex quinquefasciatus* (Say) (Diptera: Culicidae). *Tropical Biomedicine*, *23*, 208–212.

[76] Pushpanathan, T., Jebanesan, A. and Govindarajan, M. (2008). The essential oil of *Zingiber officinalis* Linn. (Zingiberaceae) as a mosquito larvicidal and repellent agent against the filarial vector *Culex quinquefasciatus* Say (Diptera: Culicidae). *Parasitology Research*, *102*, 1289–1291.

[77] Rajkumar, S. and Jebanesan, A. (2005). Repellency of volatile oils from Moschosma polystachyum and *Solanum xanthocarpum* against filarial vector *Culex quinquefasciatus* Say. *Tropical Biomedicine*, *22*, 139–142.

[78] Rani, P.U. and Osmani, Z. (1980). Insecticidal, fumigant and repellent properties of citral against housefly *Musca domestica* nebulo (Fabr.). *Pesticides*, *14*, 6-8.

[79] Renwick, J.A.A. (2001). Variable diets and changing taste in plant-insect relationships. *Journal of Chemical Ecology*, *27*, 1063-1076.

[80] Sedivy, J. and Kocourek, F. (1994). Flight activity of winter rape pests. *Journal of Applied Entomology*, *117*, 400–407.

[81] Schmutterer, H. (1990). Properties and potential of natural pesticides from the Neem tree, *Azadirachta indica. Annual Review of Entomology, 35*, 271-297.

[82] Schoonhoven, L.M. (1982). Biological aspects of antifeedants. *Entomologia Experimentalis et Applicata, 31*, 57-69.

[83] Schoonhoven, L.M., Blaney, W.M. and Simmonds, M.S.J. (1992). Sensory coding of feeding deterrents in phytophagous insects. In: E. A. Bernays (Eds.), *Insect-Plant Interactions, Vol 4* (CRC Press, pp. 59-99). Boca Raton, FL: U.S.A.

[84] Singha, D. and Singha, A.K. (1991). Repellent and insecticidal properties of essential oils against housefly, *Musca domestica* L. *International Journal of Tropical Insect Science, 12*, 487-491.

[85] Smith, C. M. (1989). *Plant resistance to insects.* (John Wiley and Sons, Inc.), New York, U.S.A.

[86] Stadler, E. (1992). Behavioral responses of insects to plnat secondary compounds. In: G. A. Rosenthal and M.R Berenbaum (Eds.), *Herbivores: Their Interaction with Secondary Plant Metabolites; Evolutionary and Ecological Processes* (Academic Press, pp. 44-88). San Diego, U.S.A.

[87] Syed, Z. and Leal, WS (2008). "Mosquitoes smell and avoid the insect repellent DEET". *Proceedings of the National Academy of Sciences of the U. S. A., 105*, 13598–13603.

[88] Tawatsin, A., Wratten, S.D., Scott, R.R., Thavara, U. and Techandamrongsin, Y. (2001). Repellency of volatile oils from plants against three mosquito vectors. *Journal of Vector Ecology, 26*, 76–82.

[89] Thompson, K.C., Rao, J. and Romeo, T. (1978). Antitick grasses as the basis for developing practical tropical tick control packages. *Tropical Animal Health and Production, 10*, 179–182.

[90] Thorsell, W., Mikiver, A.and Tunón H. (2006). Repelling properties of some plant materials on the tick *Ixodes ricinus* L. *Phytomedicine, 13*, 132–134.

[91] Trongtokit, Y., Rongsriyam, Y., Komalamisra, N. and Apiwathnasorn, C. (2005). Comparative repellency of 38 essential oils against mosquito bites. *Phytotherapy Research, 19*, 303–309.

[92] Tsitsanou, K.E., Thireou, T., Drakou, C.E., Koussis, K., Keramioti, M.V., Leonidas, D.D., Eliopoulos, E., Iatrou, K. and Zographos, S.E. (2012). *Anopheles gambiae* odorant binding protein crystal complex with the synthetic repellent DEET: implications for structure-based design of novel mosquito repellents. *Cellular and Molecular Life Science, 69*, 283–297.

[93] Tunón, H., Thorsell, W., Mikiver, A. and Malander, I. (2006). Arthropod repellency, especially tick (*Ixodes ricinus*), exerted by extract from *Artemisia abrotanum* and essential oil from flowers of *Dianthus caryophyllum. Fitoterapia, 77*, 257–261.

[94] Verma, S.K. and Singh, M.P. (1985). Antifeedant effects of some plant extracts on *Amsacta moorei* Butler. *Indian Journal of Agricultural Science, 55*, 298-299.

[95] Wimalaratne. P.D.C., Slessor, K.N., Borden, J.H., Chong, L.J. and Abate T. (1996). Isolation and identification of house fly, *Musca domestica* L., repellents from pepper tree, *Schinus molle* L. *Journal of Chemical Ecology*, 22: 49-59.

[96] Witting-Bissinger, B.E., Stumpf, C.F., Donohue, K.V, Apperson, C.S. and Roe, R.M. (2008). Novel arthropod repellent, BioUD, is an efficacious alternative to DEET. *Journal of Medical Entomology*, *45*, 891–898.

[97] Yang, P. and Ma, Y. (2005). Repellent effect of plant essential oils against *Aedes albopictus*. *Journal of Vector Ecology*, *30*, 231–234.

[98] Yano, K. (1987). Minor components from growing buds of *Artemisia capillaris* that act as insect antifeedants. *Journal of Agricultural and Food Chemistry*, *35*, 889-891.

[99] Yano, K. and Kamimura, H. (1993) Antifeedant activity toward larvae of *Pieris rapae* crucivora of phenol ethers related to methyleugenol isolated from *Artemisia capillaris*. *Bioscience Biotechnology and Biochemistry*, *57*, 129-130.

[100] Yoshida, H.A. and Toscano, N.C. (1994). Comparative effects of selected natural insecticides on *Heliothis virescens* (Lepidoptera: Noctuidae) larvae. *Journal of Economic Entomology*, *87*, 305-310.

[101] Zhang, A., Klun, J.A., Wang, S., Carroll, J. and Debboun, M. (2009). Isolongifolenone: a novel sesquiterpene repellent of ticks and mosquitoes. *Journal of Medical Entomology*, *46*, 100–106.

In: Plant Extracts
Editors: Adriana Giordano and Alfieri Costs
ISBN: 978-1-62417-534-3

Chapter 9

Essential Oils and Extracts of Medicinal Plants in the Control of Diseases in Plants

Eduardo Alves[1,*], Ricardo Borges Pereira[2], Gilvaine Ciavareli Lucas[2] and Lenita Lima Haber[1]
[1]Federal University of Lavras, Lavras, Brazil
[2]Embrapa Vegetables, Brasília, Brazil

Abstract

In recent years medicinal plants have been widely studied because they have bioactive compounds that are able to inhibit or retard the growth of microorganisms. Currently, with an increasing demand from society for healthy food and sustainable agriculture, there have been more studies with the use of oils and extracts from medicinal plants for disease control in field and post harvest, to replace pesticides. Bioactive compounds of plants are generally extracted in the aqueous form, with ethanol or other organic solvents, or by steam, when present in the essential oils. These substances are mostly secondary metabolites which have biological effects on microorganisms. In fungi, the bioactive compounds are able to inhibit and/or delay mycelial growth, cause morphological changes in the hyphae, vacuolization and disorganization of the cytoplasm, rupture of plasma and mitochondrial membranes and other organelles, as well as changes in cell permeability and the flow of its constituents. In bacteria, these compounds, besides impairing pathogenicity, cause irreversible damage to cellular structure with wall degradation; changes in cytoplasm density (loss of electro-dense material); cytoplasmic vacuolation; alterations in membrane permeability with negative effects on the metabolic activity, the inhibition of specific enzymatic processes; and protein denaturation that leads to breakage, leakage of the cytoplasm and cell death. The compounds can also act indirectly on the pathogens by activating the plant defense system, which recognizes the bioactive substances present in extracts or essential oils through membrane-anchored receptors and activates the transcription of genes

[*] ealves@dfp.ufla.br, ricardobp@cnph.embrapa.br, gilciavareli@yahoo.com.br, lenita.haber@ cnph.embrapa.br.

responsible for production of pathogenesis-related protein defense (PRPs - chitinase, β-1,3-glucanases, peroxidases, among others) and synthesis of phenolic compounds (phenol and lignin) and phytoalexins. Such effects have been confirmed in several pathosystems by molecular tools (gene expression), light microscopy and scanning electron and transmission and biochemical analyses of the activity of defense enzymes.

Introduction

The intense and indiscriminate use of agricultural pesticides has seriously damaged the environment and human health, especially due to intake of contaminated foods (Moreira & Almeida, 2008). With regard to the environment, the main problems are related to the accumulation of these products in fields, the water table and water sources, because of repeated applications of pesticides.

It is therefore very important to study alternative products, made from natural substances, for use in the management and control of pests and diseases. These studies are backed by the growing demands from society for healthier foods that are free of pesticides, especially vegetables, as these are usually consumed *in natura*. Society also wants a more sustainable form of agriculture that generates fewer impacts on the environment.

In this context, research with medicinal and aromatic plants has been intensifying greatly in recent years. A review carried out by Marques (2009) in *Science Direct* found 401 articles in journals and 95 books for the key words essential oil and phytopathogenicity, 295 articles in journals and 78 books for essential oil and insecticidal activity, 150 articles in journals and 71 books for essential oil and phytoalexins and 13 articles in journals and two books for essential oil and bioherbicide.

Products of plant secondary metabolism, the bioactive compounds from many medicinal and aromatic species have an important role as the source of raw material for food, pharmaceutical, perfumery and other industries. They are now being used in agriculture, because they present fungicidal, bactericidal, insecticidal and nematicidal activity in the field and post-harvest.

Essential oils, also known as volatile oils, ethereal oils and essences, constitute an important group within the secondary metabolites. According to the *International Organization for Standardization* they are products obtained from parts of plants – leaves, flowers and bark – generally by steam distillation. They are also products obtained by expression of the pericarp of citrus fruits, being mainly constituted by monoterpenes, sesquiterpenes and phenylpropanoids.

Besides being the source of raw material for various industries, essential oils play an important role in nature, acting in the adaptation of plants to the environment, working as chemical signals for communication between species, in protection against pathogenic microorganisms (Taiz & Zeiger, 2006; Maraschim & Vepoorte, 1999).

Essential oils are obtained principally through steam distillation, in which a steam current passes through the plant material, dragging out the substances. They can, however, also be obtained by hydro-distillation, a process in which plant material is submerged in water. In both processes, the water and essential oil vapor is condensed, and the essential oil is separated from the water, since it does not mix with water.

Other classes of substances, alkaloids and tannins among them, with potential use in the control of diseases in vegetables, can be obtained from plant extracts. However, the efficiency of their biological activity will depend on the species used, the concentration of the active principle present in the plant, the source (stem, leaves, seeds, etc), the method used to obtain it (hot or cold extraction) and the stability of the components (Kamel, 2000). Furthermore, the solvent used for extraction is also important, since due to the various polarities in existence, different classes of substances can be extracted using different solvents, for example, ethanol and water.

In Spain, products formulated from essential oils and plant extracts are already being sold. Daymsa Desarrollo Agrícola y Minero, S.A. is a pioneering company that develops, produces and sells agronutrients and phytoprotectants. In the latter group there are already 16 plant-based products on the market. Of the 16 products, four are formulated from essential oils, namely: Botania® (insecticide), formulated from a mixture of essential oils from the Lamiaceae family; Nematon EC® (nematicide), prepared from oils from species of the Pedaliaceae family; Retenol®, formulated from pine oil; and TopOut®, formulated from essential oils from species of the Euforbiaceae family.

The other phytoprotectants are formulated from pine extracts, liquid wax from species of the Buxaceae family, extracts from allioideae, solanaceae and rutaceae, pine resin, *Allium sativum* extract and formulated based on potassium salts of fatty acids, of high quality and purity.

Natural substances obtained from plant extracts and essential oils not only have the advantage of offering no risk to human health and no contribution to environmental contamination. They are also promising tools in disease control in various crops, being therefore an alternative to the use of agricultural pesticides.

Mode of Action of Bioactive Compounds on Fungi

Work carried out with plant extracts or essential oils obtained from medicinal plants has indicated that these have potential in the control of phytopathogenic fungi. Considering the complexity of these compounds, it must be admitted that there are various action mechanisms, many of which are still unknown. The mechanisms may be due to direct fungitoxic action, inhibiting mycelial growth and spore germination, or to activation of the plant's defense mechanisms, such as induction of phytoalexins and proteins related to resistance (PRPs), indicating the presence of compounds with elicitor characteristics (Stangarlin, et al., 1999). The latter are considered promising in alternative control of plant diseases (Guiraldo et al., 2004; Schwan-Estrada & Stangarlin, 2005; Santos et al., 2007; Delamare et al., 2007; Omidbeygi et al., 2007; Stangarlin et al., 2008; Alves, 2008; Pereira et al., 2008; Pinto et al., 2010).

Various studies are presented, below, to show the cutting edge of research into these compounds and their mode of action, focusing on the control of phytopathogenic fungi.

The effect on mycelial growth of fungi has been seen for compounds from various medicinal plants and different fungi. Ranasinghe et al. (2005) confirmed that essential oils of *Cinnamomum zeylanicum* (L.) (cinnamon) extracted from the bark and leaves, and *Syzygium aromaticum* (L.) (clove) presented a fungistatic effect on mycelial growth of *Colletotrichum musae*, the etiological agent of anthracnose and crown rot in banana, caused by *Lasiodiplodia theobromae* and *Fusarium proliferatum*. Azizi et al. (2006) evaluated the antifungal activity of essential oils of *Thymus vulgaris*, *Mentha piperita*, *Satureja hortensis*, *Cuminum cyminum* and *Trachyspermum copticum*, at concentrations of 250, 500 and 1000 mg.L^{-1}, on mycelial growth and spore germination of *Penicillium italicum*, *P. digitatum* and *Alternaria citri*, important post-harvest fungi in citrus.

The fungicidal action of oil of *Cymbopogon flexuosus* was verified at minimum concentrations of 0.2 μL.mL^{-1}, for *Alternaria alternata*, 0.4 μL.mL^{-1} for *Aspergillus flavus*, *A. fumigatus*, *A. niger*, *A. parasiticus*, *Cladosporium cladosporioides*, *Colletotrichum capsici*, *C. falcatum*, *Curvularia lunata*, *Fusarium cerealis*, *F. culmorum*, *F. oxysporum*, *F. udum*, *Gloeosporium fructigenum*, *Penicillium expansum*, *P. italicum*, *P. implicatum*, *P. digitatum*, *P. minio-luteum*, *P. variable* and 0.5 μL.mL^{-1} for *Botrytis cinerea*, *Helminthosporium oryzae*, *H. maydis*, *Phoma violácea* and *Rhizopus nigricans* (Shahi et al., 2003)

Souza Júnior et al. (2009) evaluated the fungitoxic effect of essential oils on *Colletotrichum gloeosporioides* isolated from yellow passion-fruit, and affirmed that the essential oils of *Lippia sidoides*, *Ocimum gratissimum*, *Lippia citriodora* and *Cymbopogon citratus* promoted total inhibition of mycelial growth, while *Psidium guayava* var. *pomifera* showed greater inhibition of *C. gloeosporioides* mycelia as concentrations increased (1, 3, 5 and 10 μL.mL^{-1}).

The effect of essential oils has also been studied on food fungi. The study developed by Viegas et al. (2005) used essential oil of cinnamon bark (*Cinnamomum zeylanicum* Breym.) and of garlic bulbs (*Allium sativum* L.), showed that this essential oil promoted inhibition of mycelial development of *Aspergillus flavus*, responsible for the synthesis of aflatoxin in peanut.

Not only fixed compounds, but also volatile ones from essential oils and plant extracts can act on fungus. Neri et al. (2006) observed significant antifungal activity of volatile compounds *trans*-2-hexenal, carvacrol, *trans*-cinnamaldehyde and citral on mycelial growth, and on germination and sporulation of *Penicillium expansum*, the etiological agent of blue mold on pear. Sukatta et al. (2008) used the inverted Petri dish method to verify the antifungal activity of clove and cinnamon oil against *Aspergillus niger*, *Alternaria alternata*, *Colletotrichum gloeosporioides*, *Lasiodiplodia theobromae*, *Phomopsis viticola* and *Rhizopus stolonifer*, post-harvest pathogens that cause grapes to deteriorate. Kishore et al. (2007) noted that essential oils of clove, cinnamon and the citral component, at a concentration of 0.01% (vol/vol) inhibited germination of conidia of *Cercospora arachidicola, Phaeoisariopsis personata* and *Puccinia arachidis* by more than 90%.

Singh et al. (2002) evaluated the antifungal activity of vapor from pure monoterpenes against four pathogenic fungi of sugarcane. Among the tested monoterpenes, they observed total inhibition of the mycelial growth of *Curvularia pallescens* in the presence of menthol at 2000 ppm and linalol at 3000 ppm and of *Colletotrichum falcatum*, using geraniol, at 3000 ppm. In post-harvest pathogens Rozwalka et al. (2008) observed total inhibition of the mycelial growth of *Glomerella cingulata* and *Colletotrichum gloeosporioides* isolated from guava fruit, in the presence of essential oil vapor from cloves. Oil from lemongrass (*Cymbopogon citratus*) presented fungitoxic action against *C. gloeosporioides* and, against *G. ingulata*, fungistatic action until the fifth day. They noted the formation of halos of inhibition when *G. ingulata* made contact with essential oils of chamomile (*Chamomila recutita*), ginger (*Zingiber officinale*) and guava leaves (*Psidium guajava*) and of *C. loeosporioides* with chamomile, guava and tagetes (*Tagetes minuta*), probably due to the presence of fixed compounds with fungitoxic action.

In a dose-dependent way, vapors from essential oils of peppermint and basil inhibited growth of the fungi *Sclerotinia sclerotiorum*, *Rhizopus stolonifer* and *Mucor* sp., which cause deterioration and major losses of peach fruits during transport, storage and sales (Edris & Farrag, 2003). Plotto et al. (2003) noted total inhibition of *Botrytis cinerea* and *Alternaria arborescens* in the presence of vapor from oil of thyme, oregano and lemongrass.

Soylu et al. (2006) reported antifungal activity of the volatile phases of essential oils of oregano (*Origanum syriacum* var. Bevanii) and thyme (*Thymbra spicata* subsp. *spicata*), from 0.3 μg.mL-1 of air upwards; of fennel (*Foeniculum vulgare*), at 0.4 μg.mL-1; of rosemary (*Rosmarinus officinalis*), at 1.2 μg.mL-1; of lavender (*Lavandula stoechas* subsp. *stoechas*), at 1.6 μg.mL-1 and of laurel (*Laurus nobilis*), at 2.0 μg.mL-1 of air, by means of total inhibition of mycelial growth of *Phytophthora infestans*. Lee et al. (2007), evaluating the inhibitory effects of volatile compounds of 39 essential oils, at concentration of 1 μL per plate, confirmed the inhibition of mycelial growth of *Botrytis cinerea* and *Colletotrichum gloeosporioides*, post-harvest pathogenic fungi, and of *Fusarium oxysporum*, *Pythium ultimum* and *Rhizoctonia solani*, soil pathogenic fungi. Essential oil of *Origanum vulgare* inhibited the mycelial growth of all five tested fungi, as did *Cuminum cyminum* and *Eucalyptus citriodora*, but not *C. gloeosporioides*. *Thymus vulgaris* suppressed the mycelial growth of *C. gloeosporioides*, *Fusarium oxysporum* and *Rhizoctonia solani,* while *Cymbopogon citratus* only worked on *Fusarium oxysporum*.

The effect of essential oils and extracts has also been observed on the germination of fungal conidia. Barrera-Necha et al. (2008) verified the efficiency of oils of *Cinnamomum zeylanicum* and *Syzygium aromaticum* at 50, 100, 150, 200 and 250 μg.mL-1, on the germination of conidia of *Colletotrichum gloeosporioides* isolated from papaya (*Carica papaya* L.), with inhibition over 98%. Pereira et al. (2011) observed that essential oils of cinnamon, citronella, lemon balm, clove, tea tree, thyme, neem and eucalyptus inhibited the germination of conidia of *Cercospora coffeicola* with increasing concentrations; and the oil of clove, cinnamon, neem, thyme and lemongrass inhibited mycelial growth of this fungus.

Tzortzakis (2009) observed a reduction in the development of colonies of *Colletotrichum coccodes*, *Botrytis cinerea*, *Cladosporium herbarum*, *Rhizopus stolonifer* and *Aspergillus niger*, in the presence of vapor of the essential oil of cinnamon (*Cinnamomum zeylanicum* L.), between 25 and 500 ppm, *in vitro*. In comparison with the control, sporulation was inhibited by up to 63% at 25 ppm and completely inhibited at 500 ppm, except for *B. cinerea*. There was reduction in the germination of spores and length of germinative tubes of *Colletotrichum coccodes*, *Botrytis cinerea*, *Cladosporium herbarum* and *Rhizopus stolonifer* in a dose-dependent way, except for *A. niger*, in which cinnamon oil at 100 ppm triggered an increase in spore germination.

Working with essential oils in the control of *Phakopsora pachyrhizi*, Medice et al. (2007) observed that essential oil of thyme worked on the germination of urediniospores on the leaf, on the formation of uredinia and on the viability of urediniospores, while oils of lemon eucalyptus, citronella and neem worked only on the size of the uredinia.

For the study of the effect of essential oils and plant extracts on the structure of fungi, scanning electron microscopy (SEM) and transmission electron microscopy (TEM) have been used. Using SEM, Zambonelli et al. (1996) observed the degeneration of hyphae and cellular leakage of *Colletotrichum lindemuthianum* and *Pythium ultimum* with thyme oil at 800 ppm, which promoted the reduction in mycelial growth, *in vitro*.

Furthermore, Bianchi et al. (1997) carried out an ultra-structural study to clarify the mode of action and effects of *Allium sativum* on phytopathogenic fungi, *in vitro*. Using SEM and TEM, they noted the collapse of hyphae of *Rhizoctonia solani* and *Colletotrichum lindemuthianum*. Hyphae of *Fusarium solani* were less damaged, but were seen to be finer than the control hyphae. A general increase in vacuolization was also observed, with a consequent reduction in the cytoplasm of the treated fungal cells. In *Rhizoctonia solani*, there was also a thickening of the cell wall and, in *Colletotrichum lindemuthianum*, an accumulation of osmophyllic bodies on the cell membrane.

In more detailed studies it was seen that the action of essential oils and their substances in contact with microorganisms promotes cell permeability, causing leakage of the constituents (Piper et al., 2001). This was proved by Rasooli et al. (2006), by means of TEM, in which essential oils of *Thymus eriocalyx* and *Thymus x-porlock* caused severe damage to the cell walls, membranes and organelles of spores of *Aspergillus niger*.

Soylu et al. (2006) used SEM and light microscopy to observe morphological alterations in the hyphae of *Phytophthora infestans* exposed to volatile and contact phases of oregano oil (*Origanum syriacum* var. *bevanii*), such as cytoplasmic coagulation, vacuolization, wrinkled hyphae and leakage of the protoplast. Sharma &Tripathi (2006) used SEM to observe not only distortion, but also thickening of the wall and reduction in the diameter of the hyphae, and absence of conidiophores in *Aspergillus niger* promoted by the essential oil *Citrus sinensis*.

Observations with TEM revealed considerable morphological alterations in the hyphae and sclerotia of *Sclerotinia sclerotiorum*, caused by essential oils of oregano (*Origanum syriacum* var. *bevanii*) and fennel (*Foeniculum vulgare*), which demonstrated efficiency in significantly inhibiting the growth of *S. sclerotiorum* in treated soil, thus increasing the survival rate of tomato seedlings in 69.8% and 53.3%, respectively (Soylu et al., 2007).

Rozwalka et al. (2010) observed the accumulation of electro-dense material caused by Eugenia caryophyllata, as well as leakage of the cytoplasmic content in some conidia of Colletotrichum gloeosporioides and of Colletotrichum musae treated with essential oils of C. citratus and Cinnamomum sp. The authors also saw changes in the cell wall and the plasmatic membrane, as well as vacuolization of the cytoplasm in conidia treated with essential oils of C. martinii, E. caryophyllata, T. vulgaris, C. citratus and Cinnamomum sp.

To understand the mode of action of the essential oil of *Cinnamomum jensenianum* in the control of *Aspergillus flavus*, Tian et al. (2012) used TEM to observe changes in the ultrastructure of the mycelium submitted to the oil. The main changes were that the cell membrane became rough and showed continuous folds in the cytoplasm, forming lomasomes. A reduction of the cytoplasmic matrix with some ruptured mitochondria, with reduced cristae and complete lysis, were also noted. However, the greatest alterations observed were intense vacuolization of the cytoplasm and detachment of the plasmatic membrane of the cell wall. To confirm the target of the oil in the plasmatic membrane, the authors studied the inhibition of the synthesis of ergosterol, the main esterol component of the cell membrane in fungi and responsible for maintaining cell function and integrity. They demonstrated that the oil considerably reduced the quantity of ergosterol, by between 56 and 92%.

In a study carried out by Tyagi & Malik (2010) on the effect of essential oil of *Cymbopogon citratus* on *Candida albicans*, by means of observations using SEM, TEM and

AFM (atomic force microscope) ultrastructural alterations were seen in the fungus, with the oil in both its liquid form and as vapor, with the latter showing greater damage. In the SEM micrographs, shrinkage, deformation and rupture of the cells could be seen. TEM micrographs showed distortion of the cell membrane, disorganization of the cytoplasm, vacuolization and variation in the thickness of the cell wall, which in some places seemed to be broken. In the AFM micrographs, the treated *C. albicans* cells presented a lower height than those of the control, 2000nm and 700 to 100nm respectively, as well as being less wrinkled than those of the control.

Pereira et al. (2011) tested the essential oils of lemongrass and citronella in the control of *Cercospora coffeicola* and using SEM observed that in all the cultivars used (Mundo Novo 379/19, Catuaí IAC 62 and Catucaí 2SL) the germination of conidia started four hours after inoculation, but that significant differences were only seen eight hours after inoculation. In the control leaves of the three cultivars studied, the conidia were in an advanced stage of germination with mycelial development on their surface. However, in the leaves treated with oils, few conidia were seen to have germinated and few germinative tubes were emitted. In some conidia, cell leakage was observed, indicated by the process of plasmolization, which indicates that these oils can be used effectively in the alternative control of *C. coffeicola.*

In the same context, Pereira et al. (2008) evaluated the effect of concentrations of coffee husk extract (CHE), essential oil of thyme (EOT) and acibenzolar-S-methyl (ASM) on the pathogen, using SEM. The authors observed that the conidia of C. *coffeicola* started germinating four hours after inoculation in all treatments, with the exception of the EOT treatment, in which germination was barely shown. In the treatments in which plants received CHE, conidia of *C. coffeicola* produced more developed germinative tubes in greater numbers, more developed mycelia and higher mycelial growth than in other treatments, 8, 16 and 48 hours after inoculation, respectively. The authors reported that in this case, a large amount of the extract was still found sticking to the leaf surface. However, in the treatments with EOT, conidia of the fungus showed low germination, reduced growth and cell leakage, suggesting that this may be a promising product in the alternative control of brown eye spot in coffee.

In the case of fungi, studies are already reaching the field and storage environments. Chu et al. (2001) observed a significant reduction in *Monilinia fructicola* (brown rot), when inoculated and fumigated with 10 $mg.L^{-1}$ of thymol before cold storage. *P. expansum* was significantly reduced, from 16% to 2%, by fumigation with acetic acid, at 6 or 10 $mg.L^{-1}$, which did not present an effect on brown rot. Ranasinghe et al. (2005) observed control of crown rot in mature ‘Embul’ bananas (Musa AAB) treated with emulsions of essential oils made from the bark and leaves of *Cinnamomum zeylanicum* (cinnamon).

Tzortzakis (2007) verified a reduction in the deterioration of strawberry (*Fragaria ananassa* Duch.) and tomato (*Lycopersicon esculentum* L.) treated with vapor of oils of *Eucalyptus globulus* L. (eucalyptus) and *Cinnamomum zeylanicum*, Blume (cinnamon). In the control of anthracnose in post-harvest mango fruits (Nam Dokmai), Duamkhanmanee (2008) observed the efficiency of essential oil of lemongrass (*Cymbopogon citratus* Stapf.) at the concentration of 4000 ppm in hot water, before and after inoculation with a suspension of *Colletotrichum gloeosporioides* conidia. These results show the practical applicability of essential oils and plant extracts in the control of plant-pathogenic fungi, and confirm that these substances are promising in the control of diseases.

Mode of Action of Bioactive Compounds on Bacteria

One of the uses of essential oils from medicinal plants is in the control of diseases. Various studies have been done in recent years to help understand the form in which these substances work on the bacterial cell, mainly focusing on human bacteria, as explained below.

Degradation of the cell wall has been confirmed by a number of authors working with bacteria exposed to the essential oil of tea-tree (*Melaleuca alternifolia* Cheel). Gardner & Peel (1991), Carson & Riley (1993, 1994, 1995), Carson et al. (1995) and Raman et al. (1995) observed that the essential oil of tea-tree has the capacity to kill a large range of micro-organisms (gram-negative and gram-positive bacteria, and yeasts), as well as to actively disinfect the cell membrane, denaturing proteins of the membrane and altering its structure, thus leading to cytoplasmic leakage, lysis and cell death. Other studies have demonstrated that some components of this essential oil, such as linalool, a-terpineol and 1,8 cineole, have the capacity to alter or penetrate lipid structures (Williams & Barry 1991; Kararli et al. 1995).

Gustafson et al. (1998) observed the loss of electro-dense material of the cells of *Escherichia coli* treated with essential oil of tea-tree, indicating the loss of cellular constituents and breakage of the cell wall. The authors believe that the coagulated material observed in some of the cells cultivated in the presence of the oil represents coagulated membranes and cytoplasmic components pushed through perforations produced in the cell wall by the action of the oil. These effects are already well known in bacteria treated with disinfectants that act on membranes, such as chlorhexidine and compounds of quaternary ammonium cations (Gardner & Peel 1991).

Blaszyk & Holley (1998), working with eugenol, another compound of this same oil, proved that it is active against pathogenic bacteria *Escherichia coli* O157:H7 and *Listeria monocytogenes*. Studies with eugenol have presented evidence indicating possible roles in membrane interactions and inhibition of specific processes or cellular enzymes (Wendakoon & Sakaguchi, 1995; Helander et al., 1998; Kwon et al., 2003; Walsh et al., 2003).

Ultee et al. (1999) observed that the hydrophobic component of carvacrol, the compound present in the essential oil of clove, interacts with the membranes of *Baccilus cereus*, altering its permeability for cations such as H^+ e K^+. This results in a bactericidal or bacteriostatic action, causing damage to the essential processes of the cell, such as inhibition of various enzymes, due to leakage of these essential ions, loss of turgidity, alteration in the synthesis of DNA and reduction in the metabolic activities. Another secondary effect of carvacrol is the loss of membrane integrity due to the disturbance of hydrophobic interactions between lipids and proteins, reduction of ATP synthesis and, finally, cell death. Carvacrol also presents proven activity against the bacteria *Listeria monocytogenes*, *Salmonella typhimurium*, *Escherichia coli* O157: H7 (Kim et al., 1995; Cosentino et al., 1999), *Pseudomonas aeruginosa* and *Staphylococcus aureus* (Lambert et al., 2001).

Gill et al. (2006) demonstrated that eugenol and carvacrol disturb the cell membranes of *Escherichia coli*, *Listeria monocytogenes* and *Lactobacillus sakei*, when submitted to the concentration of 10 mM, enough to obtain an immediate bactericidal effect. Rupture of the membrane was confirmed by the increase in extracellular ATP and the decrease in cellular ATP. Piper et al. (2001) also reported that the substances present in essential oils, when in contact with microorganisms, affect cell membrane integrity, causing leakage of the contents.

Studies with plant pathogenic bacteria have confirmed that the action mode of essential oil on these kind of bacteria are same that observed for others. Lucas et al. (2012) observed that essential oil of clove (*Syzygium aromaticum* (Linne) Merril) at a concentration of 1000 μg mL^{-1} is capable of degrading the cell wall, causing loss of electro-dense material and vacuolization of the cells of *Xanthomonas vesicatoria*, the bacterium responsible for bacterial spot in tomato.

Apart from studies that deal with the above pathogens, little research has been done on the antibacterial action mechanism of the components of essential oils from medicinal plants (Gill et al., 2006).

Essential Oils and Plant Extracts in the Activation of Plant Defense Responses

The bioactive compounds present in essential oils and extracts from medicinal and aromatic plants, such as alkaloids, terpenes, flavonoids, steroids and others (Silva et al., 2005), may also exert important functions in the plant-pathogen interaction, either by inducing plant defense responses (Guleria & Kumar, 2006; Pereira et al., 2008; Medeiros et al., 2009; Pereira et al., 2012), producing phytoalexins, increasing activity of pathogenesis-related proteins (PRPs – chitinases, β-1,3-glucanases, peroxidases, among others), or by synthesizing other biochemical and structural compounds for plant defense, such as phenols and lignins (Schwan-Estrada, 2003; Kagade et al., 2004; Balbi-Peña et al., 2006).

The phytoalexins constitute a chemically diverse group of secondary metabolites (Taiz & Zeiger, 2006). They are defined as antimicrobial compounds of low molecular weight, synthesized by plants, which temporarily accumulate in plant cells after physical, chemical or biological stresses, in response to microbial infection (Braga, 2008). These possibly represent the defense responses to pathogens that have been most studied in plants.

The increase in the synthesis of phytoalexins has been observed in various crops when submitted to treatment with essential oils and plant extracts, denoting the presence of a compound with elicitor characteristics. In sorghum, significant increases in the synthesis of phytoalexins were obtained by applying essential oil of citronella *Cymbopogon nardus* (L.) Rendle.) (Moreira et al., 2008) and hydrolates (Franzener et al., 2007). *Helietta apiculata* and *Conyza canadensis* also provoked increases in the synthesis of phytoalexins (Franzener et al., 2007). This increased synthesis of phytoalexins in sorghum had also been obtained by the application of plant extracts, such as aqueous extract of camphor (*Artemisia camphorata*), which induced the synthesis of phytoalexins 20 times greater than that observed in untreated plants of the control group (Pazuch et al., 2005).

In other plants, such as barley (*Hordeum vulgare* L.) and soybean, significant increases in the synthesis of phytoalexins were observed after applying aqueous extract of neem leaves (*Azadirachta indica* A. Juss.) (Paul & Sharma, 2002) and ginger (*Zingiber officinalis*) (Rodrigues et al., 2007), respectively.

As well as phytoalexins, some enzymes are related to plant defense mechanisms against pathogens, such as peroxidase, polyphenoloxidase, chitinase, β-1,3-glucanases, and phenylalanine ammonia-lyase (Cavalcanti et al., 2005b). These proteins are responsible for the plant's defense against pathogens, and their synthesis is induced in the host plant in response to infection by a pathogen, and to biotic or abiotic stimuli (Agrios, 2005). The activation of pathogenesis-related proteins (PRPs) has been seen in various crops after application of bioactive compounds. In lettuce, aqueous raw extract of ginger was able to increase signify-cantly the activity of the enzyme peroxidase (Rodrigues et al., 2007). In coffee seedlings essential oil of citronella (Pereira et al., 2012) and thyme (Pereira et al., 2008) provoked a significant increase in the activity of peroxidases and chitinases. In tomato seedlings, spraying with essential oil of lemongrass promoted a significant increase in activity of peroxidases, polyphenoloxidase, chitinases and β-1,3-glucanases in the leaves (Lucas, 2012). Application of ethanolic tincture of lemon balm (*Lippia Alba*) also promoted significant increases in the content of peroxidase and polyphenoloxidases in leaves of runner bean (Vigo-Schultz, 2008).

Also working to induce resistance, aqueous extract of neem leaves provoked significant increases in the activity of enzymes phenylalanine ammonia-lyase and tyrosine ammonia-lyase in barley plants (Paul & Sharma, 2002). In cucumber and tobacco, treatment with extract of *Reynoutria sachalinensis* led to increased activity of chitinase, β-1,3-glucanase, peroxidase, polyphenoloxidase and phenylalanine ammonia-lyase (Schneider & Ullrich, 1994).

As observed the studies with essential oils and plant extract to control plant pathogen and plant disease have developed in the last years, and we think could be possible in the few year farms can use this products to control disease in many crops.

References

Agrios, G. N. (2005). *Plant pathology* (5th). New York, NY: Academic.

Alves, K. F. (2008). *Controle alternativo da antracnose do pimentão com extratos vegetais.* 47p. Dissertação (Mestrado em Fitopatologia) – Universidade Federal Rural de Pernambuco, Recife.

Azizi, M.; Farzad, S.; Jafarpour, B.; Rastegar, M. F.; Jahanbakhsh, V. (2008). Inhibitory effect of some medicinal plants' essential oils on postharvest fungal disease of citrus fruits. In International Horticultural Congress, 27. Belgium. *Anais*... Belgium: Acta Horticulturae, p. 279-286.

Balbi-Peña, M. I.; Becker, A.; Stangarlin, J. R.; Franzener, G.; Lopes, M. C.; Schwan-Estrada, K. R. F. (2006). Controle de *Alternaria solani* em tomateiro por extratos de *Curcuma longa* e curcumina – I. Avaliação *in vitro*. *Fitopatologia Brasileira*, *31,* 310-314.

Barrera-Necha, L. L.; Bautista-Baños, S.; Flores-Moctezuma, H. E.; Rojas- Estudillo, A. (2008). Efficacy of essential oils on the conidial germination, growth of *Colletotrichum gloeosporioides* (Penz.) Penz. and Sacc and control of postharvest diseases in papaya (*Carica papaya* L.). *Plant Pathology*, *7*, 174-178.

Bianchi, A.; Zambonelli, A.; D'Aurelio, A. Z.; Bellesia, F. (1997). Ultrastructural studies of the effects of *Allium sativum* on phytopathogenic fungi *in vitro*. *Plant Disease*, *81*, 1241-1246.

Blaszyk, M. & Holley, R. A. (1998). Interaction of monolaurin, eugenol and sodium citrate on growth of common meat spoilage and pathogenic organisms. *International Journal of Food Microbiology*, *39*, 175-183.

Braga, M. R. (2008). Fitoalexinas. In S. F. Pascholati, B. Leite, J. R. Stangarlin, & P. CIA (Eds.), *Interação planta patógeno: fisiologia, bioquímica e biologia molecular.* (pp. 305-346). Piracicaba, SP: FEALQ.

Carson, C. F.; Cookson, B. D.; Farrelly, H. D.; Riley, T. V. (1995). Susceptibility of methicillin-resistant *Staphylococcus aureus* to the essential oil of *Melaleuca alternifloia. The Journal of Antimicrobial Chemotherapy*, 35: 421-424.

Carson, C. F. & Riley, T. V. (1993). Antimicrobial activity of the essential oil of *Melaleuca alternifolia. Letters in Applied Microbiology*, *2*, 49-55.

Carson, C. F. & Riley, T. V. (1995). Antimicrobial activity of the major components of the essential oil of *Melaleuca alternifolia. Journal of Applied Bacteriology*, *78*, 264-269.

Carson, C. F. & Riley, T. V. (1994). Susceptibility of *Propionibacterium acnes* to the essential oil of *Melaleuca alternifolia. Letters in Applied Microbiology*, *19*, 24-25.

Cavalcanti, L. S.; Brunelli, K. R.; Stangarlin, J. R. (2005). Aspectos bioquímicos e moleculares da resistência induzida. In L. S. Cavalcanti, R. M. Di Piero, P. CIA, S. F. Pascholati, M. L. V. Resende, & R. S. Romeiro (Eds.), *Indução de resistência em plantas a patógenos e insetos.* (pp. 81-124). Piracicaba, SP: FEALQ.

Chu, C. L.; Liu, W. T.; Zhou, T. (2001). Fumigation of sweet cherries with timol and acetic acid to reduce postharvest brown rot and blue mold rot. *Fruits*, *56*, 123-130.

Cosentino, S.; Tuberoso, C. I. G.; Pisano, B.; Satta, M.; Mascia, V.; ARZEDI, E.; Palmas, F. (1999). *In vitro* antimicrobial activity and chemical composition of Sardinian Thymus essential oils. *Letters in Applied Microbiology*, *29*, 130-135.

Delamare, A. P.; Moschen-Pistorellom, I. T.; Artico, L.; Atti-Serafini, L.; Echeverrigaray, S. (2007). Antibacterial activity of the essential oils of *Salvia officinalis* L. and *Salvia triloba* L. cultivated in South Brazil. *Food Chemistry*, *100*, 603–608.

Duamkhanmanee, R. (2008). Natural essential oils from lemon grass (*Cymbopogon citratus*) to control postharvest anthracnose of mango fruit. *International Journal of Biotechnology*, *10*, 104-108.

Edris, A. E. & Farrag, E. S. (2003). Antifungal activity of peppermint and sweet basil essential oils and their major aroma constituents on some plant pathogenic fungi from the vapor phase. *Nahrung: chemie, biochemie, mikrobiologie, technologie, ernaehrung*, *47*, 117-121.

Franzener, G.; Martinez-Franzener, A. da S.; Stangarlin, J. R.; Czepak, M. P.; Schwan-Estrada, K. R. F.; Cruz, M. E. S. (2007). Atividades antibacteriana, antifúngica e indutora de fitoalexinas de hidrolatos de plantas medicinais. *Semina: Ciências Agrárias*, *28*, 29-38.

Gardner, J. F. & Peel, M. M. (1991). Principles of chemical disinfection. In: *Introduction to sterilization, disinfection and infection control.* (2nd, pp. 170-192). London: Churchill Livingstone.

Gill, A. O. & Holley, R. A. (2006). Disruption of *Escherichia coli*, *Listeria monocytogenes* and *Lactobacillus sakei* cellular membranes by plant oil aromatics. *International Journal of Food Microbiology*, *108*, 1-9.

Guiraldo, N.; Ambrosano, E. J.; Mendes, P. C. D.; Rossi, F.; Avéralo, R. A. (2004). Medidas de controle de doenças em sistema agroecológicos. *Summa Phytopathologica*, *30*, 153-156.

Guleria, S. & Kumar, A. (2006). *Azadirachta indica* leaf extract induces resistance in sesame against Alternaria leaf spot disease. *Journal of Cell and Molecular Biology*, *5*, 81-86.

Gustafson, J. E.; Liew, Y. C.; Chew, S.; Markham, J.; Bell, H. C.; Wyllie, S. G.; Warmington, J. R. (1998). Effects of tea tree oil on *Escherichia coli*. *Letters in Applied Microbiology*, *26*, 194-198.

Helander, I. M.; Alakomi, H. L.; Latva-Kala, K.; Mattila-Sandholm, Y.; Pol, L.; Smid, E. J.; Gorris, L. G. M.; Wright, A. von. (1998). Characterization of the action of selected essential oils components on Gram-negative bacteria. *Journal of Agricultural and Food Chemistry*, *46*, 3590-3595.

Kagade, S.; Marimuthu, T.; Thayumanavan, B.; Nandakumar, R.; Samiyappan, R. (2004). Antimicrobial activity and induction of resistance in rice by leaf extract of *Datura metel* against *Rhizoctonia solani* and *Xanthomonas oryzae* pv. *oryzae*. *Physiological and Molecular Plant Pathology*, *65*, 91-100.

Kamel, C. (2000). A novel look at a classic approach of plant extracts. *Feed Mix*, *9*, 19-24.

Kararli, T. T.; Kirchhoff, C. F.; Penzotti, S. C. (1995). Enhancement of transdermal transport of azidothymidine (AZT) with novel terpene and terpene-like enhancers- *in vivo in vitro*correlations. *Journal of Controlled Release*, *34*, 43-51.

Kim, J.; Marshall, M. R.; Wei, C. I. (1995). Antibacterial activity of some essential oil components against five foodborne pathogens. *Journal of Agricultural and Food Chemistry*, *43*, 2839-2845.

Kishore, G. K.; Pande, S.; Harish, S. (2007). Evaluation of essential oils and their components for broad-spectrum antifungal activity and control of late leaf spot and crown rot diseases in peanut. *Plant Disease*, *91*, 375-379.

Kwon, J. A.; Yu, C. B.; Park, H. D. (2003). Bactericidal effects and inhibition of cell separation of cinnaic aldehyde on *Baccilus cereus*. *Letters in Applied Microbiology*, *37*, 61-65.

Lambert, R. J. W.; Skandamis, P. N.; Coote, P. J.; Nychas, G. J. E. (2001). A study of the minimum inhibitory concentration and mode of action of oregano essential oil, thymol and carvacrol. *Journal of Applied Microbiology*, *91*, 453-462.

Lee, S. O.; Choi, G. J.; Jang, K. S.; Lim, H. K.; Cho, K. Y.; Kim, J. C. (2007). Antifungal activity of five plant essential oils as fumigant against postharvest and soilborne plant pathogenic fungi. *Plant Pathology*, *23*, 97-102.

Lucas, G. C.; Alves, E.; Pereira, R. B.; Zacaroni, A. B.; Perina, F. J.; SOUZA, R. M. (2012). Indian clove essential oil in the control of tomato bacterial spot. *Journal of Plant Pathology*, *94*, 45-51.

Lucas, G.C. (2012). *Óleos essenciais no controle da pinta preta do tomateiro*. 92p. Tese (Doutorado em Fitopatologia) – Universidade Federal de Lavras, Lavras.

Maraschin, M. & Vepoorte, R. (1999). Engenharia do metabolismo secundário. *Biotecnologia, Ciência e Desenvolvimento*, *23*, 24-28.

Marques, M. O. M. (2009). Uso de óleos essenciais na agricultura. In Simpósio De Brasileiro De Óleos Essenciais, 5. Rio de Janeiro. *Anais...* Rio de Janeiro: Anais do V Simpósio de Brasileiro de Óleos Essenciais.

Medeiros, F. C. L.; Resende, M. L. V.; Medeiros, F. H. L.; Zhang, H. M. Pare, P. (2009). Defense gene expression induced by a coffee-leaf extract formulation in tomato. *Physiological and Molecular Plant Pathology*, *74*, 175-783.

Medice, R.; Alves, E.; Assis, R.T.; Júnior, M.R.G.; Lopes, E.A.G.L. (2007). Óleos essenciais no controle da ferrugem asiática da soja *Phakopsora pachyrhizi* Syd. & P. Syd. *Ciência e Agrotecnologia*, *31*, 83-90.

Moreira, C. G. A.; Schwan-Estrada, K. R. F.; Bonaldo, S. M.; Stangarlin, J. R.; Cruz, M. E. S. (2008). Caracterização parcial de frações obtidas de extratos de *Cymbopogon nardus* com atividade elicitora de fitoalexinas em sorgo e soja e efeito sobre *Colletotrichum lagenarium*. *Summa Phytopathologica, 34*, 332-337.

Moreira, M. I. F. & Almeida, J.R. de S. Uso de Agrotóxicos, seus efeitos para a saúde e o ambiente e o uso de outras alternativas. 2008/01/01. Available from: http://www. Web-artigos.com/articles/10698/1/uso-de-agrotoxicos-seus-efeitos-para-a-saude-e-o-ambiente-e-o-uso-de-outras-alternativas/pagina1.html.

Neri, F.; Mari, M.; Brigati, S. (2006). Control of *Penicillium expansum* by plant volatile compounds. *Plant Pathology*, 55, 100-105.

Omidbeygi, M.; Barzegar, M.; Hamidi, Z.; Naghdibadi, H. (2007). Antifungal activity of thyme, summer savory and close essential oils against *Aspergillus flavus* in liquid medium and tomato paste. *Food Control, 18*, 1518-1523.

Paul, P. K. & Sharma, P. D. (2002). *Azadirachta indica* leaf extract induces resistance in barley against leaf stripe disease. *Physiological and Molecular Plant Pathology, 61*, 3-13.

Pazuch, D.; Stangarlin, J. R.; Franzener, G.; Schiwan-Estrada, K. R. F.; Martinez-Franzener, A. S.; ASSI, L. (2005). Efeito do congelamento e da incorporação de antioxidantes na atividade antifúngica e indutora de fitoalexinas do extrato aquoso de *Artemisia camphorata* (cânfora). In Encontro Anual de iniciação Científica – EAIC, 14. Guarapuava, *Resumos...* Guarapuava: UNICENTRO.

Pereira, R. B.; Alves, E.; Ribeiro Júnior, P. M.; Resende, M. L. V.; Lucas, G. C.; Ferreira, J. B. (2008). Extrato de casca de café, óleo essencial de tomilho e acibenzolar-S-metil no manejo da cercosporiose-do-cafeeiro. *Pesquisa Agropecuária Brasileira, 43*, 1287-1296.

Pereira, R. B.; Lucas, G. C.; Perina, F. J.; Resende, M. L. V.; Alves, E. (2022). Potential of essential oils for the control of brown eye spot in coffee plants. *Ciência e Agrotecnologia, 35*, 115-123.

Pereira, R. B.; Lucas, G. C.; Perina, F. J.; Ribeiro Júnior, P. M.; Alves, E. (2002). Citronella essential oil in the control and activation of coffee plants defense response against rust and brown eye spot. *Ciência e Agrotecnologia, 36*, 383-390.

Pinto, J. M. A.; Souza, E. A.; Oliveira, D. F. (2010). Use of plant extracts in the controle of common bean anthracnose. *Crop Protection, 29*, 838-842.

Piper, P.; Calderon, C. O.; Hatzixanthis, K.; Mollapour, M. (2001). Weak acid adaptation: the stress response that confers resistance to organic acid food preservatives. *Microbiology, 147*, 2635-2642.

Plotto, A.; Roberts, D. D.; Roberts, R. G. Evaluation of plant essential oils as natural postharvest disease control of tomato (*Lycopersicon esculentum*). (2003). *Acta Horticulturae, 62*, 737-745.

Raman, A.; Weir, U.; Bloomfield, S. F. (1995). Antimicrobial effects of tea tree oil and its major components on *Staphylococcus aureus*, *Staph. epidermidis* and *Propionbacterium acnes*. *Letters in Applied Microbiology, 21*, 242-245.

Ranasinghe, L.; Jayawardena, B.; Abeywickrama, K. (2005). An integrated strategy to control post-harvest decay of Embul banana by combining essential oils with modified atmosphere packaging. *International Journal of Food Science and Technology, 40*, 97-103.

Rasooli, I.; Rezaei, M. B.; Allameh, A. (2006). Growth inhibition and morphological alterations of *Aspergillus niger* by essential oils from *Thymus eriocalyx* and *Thymus x-porlock*. *Food Control*, *17*, 359-364.

Rodrigues, E.; Schwan-Estrada, K. R. F.; Fiori-Tutida, A. C. G.; Stangarlin, J. R. Cruz, M. E. S. (2007). Fungitoxicidade, atividade elicitora de fitoalexinas e proteção de alface em sistema de cultivo orgânico contra *Sclerotinia sclerotiorum* pelo extrato de gengibre. *Summa Phytopathologica, 33*, 124-128.

Roswalka, L. C. (2010). *Óleos essenciais: ação sobre Colletotrichum gloeosporioides e Colletotrichum musae, associados ou não a película de fécula de mandioca no controle da antracnose em goiaba*. 192p. Tese (Doutorado em Fitopatologia) – Universidade Federal de Lavras, Lavras.

Rozwalka, L. C.; Lima, M. L. R. Z. C.; May-De Mio, L. L.; Nakashima, T. (2008). Extratos, decoctos e óleos essenciais de plantas medicinais e aromáticas na inibição de *Glomerella cingulata* e *Colletotrichum gloeosporioides* de frutos de goiaba. *Ciência Rural*, *38*, 31-36.

Santos, F. S.; Souza, P. E.; Resende, M. L. V.; Pozza, E. A.; Ribeiro JÚNIOR, P. M.; Miranda, J. C.; Manerba, F. C. (2007). Efeito de extratos vegetais no progresso de doenças foliares do cafeeiro orgânico. *Fitopatologia Brasileira*, *32*, 59-63.

Schneider, S. & Ullrich, W. R. (1994). Differential induction of resistance and enhanced enzyme activities in cucumber and tobacco caused by treatment with various abiotic and biotic inducers. *Physiological and Molecular Plant Pathology*, *45*, 291-304.

Schwan-Estrada, K. R. F. (2003). Potencial de extrato e óleos essenciais de vegetais como indutores de resistência - plantas medicinais. In Reunião Brasileira Sobre Indução De Resistência Em Plantas Contra Fitopatógenos, 1. São Pedro. *Anais...* São Pedro: Escola Superior de Agricultura Luiz de Queiróz /USP.

Schwan-Estrada, K. R. F. & Stangarlin, J.R. (2005). Extratos e óleos essenciais de plantas medicinais na indução de resistência. In L. S. Cavalcanti, R. M. Di Piero, P. Cia, S. F. Pascholati, M. L. V. Resende, & R. S. ROMEIRO. (Orgs). *Indução de resistência em plantas a patógenos e insetos*. (pp. 125-138). Piracicaba, SP: FEALQ.

Shahi, S. K.; Patra, M.; Shukla, A. C.; Dikshit, A. (2003). Use of essential oil as botanical-pesticide against post harvest spoilage in *Malus pumilo* fruit. *BioControl*, *48*, 223-232.

Sharma, N. & Tripathi, A. (2006). Fungitoxicity of the essential oil of *Citrus sinensis* on post-harvest pathogens. *World Journal of Microbiology and Biotechnology*, *22*, 587-593.

Silva, M. C.; Guerra–Guimarães, L. & Nicole, M. (2005). Cytological and biochemical mechanisms involved in coffee leaf rust resistance. In L. Zambolim, E. M. Zambolim, & V. M. P. Várzea. (Eds). *Durable Resistance to Coffee Leaf Rust*. (pp. 249-284). Viçosa, MG: UFV.

Singh, G.; Singh, O. P.; Rao, G. P.;Sharma, S. R. (2002). Studues on essential oil, part 37: The vapour action of essential oils and monoterpenoides against pathogenic Fungi. *Sugar Technology Reviews*. *4*, 69-71.

Souza Júnior, I. T.; Sales, N. L. P.; Martins, E. R. (2009). Efeito fungitóxico de óleos essenciais sobre *Colletotrichum gloeosporioides*, isolado do maracujazeiro amarelo. *Biotemas*, *22*, 77-83.

Soylu, E. M.; Soylu, S.; Kurt, S. (2006). Antimicrobial activities of the essential oils of various plants against tomato late blight disease agent *Phytophthora infestans*. *Mycopathologia*, *161*, 119-128.

Soylu, S.; Yigitbas, H.; Soylu, E. M.; Kurt, S. (2007). Antifungal effects of essential oils from oregano and fennel on *Sclerotinia sclerotiorum*. *Journal of Applied Microbiology, 103*, 1021-1030.

Stangarlin, J. R.; Schwan-Estrada, K. R. F.; Cruz, M. E. da S.; Nozaki, M. H. (1999). Plantas medicinais e controle alternativo de fitopatógenos. *Revista Biotecnologia Ciência & Desenvolvimento, 2*, 16-21.

Stangarlin, J. R.; Kuhn, O. J.; Schwan-Estrada, K. R. F. (2008). Controle de doenças de plantas por extratos de origem vegetal. *Revisão Anual de Patologia de Plantas, 16*, 265-304.

Sukatta, U.; Haruthaithanasan, V.; Chantarapanont, W.; Dilokkunanant, U.; Suppakul, P. (2008). Antifungal activity of clove and cinnamon oil and their synergistic against postharvest decay fungi of grape *in vitro*. *Natural Science, 42*, 169-174.

Taiz, L. & Zeiger, E. (2006). Metabólitos secundários e defesa vegetal. In L. Taiz & E. Zeiger. (Eds.), *Fisiologia Vegetal*. (3th, pp. 309-332). Artmed.

Tian, J.; Huang, B.; Luo, X.; Zeng, H.; Ban, X.; He, J.; Wang, Y. (2012). The control of *Aspergillus flavus* with *Cinnamomum jensenianum* Hand.-Mazz essential oil and its potential use as a food preservative. *Food Chemistry, 130*, 520–527.

Tyagi, A. K. & Malik, A. (2010). In situ SEM, TEM and AFM studies of the antimicrobial activity of lemon grass oil in liquid and vapour phase against *Candida albicans*. *Micron, 41*, 797–805.

Tzortzakis, N. G. (2009). Impact of cinnamon oil-enrichment on microbial spoilage of fresh produce. *Innovative Food Science and Emerging Technologies, 10*, 97-102.

Tzortzakis, N. G. (2007). Maintaining postharvest quality of fresh produce with volatile compounds. *Innovative Food Science and Emerging Technologies, 8*, 111-116.

Ultee, A.; Kets, E. P.W.; Smid, E. J. (1999. Mechanisms of action of carvacrol on the food-borne pathogen *Bacillus cereus*. *Applied and Environmental Microbiology*, 65, 4606-4610.

Viegas, E. C.; Soares, A.; Carmo, M. G. F.; Rossetto, C. A. V. (2005). Toxicidade de óleos essenciais de alho e casca de canela contra fungos do grupo *Aspergillus flavus*. *Horticultura Brasileira, 23*, 915-919.

Vigo-Schultz, S. C. (2008). *Avaliação da indução de resistência no controle do crestamento bacteriano comum do feijão vagem*. 78 p. Tese (Doutorado em Fitopatologia) - Universidade Estadual Paulista, Faculdade de Ciências Agronômicas, Botucatu.

Walsh, S. E.; Maillard, J. Y.; Russel, A. D.; Catrenich, C. E.; Charbonneau, D. L.; Bartolo, R. G. (2003). Activity and mechanisms of action of selected biocidal agents on Gram-positive and negative bacteria. *Journal of Applied Microbiology, 94*, 240-247.

Wendakoon, C. N. & Sakaguchi, M. (1995). Inhibition of amino acid decarboxylase activity of *Enterobacter aerogenes* by active components in spices. *Journal of Food Protection, 58*, 280-283.

Williams, A. C. & Barry, B. W. (1991). Terpenes and the lipid-protein-partitioning theory of skin penetration enhancement. *Pharmacology Research, 8*, 17-24.

Zambonelli, A.; D'aulerio, A. Z.; Bianchi, A.; Albasin, A. (1996). Effects of essential oils on phytopathogenic fungi in vitro. *Journal of Phytopathology, 144*, 491-494.

In: Plant Extracts
Editors: Adriana Giordano and Alfieri Costs
ISBN: 978-1-62417-534-3

Chapter 10

Herbal Therapies against Common Dermatological Disorders

Mahomoodally Mohamad Fawzi [*] ***and Ziaad Dil Hossain***
Department of Health Sciences, Faculty of Science,
University of Mauritius, Réduit, Mauritius

Abstract

Dermatological disorders (DD) have been projected to amount to as high as 34% of all occupational diseases. Amplification of chronic, degenerative and malignant diseases, economic factors and limitations of conventional medicines such as antibiotics resistance have encouraged renewed public interest in herbal therapies (phytotherapies), which is classified as complementary and alternative medicines (CAM) to treat and/or manage common DD. It is of no denying that herbal extracts have proved to be key sources of novel templates for the pharmaceutical armamentarium. Interestingly, herbal extracts geared towards skin disorders, are currently gaining popularity among patients and to a lesser degree among physicians. For example, ancient Egyptians used the fruits and leaves of the bishop's weed (*Ammi majus*) to treat vitiligo, a skin condition characterised by a loss of pigment. Recently, β-methoxypsoralen, a drug produced from this plant which showed promising potential to treat psoriasis and other skin disorders, as well as T-cell lymphoma has been developed. Additionally, herbal medicines for DD have proved to be of great importance in the treatment and/or management of HIV-associated dermatological problems. Since such patients already require panoply of conventional medicines, topical herbal therapies have proved to be of great significance in avoiding unnecessary drug-drug interactions which can set up harmful allergic reactions. Consequently, an advanced investigation should be geared towards traditionally used medicinal plants with the aim of identifying any novel molecule that can be used as alternative remedies against common DD. However, appropriate standardization and exact mechanism(s) of action must be considered for effective and safe uses of such remedies. This present book chapter endeavors to present an overview of the prevalence of common DD and highlights the potential of plant extracts in the management and/or treatment of common human DD.

[*] E-mail address: f.mahomoodally@uom.ac.mu; mmfawzi@gmail.com.

Introduction

Dermatological disorders (DD) affect all ages from the neonate to the elderly and cause harm in a plethora of ways. It has been estimated that DD amount to as high as 34% of all occupational diseases [1]. Amplification in chronic and malignant diseases that are difficult to treat, high cost of new technologies, increasing suspicions of patients on current treatment methods and the high prevalence of side-effects and antibiotic resistance experienced with conventional medicines has led to an increasing public interest in complementary and alternative medicines (CAM) [2].

CAM include an extensive scope of forms, such as herbal therapy, zootherapy, spiritual healing and healing with hands, for instance, acupuncture, massage and so on [1-3]. Several diverse lines of evidence indicate that medicinal plants represent the oldest and most widespread forms of medication. Until the last century, most medicines were derived directly from natural sources such as plant or animal products. Globally, about 85% of all medications for primary health care originate from plants [1]. Indeed, phytotherapy has provided a source of inspiration for novel drug compounds, as plant-derived medicines have made large contributions to human health and well-being. Their role in the development of new drugs could be either by serving as a natural blueprint, or directly as phytomedicine to be used in the treatment or management of diseases. Many commercially proven drugs used in conventional medicine were initially used as crude forms in folklore healing methods [3]. The traditional practice of topically treating DD with plant-derived therapeutic preparations predates the cultures of ancient Egypt and remains vital today in our life [4]. However, contradictorily to general beliefs, CAM are not utterly safe. In fact, the list of undesired side-effects of these traditional remedies is long, for instance, many of the plants employed contain pharmacologically active substances but potentially toxic components also which are capable of adversely interacting with other drugs [5]. Furthermore, The Beijing Declaration, published by the WHO (2008), calls for a partnership between conventional medicines and CAM to help bridge the equity gap in public health and highlights the importance of research to support the development of CAM in delivering appropriate, safe and effective treatments [6].

CAM, particularly are vital components of human health care systems in many developing countries as well as in developed countries [7]. Although conventional medicine may be available in many areas of the world, these CAM have often been used for various historical, cultural and ecological reasons [8]. The uses of CAM are widespread throughout Asia including China, India, Japan, Pakistan, Sri Lanka and Thailand. In Japan, herbal medicinal preparations are more in demand than mainstream pharmaceutical products [9]. 60% to 70% of conventional doctors in Japan prescribe CAM for their patients. In Malaysia, traditional forms of Malay, Chinese and Indian medicines are used extensively [10]. China is the leading country for incorporating CAM into a modern health care system. In this country, CAM accounts for around 40% of all health care delivered and are used to treat roughly 200 million patients annually. According to a survey, almost 7,300 plants have been used in Traditional Chinese Medicine [11].

Many people in the United States are turning to herbal medicine to treat their ills. It has been estimated that up to 50% of the prescriptions dispensed in the United States may contain one or more natural product drugs [12]. Over 20,000 herbal and other natural products are available in the United States [13]. In Europe, some 1,500 species of medicinal and aromatic plants are widely used in Albania, Bulgaria, Croatia, France, Germany, Hungary, Poland, Spain, Turkey, and the United Kingdom [14]. Germany and France together represented 39% of the global retail markets [15]. In 1999, among the 25 best selling drugs in the world, 30% came from natural products (WHO 2001).

According to the WHO, about three quarters of the world's population involving 80% from the developing countries relies on CAM for primary health care needs and most of this treatment involves the use of plant extracts or their active components [6, 16, 17]. In some countries, the use of medicinal plants is often associated with witchcraft and superstition, because people did not have the scientific insight to explain and predict the curative action of plants. One example of such an irrational concept is the Doctrine of Signatures, elements of which are found in many of the healing cultures of the world. It is based on the assumption that the appearance of plants may give clues to their medicinal properties-it is interpreted as God's signature on the plant [18].

Today, the interest in CAM has increased for reasons such as an amplification in chronic, degenerative and malign diseases that are difficult to treat, in parallel with an increase in population lifespan, the high cost of new technologies, the difficulties many patients have in accessing these opportunities, physicians not being able to allocate adequate time to their patients, suspicions of current treatment methods and fear of probable side effects [2]. In most societies today, conventional and traditional systems of medicine occur side by side in a complementary way. The former treats serious acute conditions while the latter is used for chronic illnesses, to reduce symptoms and improve the quality of life in a cost-effective way [18]. There is also a general belief that CAM are safe and more readily acceptable by the body. Medicinal plant research is directed at verifying ethno-medical claims by herbalists with the ultimate aim of isolating active compounds and standardizing the crude extracts used as CAM. There is a great diversity in the plants used in traditional medicine systems and the diversity varies with local cultures and traditions [19].

Ayurveda is the most ancient of all medicinal traditions and probably older than the Traditional Chinese Medicines. It is actually a practical and holistic set of guidelines to maintain balance and harmony in the system. Ayurveda is derived from the Indian words '*Ayar*' (Life) and '*Veda*' (Knowledge or Science) and hence means the Science of Life. Following the system is believed to ensure a long life and is considered to be the instrument for achieving righteousness *(dharma)*, wealth *(artha)* and happiness *(sukha)*. The first school to teach Ayurvedic medicine was at the University of Banaras in 500 BC and the great Samhita (or encyclopedia of medicine) was written. 700 years later, another great enclyclopedia was written and these two together form the basis of the Ayurveda [18].

Chinese Traditional Medicine is believed to be more than 5000 years old and is based on two separate theories about the natural laws that govern good health and longevity; namely *Yin* and *Yang*. The most complete reference to Chinese herbal prescription is the Modern Day Encyclopedia of Chinese *material medica* published in 1977. It lists nearly 6000 drugs out of which 4800 are of plant origin [18].

Despite the increasing use of factory-made synthetic drugs, natural organic healing materials have persisted as the 'treatment of choice' for a multitude of health problems in populations throughout the world [20]. The ancient use of plants for healing purposes forms the origin of much of modern medicine. Many conventional drugs originate from plant sources: a century ago, most of the few effective drugs were plant based. Examples include aspirin (from willow bark), digoxin (from foxglove), quinine (from cinchona bark), and morphine (from opium poppy) [21]. Several new small molecules of natural product-derived drugs have been introduced into therapy in western countries in recent years, including acarbose, artemether, capsaicin, docetaxel, dronabinol, galanthamine, irinotecan, paclitaxel, tacrolimus, and topotecan. This trend is likely to continue in the future, at least for the treatment of disease states such as cancer and infectious diseases. In a statistical survey, it was pointed out that the origin of 30,000 bioactive natural products could be divided between animals (13%), bacteria (33%), fungi (26%), and higher plants (27%) [12].

The knowledge of herbal remedies, developed through trial and error over the centuries, is being used as guide to lead the chemists towards different classes of compounds. 25% of all medical prescriptions are based on substances derived from plants or plant-derived synthetic analogues [18]. It is estimated that there are 250,000 to 500,000 species of higher plants on earth. Only a small proportion (1-10%) of these is used as foods by both humans and other animal species. It is possible that even more are used for medicinal purposes [22]. However, only a relatively small percentage (5-15%) has been systematically investigated for the presence of bioactive compounds. Plants offer the scientist searching for novel bioactive compounds the added advantage of ethnobotanical observations, since many species are used as CAM. It has been estimated that nearly 75% of about 120 biologically active plant derived substances used throughout the world were discovered by following up on leads from traditional medicine [12].

1. The Skin as a Protective Barricade

The skin is primarily an organ of protection. As the body's first line of defense, the skin is continuously subjected to potentially harmful environmental agents, including solid matter, liquids, gases, sunlight, and microorganisms [24]. Although it has remarkable properties that allow for a continuous cycle of healing, shedding and cell regeneration, the skin is subjected to a very wide range of disorders [25].

2. Prevalence of Common Dermatological Disorders - An Overview

Some of the most common DD are acne, eczema, xerosis, psoriasis, contact dermatitis, skin cancers or precursors to skin neoplasms, herpetic infections and warts. In less developed parts of the world, infections predominate as health considerations are different and data are even harder to collect [26]. It has been reported that the prevalence of common DD varies from one nation to another. Acne is thought to be the most common DD in the first world nations, although prevalence and severity of the disease varies genetically across racial and ethnic groups. Papular lesions together with acne hyperpigmented macules and acne scars are reported to be common in the first world nations and widespread amongst Asians. Eczema, in turn, is reported to be an increasing problem in western nations and is affected by genetic, cultural and environmental factors. An increased risk of eczema was also seen in Asian and black populations in Australia. Psoriasis is found to occur in 1-3% of population worldwide. Aging is another skin disorder that affects all populations [26].

2.1. Infectious Skin Disorders

The skin is very prone to attack by panoply of microorganisms. Normally the skin flora, sebum, immune responses and other protective mechanisms guard the skin against infection. Depending on the virulence of the infecting agent and the competence of the host's resistance, infections may result [24].

Bacterial Infection

Bacteria are considered normal flora of the skin. Most bacteria are not pathogenic; however, when pathogenic bacteria invade the skin, superficial or systematic infections may develop. Bacterial infections are classified as primary, or superficial (e.g. impetigo) and secondary, or deep (e.g. infected ulcers). The most common bacterial infections include impetigo, *folliculitis*, furuncles and carbuncles [24]. Impetigo is an acute bacterial infection that occurs superficially on the skin as serous and purulent vesicles that later rupture and forms a golden crust. The causative organisms include β- hemolytic *Streptococci* and coagulase-positive *Staphylococci*. *Folliculitis* is also a bacterial infection of the skin originating within the hair follicle. *Staphylococci* are the usual causative organisms. Furuncles frequently develop from a preceding *Staphylococcal folliculitis* and are usually located in body areas containing hair follicles. Carbuncles are larger *Staphylococcal* abscesses that drain through various points [27].

Fungal Infections

Fungal infections of the skin can be superficial, intermediate or deep. Some are opportunistic and affect a susceptible host while some are truly pathogenic and can infect a healthy person. The superficial infections are called *dermatophytoses* and they are commonly known as *Tinea*, or ringworm. Different forms of *Tinea* affect different body areas. *Tinea* can affect the body (*Tineacorporis*), scalp (*Tineacapitis*), beard (*Tineabarbae),* hands (*Tineamanus*) feet (*Tineapedis*), nail (*Tineaunguium*) or groin and upper parts of the thigh (*Tineacruris*). Individual species of three genera have been identified as the invading fungi in most forms of *Tinea*: *Microsporum*, *Epidermophyton*, and *Trychophyton*. Intermediate fungal diseases invade both the superficial and deeper tissues, *moniliasis* caused by *Candida albicans* being an example. Deep fungal infections involve the epidermis, dermis and subcutis. Infections that are typically superficial may exhibit deep involvement in immunosuppressed individuals [28].

Viral Infections

Viruses are intracellular pathogens. They rely completely on live cells for reproduction. The viruses seen in skin lesion disorders tend to be DNA-containing viruses. Viruses invade the keratinocyte, begin to reproduce and cause cellular proliferation or cellular death. The rapid increase in viral skin diseases has been attributed to the use of birth control medications and corticosteroid drugs, which have immunosuppressive property, and the use of antibiotics, which alter the bacterial flora of the skin [24]. *Herpes simplex*, *Varicella* (also known as chickenpox), *Herpes zoster*, Warts (*Verrucae*), *Rubeola* (Measles), and *Rubella* (also known as German measles) are some of the most common infectious skin disorders of viral origin [27].

2.2. Inflammatory and Allergic Dermatological Disorders

The inflammatory skin diseases are usually localized to the skin and are rarely associated with a specific internal disease. They produce marked variations in normal skin, usually papulosquamous in nature. Inflammation and erythema are common.

Acne, *Lichen planus*, psoriasis, and *Pityriasisrosea*, are among the most common DD of this type. Allergic skin responses involve the body's immune system and are caused by hypersensitivity reactions. They include contact dermatitis, atopic and cummular eczema and drug reactions [28].

Acne vulgaris is a common, chronic, inflammatory disease of the sebaceous glands and hair follicles of the skin, also known as pilosebaceous ducts. *Lichen planus* on the other hand is a relatively common chronic, pruiritic disease involving inflammation and papular eruption of the skin and mucous membranes [28]. Psoriasis is another chronic inflammatory skin disorder characterized by variably pruritic rashes [29]. *Pityriasisrosea* is a self-limiting rash seen in adolescents and young adults. Eczema describes an acutely inflamed weeping skin with vesicles. It is synonymous with the term 'dermatitis' and the two words are interchangeable [30]. Itching may be severe and secondary infections are common [24].

2.3. HIV-Associated Skin Disorders

HIV infection commonly causes significant DD. A rash may even be the presenting feature of the underlying HIV infection. It is estimated that 90% of the HIV positive patients will suffer with a mucocutaneous disorder during the illness. It is also estimated that up to 30% of people with AIDS will suffer from three different dermatoses. These rashes can often be clinically atypical and difficult to diagnose. On top of this, many of the skin problems are resistant to standard treatments. The dermatoses may be arbitrarily divided into cutaneous and opportunistic infections, inflammatory dermatoses, autoimmune dermatoses, drug rashes, cutaneous tumours and specific HIV dermatoses [30].

2.4. Prevalence of Skin Disorders

Skin diseases have a high prevalence throughout the world. In developing countries, infectious diseases are more common, whereas in developed countries inflammatory disorders are more common [30]. Atopic dermatitis now affects 15 to 20% of children in developed countries and the prevalence in cities in developing countries undergoing rapid demographic changes is quickly following suit. Atopic eczema affects between 5 and 20% of children aged up to 11 years at one time or other [31].

A research made on primary school children in Turkey revealed that infectious skin diseases were frequently observed: *Pediculosis capitis* (9.4%), scabies (2.2%), viral skin diseases (3.8%), and fungal infection (0.7%) [32]. A similar study made in Aman, Jordan showed an overall prevalence of skin disorder to be 19.2% [33]. A study made in rural Tanzania also reported that 34.7% of 800 villagers had one or more skin diseases, the most common of which were *Tinea capitis, Tinea corporis*, scabies, acne and eczema [34].

Although limited studies have been made on the prevalence of skin diseases in Ethiopia, the available records are indicative of the high prevalence of this problem in the country. A study made at Black Lion hospital indicated that skin diseases are among the leading causes of hospital visits. The most common skin diseases were allergic skin diseases (25.5%), infections (25.4%), photo-dermatoses (22.9%), followed by papulosquamous diseases (11.4%) and pruritus of unknown origin (3.3%). The pattern of skin diseases observed in this analysis shows that allergic and infectious causes account for three-quarter of skin problems [35].

The prevalence study among school children in South West Ethiopia revealed that 80.4% of 112 children included in the study were found to have one or more skin diseases. Infestations were the most prevalent skin pathology (81.2%) followed by fungal infection [36]. A similar study made in Northwest Ethiopia showed a prevalence rate of 49.2%. *Tinea capitis*, impetigo, infected wounds and verrucae were the most common skin infections [37]. Another study made at 5 elementary schools in Dembia district, North Ethiopia indicated that infectious skin disorders are very common among the children (5 to 15 years old). Only 16% of all children were found to be free of infectious skin diseases [38]. A study on the prevalence of skin diseases in two different rural communities in Southwestern Ethiopia also showed that parasitic infestations (46%) were the commonest complaints among the case studies followed by bacterial and fungal infections (33%) [39]. A related study made at Kazanchi's Health Center in Addis Ababa also reported that bacterial, fungal and viral infections were dominating, comprising 19.4%, 18.5% and 6.5% of the cases [37].

The frequency of skin problems in patients with HIV infection and AIDS is well known. As HIV disease progresses, these skin manifestations may become more severe and varied. In advanced immunosuppression, opportunistic pathogens may present as atypical cutaneous lesions [40, 41]. Cutaneous and mucosal complications eventually occur in nearly all individuals (up to 92%) with HIV infections. A cutaneous eruption is present in approximately 75% of cases. More than 90% of patients will have oral candidiasis at some stage of their disease. *Staphylococcus aureus* is the most common (50%) cutaneous and systemic bacterial pathogen in HIV infected adults. Superficial fungal infections with dermatophytes are also very common [40, 42].

Although generally less well known than most opportunistic infections, some of the noninfectious cutaneous disorders associated with HIV disease are remarkably prevalent. Seborrhetic dermatitis is seen in up to 85% of all HIV-infected individuals. Psoriasis develops in 5% of patients with HIV infection, which is a much higher incidence than the 1 to 2 % reported for the general population. *Pityriasis rubrapilaris* may develop in 20 to 30% of HIV infected patients. Cutaneous drug reactions are the most common manifestations of drug hypersensitivity. A recent study of 684 HIV-infected patients in Boston revealed that 79% had one or more dermatologic diagnosis, 188 of which included cutaneous reactions to drugs [43, 44]. The skin has many essential functions, including protection, thermoregulation, immune responsiveness, biochemical synthesis, sensory detection, and social and sexual communication. Therapy to correct dysfunction in any of these activities may be delivered topically, systemically, intra-lesionally or through ultraviolet radiation.

3. Conventional Approaches

3.1. Treatment of Infectious Dermatological Disorders

Bacterial Infections

Antibiotics are used empirically for the treatment of bacterial skin infections with consideration for resistance patterns. Current antibiotics recommendations include penicillinase-resistant penicillins, azithromycin, clarithromycin, amoxycillin-clavulanic acid, fluoroquinolones and cephalosporins [28, 45]. Systemic antibiotics including tetracycline, doxycycline, minocycline, erythromycin, trimethoprim-sulfamethoxazole, and clindamycin are commonly used medications in most bacterial infections. Topical antibiotics are commonly used in acne and superficial secondary infections. Superficial cutaneous *Staphylococcus* infections are treated with mupirocin [28] Neomycin and gentamycin are active against gram-negative organisms, including *E. coli*, *Proteus*, *Klebsiella*, and *Enterobacter.* Gentamycin is also active against *Staphylococci* and beta-hemolytic *Streptococci* [46].

Fungal Infections

The choice of an antifungal agent is dependent on the site involved, the extent of clinical infection, the age and general health of the patient and concomitant drug therapy. Superficial infections of the skin and nails may be managed either topically or systemically [28]. The treatment of superficial fungal infections caused by dermatophytic fungi may be accomplished with topical antifungal agent (e.g. clotrimazole, miconazole, econazole, ketoconazole, oxiconazole, sulconazole, ciclopiroxolamine, naftifine, terbinafine, and tolnaftate) or with orally administered agents (i.e. griseofulvin, terbinafine, ketoconazole, fluconazole, and itraconazole). Superficial infections caused by candida species may be treated with topical applications of clotrimazole, miconazole, econazole, ketoconazole, oxiconazole, ciclopiroxolamine, nystatin, or amphotericin B [28].

Viral Infections

Few effective medications are available for viral diseases. Acyclovir, famciclovir and valacyclovir frequently are used systemically to treat herpes simplex and varicella infections. Acyclovir and penciclovir are available for topical use in treating mucocutaneous herpes simplex. Podophyllin (25% solution) and podofilox (0.5% solution) are used to treat moist warts of condyloma. The immune response modifier imiquimod, which induces interferon production, is approved for treatment of condyloma and is efficacious for the treatment of verrucae, *molluscum contagiosum*, and superficial basal cell carcinomas [47].

3.2. Treatment of Inflammatory and Allergic Skin Disorders

Urticaria

Urticaria is a common reaction pattern characterized by the presence of itchy or 'burning' edematous swellings (wheals) occurring anywhere in the body [48]. Antihistamines are still the mainstay of treatment for chronic urticaria, although they tend to be more effective in suppressing itching than healing. Oral antihistamines (H_1 blockers) are the most important part of treating the idiopathic cases. Angioedema of the mouth and throat may require urgent treatment with intravenous steroids and subcutaneous adrenaline [28, 30].

Eczema

The basis of direct treatment of eczema is to suppress the symptoms and control or prevent complications. Anti-inflammatory topical steroids are the drugs of choice. In children, when the eczema is very active, stronger steroids such as betamethasone valerate may be required. Normally, clobetasone or mometasone may be adequate for treating areas other than the face; 1% hydrocortisone is the main steroid for the face [49]. Acute flares are often induced by *Staphylococcal* super infection. Systemic antibiotics (flucloxacillin or erythromycin) should be used. As second line of treatment, corticosteroids should be used very seldom. Ultraviolet light and photo chemotherapy with psoralen and long wavelength ultraviolet irradiation may be used in older children and adults. When second line treatment fails or is not suitable, further treatment options include cyclosporin A and azathioprine [49, 50].

Psoriasis

The selection of therapy for psoriasis is multifactorial. The overall health status of the patient and the percent of body surface area involved should be taken into account. Topical therapy for psoriasis includes multiple options the first of which are emollients to soften and moisturize psoriatic plaques. Topical keratolytic agents, formulated with urea or salicylic acid, also are useful in the treatment of localized or limited psoriasis. Topical steroids are the mainstay of treatment for localized psoriasis. A vitamin D analog, calcipotriene, is useful for the topical treatment of psoriasis, as a solution, an ointment or a cream. Anthralin and the topical retinoid tazarotene also are beneficial. The use of systemic medications for the treatment of psoriasis may be indicated by the extent or severity of the disease. Some cancer chemotherapeutic agents have been used with good results in psoriasis, especially methotrexate, thioguanine and hydroxyurea [47].

Acne

Treatment of acne is aimed at decreasing sebum production, decreasing bacteria, normalizing duct keratinization or decreasing inflammation. Topical treatment of acne involves the use of retinoids and antimicrobials. Tretinoin (Acutane®) is a comedolytic agent that loosens existing comedones and prevents the formation of new ones. When combined with topical retinoids, treatment with antimicrobials is more effective than if either medication is used alone. Commonly used topical antimicrobials include erythromycin, clindamycin and benzoyl peroxide. Systemic therapy in acne is most often reserved for inflammatory lesions. Systemic treatment includes retinoids, antibacterials and hormones [28].

3.3. Limitations of Conventional Approaches

Conventional medications are often unsatisfactory for many patients with chronic skin disorders such as eczema because of adverse effects on long-term use [51]. Chronic use of very high potency topical glucocorticoids in the treatment of eczema can lead to cutaneous atrophy and often accompanied by loss of effectiveness. In patients with chronic atopic eczema, therapy with systemic glucocorticoids will generally clear the skin only briefly, but cessation of the systemic therapy will invariably be accompanied by return, if not worsening, of dermatitis. The side effects of daily doses of systemic glucocorticoids preclude chronic use in virtually all patients with atopic dermatitis, and the efficacy of alternate-day regimens in this disease is limited [28]

Most patients with localized plaque-type psoriasis can be managed with mid-potency topical glucocorticoids, but their long-term use is often accompanied by loss of effectiveness. The use of ultraviolet A (UV-A) radiation and either oral or topical psoralens is also effective for the treatment of psoriasis but the photosensitizing potential of psoralens and unknown long term toxicity may limit the use of this therapy. The systemic retinoid etretinate is effective in some patients with severe psoriasis, but is a potent teratogen with an extremely long tissue half-life, thus precluding its use in women of child-bearing age [29].

Dermatologists rely heavily on antibiotics to treat their patients for a variety of conditions. But treating bacterial pathogens with antibiotics has suffered enormous setbacks since the introduction of these drugs. In 1950, the antibiotic arsenal included penicillin, streptomycin and sulfonamides followed by tetracyclines, chloramphenicol and erythromycin. Today, bacterial resistance has compromised the effectiveness of these antimicrobials. Until the last decade, the development of new antibiotics has roughly kept pace with the continuing evolution of bacterial resistance to antimicrobials. Today, the problem of resistance has become far more serious, escalating to crisis proportions [52].

A study conducted in Latin America hospitals to evaluate the frequency of occurrence and antimicrobial susceptibility of bacterial isolates collected from patients with skin and soft tissue infections indicated that 31% of *S. aureus* isolates were resistant to methicillin [53]. In a similar study made at 20 hospitals in 12 different European countries, methicillin resistance in *S. aureus* (MRSA) averaged 22% across Europe, only slightly less than that in isolates derived from blood. Topical antibiotics are widely used but resistance has been encountered with tetracycline, gentamycin and fusidic acid. Systemic toxicity and contact allergy is also a possibility, for example the topical use of neomycin in patients with extensive skin damage may result in deafness [25].

4. Phytopharmaceutical Approaches

The traditional practice of topically treating DD with plant-derived therapeutic preparations predates the cultures of ancient Egypt and remains vital today in our life [4]. For example, ancient Egyptians used the fruits and leaves of the bishop's weed (*Ammi majus*) to treat vitiligo, a skin condition characterised by a loss of pigment. Recently, β-methoxy-psoralen, a drug produced from this plant which showed promising potential to treat psoriasis and other skin disorders, as well as T-cell lymphoma. Alternative therapies are becoming increasingly popular, particularly among patients with chronic skin disorders such as eczema, which are not cured by standard treatments [54].

Recently, the use of topical use of herbs (topical herbal remedies or THRs) is exploding, making it the fastest growing segment of a herbal market projected to be more than $23 billion in 2009. Fifty percent of plastic surgery patients use over-the-counter herbal medications. THRs promise to treat a variety of skin ailments and promote skin rejuvenation making them particularly attractive to baby-boomers. Patients strongly trust in the 1000-year tradition that makes them believe that THRs are effective and that since they are of 'natural' origin, they are 'harmless'[68].

The American Indians used roots of *Polemonium caeruleum* in prescriptions for piles and to treat eczema. The root of *Sanguinaria candensis* is used as a local application in chronic eczema. *Centaurium erythraea* is a highly praised plant by American Indians as a remedy for wound, eczema and all types of sores. *Trifolium pratense* has long been recognized amongst herbalists as a useful remedy in eczema and psoriasis [55]. *Thuja accidentalis* has been used successfully for the treatment of psoriasis, rheumatism and for warts. *Smilax aristolochia efolia* has anti-inflammatory properties, which would further relieve the discomfort of psoriasis. *Stellaria media* contains saponin glycosides, coumarins and hydroxycoumarins, flavonoids, carboxylic acids, triterpenoids and Vitamin C. These constituents give the plant antipruritic, vulnerary, emollient and antirheumatic properties, which when used in an ointment or poultice is excellent for eczema, psoriasis, ulcers and boils. *Anaphalis margaritaceae* is used by American Indians for rheumatism, burns, sores, bruises, swellings and psoriasis. Interestingly a related species found in Nepal, *Anaphalis triplinervis* is used as an antiseptic on wounds. Numerous other plant species including *Artemisia tridentate*, *Trillium erectum*, *Grindelia robusta*, *Clematis virginiana* and *Eupatorium perfoliatum* are used worldwide for the treatment of wounds, cuts and other skin problems [55].

In the hot and humid lowlands of Mexico, half of all recorded species of plants (49%) are used to treat DD, and about one fifth of the use-reports fall into this group. The fresh leaf sap from *Hamelia patens* is applied to stop the bleeding of a wound. The closing of a wound is accelerated by dried leaf powder of *Solanum torvum*, *Solanum rudepannum* or the bark extract of *Heliocarpus* species. *Tithonia diversiflora* is used for scabies and for washing wounds. The use of this species for these purposes can presumably be explained by the presence of sesquiterpene lactones, which are potent inhibitors of inflammatory processes and have antibacterial effects [56].

The African continent is a rich source of plants used for treating various skin diseases. The juice from the aerial parts of *Borreria verticillata* is used in west coast of Africa to treat various skin diseases including eczema. Another species *B. campacta* is used similarly in Southern Africa. *B. natalensis* is used by the Zulus for the treatment of leprosy and furuncles. *Anogeissus leiocarpus*is used in Nigeria for the treatment of skin diseases and the itch of psoriasis. The powdered bark is applied to wounds, sores, boils, cysts, diabetic ulcers and acne with good results. *Andira inermis* is also used to treat psoriasis and ringworm of the scalp. Extracts of *Ammivis naga* have been used to treat psoriasis, where khellin is the major component responsible for the beneficial action [57]. *Plumbago zeylanica* is a plant, which is extremely popular throughout Africa and Asia as a remedy for parasitic skin diseases, especially leprosy, scabies, acne vulgaris, sores and leg ulcers. The root contains plumbagin, which is responsible for its antimicrobial activity. Plumbagin is lethal to a wide spectrum of pathogenic bacteria and fungi at very low concentrations. The Negritos of Philippines use *Artemisia vulgaris* or *Artemisia absinthus* in fermentations for skin diseases including ulcerative sores and wounds. The leaves are used in eczema, herpes andpurulent scabies. *Artemisia herba-alba* is extremely popular throughout much of North Africa; being used for a variety of skin problems such as burns, wound, abscesses, excessive bleeding, herpes lesions, acne and curbuncles. *Artemisia afra* is also popular for skin ailments. It is used externally for boils, carbuncles, and large acne pimples. A poultice of the leaf is applied locally to relieve neuralgia, to the swelling in mumps, and to any glandular or skin inflammation [57].

Table 1. Allergic skin reactions to herbal remedies and aromatherapy oils [58]

Remedy	Indications	Application	Type of side-effects
Aloe vera	Various effects	Topical	Severe and prolonged allergic dermatitis
Aristochol	Cholagogue	Oral	Allergic contact dermatitis
Balsam of Peru	Cosmetic wound healing	Topical	Allergic contact dermatitis
Black curmin oil	Acne, eczema and others	Topical	Allergic contact dermatitis
Camomile	Anti-inflammatory	Topical	Allergic contact dermatitis
Camphor	Various	Topical	Erythematous, papulousoedematous, puriginouseruptions
Cedarwood oil	Aromatheraphy	Topical	Allergic contact dermatitis
Chinese herbal mixtures	Various	Topical	Allergic contact dermatitis
Citrus hystrix	Insect bite	Topical	Allergic contact dermatitis
Curcumin	Various	Topical	Allergic contact dermatitis
Echinaceae	Immune stimulant	Oral	Generalised uticaria anaphylaxis
French marigold	Aromatherapy	Topical	Allergic dermatitis
Garlic	Lipid lowering	Oral	Urticaria, angioderma
Inulahelenium	Massage oil	Topical	Allergic contact dermatitis
Jasmine	Aromatherapy oil	Topical	Contact allergy (Oedematous, exudative and erosive erythema)
Kava	Anxiolytic	Oral	Contact dermatitis
Lavender	Aromatherapy oil	Topical	Allergic contact dermatitis (Oedematous, exudative and erosive erythema)
Mai Meu Dong Tang	Chinese health drink	Oral	Stevens-Johnson syndrome
Olive oil	Massage oil	Topical	Allergic contact dermatitis
Paprika/capsaicin	Mucoskeletal pain	Topical	Dermatitis anaphylaxis
Peppermint oil	Aromatherapy	Topical	Allergic contact dermatitis (erythematouseruptions)
Rosewood	Aromatherapy oil	Topical	Contact allergy (oedematous exudative and erosive erythema)
Tea tree oil	Aromatherapy/cosmetics	Topical	Allergic contact dermatitis
Various aromatherapy oil mixtures	Aromatherapy	Topical or airborne	Allergic contact dermatitis
Ylangylang	Aromatherapy	Topical	Allergic contact dermatitis (erythematouseruptions)

The most common cutaneous side-effect of topical herbal preparations is allergic contact dermatitis. *Aloe vera*, for instance, is considered a weak sensitizer. Likewise, the allergologic relevance of other plants (arnica, chamomile, propolis, tea tree and marigold) widely utilised has also been documented. Plants do not only contain potential allergens, they may also incorporate potent irritants [5]. Ernst, in his work [58] summarises the reported side effects encountered with herbal drugs in dermatology (Table 1).

4.1. Non-Herbal Complementary and Alternative Medicines

Complementary and alternative medicine against common dermatological disorders also includes non-herbal therapies such as zootherapy, homeopathy, acupuncture, aromatherapy and other modalities [59].

Zootherapy, Industrial and Mineral Remedies

Zootherapy is the healing of human diseases by the use of therapeutics obtained or ultimately derived from animals. Prehistoric societies made extensive use of animals and their products as food, to make tools and weapons as well as clothes from animal skins and fur [60]. Despite its prevalence in folklore medicine throughout the world, the phenomenon of zootherapy has been largely been unutilized in comparison to the medicinal plant therapies. The sources of animal-derived remedies include humans, pigs, bee, chicken and slug mucous. Remedies of mineral and industrial origin are also important and include sources such as copper sulphate, salt, soot, old gas among others. They are made use of in the treatment of burn wounds, furuncle, facial wrinkles, abscess, broken bones, bruises, chapped skin, conjunctivitis, wart, acne and dermatitis [61].

Homeopathy

Homeopathy is a treatment method using diluted doses of substances that are normally capable of causing disease in a healthy person. Homeopathy treats a person as a whole and focuses on the patient as a person as well as his pathological condition. Homeopathic medicines are usually selected after a full individualized assessment and case-analysis which includes the medical history of the patient, his physical and mental constitutions among others [62]. The psychological, physical and psychosomatic symptoms and effects of chronic skin diseases are inseparable. Individualized homeopathic treatment can yield a good response in patients with chronic skin disease; therefore the holistic approach used in homeopathy may be a fruitful strategy in parallel with conventional treatments [63].

Acupuncture

Acupuncture therapy refers to a treatment method which involves inserting needles along specific pathways or meridians. The placement varies with the disease being treated. It is sometimes used in conjunction with heat, acupressure or electric stimulation. Over the years acupuncture has certainly been tried for skin disease but unfortunately without consistent success. Most skin diseases are associated with itching which is not only irritating for the patient but may also be stressful. Acupuncture is thus usually used to calm the patient before or during treatment. Acupuncture has been reported to have yielded positive results in the treatment of some skin diseases such as chronic eczema, psoriasis, acne, vitiligo, warts and wrinkles [64].

Mind-Body Healing

Mind-influencing therapies that can enhance treatment of skin disorders include standard psychotropic drugs, alternative herbs and supplements, the placebo effect, suggestion, cognitive behavioural methods and hypnosis [65].

Yoga and Meditation Healing

Kapalbhati is a meditation technique that should be performed in the morning on an empty stomach. It involves sitting in a particular position and carrying out forceful inhalations and exhalations which provide aeration to blood circulation. *Kapalbhati* has been shown to be extremely beneficial in skin treatments such as skin psoriasis, eczema, allergies, leucoderma and vitiligo [62].

Hypnosis

Hypnosis is a CAM that has been used since ancient times to treat medical and dermatological problems. A wide spectrum of dermatological disorders may be improved or cured using hypnosis, including urticaria, acne, atopic dermatitis, psoriasis and vitiligo [66]. *Spiritual healing:* It has also been reported that spiritual healing and prayers are CAM that are used in treating dermatological disorders such as warts, acne and fungal infections. However it is argued that spiritual healing and prayers may only represent a placebo effect [67].

Conclusion

Complementary and alternative medicines, particularly herbal extracts, have been used since time immemorial for the treatment and management of many diseases including infectious and non-infectious diseases as well as skin disorders. Plants may represent a large source of chemicals which can be of great therapeutic value. The widespread practice of traditional medicines in developing countries as well as in developed countries together with the rapidly increasing demand for alternative medicinal remedies largely encourages research and development in the field of CAM. Further motivation is brought about by the Beijing Declaration published by the World Health Organization in 2008 which calls for an integration of traditional remedies in modern medicine. The outcomes of conventional medicine are not always satisfactory in the treatment of DD. Microbial resistance, numerous side-effects associated with conventional medicine, the fear of using corticosteroids as a remedy and overall failure of conventional remedies contribute to such patients' complaints. This is why large number of people are now turning towards traditional remedies. Medicinal plants are highly expected to yield fruitful results in the discovery of new therapeutic molecules, mainly those to be used against resistant microorganisms, in inflammatory and allergic skin disorders and as cosmetics. Herbal medicines as a remedy for skin disorders also prove to be of great importance in the treatment of HIV-associated dermatological problems. Since such patients already require a large number of conventional medicines, herbal therapies prove to be of great value in avoiding unnecessary drug-drug interactions which can set up harmful allergic reactions. An advanced investigation on the less commonly used medicinal plants with the aim of identifying any novel molecule that can be used as remedies against dermatological disorders.

References

[1] Abbasi, A.M., Khan, M.A., Ahmad, M., Zafar, M., Jahan, S., Sultana, S., 2010. *Journal of Ethnopharmacology,* 128(2), 322-335.

[2] Eser, I., Khorshid, L., Demir, Y. And Denat, Y., 2010. *International Journal of Human Sciences,* 7 (1).

[3] Dagne, E., 1996. *Proceedings of the workshop on development and utilization of herbal remedies in Ethiopia, Nazareth*, 40-41.

[4] Jones, M., Schmitz, F., Fluit, A., Acar, J., Gupta, R. And Verhoef, J., 1999. *European Journal of Clinical Microbiological Infectious Diseases*, 18(6), 403-408.

[5] Corazza, M., Borghi, A., Lauriola, M.M. and Virgilli, A., 2009. *Journal of European Academy of Dermatology and Venerology,* 23, 1298-1303.

[6] Picking, D., Youngerb, N., Mitchellc, S. And Delgodaa, R., 2011. *Journal of Ethnopharmacology,* 137(1), 305-311.

[7] Kartal, M., 2007. *Phytotherapy Research,* 21, 113-119.

[8] Mutheeswaran, S., Pandikumar, P., Chellappandian, M. And Ignacimuthu, S., 2011. *Journal of Ethnopharmacology*, 137(1), 523-33.

[9] Baquar, S. R., 1995. *Traditional Medicine in Africa*, Issaq, S. (Editor), East Africa Educational Publishers Ltd., Nairobi, pp. 141-142.

[10] Der, A. H., Kratz, A. M. And Riedlinger, J. E., 2000. *Remington, The Science and Practice of Pharmacy*, 20th Edition, Gennaro, A. R. *et.al.* (Editiors), Lippincott Williams and Wilkins, USA, p. 1769.

[11] Ming, K. J., Khang, G. N., Sai, C. L. And Fatt, C. T., 2003. *Acta Pharmacologica Sinica*., 24(1), 7-21.

[12] Ernst, E., 1999. *Pharmaceutical News, Focus on Contemporary Medicine*, Special Issue, p. 17.

[13] Douglas, K. A., 2001. *Journal of Pharmacy and Pharmacology*, 53, 135-148.

[14] Duke, J. A., 1993. *New Crops*, Janick, J. and Simson, E. (Editors), Wiley, New York, pp. 664-669.

[15] Lanfranco, G., 1992. *Insula*, No.1, pp. 34-35.

[16] Namsaa, N.D., Tag, H., Mandalb, M., Kalita, P. And Dasc, A.K., 2009. *Journal of Ethnopharmacology,* 125, 234-245.

[17] Wei Huang, T.H., Kota, B.P., Razmovski V., and Roufogalis, B.D., 2005. *Basic & Clinical Pharmacology & Toxicology,* 96, 3-14.

[18] GURIB-FAKIM, A., 2006. *Molecular Aspects of Medicine*, 27 (1), 1-93.

[19] Abubakar, M.S., Musab, A.M., Ahmeda, A. And Hussaini, I.M., 2007. *Journal of Ethnopharmacology*, 111, 625-629.

[20] Halberstein, R.A., 2005. *Annals of Epidemiology,* 15, 686-699.

[21] Tyler, V. E., Brady, L. E., And Robbers, J. E., 1976. *Pharmacognosy*, 7th Edition, Lea and Febiger, Philadelphia, pp. 1-3.

[22] Taylor, L., 2000. Raintee Nutrition Inc., pp. 1-5.

[23] Bensouilah, J and Buck, P., 2006. *Aromatherapy.* Abington, U.K: Radcliffe Publishing Company.

[24] Mitscher, L. A., Drake, S., Gollapudi, S. R. And Okwute, S. K., 1987. *Journal of Natural Products*, 50(6), 1025-1040

[25] Simandi, C., 1990. *Pathophysiology, Concepts of Altered Health States*, 3rd Edition, Porth, C. M. (Editor), J. B. Lippincott Company, USA, pp. 108-140.

[26] Kimball, A.B., 2008. *Journal of Investigative Dermatology Symposium Proceedings*, 13, 2-5.

[27] Martindale, *The Extra Pharmacopoeia*, 31st Edition, the Royal Pharmaceutical Society, Reynolds, J. E. F. (Editor), London, 1996, pp. 160-163.

[28] Polotsky, H., 2009. *American College of Physicians, ed. General Internal Medicine.* United States of America: Clerkship Directors in Internal Medicine, 128-132.

[29] Barbara, L. B. And Rosendahl, P. P., 1988. *Pathophysiology, Adaptations and Alterations in Function*, 2nd Edition, Scott, Foresman and Company, London, pp. 683 693.

[30] Thomas, J. L. And Robert, A. S., 1991. *Harrison's Principles of Internal Medicine*, 12th Edition, Wilson, J. D., *et.al* (Editors), McGraw-Hill, Inc., New York, p. 309.

[31] Williams, H., 2002. *British Medical Journal*, 324, 1534-1535.

[32] Fennesy, M., Coupland, S., Popay, J. And Naysmith, K., 2000. *Journal of Epidemiology and Community Health*, 54: 581-589.

[33] Inanir, I., Sahin, M., Gunduz, K., Turel, A. And Ozturkan, S., 2002. *Pediatric Dermatology*, 19(4): 307-311.

[34] Shakkoury, W. And Abu-Wndi, E., 1999. *Archives of Dermatology*, 5(5): 955-959.

[35] Satima, F. T., Mc Birde, S. R. And Leppard, B., 1999. *Archives of Dermatology*, 134(11), 1050-1055.

[36] Sheibeshi, D., 2000. *International Journal of Dermatology*, 39(11), 822-825.

[37] Figueroa, J. I., Fuller, L. C., Abraha, A. And Hay, R. J., 1998. *International Journal of Dermatology*, 37(10), 752 758.

[38] Dagnew, M. B. And Erwin, G., 1991. *Tropical Geographical Medicine*, 43(1- 2), 152-155.

[39] Dagnew, M. B. And Gunther, E., 1990. *Dermatologische Monatsschrift*, 176(4), 219-223.

[40] Chism, J., 1993. *Newsletter of Experimental Aids Therapies*, 7(6), 1-3.

[41] Hiletework, M., 1998. *Ethiopian Medical Journal*, 36(4), 245-54.

[42] Darley, C., 1998. *Aids Patient Care*, 12(11), 849-842.

[43] Tappers, J. W., Perkins, B. A. And Gerger, T. G., 1995. *Clinical Microbiological Reviews*, 8, 440-450.

[44] Cockerell, C. J., 1995. *Journal of Clinical Dermatology*, 9, 531-541.

[45] Wyatt, E. L., Sutter, S. H. And Drake, L. A., 2001. *Goodman and Gilman's the Pharmacological Basis of Therapeutics*, 10th Edition, Hardman, J. G. And Limbird, L. E. (Editors), Mcgraw-Hill, USA, Pp. 1795-1815.

[46] Stulberg, D. L., Penrod, M. A. And Blatny, R. A., 2002. *American Family Physician*, 66(1), 119-124.

[47] Goodman, D. S., Tepletts, E. D. And Wishner, A., 1987. *Journal Of American Academy Of Dermatology*, 16,485-506.

[48] James, S. D., 1998. *National Medicines and Information Center Bulletin*, 4(4), 9-11.

[49] Greaves, M. W., Sabroe, R. A., 1998. *British Medical Journal*, 316, 1147-1150.

[50] Friedmann, P. S. 1998. *British Medical Journal*, 316,1226-1228

[51] Charman, C., 1999. *British Medical Journal*, 318, 1600-1603

[52] Worm, M. And Henz, B., 2000. *Dermatology*, 201(3): 191-195.

[53] Fingold, D. S., 1999. Apua Newsletter, 17(3): 4-5.
[54] Naguib, Y., 2002. *Vitamin Retailer*, Pp. 1-7.
[55] Keane, F. M., Munn, S. E., Vivier, A. W., Tylor, N. F. And Higgins, E. M., 1999. *British Medical Journal*, 318: 363-564.
[56] Dweck, A. C., 1996. *Ethnobotanical Use of Plants, Part 4*: American Continent, Article for Cosmetics And Toiletries, Pp. 1-5.
[57] Leonti, M., Vibrans, H., Sticker, O. And Hinrich, M., 2002. *Journal Of Pharmacy And Pharmacology*, 53: 1653-1668.
[58] Ernst, E., 2000. *British Journal Of Dermatology*. 143, 923-929
[59] Vender, R. B., 2002. *Skin Therapy Letter,* 7(2).
[60] Efraim, L., 2003. *Journal of Ethnopharmacology*, 85:107-118.
[61] Quave, C.L, Pieroni, A., Bennet, B.C., 2008 *Journal of Ethnobiology and Ethnomedicine*: 4(5).
[62] Soni, P., Patidar, R., Soni V., Soni, S., 2010. *International Journal of Pharmaceutical and Biological Archives*, 1(3): 220-227.
[63] Intamura, R., 2007. *Complementary Therapies in Medicine,* 15(2): 115-120.
[64] Rosted, P., 1992. *Acupuncture in Medicine*, 10: 66-68.
[65] Shenefelt, P.D., 2011. *International Journal of Dermatology*, 50(11): 1309-1322.
[66] Shenefelt, P.D., 2000. *Archives of Dermatology*, 136930: 393-399.
[67] Gonul, M., Gul, U., Çakmak, S.K, Kılıç, S., 2009. *International Journal of Dermatology,* 48: 639-644.
[68] Krajewski A, Garg M., Chandawarkar R.Y. 2010. Topical herbal remedies: Research opportunities for plastic surgeons. *Journal of Plastic, Reconstructive & Aesthetic Surgery* 63, 896e905.

In: Plant Extracts
Editors: Adriana Giordano and Alfieri Costs

ISBN: 978-1-62417-534-3

Chapter 11

Polyphenols as a Source of Antimicrobial Agents against Human Pathogens

*Allan de Jesus dos Reis Albuquerque, Patrícia Maria de Freitas e Silva, Ana Lúcia Furtado de Almeida Cavalcante and Fábio Correia Sampaio**

Federal University of Paraíba, Campus I, Health Science Center, Laboratory of Oral Biology, João Pessoa, Brazil
State University of Paraíba, Campus I, Bodocongó, Pharmaceutical Department, Campina Grande, Brazil

Abstract

Polyphenols or phenolic compounds are groups of secondary metabolites that are frequently found in extracts of medicine plants. Many medicine plants have been tested in clinical and experimental studies as antimicrobials. As least for some species, this biological activity can be partially explained by the presence of phenolic compounds. These compounds comprise a wide variety of molecules that are usually divided into flavonoids and nonflavonoids structures and sequentially into several sub-groups of compounds depending on the oxidation state. Apart from this clear chemical classification, the literature is rather confusing in presenting the health effects of polyphenols. The major aim of this review is to present an updated material on the effects of polyphenols acting as antimicrobial agents against human pathogens. There is a wide variety of proposals for using phenolic compounds as antimicrobial agents but very little is discussed about the real potential of these molecules when incorporated in medicines and surface cleaning agents. The antibacterial activity of these molecules against

* Federal University of Paraíba, Campus I, Health Science Center, Laboratory of Oral Biology, João Pessoa, Brazil, 58051-900. State University of Paraíba, Campus I, Bodocongó, Pharmaceutical Department, Campina Grande, 58429-500.

pathogens of the oral cavity is particularly evidenced since limitations on the use of phenolic compounds in the oral cavity can exist (e.g. stains of the teeth and restorations). Examples of polyphenol-rich medicine plants that can show antibiotic activity against resistant bacteria are presented. The synergism and antagonism effects of the combination of phenolic compounds and other drugs are also discussed due to the fact that this is of upmost interest for health professionals. Finally, some medicine plants are presented as source of antibacterial phenolic compounds but framing these plants in the perspective of advantages and disadvantages for promoting health in humans.

1. Introduction

According to World Health Organization (WHO) medicine plant is "any plant that has, in one or more organs, substances that can be used for therapeutic purposes or which can be semi-synthetic drug precursor" (WHO, 2005). Nowadays, about half of the drugs used in therapy are originated, directly or indirectly, from natural products, especially from medicine plants that remain an important source of drugs (Carvalho et al., 2007).

Among the activities of the many products of vegetable origin, antimicrobial activity (antibacterial, antiviral, antifungal) can be regarded as a very important one particularly at this specific period of human history due to bacteria resistance that are constantly imposing new scientific challenges. In spite of great development in drug therapy, infectious diseases remain as most common reasons for mortality in many developing countries. Therefore, the search for new antimicrobial agents is becoming a hot topic since resistant microorganisms and fatal opportunistic infections associated with AIDS, cancer chemotherapy and transplantation are registered every year (Penna et al., 2001). For centuries, antimicrobial properties of substances in extracts and essential oils produced by plants are known empirically (Jansen et al, 1987).

Several studies have been developed to discover new antimicrobial agents from plant extracts and other natural products to be used in pharmaceutical products and cosmetics but few were clinical trials (Michelin et al., 2005; Ostrosky et al., 2008). Researches on antimicrobial activity of extracts and essential oils of native plants have been reported in many countries that have a diverse flora and a rich tradition of using medicine plants as antimicrobial agents, such as Brazil, Cuba, South Africa, Rwanda, India, Mexico and Jordan (Araújo, 2010). Many of these medicine plants are sources of active phytochemicals such as polyphenols.

The term polyphenols or phenolic compounds refers to a large group of molecules found in vegetables, fruits, cereals, tea, coffee, cocoa, wine, fruit juice and soy. In plants, they have photoprotection function, protection against microorganisms and insects, and are responsible for pigmentation and some organoleptic characteristics of food. The polyphenols are important anti-oxidants in plants and they are also responsible for the color and partially respond for the taste and flavor of many plants, and fruits such as apples and grapes. The polyphenols are effective hydrogen donors and therefore are regarded as potent antioxidant phytochemicals.

This antioxidant ability can be related to the position of the hydroxyl groups in their chemical structure.

2. Classification

In plants, polyphenols minimize the effects of UV radiation against abiotic agents and protection against herbivory (Daglia, 2012). Polyphenols are usually solid, crystalline, toxic, caustic and with low solubility in water. Many of these substances are classified as natural antioxidants and have therapeutic properties, being present in food and medicine plants (Boudet, 2007).

Polyphenols are characterized by the presence of several phenol groups (i.e., aromatic rings with hydroxyls) which derive from L-phenylalanine (Boudet, 2007). Its basic structure is rather simple and it is presented in figure 1.

Polyphenols are usually natural substances of vegetable origin, which have one or more hydroxyl groups linked to an aromatic ring. They are, therefore, phenols but may have one or more hydroxyl groups and more than one aromatic ring. Based on its structure and considering the way polyphenolic rings bind to each other, they can be classified as nonflavonoids and flavonoids. But more detailed classification is also accepted (Figure 2).

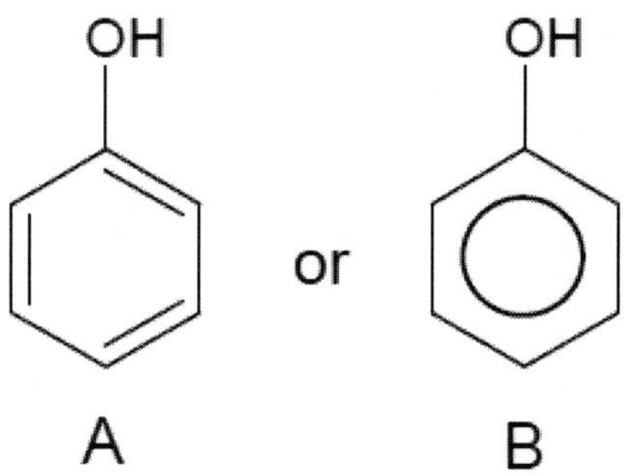

Figure 1. Basic structure of polyphenols and the chemical structure showing the aromatic ring (benzene) which carries a hydroxyl group. The right chemical structure (with a solid circle that indicates hybrid benzene) is an alternative way to show the resonance feature of double ligants of the chemical structure.

Table 1. Classification of polyphenols based on the number of carbons atoms

Atoms number	Basic carbon skeleton	Phenolic compounds
6	C_6	Simple Phenols and Benzoquinones
7	C_6 - C_1	Phenolic acids
8	C_6 – C_2	Acetophenones and phenyl acetic acids
9	C_6 - C_3	Hydroxycinnamic acids (hydroxycinnamates), phenylpropanoids
10	C_6 - C_4	Naphthoquinones
13	C_6 - C_1- C_6	Xanthone
14	C_6 - C_2- C_6	Stilbenes and anthraquinone
15	C_6 - C_3- C_6	Flavonoids and isoflavonoids
18	$(C_6 - C_3)_2$	Lignans and neolignans
30		Biflavanoids
N	$(C_6 - C_3)_n$ $(C_6)_6$ $(C_6\text{-}C_3 - C_6)_n$	Lignin, catechol melanins and flavolans

Source: Adapted from Castillo and collaborators (2012).

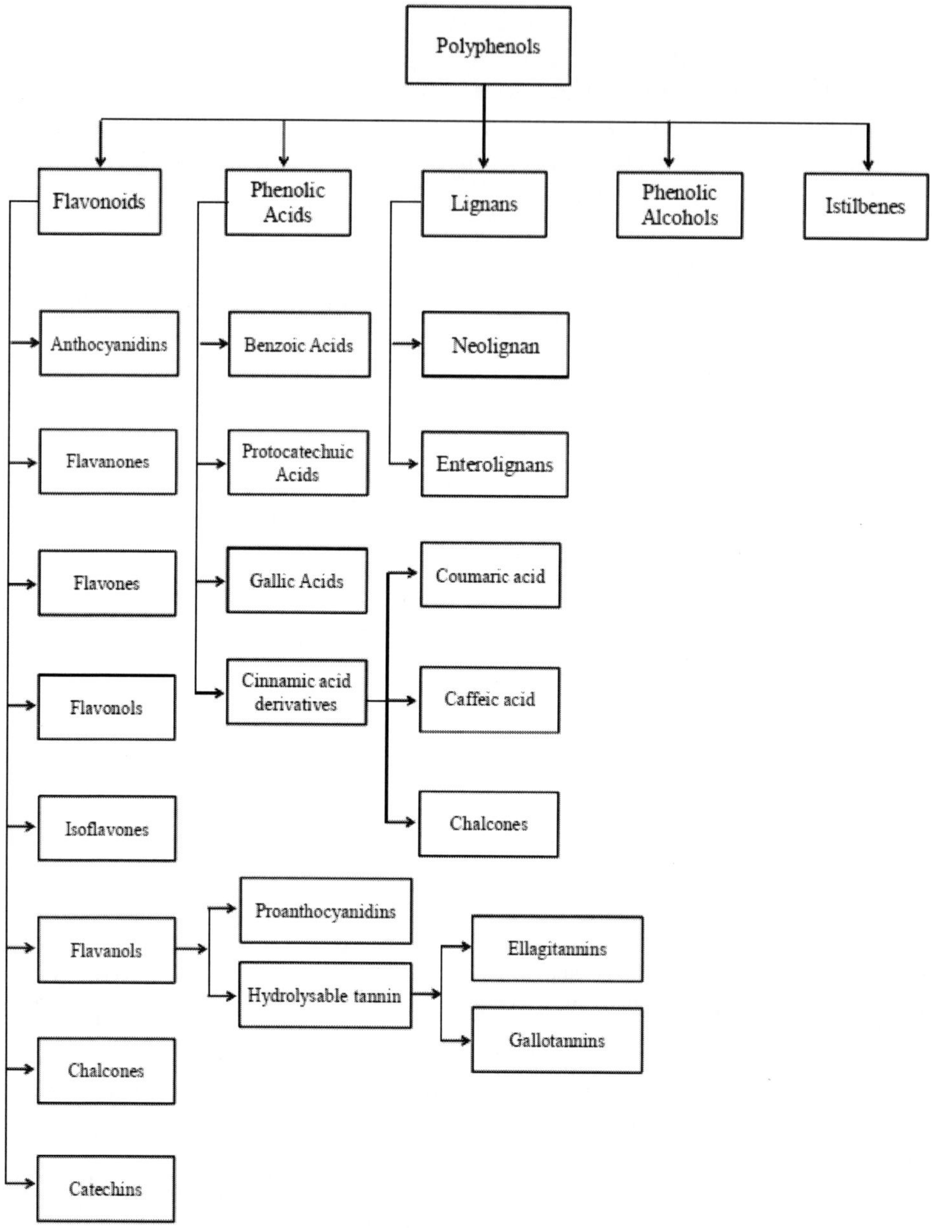

Figure 2. Classification of polyphenols and some derivates.

In general, the most frequent pattern to be considered for polyphenols classification is the number of carbon atoms in their main structure. A list of some polyphenols according to this criterion is presented in table 1. Polyphenols are polyhydroxylated phytochemicals which have common structures with uncountable variations in the other parts of the molecule. Hence, the possibilities of chemical combination are so many that eight thousand compounds have already been isolated. The researchers on this field must bear in mind that selecting one single molecule for testing its antimicrobial activity can leave behind several compounds of the same group to be tested in the future.

In order to simplify the description of these phytochemicals the polyphenols can be subdivided in three main subclasses: the flavonoids, the phenolic acids, and the stilbenoids. The former are by large the most studied subclass.

Flavonoids: This polyphenol group is characterized as containing two or more aromatic rings, each bearing one or more phenolic hydroxyl groups, and connected by a carbon bridge. Taking into account this structure the flavonoids are distributed into 14 classes (Simões et al., 2007). These phytochemicals are of upmost interest due to the fact that may have antimicrobial activity against resistant bacteria. Recent studies have partially clarified the mechanism of action of flavonoids for their direct antibacterial activity, synergism with antibiotics, and suppression of bacterial virulence. A major mechanism is related to the damage of the cytoplasmic membrane possibly by generating hydrogen peroxide (Sato et al., 2000). Other mechanisms are connected to inhibition of cell wall synthesis and inhibition of cell membrane synthesis (Bae et al., 1999; Kim et al., 1999).

In addition, it is also suggested that flavonoids may interfere with specific intracellular or surface enzymes and many bacterial virulence factors, such as toxins, enzymes, and signal receptors (Cushnie and Lamb, 2011). This evidence opens possibilities to the development of medicines able to have synergic effect with antibiotics creating a wide field of research in health sciences.

Phenolic acids: Phenolic acids are a subclass of phenolic compounds, encompassing hydroxybenzoic acids (C6–C1structures) and hydroxycinnamic acids (C6–C3 structures) (Table 1) (Dai and Mumper, 2010). Phenolic acids are usually divided in two main subgroups: benzoic acids (7 carbon atoms) and cinnamic acids (9 carbon atoms) (Fresco et al., 2006). The majority of the isolated compounds are gallic acid, chlorogenic acid, caffeic acid, ferulic acid, p-coumaric acid, and gentisic acid.

Phenolic acids are frequently recognized as direct antioxidants. However, these phytochemicals may also exhibit indirect antioxidant effects through the induction of endogenous protective enzymes, and beneficial regulatory effects on signaling pathways (Stevenson and Hurst, 2007). The relative low solubility of phenolic compounds in apolar media may be seen as a drawback when considering their use as antioxidants for stabilizing unsaturated lipids in formulated complex emulsified systems (Reis et al., 2010). Considering their chemical structure, the numbers of hydroxyl groups bonded to the aromatic ring and their positions are probably important factors influencing the activity of phenolic acids (Cueva et al., 2010).

Phenolic acids have attracted much attention as antioxidant compounds than for their antibacterial activity. These phytochemicals have antimicrobial activity and hold promise for application as preservatives in food and food-packing materials. The current knowledge on structure–function relationships of the antimicrobial activity of phenolic acids does not account for this diversity of compounds (Sánchez-Maldonado et al., 2011). A complex situation for studying phenolic acids is due to the fact that these acids can be originated from different sources: directly from dietary items; from endogenous sources and from the microbial metabolism of other dietary phenolic compounds.

Stilbenoids: this subclass is regarded as the smallest among polyphenols. Stilbenoids comprise polyhydroxylated stilbenes and the main representative example is the compound called resveratrol. The *trans*-resveratrol gained much of its popularity due to the anticancer activity related to the potent antioxidant activity combined also because it is present in grapes and wines.

However, studies show low bioavailability of this compound and low concentration in wine (Cowan, 1999). Resveratrol is a phytoalexin produced naturally by several plants when under attack by bacteria or fungi. Thus, it is suggested that these phytochemicals can be an interesting source of antimicrobial molecules (Eloff et al., 2005).

3. Health Effects

Polyphenols are secondary metabolites found in aerial parts of higher plants. They have received much attention from the scientific community for its numerous biological effects, such as free radical scavenging activity (antioxidant), modulation of some specific enzymes activity, inhibition of cell proliferation as well as its potential as anti-allergic, anti-inflammatory, antihypertensive, antimicrobial and antineoplastic (Haslam et al., 1989; Manach et al., 2004; Nagori et al., 2010; Daglia, 2012).

The first observations of polyphenols activity were their powerful antioxidant effect. For *in vitro* experiments it was shown that they were capable of scavenging a wide range of reactive oxygen, nitrogen, and chlorine species, such as superoxide anion, hydroxyl radical, peroxyl radicals, hypochlorous acid and peroxynitrous acid (Petti and Scully, 2009). Polyphenols high consumption has been associated with decreased cancer risk, cardiovascular diseases and neurodegenerative diseases. Many of these biological activities were evaluated *in vitro* with pure enzymes, cultured cells or isolated tissues (Cowan, 1999; Petti and Scully, 2009; Cueva et al., 2010).

Polyphenols antimicrobial activity found in plant foods and medicine plants has been extensively investigated against a wide range of microorganisms (Cowan, 1999; Daglia, 2012). Among polyphenols, flavonols, tannins and flavan-3-ol (sometimes referred to as flavanols) received more attention due to its broad spectrum and antimicrobial activity when compared with other polyphenols and the fact that most of them are able to suppress microbial virulence factors (such as inhibition of biofilm formation, reduction of receptor ligands adhesion, and bacterial toxins neutralization) and show synergism with antibiotics (Aron and Kennedy, 2008; Daglia, 2012). Interestingly all these biological activities have a great impact in general and oral health.

A challenge to face at this state of the art in Pharmacognosy is to evaluate all the available molecules against different microorganisms and applying specific and appropriate techniques. Table 2 presents a list of some families of polyphenols that have been already tested for their antimicrobial activity.

Taking into account the broad-spectrum antibacterial activity of polyphenols, these phytochemicals can be a source of drugs against many infection diseases.

Recent studies have shown that flavonoids can reach an impressive low minimum inhibitory concentration (MIC) < 0.6 μg/mL. The MIC is the lowest level of the substance that can interfere in bacterial growth and at this level only antibiotics can reach such efficacy (Cushnie and Lamb, 2011).

Moreover, it must be pointed out that although the focus of this chapter is polyphenols as antimicrobial agent and bacteria-related-diseases, most polyphenols are regarded as micronutrients of our diet and plays a relevant role in the prevention of degenerative diseases (Manach et al., 2004; 2005).

Table 2. List of some polyphenols compounds related to their respective activity against some microorganisms (bacteria, fungi and virus)

Classification of Polyphenols		Antimicrobial activity			References
		Bacteria	Fungi	Virus	
Flavonoids		*V.cholerae, S.mutans, C.jejuni, C.perfringens, E.coli, B.cereus, H.pylori, S.aureus, Lacidophilus, P.oralis, P.gengivalis, P. melaninagem, F.nucleatum, C.pneumoniea.*	*C.albicans, M. gypseum, T. mentagrophytes, T.rubrum.*	Adenovirus, Enterovirus, Flavivirus, HIV.	(Cowan et al., 1999)
Phenolic Acids		*S.aureus, L. brevis, L. fermentum, Listeria monocytogenes, E. coli, Pseudomonas aeruginosa, Bacillus cereus, B. megaterium, B. subtilis, E. coli, M. tuberculosis.*	*C. albibans, C. krusei, C. parapsilosis, C. tropicalis*	HIV.	(Cowan et al., 1999; Fernández et al., 1996; Cueva et al., 2010).
Coumarin		*Streptococcus mutans, S. aureus.*	*Candida albicans.*	-	(Bose et al., 1958; Khanet al., 2004)
Lignans		*P. acnes, S. aureus, Enterococcus faecalis, E. coli, M. tuberculosis.*	-	HIV.	(Cowan et al., 1999; Favela-Hernández et al., 2012)
Phenolic Alcohols		*S. aureus, M. tuberculosis.*	-	-	(Cowan et al, 1999)
Tannins	Condensed	*S.mutans, E.coli, S.aureus.*	*Candida albicans.*	Influenza A, Herpes simplex (type 1), HIV.	(Cowan et al., 1999; Ishida et al., 2006)
	Hydrolyzed	*Salmonella, Staphylococcus, Helicobacter, E.coli, Bacillus, Clostridium, Campylobacter, Lysteria.*	*Candida parapsilosis.*	Epstein-Barr vírus,Herpes virus 1 e 2, HIV.	(Cowan et al., 1999)
Istilbens		*S. aures, P. aeruginosa, P. syringae Bacillus subtilis.*	*A niger, B. cinerea, C.herbarum M. aucupariae. A.niger, B cinerea, C. herbarum*	-	(Eloff et al., 2005; Aslam et al., 2009)

4. Antimicrobial Activity

Antimicrobial properties of some classes of polyphenols have been proposed either for the development of new food preservatives to prevent synthetic preservatives, or for developing new therapies for treating several different microbial infections, given the increase in microbial resistance against conventional antibiotic therapy (Daglia, 2012).

The experimental protocol is a key point for comparing studies that tested polyphenols against bacteria and fungi.

The majority of the *in vitro* protocols were developed for testing inorganic drugs that are frequently of a very low level of impurities, colorless, with polar features (i.e., soluble in many polar solvents such as water). As a result, some methodological items are important before understanding the results of an antimicrobial assay (Percival et al., 2006). The extraction procedures of the phytochemicals are of upmost relevance since a high temperature may interfere in some constituents as well as the polarity of the solvents (Cowan, 1999). It is rather common that Gram positive and Gram negative bacteria differ in susceptibility to antimicrobials of natural origin (Brondani et al., 2003).

In addition, the bacterial tests must consider the conditions of growth. For instance, the minimum inhibitory concentration (MIC) of most oral antimicrobials are increased in the magnitude of ten to hundred folds when oral bacteria are cultivated as single or multispecies biofilms (Sampaio et al., 2009).

The combination of drugs is also a relevant topic. There is a great difference when testing an isolated phytochemical of polyphenol and the extract of a medicine plant. The former is a drug with known chemical characteristics such as molecular weight. The extract is a mixture of compounds that may act by synergism or additive effect and may have less activity due to antagonism effect of its own constituents (Cowan, 1999).

The glycosylation of polyphenols in nature helps to protect these molecules from oxidation, extending their half-life. This is an important step for the plant.

However, glycosylation can increase solubility of the molecule. Moreover, glycosylation of a polyphenol can affect its biological activity. The glycosylation is the chemical reaction in which a carbohydrate (glycosyl donor) is attached to a hydroxyl or other functional group of a target molecule. In summary, glycosylation of polyphenols can weaken their binding affinities (Xiao and Kai, 2012).

In addition, glycosyl groups can react with some acids, forming an ester such as glucose 1-phosphate which in some environment can be a nutritional source of energy for bacteria. These chemical observations may provide strange or unexpected results when testing polyphenols under laboratorial conditions (Cavalcante, 2010).

It is important to bear in mind that antimicrobial research may focus in different perspectives when comes to consider natural products as source of antimicrobials. For instance, increases in antibacterial activity can be accompanied by decreases in selectivity and this is not interesting for drugs developed for target pathogens. Future structure–activity studies should therefore perform cytotoxicity testing in parallel with antibacterial testing (Cushnie and Lamb, 2011).

5. Polyphenols and Oral Health

The oral mouth is generally regarded as one single entity, but it comprises several compartments when comes to the study of antimicrobials. For instance, saliva is a key component for bacterial development that is frequently neglected when testing antimicrobial activity for oral health (Guggenheim et al., 2001; Guggenheim et al., 2004).

The mouth is warm and moist, and is able to support the growth of a distinctive collection of microorganisms and can be easily colonized by microorganisms. In fact, the variety of these cells in the mouth is so high that it can reach half of the 700 species of commensal microorganisms that can be found in the whole human body. (van der Reijden et al., 2001; Marsh et al., 2011). This great amount of bacteria is present in the saliva and oral fluids and mainly on tooth surfaces and tissues such as on the tongue. Although the same bacteria species are regularly found at the same site, the oral microorganisms can get adapted to multiple environmental conditions. Variations in pH, nutritional sources, oxygen availability and humidity can modulate the colonization and affect the ecology within these microenvironments. It is important to bear in mind that metabolic activity can be maintained even during low consumption of food by the host. Salivary glycoproteins can be degraded under such conditions. During high consumption of fermentable carbohydrates the metabolic activity of the oral microorganisms is enhanced considerably (Marsh et al., 2011).

The oral microorganisms do not adhere directly to tooth surfaces. The acquired pellicle (a glycoprotein layer) plays an important role in bacterial colonization since it can contribute for the first colonizers which are streptococci species.

Taking into account the microorganisms-related-diseases (e.g. dental caries, periodontal diseases, halitosis), a poor oral health can be regarded a disequilibrium on the microbial biofilms present in the oral mouth. It is very important to control any oral disequilibrium because oral streptococci are frequently associated with infective endocarditis. Moreover, periodontal disease can increase the risk for cardiac infarction, pneumonia and many other conditions (Genco et al., 2002).

The most effective antimicrobial used for controlling oral biofilms is chlorhexidine gluconate a cationic bis-biguanide biocide with low mammalian cell toxicity and broad-spectrum antibacterial activity (Guggenheim and Meier et al., 2011).

The major mechanism of action of chlorhexidine is membrane disruption but inhibition of proteolytic and glycosidic enzymes in the bacteria may also take place. In spite of its great effectiveness this substance shows many side effects (e.g.: tooth stain, bad taste and bacteria resistance). Since seventies, the use of chlorhexidine mouth washes is recommended for short periods only (Flötra et al., 1971). As a result, an alternative antimicrobial for chlorhexidine became a focus for many microbiologists.

Oral pathogens virulence can be inhibited or conversely strengthened by dietary factors, use of antimicrobials, combination of medicines and systemic conditions of an individual. Dental caries, periodontal diseases and halitosis are three major oral problems that are directly related to presence of bacteria. Oral candidiasis is also frequently observed in some patients and polyphenols can be a therapeutic strategy for controlling the infection (Vasconcelos et al., 2003).

Dental caries: The assumption that polyphenols has a positive effect on cariogenic bacteria (*Streptococcus mutans* and *Streptococcus sobrinus*) is not recent (Elvin-Lewis et al., 1982; Taniguchi et al., 1993; Matsumoto et al., 1999). Several *in vitro* studies have indicated that catechins and flavonoids can prevent the development of a cariogenic biofilm. Among the sources of active anti-caries polyphenols are: cocoa beans (*Theobroma cacao* L.), apple, large cranberry (*Vaccinium macrocarpon* Ait.), pomegranate (*Punica granatum* L.), green tea (*Camellia sinensis* L.), cashew tree bark (*Anacardium occidentale* L.); propolis (Vasconcelos et al., 2003; Percival et al., 2006; Pereira et al., 2006; Cavalcante, 2010). Recently, a review on this topic highlighted the possible biological activities of polyphenols on oral bacteria: a) a direct effect against *S. mutans* (possibly in cell wall); b) inhibiting the adherence of bacterial cells to the tooth surface due an interaction with microbial membrane proteins and c) the inhibition of glucosyltransferase and amylase affecting cell metabolism (Ferrazzano et al., 2011). Although more *in vivo* studies are necessary for providing conclusive evidence on this subject, the pathways for future studies have been extensively discussed by other authors (Petti and Scully, 2009; Jeon et al., 2011). It is true that there is a big gap between the *in vitro* studies starting point and the final medicine ready to be prescribed.

After extensive evaluation of *in vitro* studies, it has been established that a minimum inhibitory concentration (MIC) acceptable for antibacterial isolated natural compounds should be below 100 μg/mL (Ríos et al., 1988; Ríos and Recio, 2005). For antifungal activity, it is taken 500 μg/mL as a threshold concentration for a good MIC (Sartoratto et al., 2004). These values are far below most toxic doses for many phytochemicals. However, the effectiveness of the compounds must be tested in biofilm environment and under clinical influence of saliva and other organic constituents (Marsh, 2004; Sampaio et al, 2009; Jeon et al., 2011).

Another point to be considered are the limitations of this effect on a regular basis. For instance, polyphenols is frequently present in very low concentration to be regarded an effective antimicrobial agent in the oral mouth. Moreover, when present in the oral mouth it can be inactivated by some salivary proteins or enzymes. Another important point to consider is the drawback of having similar undesirable side effects as chlorhexidine mouth wash. A simple experimental evaluation of resin base dental material exposed to pomegranate extract diluted to MIC for *S. mutans* at 65 μg/mL showed an evident reduction in the original color of the material (data not shown). Due to possible influence of these factors on polyphenols effectiveness against oral cariogenic bacteria, some authors regard these compounds as moderate antimicrobials (Petti and Scully, 2009). This cannot be considered a failure because a moderate effect might be preferable than a total killing agent. One must bear in mind that modern assumption on cariology is that antibacterial agents must bring back the biofilm equilibrium and not its complete depletion (Marsh et al., 2011).

The Brazilian medicine plants *Ziziphus joazeiro* Mart. (juá), *Psidium guajava* L. (guava) *Caesalpinia pyramidales* Tul. (catingueira), *Aristolochia cymbifera* (cipó mil homens) *Caesalpinia ferrea* Mart. (jucá) and the possible exotic plant *Cocos nucifera* L. (coco), showed antimicrobial activity against important pathogens such as *Streptococci* sp, *Escherichia coli*, *Salmonela* spp, *Shigella* spp, *Proteus* spp e *Pseudomonas aeruginosa.* For some studies, bacteria was cultivated in biofilms providing more data for future clinical trials and the development of drugs (Alviano et al., 2008; Carvalho *et al.,* 2002; Sampaio et al., 2009). Other promising plants candidates as source of active polyphenols for dental caries are *Punica granatum L.* (pomegranate) and *Anacardium occidentale* L. (cashew) due to their potent antimicrobial activity against *Streptococci* sp and availability of plant material (Pereira et al., 2006; Cavalcante, 2010).

Periodontal diseases: Periodontitis is an inflammatory disease of the periodontium that affects around 70% of adult population of many countries including United States (Genco et al., 2002). Although these figures are questioned by differences in diagnostic tools, there is an overall consensus that it is neglected in many health systems (Dye, 2012). The disease is characterized by a progressive destruction of the tissues supporting the tooth and its primary etiology is an ill-defined series of microbial infections basically Gram negative types such as *Bacteroides forsythus*, *Porphyromonas gingivalis, Prevotella intermedia* and *Fusobacterium nucleatum*. The progression of the disease is frequently periodic with episodes of rapid tissue destruction but variations in patterns of disease activity are observed. The inflammatory process can be related to smoking, diabetes and other general conditions that can increase the oxidative stress. Providing antioxidants from dietary sources might control periodontal disease development.

Although chlorhexidine 0.12% is the most frequent antimicrobial indicated for periodontitis, polyphenols have been tested with some relative fair success (Petti and Scully, 2009; González et al., 2012). Catechins from wine, green tea and cranberry showed antimicrobial activity against *Porphyromonas gingivalis* and *Prevotella intermedia* (Ho et al., 2001).

A randomized clinical trial with a low concentration of pomegranate in toothpastes showed a strong reduction in gingivitis which is a general inflammation of the gums in adolescents (Pereira, 2002). Nevertheless, like in dental caries studies, the majority of the surveys are laboratorial type and few are clinical ones. So far, moderate evidence can be accounted for polyphenols controlling periodontal diseases (Petti and Scully, 2009).

Halitosis: Halitosis is generally defined as breath that is very offensive to others. It is a multifactorial condition that is caused by the interplay of several reasons such as: bacterial coating of tongue, periodontal disease, systemic disorders and different types of food (Kleinberg and Codipilly, 1999). Most cases of halitosis are not related to stomach problems as many researchers believed. Actually, halitosis is frequently related to Gram negative bacteria in the oral cavity mainly on the dorsum of the tongue. These microorganisms (*Treponema denticola*, *Porphyromonas gingivalis* and *Tannerella forsythia*) can produce volatile sulfur compounds which are responsible for the unpleasant breath (Veloso, 2006).

The majority of the studies with natural products for controlling halitosis have tested essential oils due to the volatile characteristic of these compounds. However, this strategy can be just a masking agent capable to decrease odor (Veloso, 2006). Since the presence of these Gram negative microorganisms is responsible for producing bad breath, any type of treatment approach that has impact in the local flora has the potential of reducing halitosis. Thus, evaluation of medicinal plants, some of those being rich in polyphenols, were tested in a double blind clinical trial. It could be concluded that hydrolysable tannins were able to reduce the malodor but no organic compound showed good results in a similar level as chlorhexidine (Veloso, 2006).

Recently, it was observed that green tea, which is rich in catechins, can modify odorant sulfur components which indicate a mask effect. So far, it is unclear if polyphenols can effectively reduce halitosis and what are the main mechanisms of action of these compounds, if there are some (Narotzki et al., 2012; Varoni et al., 2012).

Oral candidiasis: Candidiasis is a fungal infection of the most superficial part of the lining of the mouth that is generally caused by *Candida albicans* (Vasconcelos et al., 2003). This condition causes either a red or white (sometimes both) lesions in the oral mucosa and for most patients the infection is quickly cured by antifungal agents that are commonly used. However, low immune conditions can aggravate candida infections and poor oral hygiene of dentures can also be a problem.

There are several reports about the potential antifungal activity of polyphenols for treating oral candidiasis (Vasconcelos et al., 2003; Tamura et al., 2011). Nevertheless, much research is needed for the understanding of these compounds on fungal metabolism. Moreover, there is a lack in clinical trials that could indicate the magnitude of the effect of polyphenols on oral candidiasis. One of the few clinical trials on this topic evaluated the extract of pomegranate in patients with denture stomatitis. The results were promising in spite of the fact that clinical signs reduced in much slower speed than conventional therapies with synthetic antifungal drugs (Vasconcelos et al., 2003). More randomized clinical trials are needed before the final conclusion can be drawn on efficacy of pomegranate fruit extract, but the preliminary results are promising for this as well as for other phenolic compounds (Casaroto and Lara, 2010; Varoni et al., 2012). Finally, many reports of the promising effect of some phenolic compounds on oral health can be found in the scientific literature. The oral conditions investigated included caries, gingivitis, periodontal disease, candidiasis, oral aphtae, oral mucositis, oral lichen planus, leukoplakia and oral cancer. Many stilbenes, flavonoids and proanthocyanidins and crude extract rich in a mixture of polyphenols were investigated (Cavalcante, 2010; Petti and Scully, 2009; Varoni et al., 2012). Hopefully, future research will provide data on efficacy, effective doses and limitations for using these compounds in the oral mouth.

6. Combining Polyphenols with Drugs

Three types of interactions between drugs can be observed when evaluating efficacy: additive, synergistic, and antagonistic. This classification is valid for plant-antibiotic and plant-plant combination. The additive interaction takes place when the effect of two drugs/chemicals is equal to the sum of the effect of the two drugs/chemicals taken separately. In general, this effect means that both drugs have the same mechanism of action or have the same therapeutic effect. In other words, these drugs are operating separately. Synergistic interaction implies that the effect of two drugs/chemicals taken together is greater than the sum of their separate effect at the same doses. Antagonistic interaction means that the effect of two drugs/chemicals is lower than the sum of the effect of the two drugs taken independently of each other. This can be interpreted as an increment on the efficacy of one drug upon the other due to blocking of target molecules and/or the effect of the second drug increasing excretion of the first drug (Lin et al., 2008; Cavalcante, 201).

During the last decades, several reports indicated that different bacteria strains are resistant against almost all clinically available antibiotics (Betoni et al., 2006; Gootz, 2010; Abreu et al., 2012). Certainly, the wide use and misuse of antibiotics have led to a situation of spreading resistant bacteria.

Some studies have suggested that a combination of plant extract and antibiotics can increase antimicrobial activity (Betoni et al., 2006; Abreu et al., 2012). Conversely, this plant-antibiotic combination can result in antagonism effect (Cavalcante, 2010). It is important to point out that both situations might be interesting as long as side effects of antibiotics are minimized and its efficacy is increased. An additive effect of antimicrobial activity is also expected when combination of plant-drug is proposed.

The plant-antibiotic interaction can be evaluated for direct physical-chemistry changes of both drugs before intake or absorption (e.g. plant infusion inhibiting bioavailability of a drug) and for pharmacokinetics effect of one of the drugs modulating the activity of the other.

The antibiotic effect can be increased (additive, synergistic effect) or reduced (antagonism) by the presence of polyphenols (Lin et al., 2008). Additive, synergistic, and/or antagonistic interaction of plant-antibiotics upon oral bacteria can indicate the potential effect of plant polyphenols on antibiotic activity.

In a preliminary study carried out with some polyphenol-rich medicine plants from Brazil, the effect of plant extract on the antibiotic activity was measured (Cavalcante, 2010). First the minimum inhibitory concentration (MIC) of antibiotics and plant extracts were determined separately using microdilution techniques (Andrews, 2001). The combination of plant extract-antibiotics were evaluated taking into account five different doses (5 up to 25 μg/mL) of the plant extract and a fixed dose equal to the MIC of the antibiotics for each bacteria (Table 3). The plants tested are known to be rich in polyphenols. Some are ingested after infusion or bark of fruit peels such as pomegranate. Interesting is the fact that these plants are sometimes used as the liquid vehicle (e.g. tea, juices) for ingesting medicines (antibiotics).

Hence, the potential effect of this combination is real either by direct interaction during ingestion or later effect during pharmacokinetics of the drug. This can be illustrated with the pomegranate example. Pomegranate juice is becoming a popular drink in many countries but most consumers are not aware that high blood pressure medications may be less effective when in combination with pomegranate fruit whereas the pomegranate polyphenols may enhance antibiotic activity (Table 3). This effect of pomegranate polyphenols was also observed in a previous study testing different antibiotics against resistant bacteria (Braga et al., 2005).

An additional result presented in table 3 was the difference in bacteria behavior when exposed to the plant-antibiotics combination. In most cases, additive effect was observed which is promising for polyphenols since these phytochemicals can be used to enhance antibiotic activity.

Finally, the observation that plant extracts can interfere on antibiotic effect is a good indication that most medicines cannot be taken with tea, juices and other types of drinks that contains some amount of polyphenols.

Conclusion

There is enough evidence that polyphenols can reduce oral bacterial growth rate and interfere on their ability to adhere to tooth surfaces and mucosa tissue. The effective concentration is still a matter of discussion as well as the strategies to increase substantivity (i.e., permanence of a drug in the oral mouth) and reduce side effects (staining of teeth and bad taste). Slow delivery devices of polyphenols can be an interesting strategy to maintain the effective concentration of these phytochemicals in the oral mouth. Certainly the effective concentration will vary since thousands of polyphenols compounds can be isolated from medicine plants.

Future research may focus on delivery systems (nanotechnology) and combination with other polyphenols or drugs taking the advantage of some polyphenols for increasing effectiveness when delivered in combination.

On the same trend, undesirable antagonism effect of polyphenols on antibiotics needs to be investigated, especially applying *in vivo* experimental models. Finally, a very important issue is related to the lack of knowledge on structure-activity particularly among flavonoids.

Table 3. Additive, synergistic, and/or antagonistic interaction of plant-antibiotic upon oral bacteria. The combination of plant extract-antibiotics were measured taking into account five different doses (5 up to 25 μg/mL) of the plant extract and a fixed dose that was equal to the MIC (minimum inhibitory concentration) of the antibiotic for each bacteria. Dark squares (□) indicate that additive/synergism as well as antagonism interaction were both observed at least once in one when different concentrations of a plant extract was included in the broth media. Grey squares (▒) indicate that only one effect, additive/synergism or antagonism interaction affected the antimicrobial activity of the antibiotic. White squares (□) indicate no effect of the plant extract on the MIC of the antibiotic. MIC values were previously determined by microdilution techniques. Plants (popular names and part of the plant used in brackets) were selected after Ethnobotanical research carried out in Brazil: A=*Schinus terbinthifolius* Raddi (aroeira, bark); B=*Abarema cochliacarpos* {(Gomes) Barneby and JW Grimes} (barbatimão, bark); C=*Anacardium occidentale* Linné (cashew, bark); D=*Ziziphus joazeiro* Mart. (juazeiro, bark); E=*Punica granatum* Linné (pomegranate, fruit)

Microorganisms	Antibiotics														
	Ampicillin					Amoxicillin					Cephalexin				
	A	B	C	D	E	A	B	C	D	E	A	B	C	D	E
S. mutans ATCC700610	Dark	White	Grey	Grey	Grey	Grey	White	Grey	Grey	Grey	Dark	White	Grey	Grey	Grey
S.mutans ATCC25175	Dark	Grey	Grey	Grey	Dark	Dark	Dark	Dark	Grey	Dark	Dark	White	Grey	Grey	Dark
S.salivarius ATCC7073	White	Grey	Grey	Grey	Grey	Dark	Dark	Dark	Dark	Dark	Grey	White	Grey	Grey	Grey
S. oralis ATCC 10557	White	White	Dark	Dark	Dark	Dark	Grey	Dark	Dark	Grey	Dark	White	Dark	Dark	Grey
S. aureus ATCC6538	Dark	Grey	Dark	Dark	Dark	White	Grey	Grey	Grey	Dark	Dark	Grey	Dark	Dark	Dark

References

Abreu, A. C., McBain, A. J., Simões, M. Plants as sources of new antimicrobials and resistance-modifying agents. *Natural Product Reports*, 2012, 7, 29,1007-1021.

Andrews, J. M. Determination of minimum inhibitory concentrations. *Journal of Antimicrobial Chemotherapy,* 2001, 48, Suppl., 1:5-16.

Alviano, W. S., Alviano, D. S., Diniz, C. G., Antoniolli, A. R., Alviano, C. S., Farias, L. M., Carvalho, M. A., Souza, M. M., Bolognese, A. M. In vitro antioxidant potential of medicinal plant extracts and their activities against oral bacteria based on Brazilian folk medicine. *Archives of Oral Biology*, 2008, 53, 6, 545-52.

Araújo, N. R. R. Evaluation of *in vitro* antimicrobial activity of plant extracts on microorganisms related to oral mucositis. Dissertation (Master degree). Graduate Post-graduation Program in Pharmaceutical Sciences, Institute of Health Sciences, Federal University of Pará, Belém-PA (Brazil), 2010, 99 p. [Portuguese].

Aron, P. M., Kennedy, J. A. Flavan-3-ols: Nature, occurrence and biological activity. *Molecular Nutrition and Food Research*, 2008, 52, 79-104.

Aslam, S. N., Stevenson, P. C., Kokubun, T., Hall, D. R. Antibacterial and antifungal activity of cicerfuran and related 2-arylbenzofurans and stilbenes. *Microbiological research*, 2009,164,2,191-5.

Bae, E. A., Han, M. J., Kim, D. H. In vitro anti-Helicobacter pylori activity of some flavonoids and their metabolites. *Planta Med.*, 1999, 65, 442-3.

Betoni, J. E., Mantovani, R. P., Barbosa, L. N., Di Stasi, L. C., Fernandes Junior, A. Synergism between plant extract and antimicrobial drugs used on Staphylococcus aureus diseases. *Memórias do Instituto Oswaldo Cruz*. 2006, 101,4,387-90.

Bose, P. K. On some biochemical properties of natural coumarins. *Journal of the Indian Chemistry Society*, 1958, 58, 367–375.

Boudet, A. M. Evolution and current status of research in phenolic compounds. *Phytochemistry*, 2007, 68, 2722–2735.

Braga, L. C., Leite, A. A., Xavier, K. G., Takahashi, J. A., Bemquerer, M. P., Chartone-Souza, E., Nascimento, A. M. Synergic interaction between pomegranate extract and antibiotics against Staphylococcus aureus. *Canadian Journal of Microbiology*, 2005,51,7,541-547.

Brondani, D. J., Silva-Filho, A. A., Leite, A. C. L., Nascimento, C. R. M., Rolin-Neto, P. J., Bieber, L. W. Síntese e Avaliação da Atividade Antimicrobiana de Análogos da Primina 5 e 6 Alquil-Substituídos. *Latin American Journal of Pharmacy*, 2003, 22, 3, 217-221. [Portuguese]

Carvalho, A. A. T., Sampaio, F. C., Melo, A. F. M., Sena, K. X. F. R., Chiappeta, A. A., Higino, J. S. Atividade Antimicrobiana in vitro de Extratos Hidroalcoólicos de Psidium guajava L. sobre Bactérias Gram-Negativas. *Acta Farmaceutica Bonaerense*, 2002, 21, 4, 255-258.

Carvalho, A. C. B., Nunes, D. S. G., Baratelli, T. G., Nur, S. M. S. A. Q. S., Netto, E. M. Aspectos da legislação no controle dos medicamentos fitoterápicos. *TandC Amazônia*, 2007, 5, 11, 26-32 [Portuguese]

Casaroto, A. R., Lara,V. S. Phytomedicines for Candida-associated denture stomatitis. *Fitoterapia*, 2010, 81, 5, 323-328.

Castillo, F., Hernández, D., Gallegos, G., Rodríguez, R., Aguilar, C. N. Antifungal Properties of Bioactive Compounds from Plants. In: Dharumadurai Dhanasekaran (Ed.), *Fungicides for Plant and Animal Diseases,* Rijeka, Croatia: InTech, 2012, 81-106.

Cavalcante, A. L. F. A. Medicine Plants and oral health in the National Health System (SUS): Ethnobotanical study and risk factors for medicaments interaction. 2010. Dissertation (Master degree). Post-graduation program in Dentistry – Health Science Center, Federal University of Paraiba, João Pessoa (Brazil), 2010, 206 p. [Portuguese].

Cowan, M. M. Plant Products as Antimicrobial Agents. *Clinical Microbiology Reviews*, 1999, 12, 4, 564-582.

Cueva, C., Moreno-Arribas, M. V., Martín-Álvarez, P. J., Bills, G., Vicente, M. F., Basilio, A., López, C. R, Requena, T., Rodríguez, J. M., Bartolomé, B. Antimicrobial activity of phenolic acids against commensal, probiotic and pathogenic bacteria. *Research in Microbiology*, 2010, 161, 372-382.

Cushnie, T. P. T., Lamb, A. J. Recent advances in understanding the antibacterial properties of flavonoids. *International Journal of Antimicrobial Agents*, 2011, 38, 99-107.

Daglia, M. Polyphenols as antimicrobial agents. *Current Opinion in Biotechnology*, 2012, 23, 174–181.

Dai, J., Mumper, R. J. Plant phenolics: extraction, analysis and their antioxidant and anticancer properties. *Molecules*, 2010, 15, 7313–7352.

Dye, B. A. Global periodontal disease epidemiology. *Periodontology* 2000. 2012, 58, 1, 10-25.

Eloff, J. N., Famakin, J. O., Katerere, D. R. P. Isolation of an antibacterial stilbene from *Combretum woodii* (Combretaceae) leaves. *African Journal of Biotechnology*, 2005, 4-10, 1167-1171.

Elvin-Lewis, M. The therapeutic potential of plants used in dental folk medicine. *Odonto-stomatologie Tropicale*, 1982, 5, 3, 107-17.

Favela-Hernández, J. M. J., García, A., Garza-González, E., Rivas-Galindo, V. M., Camacho-Corona, M. R. Antibacterial and Antimycobacterial Lignans and Flavonoids from Larrea tridentata. *Phytotherapy Research*. 2012, 1-4.

Fernández, M. A., García, M. D., Sáenz, M. T. Antibacterial activity of the phenolic acids fraction of Scrophularia frutescens and Scrophularia sambucifolia. *Journal of Ethnopharmacology*, 1996, 53, 11-14.

Ferrazzano, G. F., Amato, I., Ingenito, A., Zarrelli, A., Pinto, G., Pollio, A. Plant Polyphenols and Their Anti-Cariogenic Properties: A Review. *Molecules,* 2011, 16, 1486-1507.

Flötra, L., Gjermo, P., Rölla, G., Waerhaug, J. Side effects of chlorhexidine mouth washes. *Scandinavian Journal of Dental Research*, 1971, 79, 2, 119-25.

Fresco, P., Borges, F., Diniz, C., Marques, M. P. New insights on the anticancer properties of dietary polyphenols. *Medicinal Research Reviews*, 2006, 6, 26, 747-66.

Genco, R., Offenbacher, S., Beck, J. Periodontal disease and cardiovascular disease: epidemiology and possible mechanisms. *Journal of the American Dental Association*. 2002, 133, Suppl.:14S-22S.

González, O. A., Escamilla, C., Danaher, R. J., Dai, J., Ebersole, J. L., Mumper, R. J., Miller, C. S. Antibacterial effects of blackberry extract target periodontopathogens. *Journal of Periodontal Research*, 2012, 1-7.

Gootz, T. D. The global problem of antibiotic resistance *Critical Reviews in Immunology*. 2010,30,1,:79-93.

Guggenheim, B., Giertsen, E., Schüpbach, P., Shapiro, S. Validation of an in vitro biofilm model of supragingival plaque. *Journal Dental Research*, 2001, 80, 1, 363-370.

Guggenheim, B., Guggenheim, M., Gmür, R., Giertsen, E., Thurnheer, T. Application of the Zürich biofilm model to problems of cariology. *Caries Research,* 2004, 38, 3, 212-222.

Guggenheim, B., Meier, A. In vitro effect of chlorhexidine mouth rinses on polyspecies biofilms. *Schweiz Monatsschr Zahnmed*, 2011, 121, 5, 432-441.

Hara, K., Ohara, M., Hayashi, I., Hino, T., Nishimura, R., Iwasaki, Y., Ogawa, T., Ohyama, Y., Sugiyama, M., Amano, H., et al. The green tea polyphenol (-)-epigallocatechin gallate precipitates salivary proteins including alpha-amylase: biochemical implications for oral health. *European Journal of Oral Sciences,* 2012, 120, 132-139.

Haslam, E., Liley, T. H., Cai, I., Martin, R., Magnolato, D. Traditional herbal medicines: the role of polyphenols. *Planta Medica*, 1989, 55, 1-8.

Ho, K. Y., Tsai, C. C., Huang, J. S., Chen, C. P., Lin, T. C., Lin, C. C. Antimicrobial activity of tannin components from Vaccinium vitis-idaea L. *Journal of Pharmacy and Pharmacology*, 2001, 53,187-191.

Ishida, K., de Mello, J. C., Cortez, D. A., Filho, B. P., Ueda-Nakamura, T., Nakamura, C. V. Influence of tannins from Stryphnodendron adstringens on growth and virulence factors of Candida albicans. *Journal of Antimicrobial Chemotherapy,* 2006, 58, 5, 942-949.

Jansen, A. M., Scheffer, J. J. C, Baerheim Svendsen, A. Antimicrobial activity of essencial oils: a 1976-1986 literature review. Aspects of test methods. *Planta Medica*, 1987, 40, 395-398.

Jeon, J. G., Rosalen, P. L., Falsetta, M. L., Koo, H. Natural products in caries research: current (limited) knowledge, challenges and future perspective. *Caries Research*, 2011, 45, 3, 243-263.

Khan, K. M., Saify, Z. S., Khan, M. Z., Zia-Ullah, C. I. M., Atta-Ur-Rahman, P. S., Chohan, Z. H., Supuran, C. T. Synthesis of coumarin derivatives with cytotoxic, antibacterial and antifungal activity. *Journal of Enzyme Inhibition and Medicinal Chemistry*, 2004, 19, 4, 373-379.

Kleinberg, I., Codipilly, M. Modeling of the oral malodor system and methods of analysis. *Quintessence International*, 1999, 30, 5, 357-69.

Kim, D. H., Bae, E. A., Han, M. J. Anti-Helicobacter pylori activity of the metabolites of poncirin from Poncirus trifoliata by human intestinal bacteria. *Biological and Pharmaceutical Bulletin*, 1999, 22, 422–44.

Lin, R. D., Chin, Y. P., Hou, W. C., Lee, M. H. The effects of antibiotics combined with natural polyphenols against clinical methicillin-resistant Staphylococcus aureus (MRSA). *Planta Medica*, 2008,74,8, 840-846.

Manach, C., Scalbert, A., Morand, C., Rémésy, C., Jiménez, L. Polyphenols: food sources and bioavailability *American Journal of Clinical Nutrition*, 2004, 79, 727-747.

Manach, C., Williamson, G., Morand, C., Scalbert, A., Rémésy, C. Bioavailability and bioefficacy of polyphenols in humans. I. Review of 97 bioavailability studies. *American Journal of Clinical Nutrition*, 2005, 81 (suppl.), 230S– 242S.

Marsh, P. D. Dental plaque as a microbial biofilm. *Caries Research*, 2004, 38,3,204-211.

Marsh, P. D., Moter, A., Devine, D. A. Dental plaque biofilms: communities, conflict and control. *Periodontol.* 2000, 2011,55,1,16-35.

Matsumoto, M., Minami, T., Sasaki, H., Sobue, S., Hamada, S., Ooshima, T. Inhibitory effects of oolong tea extract on caries-inducing properties of mutans streptococci. *Caries Research*, 1999, 33,6, 441-445.

Michelin, D. C., Moreschi, P. E., Lima, A. C., Nascimento, G. G. F., Paganelli, M. O., Chaud, M. V. Avaliação da atividade antimicrobiana de extratos vegetais. *Revista Brasileira de Farmacognosia,* 2005, 15,4, 316-320.

Nagori, K., Singh, M. K., Dewangan, D., Verma, V. K., Tripathi, D. K. Anti-inflammatory activity and chemo profile of plants used in traditional medicine: A review. *Journal of Chemical and Pharmaceutical Research*, 2010, 2,5, 122-130.

Narotzki, B., Reznick, A. Z., Aizenbud, D., Levy, Y. Green tea: a promising natural product in oral health. *Archives in Oral Biology*, 2012, 57,5, 429-35.

Ostrosky, E. A., Mizumoto, M. K., Lima, M. E. L., Kaneko, T. M., Nishikawa, S. O., Freitas, B. R. Métodos para avaliação da atividade antimicrobiana e determinação da Concentração Mínima Inibitória (CMI) de plantas medicinais. *Revista Brasileira de Farmacognosia*, 2008, 18(2), 301-307.

Penna, C., Marino, S, Vivot, E., Cruañes, M. C., Muñoz, J. D., Cruañes, J., Ferraro, G., Gutkind, G., Martino, V. Antimicrobial activity of Argentine plants used in the treatment of infectious diseases. Isolation of active compounds from *Sebastiania brasiliensis*. *Journal of Ethnopharmacology,* 2001, 77, 37-40.

Percival, R. S., Devine, D. A., Duggal, M. S., Chartron, S., Marsh, P. D. The effect of cocoa polyphenols on the growth, metabolism, and biofilm formation by Streptococcus mutans and Streptococcus sanguinis. *European Journal of Oral Sci*ences, 2006, 114, 4, 343-348.

Pereira, J. V., Sampaio, F. C., Pereira, M. S. V., Melo, A. F. M., Higino, J. S., Carvalho, A. A. T. In vitro antimicrobial activity of an extract from Anacardium occidentale Linn on Streptococcus mitis, Streptococcus mutans and Streptococcus sanguis. *Odontologia Clínico-Científica*, 2006, 5, 2, 137-141. [Portuguese]

Pereira, J. V. Estudos com o extrato da Punica granatum Linn. (Romã): Efeito antimicrobianao in vitro e avaliação clínica de um dentifrício sobre microorganismos do biofilme dental. Thesis (Doctorate degree). Thesis (Doctorate degree). (Post-graduation in Dentistry – UFPB-UFBA) – Federal University of Paraiba, (Brazil). 2002, 97. [Portuguese]

Petti, S., Scully, C. Polyphenols, oral health and disease: A review. *Journal of Dentistry*, 2009, 37, 413-423.

Van der Reijden, W. A., Dellemijn-Kippuw, N., Stijne-van New, A. M., de Soet, J. J., van Winkelhoff, A. J. *Journal of Clinical Periodontology*, 2001, 28, 686-691.

Reis, B., Martins, M., Barreto, B., Milhazes, N., Garrido, E. M., Silva, P., Garrido, J., Borges, F. Structure-property-activity relationship of phenolic acids and derivatives. Protocatechuic acid alkyl esters. *Journal of Agricultural and Food Chemistry*, 2010, 58, 6986-6993.

Ríos, J. L., Recio, M. C., Villar, A. Screening methods for natural products with antimicrobial activity: a review of the literature. *Journal of Ethnopharmacology*, 1988, 23, 2-3,127-49.

Ríos, J. L., Recio, M. C. Medicinal plants and antimicrobial activity. *Journal of Ethnopharmacology*, 2005, 22, 100,1-2,80-4.

Sampaio, F. C., Pereira, MdoS., Dias, C. S., Costa, V. C., Conde, N. C., Buzalaf, M. A. In vitro antimicrobial activity of Caesalpinia ferrea Martius fruits against oral pathogens. *Journal of Ethnopharmacology*, 2009, 124, 2, 289-294

Sánchez-Maldonado, A. F., Schieber, A., Gänzle, M. G. Structure–function relationships of the antibacterial activity of phenolic acids and their metabolism by lactic acid bacteria. *Journal of Applied Microbiology*, 2011, 111, 1176-1184.

Sato, Y., Suzaki, S., Nishikawa, T., Kihara, M., Shibata, H., Higuti, T. Phytochemical flavones isolated from *Scutellaria barbata* and antibacterial activity against methicillin-resistant *Staphylococcus aureus*. *Journal of Ethnopharmacology*, 2000, 72, 483–488.

Sartoratto, A., Machado, A. L. M., Delarmelina, C., Figueira, G. M., Duarte M. C. T., Rehder, V. L. G. Composition and antimicrobial activity of essential oils from aromatic plants used in Brazil. *Brazilian Journal of Microbiology,* 2004, 35, 275-280.

Simões, C. M. O., Schenkel, E. P., Gosmann, G., Mello, J. C. P., Mentz, L. A., Petrovick, P. R. Farmacognosia: da planta ao medicamento. 6.ed. Porto Alegre/Florianópolis-Brasil: Editora da Universidade UFRGS/Editora da UFSC-Brasil; 2007.[Portuguese and Spanish]

Stevenson, D. E., Hurst, R. D. Polyphenolic phytochemicals - just antioxidant or much more? *Cellular and Molecular Life Sciences,* 2007, 64, 2900-2916.

Tamura, M., Saito, H., Kikuchi, K., Ishigami, T., Toyama, Y., Takami, M., Ochiai, K. Antimicrobial Activity of Gel-Entrapped Catechins toward Oral Microorganisms. *Biological and Pharmaceutical Bulletin,* 2011, 34,5, 638-643.

Taniguchi, M., Kubo, I. Ethnobotanical drug discovery based on medicine men's trials in the African savanna: screening of east African plants for antimicrobial activity II. *Journal of Natural Products*, 1993, 56, 9, 1539-1546.

Varoni, E. M., Lodi, G., Sardella, A., Carrassi, A., Iriti, M. Plant polyphenols and oral health: old phytochemicals for new fields. *Current Medicinal Chemistry,* 2012, 19, 11, 1706-20.

Vasconcelos, L. C., Sampaio, M. C., Sampaio, F. C., Higino, J. S. Use of Punica granatum as an antifungal agent against candidosis associated with denture stomatitis. *Mycoses*, 2003,46,5-6,192-196.

Veloso, D. J. Ação de fitoterápicos antimicrobianos sobre microorganismos bucais produtores de compostos sulfurados voláteis: estudo in vitro. Thesis (Doctorate degree). (Post-graduation in Dentistry – UFPB-UFBA) – Federal University of Paraiba, (Brazil). 2006, 125. [Portuguese]

WHO. World Health Organization. National policy on traditional medicine and regulation of herbal medicines. Report of a WHO global survey. WHO: Geneva; 2005

Xiao, J., Kai, G. A review of dietary polyphenol-plasma protein interactions: characterization, influence on the bioactivity, and structure-affinity relationship. *Critical Reviews in Food Science and Nutrition*, 2012, 52, 1, 85-101.

In: Plant Extracts
Editors: Adriana Giordano and Alfieri Costs
ISBN: 978-1-62417-534-3

Chapter 12

New Trends in Plant Extract Use for Food-Borne Pathogenic Bacteria Control

***J. Ruiz-Martínez*[1], *J. E. Wong-Paz*[1], *F. Castillo-Reyes*[2], *F. D. Hernández-Castillo*[3], *V. M. González*[1], *C. N. Aguilar*[1] *and R. Rodríguez-Herrera*[1*]**

[1]Food Research Department, School of Chemistry, Universidad Autónoma de Coahuila, Saltillo, Coahuila, México
[2]Campo Experimental Saltillo, INIFAP. Carretera Saltillo-Zacatecas, Colonia Hacienda de Buenavista, Saltillo, Coahuila
[3]Department of Agricultural Parasitology, Universidad Autónoma Agraria Antonio Narro, Buenavista, Saltillo, Coahuila, México

Abstract

Food-borne pathogenic bacteria are one of the major public concerns worldwide. Prevention of pathogenic and spoilage microorganisms in food is usually achieved by using synthetic chemical preservatives, but some of them are responsible for many carcinogenic and teratogenic damages, as well as residual toxicity, together with growing concern of microbial resistance toward conventional preservatives. All of the above are concerns that have originated pressure on the food industry, for progressive removal of chemical preservatives and adoption of natural alternatives, to obtain its goals concerning safe food with longer shelf live. Many plant derived products such as spices, fruit preparations, vegetal preparations or extracts have been used for centuries for preservation and extension of shelf life of foods. In this chapter, trends on food-borne bacteria control using plant extract are mentioned. Today, scientific research reveals that chemicals from plants not only have an effect against a particular disease, but also antioxidant property from plant extracts gives a beneficial effect to human health.

[*] Corresponding author: rrh961@hotmail.com.

Considerable research on antimicrobial plant extracts has been reported, moreover, different mixtures of plant or by-product extracts have been used for food-borne bacteria control. Currently, there is a tendency to create natural products and environmentally friendly. In addition, the use of different organic solvents allowed in food, which are in concordance with the context of organic production systems, as well as the use of green technologies for plant and by-product extracts obtaining is also encouraged.

Keywords: Organic solvents, synergistic effects, by-products utilization, green technologies

Introduction

Food-borne pathogen microorganisms are one of the most important concerns of consumers worldwide (Tayel and El-Tras, 2010) and are responsible of a great number of plant, animal and human diseases. A large variety of microorganisms can cause food contamination so it is a main concern in food industry (Natta *et al.*, 2008). Microorganisms, specifically bacteria, are the main cause of disease by contaminated food consumption, on the other hand, international food trade and raw materials have increased the risk of pathogenic bacteria dispersion from production to consume sites (Careaga *et al.*, 2003). Nowadays, food-borne bacteria are a huge problem in livestock producing countries (porcine, bovine, poultry and aquaculture).

Bacteria is the oldest form of life on Earth, they survive and thrive in the most severe ecosystems, springs, acid wells, cracks in the ground, airless, no light and temperatures above 100 °C. These microorganisms play a main role in nature and humanity, the presence of bacterial flora is essential, even though some of them are pathogens. In addition, they have an important role in the industry so they allow developing important progress in research, specifically cellular physiology and genetics. Microbial pathogenicity has been defined as the biochemical mechanisms by which microorganisms cause virulence and disease. Not all microorganisms have the same probability of causing infection, the presence or successful multiplication on or inside the host, and subsequently disease, which is used to describe an infection that causes significant harm in the host. In contrast, the presence of bacteria that do not cause any damage in the organism is known as colonization. If the host defenses are in good condition, a person can be infected by bacteria able to cause disease and be an asymptomatic carrier, this duo infection-disease, depends on pathogen and host.

Virulence is the measure of pathogenicity and it is related directly to the ability of an organism to cause disease despite host resistance mechanisms. Bacterial virulence factors can be encoded on chromosomal, plasmid, transposon or bacteriophage DNA, they help bacteria to invade the host, cause disease and evade host defenses. Among them are adherence factors, invasion factors, capsule formation, as well as endotoxin and exotoxins production (Peterson, 1996). It is convenient to make sure the difference between primary pathogen, which is the microorganism that causes infection and disease when it gets into a no-immune host and opportunist pathogen, that is one that rarely causes disease in healthy human, but it does in a organism where defense system is altered, and they are often fatal (Ryan and Ray, 2004). Many scientific articles point out how to avoid their presence, its sanitary and socioeconomic consequences, even so, this kind of bacteria are often not well known. In the present article, it is shown an analysis about tendencies on methods and new organic and sustainable alternatives to control food-borne pathogen bacteria.

New Tendencies

Prevention of microbial food contamination is usually achieved through synthetic chemicals, some of them responsible or associated to cancer or teratogenic damages, as well as production of toxic residues, so there is a recent concern about microorganism resistance to conventional preservatives (Pundir *et al.*, 2010). Nowadays the continuous dispersion of multi-drug resistant pathogens has become a serious problem of public health and a major concern for physicists (Iwalokun *et al.,* 2004). However, at the same time, food industry requires food free or with low preservative content which may be toxic for human, which has force food industry to adopt natural alternatives for obtaining its goals related to safety and longer shelf life of food products (Agatemor, 2009). Many vegetable products such as spices, fruit and vegetables preparations and extracts, have been used for centuries to preserve and extend shelf life of food (Pundir *et al.*, 2010). Despite of their interesting properties, plants have been barely exploited as antimicrobial agent sources, due to the lack of information about structure and mode of action of some phytochemicals, also most of them have been obtaining using processes based in conventional extraction methodologies such as Soxhlet, maceration, infusion, which sometimes are not environmentally friendly. These technologies called conventional, have the disadvantages of consuming high amounts of toxic solvents (Tamalli *et al.*, 2012) and long extraction periods (Egizabal *et al.*, 1998; Hao *et al.*, 2002; Yan *et al.*, 2010). Some of these techniques may require from 1 to 24 hours for phytochemicals extraction (Raman and Gaikar, 2002).

In addition, some of these techniques are complex (Ballard *et al.*, 2010; Esclapez *et al.*, 2011) or requires exposition of samples to high temperatures, which can significantly affect the quality of the extract incorporating the risk of thermal degradation of termolabile active compounds that could be of interest (Egizabal *et al.*, 1998; Chan *et al.*, 2011; Routray and Orsat, 2011). In addition, energy prices and the impulse to reduce CO_2 industry emissions, has emerged a new challenge to find new technologies or processes aimed to reduce energy consumption to satisfy legal requirements of emissions, ensuring or improving control and quality of process-product and also to cost reduction. For example, extractions using existing technologies often require more than 70% of total energy process used in food, chemical and pharmaceutical industries (Perino *et al.*, 2011). By these reasons, research for alternative extraction technologies is going to be one of the innovation subjects that could contribute to a sustainable growth of chemical and food industries. Nowadays, there is a tendency to create eco-friendly products (Lira-Saldivar *et al.* 2003), where multiple plant extracts used in traditional medicine could be employed to eliminate food-borne bacteria and pathogen growth (Tayel and El-Tras, 2010). At the present time, exist considerable research about anti-microbial property of vegetal extracts (Sharma and Hashinaga, 2004; Du Pont *et al.*, 2006). However, there are very few reports about obtaining plant extract using organic solvents (Castillo-Godina *et al.*, 2010). On the other hand, there is a new tendency of testing mixtures and/or more vegetal extracts, obtained from plants or agro-industrial wastes against food borne bacteria so they represent a big combination of active compounds product of plant metabolism.

Plant Extracts

It is well known that, some plants extracts have antimicrobial activity and have been used as typical remedies even before that the chemical medicine existed. Today, scientific research reveals that phytochemicals not only have effect against a particular disease, but also the anti-oxidant effect of vegetal extracts gives beneficial action for human health (Puangpronpitag and Sittiwet, 2009). In addition, the perception of being a less incidence of adverse reactions of plant preparations compared to synthetic pharmaceutical and reduced cost due to vegetal extracts preparation (Cock, 2008), has become this alternative into an attractive and potential way to obtain active compounds for food-borne bacteria control (Puangpronpitag and Sittiwet, 2009; Pundir *et al.*, 2010). Some semidesertic plants from such as creosote bush *(Larrea tridentata)*, tar bush *(Flourensia cernua)* and prickly pears cactus (*Opuntia ficus-indica)* have high amounts of components that have a phenol group and a hydroxyl group or an aromatic ring. These substances are classifies as phenolic compounds and have been considered as responsible of diverse biological activities. These antimicrobial phytochemicals probably evolved as part of defense mechanisms of host against microbial invasion (Agatemor, 2009), acquiring a very advanced biological specialization. Various authors report antifungal and antimicrobial properties of some plants, relating its activity principally to its polyphenolic content and against phytopathogens (López *et al.,* 2005; Guerrero *et al.*, 2007; Castillo *et al.*, 2010; Osorio *et al.*, 2010; Jasso de Rodriguez *et al.*, 2011).

There are reports of using multiple chromatographic techniques such as Thin Layer Chromatography (TLC), Column Chromatography, High Performance Liquid Chromatography (HPLC) and other spectroscopic techniques such as Mass Spectroscopy (MS) in order to identify the phytochemical compounds in these plant extracts. Ruiz-Martínez *et al.* (2011), report the presence of polyphenolic compounds in extracts obtained with etanol and water, using semidesertic plants like creosote bush, (*Larrea tridentata*), tar bush (*Flourensia cernua*), pecan nut husk (*Carya illinoinensis*), yucca palm (*Yucca spp.*) and oregano (*Lippia spp.*), being identified the presence of ellagic acid, cumaric acid, tannic acid, gallic acid, catechin, quercetin, methyl gallate, cinnamic acid, 2-hydroxycinnamic acid and resorcinol, through those techniques mentioned above. Those results confirmed previous studies that probed the presence of flavonols (Wollenweber and Dietz, 1981), sesquiterpenoids (Téllez *et al.,* 2001), monoterpenoids, acetylenes, *p*-acetophenone, benzopyrans and benzofurans, in extracts obtained from *Fluorensia cernua.* Some of the chemical compounds of creosote bush are widely used in human diet. It contents volatile oils, esters of fatty, sterols and other hydrocarbons such as monoterpenes (Mabry *et al.,* 1977; Waller and Gisvold, 1945) and multiple polyphenols like tannins, gallic acid and catechin. Oregano essential oil is constituted mainly by thymol and carvacrol and also for some phenolic acids and flavonoids (Kalemba and Kunicka, 2003). In pecan, it can be found different active compounds depending of the analyzed plant part, leaves are rich in naphthoquinones, tannins, inositol, flavonoid derivatives, phenolcarboxylic acids and ascorbic acid, pericarp area contains fatty acids, tannins and naphthoquinones; tegmental area is rich in polyphenols and tannins, while cotyledons are rich in unsaturated fatty acids (Raissouni, 2005).

Synergistic Effects

Likewise, there is a poor knowledge about the role that play polyphenols individually, and the chance that they could be interacting synergistically, enhancing the effect. Actually, vegetal extracts are being exhaustively studied individually, but it is the combined use of extracts the one that is not very well explored about the effect over growth of food-borne pathogen bacteria, which could give major antimicrobial effects than those achieved with only one extract (Viveros-Folleco and Castaño-Zapata, 2006) and could be also an alternative against resistance generated by microorganism in the future.

Organic Solvents

In the process to obtain a plant extract and determine its antimicrobial properties, most of those extractions have been done with unfit solvents according to organic agriculture, among which are methanol, ethanol, hexane, chloroform and diethyl ether (Guerrero *et al.*, 2007, Lira *et al.*, 2007; Jasso de Rodríguez *et al.*, 2007). For this reason, it is necessary the searching of new organic-alternative solvents, that allow to extract more and different phytochemicals with the same or better antimicrobial properties, besides that they are recognized as safe to be used under organic production systems (Castillo *et al.*, 2010). It is recently being studied the use of more eco-friendly solvents that could be adapted for a more sustainable large scale production, among these new alternatives of solvents are lanolin, and cocoa butter (Méndez *et al.*, 2012), ethanol and also water, that have all shown excellent properties in bioactive compound recovery with antimicrobial properties (Castillo *et al.*, 2010; Méndez *et al.*, 2012).

As the main principle consists on the solid phase (vegetal tissue) and solute (bio composites) separation, this happens when plant material is in contact with liquid phase, which allows separation of compounds from their original structure. Solvents applied are organic and inorganic type, which allows that each solvent produce extracts with a specified composition (Zarnowski and Suzuki, 2004), all this according to its polarity and starting point. Among the solvents used in the extraction are hexane, acetone, ether, chloroform, methanol, ethylene, propylene, methane, propane (Gamboa *et al.*, 2003; Lenis, *et al.*, 2007), those solvents are not suitable to treatment during food conservation. It has lately grown the interest of supercritical and subcritical extraction using carbon dioxide as solvent, this gas is an ideal option due to it is neither toxic nor explosive and it is easy to remove from extracted products. When using supercritical and subcritical extraction with carbon dioxide, there are not residues in the extract by its specific managing properties, especially in those that have essential oils and lipids. In Table 1 are showed antioxidant extraction conditions (temperature and pressure) using some vegetal species and supercritical CO_2. A good alternative solvent is the one that allows obtaining bioactive compounds and also presents harmless features to human consume and null reaction during metabolism of food treated with them. In this way, Méndez *et al.* (2012) using extracts from Mexican semi-desert plants obtained with ethanol, lanolin, cocoa butter and water, found that there was a good *in vitro* inhibitory response of food-borne bacteria (Figure 1 and 2), by this reason these alternative solvents are a good option to be used in agricultural organic systems.

Table 1. Natural antioxidants extracted by supercritical CO_2 (Velasco *et al.,* 2008)

Vegetal tissue	Active principle	Extract conditions	
		T (°C)	P (Bar)
Pistachia vera (Pistachio shell)	Antioxidant	35-60	101-355
Zingiber officinale R. (Rizoma Ginger)	Gingerol	20-40	150-200
Zingiber officinale R. (Rizoma Ginger)	Gingerol	20-40	100-300
Zingiber officinale R. (Rizoma Ginger)	Gingerol	20-35	200-250
Rosmarinus officanalis L. (Rosemary)	Antioxidant	40-60	100-400
Coriander sativum (Coriander seed)	Antioxidant	58-85	116-280
Hierochloe odorata (Peppermint)	Antioxidant	40	250-350

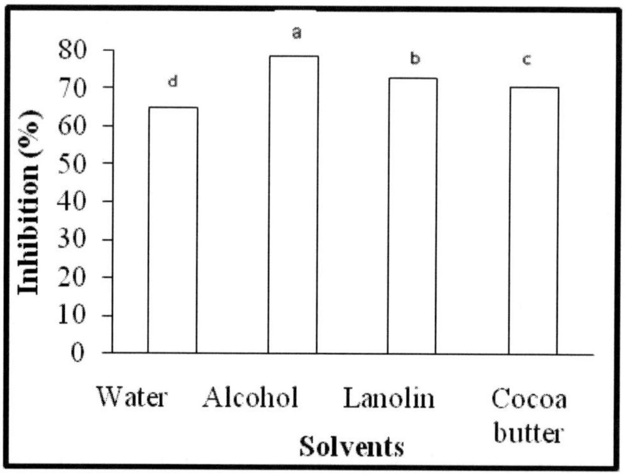

Figure 1. Antimicrobial activity solvent effect of different extracts (Méndez *et al.,* 2012).

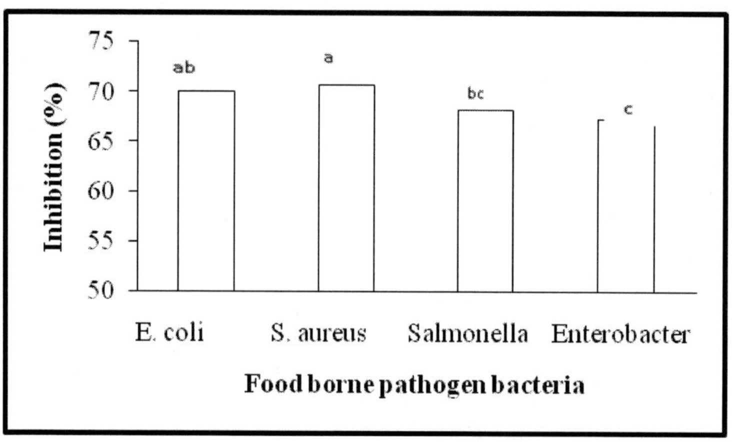

Figure 2. Percentage of inhibition of bacterial growth due to the effect of plant extracts (Méndez et al., 2012).

By-products Utilization

Agricultural and agro-industrial residues and wastes are potential sources for products of industrial interest, despite of that, only a small fraction of these sources are natural (Yemis and Mazza, 2012). These sources have a large number of compounds of interest, with multiple biological properties. A clear example of this are polyphenols, that can be found naturally in plant tissues and are necessary for pigmentation, growth, reproduction, pathogen resistance and many other functions. Martínez *et al.* (2010) achieved to extract polyphenolic compounds by microwave technology, using pomegranate peel and even though yields were not low, compared to other extraction methods, yields may improve if this methodology is optimized. A by-product that is generated in large quantities and has little use is the coffee pulp (*Coffea arabica*) and (*Coffea canephora*), which have few applications, among these; it is used as fuel and fertilizer, livestock feed or composting. Although, for manufacture of these products is used only a fraction of the available amount of waste, and these manufacture processes are not technically very efficient (Pandey et al., 2000). Some studies have focused on application of waste coffee bio-process (Pandey *et al.,* 2000; Fan *et al.*, 2000; Soares *et al.*, 2000). Coffee pulp is rich in nutrients, but it also contains compounds such as caffeine, tannins and polyphenols, which have anti-nutritional effects on animals and avoid its intensive use as feed (Ramírez, 1988, González *et al.*, 1998). An alternative use of these by-products is obtaining antimicrobial extracts. In this field, it has been reported the use of different concentrations (500 and 1000 mg / L) of phenolic compounds as antimicrobials inhibiting the growth of bacteria in wine, results showed that p-cumaric acid, ferulic and caffeic acid, inhibit bacteria growth (Stead, 1993).

Olive leaves are an abundant by-product deriving from olive tree cultivation and olive mills. The use of this residue is limited to animal feed and phytotherapy (Martin-García *et al.*, 2003). Olive leaves contain high quantities of phenol substances, very similar to those present in olives and their derived products (De Leonardis *et al.*, 2008). There is scientific evidence that olive leaf polyphenols have antimicrobial properties (Bisignano *et al.*, 1999), these authors evaluated *in vitro* susceptibility of five ATCC standard bacterial strains (*Haemophilus influenzae* ATCC 9006, *Moraxella catarrhalis* ATCC 8176, *Salmonella typhi* ATCC 6539, *Vibrio parahaemolyticus* ATCC 17802 and *Staphylococcus aureus* ATCC 25923) and 44 fresh clinical isolates (*Haemophilus influenzae*, eight strains, *Moraxella catarrhalis*, six strains, *Salmonella* species, 15 strains, *Vibrio cholerae*, one strain, *Vibrio alginolyticus*, two strains, *Vibrio parahaemolyticus*, one strain, *Staphylococcus aureus*, five penicillin-susceptible strains and six penicillin-resistant strains) to two olive (*Olea europaea*) secoiridoides, oleuropein (the bitter principle of olives) and hydroxytyrosol (derived from oleuropein by enzymatic hydrolysis and responsible for the high stability of olive oil). The minimum inhibitory concentrations (MICs) calculated in their study were evidence of the broad antimicrobial activity of hydroxytyrosol against these bacterial strains (MIC values between 0.24 and 7.85 $\mu g\ mL^{-1}$ for ATCC strains and between 0.97 and 31.25 $\mu g\ mL^{-1}$ for clinically isolated strains).

Furthermore, oleuropein also inhibited (although to a much lesser extent) the growth of several bacterial strains (MIC values between 62.5 and 500 $\mu g\ mL^{-1}$ for ATCC strains and between 31.25 and 250 $\mu g\ mL^{-1}$ for clinical isolates); oleuropein was ineffective against *Haemophilus influenzae* and *Moraxella catarrhalis*. The authors concluded that in addition to the potential employment of their active principles as food additives, or in integrated pest-management programs, *Olea europaea* can be considered a potential source of promising antimicrobial agents for treatment of intestinal or respiratory tract infections in man.

On the other hand, guava leaves contain tannins, pentacyclic triterpenoid guajanoic acid, uvaol, oleanolic acid, ursolic acid maslinic acid, volatile oils, triterpenoids, flavonoids (Meckes *et al.*, 1996; Begum *et al.*, 2004), guaijavarin, isoquercetin, hyperin, quercitrin, quercetin 3-Ogentobioside, and quercetin (Lozoya *et al.*, 1990). This by-product has been used for treatment of several human diseases (Goncalves *et al.*, 2005), the mode of action and activities of guava leaves extracts against various food pathogens were studied by Henie, *et al.*, (2009). These authors studied the killing kinetics, viability and cell leakage of *Kocuria rhizophila, Salmonella typhimurium, Listeria monocytogenes* and *Escherichia coli* O157:H7, measured after exposure to guava methanolic extracts (GME). Their results revealed a significantly higher ($p \leq 0.05$) release of bacterial nucleic acids, K^+ ions and protein than that of untreated microbes, indicating disruption of the bacterial membrane. GME caused a significantly higher ($p \leq 0.05$) release of RNA in gram-negatives compared to gram-positives. GME caused a relatively small but significant release of pyrines and pyrimidines in all tested organisms. GME probably disrupted the integrity of the Gram-negative microorganism lipopolysaccharide (LPS) layer. Unlike all the other tested microorganisms, *E. coli* O157:H7, demonstrated the lowest protein leakage, the highest K^+ leakage, the highest pyrines and pyrimidines leakage within the first 10 min of extract exposure, but the lowest after 30 minutes, which may indicate their good homeostasis ability or adaptability. Understanding the mode of action of flavonoids, present in high amounts in guava leaves extract, may help to develop an alternative biodegradable and safe, antimicrobial product for food and medicine, and as a by-product of the guava industry. Obtaining of bioactive plant compounds from industrial wastes, represent the first step to obtain raw extract from plants and it has being realized since many years from conventional methods. Some of these methods are both conventional and traditional such as: maceration, percolation, Soxhlet, reflux and infusion. Soxhlet extraction is the most conventional method used and consists in a simple distillation process that is repeated. Sources of extracts from by-product are still limited, but studies on this source of natural bactericides increase as increase the use of organic fruit and vegetables, organic packaging and pre-cut foods that require the application of microbial natural products safe for consumers.

Green Technologies

As it was mentioned above, plants have different compounds that are widely applied in food, pharmaceutical and cosmetic industry such as lipids, fat, fragrances, pigments, flavors and phytochemicals as polyphenolic compounds and terpenes. Some of these compounds are bioactive against multiple harmful microorganisms on the quality and plant growth. Separation of these compounds is done principally by lixiviation, applying some techniques to solute extraction of the solid matrix, among them, Soxhlet, hydro-distillation, maceration, which are called conventional techniques. In recent years, new alternative technologies have been developed in order to obtain plant extract, also known as "green techniques", which have come to revolutionize the extraction methods of bioactive compounds (Ballard *et al.*, 2010) due to the fact that this type of novel techniques allow to recover a bigger quantity of bioactive compounds of interest, in shorter periods of extraction (reduction hours to seconds), rising quality of extracts with less cost of processing, or extracting compounds that are hard to obtain through conventional techniques (Martins *et al.*, 2010; Yan *et al.*, 2010). Among these techniques are: ultrasonic assisted extraction, ohmic heating, high pressure, supercritical fluid extraction and microwave assisted extraction (Xia *et al.*, 2011, Hao *et al.*, 2002; Kwon *et al.*, 2003; Wang *et al.*, 2010; Chan *et al.*, 2011; Perino *et al.*, 2011; Tamalli *et al.*, 2012). Green techniques have the advantage of being more efficient and eco-friendly, because these techniques use less solvents and energy, consume less extraction time and are simpler, compared to conventional techniques and there is less degradation of thermo-labile compounds, better products with less sot and higher yields. An example for this are microwave assisted extraction, supercritical fluid extraction, high pressure extraction, ultrasound and others (Hao *et al.*, 2002; Kwon *et al.*, 2003; Wang *et al.*, 2010; Chan *et al.*, 2011; Perino *et al.*, 2011; Tamalli *et al.*, 2012). Microwave assisted extraction (MAE) has been well accepted as a powerful potential alternative for organic extractions of bioactives from vegetal samples (Yan *et al.*, 2010).

MAE is a relatively well known technique; the first works about this were published more than twenty years ago. However, microwave assisted extraction is a innovative technique that has proved to be at the present, very useful for quick extraction and obtaining compounds of high quality (Perino *et al.*, 2011; Shao *et al.*, 2012; Tamalli *et al.*, 2012). MAE is really fast, so it can transfer heat to the solvent and in this way penetrates directly in the sample, heating instantly the solvent trapped between sample pores. The main advantage of MAE is that it uses 95% of supplied energy, and therefore, helps to reduce environmental pollution with shorter extraction times and less CO_2 emissions to atmosphere (Wang *et al.*, 2007; Perino *et al.*, 2011; Yemis and Mazza, 2012). During microwave heating, hydrogen bonds rupture promoted by dipole rotation of molecules is which produces heat release. Also exists migration of dissolved ions which increases solvent penetration into the matrix; therefore, it facilitates target compounds recovery. During microwave processing, a considerable amount of pressure builds up inside of the cell, which modifies the physical properties of the biological tissue, increasing the biological matrix porosity, solvent penetration into the matrix, and releasing of compounds internally. Pressure effect also takes to cell inflammation that generally causes cell rupture. This technique allows an exact control of extraction parameters (time, temperature, potency), allowing to reach high temperatures and pressure. The main limiting which does not allow the common use of this technique is the equipment cost (Pérez *et al.*, 2008) speaking of this, microwave equipment is simpler and cheaper than the one used for supercritical fluid extraction (Hao *et al.*, 2002) and also allows a better recovery of compounds. Martins *et al.* (2010) did to optimize nordihydroguaiaretic acid obtaining from *Larrea tridentata*.

Another promising alternative technique of extraction is using supercritical fluids. This term describes the behavior of certain substances that do not condense or evaporates above certain temperature, but it only exists as fluids. Supercritical fluid extraction (SFE) generally applies carbon dioxide as solvent (alone or with other co-solvents) under temperatures and pressures near of above its critical point. This technique eliminate all negative effects over health and environment caused by the use of regular organic solvents, in this technique, solvent can be removed very easily. The main disadvantages of solvents used in this technique are that they require high pressures and a very strict control of system temperature (Pérez et al., 2008). High pressures (100-800 MPa) can easily interrupt electrostatic and hydrophobic interactions between molecules and increase mass transfer and permeability speed of cells, leading to an increased release of compounds with functional properties (He *et al.*, 2011). This is a method, time of extraction can be decreased significantly, even less than

50% of the time required in most of other methods. It is a clean method, considered eco-friendly because it uses carbon dioxide as solvent. This method is very efficient to non-polar compounds extraction, however, with this method, one could use some solvents such as ethanol, methanol, acetonitrile, acetone, water, ethyl ether and dichloromethane in order to increase solubility of polar compounds (Routray and Orsat, 2011). Ultrasound assisted extraction (20-100 kHz) offers a high reproducibility in shorter times, it has reduce consume of solvent, increases yields and uses little amount of energy. It is based in the fact that ultrasonic energy induces a phenomenon called cavitation, on which bubble formation occurs with negative pressure and they swell until they implode. By this process, shock waves form microjets of liquid and released energy causes wall cells rupture and allows a high diffusion of bioactive compounds through the cell, accelerating extraction process (Kellner *et al*, 2004; Pérez *et al.*, 2008; Vilkhu *et al.*, 2008; Khan *et al.*, 2010; Peng *et al.*, 2010; Routray and Orsat, 2011). There are some troubles when using ultrasound technique with a simple solid, because there is a chance that product of interest in the sample could be decomposed if it gets trapped inside a cavitation bubble and then it implodes. The UAE is highly recommended for solid samples that contain volatile and semi-volatile organic compounds; this technique can also be performed at room temperature and regular pressure, which facilitates thermolabile compound extraction. One of UAE advantages is that is faster than a conventional Soxhlet extraction and it allows recovery of most used sample and equipment is relatively simple and inexpensive (Pérez *et al.*, 2008).

Conclusion

In summary, we can conclude that the new tendencies in the food borne bacteria control are focused in replacing synthetic preservatives with plant extracts which have been used individually or could be used as a mixture. Besides, the new technologies and extraction solvents fit the context of organic and sustainable production of food, where some of these technologies are faster, less expensive or more eco-friendly than the traditional ones.

Acknowledgments

This work was supported by the Department of Food Science and Technology, University of Coahuila. J. R. M., J.E.W.P., F.C.R., V.M.G., want to thank to CONACYT for all financial support during their postgraduate studies.

References

Agatemor, C. (2009). Antimicrobial activity of aqueous and ethanol extracts of nine Nigerian spices against four food borne bacteria. *EJEAF Che.*, 8:195-200.

Ballard, T. S., Mallikarjunan, P., Zhou, K., O'Keefe, S. (2010). Microwave-assisted extraction of phenolic antioxidant compounds from peanut skins. *Food Chemistry.* 120:1185-1192.

Begum, S., Hassan, S.I., Ali, S.N., Siddiqui, B.S. (2004). Chemical constituents from the leaves of Psidium guajava. *Natural Product Research* 18 (2):135-40.

Bisignano, G., Tomaino A., La Cascio R., Crisafi G., Uccella N., Saija A., (1999). On the *in vitro* antimicrobial activity of oleuropein andhydroxytyrosol. *J. Pharm. Phrmacol.,* 51: 971-974.

Careaga, M., Fernández, E., Dorantes, L., Mota, L., Jaramillo, M. E., Hernandez-Sanchez, H. (2003). Antibacterial activity of capsicum extract against Salmonella typhimurium and Pseudomonas aeruginosa inoculated in raw beef meat. *Int. J of Food Microbiol,* 83: 331-335.

Castillo, F., Hernández, D., Gallegos, G., Méndez, M., Rodríguez, R., Reyes, A., Aguilar, C. N. (2010). In vitro antifungal activity of plant extracts obtained with alternative organic solvents against Rhizoctonia solani Kühn. *Industrial Crops and Products.* 32: 324-328.

Castillo-Godina, R., (2008). Antibacterial and antifungal activities of polyphenol compounds from typical Mexican semi-desert plants. BSc. Thesis. Universidad Autónoma de Coahuila campus Saltillo. México p.60 (in Spanish).

Castillo-Godina, R., Rodríguez-Herrera, R., Ventura-Sobrevilla, J., Belmares-Cerda, R., Aguilar-González, C. N. (2010). Antibacterial and antioxidant activities of polyphenolic extracts from Mexican semiarid plants pre and post-fermentation. *Innovations in Food Science and Food Biotechnology Developing Countries.* Regalado González C. and García Almendárez B. E.. AMECA, Querétaro, México. Pág. 229-244.

Chan, C. H., Yusoff, R., Ngoh, G., Kung. F. W. (2011). Microwave-assisted extractions of active ingredients from plants – A review. *Journal of Chromatography A.* 1218(37):6213-6225.

Cock, I. E. (2008). Antibacterial activity of selected Australian native plant extracts. *Int. J. Microbiol.* 4:2.

De Leonardis, A., Acetini A., Alfano G., Macciola V., Ranalli G., (2008). Isolation of a hydroxytyrosol rich extract from olive leaves (Olea Europaea L.) and evaluation of its antioxidant properties and bioactivity. *Eur. Food Res. Technol.,* 226: 653-659.

Du Pont. (2006). Sigatoka negra y amarilla. Técnicas mejoradas para manejo e identificación. *Du Pont Latín América.* Coral Gables: Florida. 17 p.

Egizabal, A., Zuloaga, O., Etxebarria, N., Fernandez, L. A., Madariaga, J. M. (1998). Comparison of microwave-assisted extraction and Soxhlet extraction for phenols in soil samples using experimental designs. *Analyst.* 123:1679-1684.

Esclapez, M. D., García, J. V., Mulet, A., Cárcel, J. A. (2011). Ultrasound-assisted extraction of natural products. *Food Eng. Rev.* 3:108-120.

Fan, L., Pandey, A., Soccol, C. R. (2000). Solid state cultivation: an efficient method to use toxic agro-industrial residues. *J. Basic. Microbiol.,* 40(3), 187-197.

Gamboa, R., Hernández F., Guerrero E., Sánchez-Villareal L., López R., Jiménez F., Lira-Saldivar R. (2003), "Antifungal effect of Larrea tridentata extracts on Rhizotocnia solani Kühn and Phytophtora infestans Mont. (De Bary)", *International Journal of Experimental Botany PHYTON,* 21(1):13-18.

Goncalves, J.L.S., Lopes, R.C., Oliveira, D.B., Costa, S.S., Miranda, M.M.F.S., Romanosa, M.T.V., Santos, N.S.O., Wigg, M.D. (2005). In vitro anti-rotavirus activity of some medicinal plants used in Brazil against diarrhea. *Journal of Ethnopharmacology* 99: 403–407.

González, N., Ramírez-Martínez, J. R., Aldana, J. O., Ramos-Niño, M. E., Clifford, M. N., Pérekar, S., et al. (1988). Isolation, characterization and determination of biological activity of coffee proanthocyanidins. *J. Sci. Food Agric.,* 77: 368-372.

Hao, J., Han, W., Huang, S., Xue, B., Deng, X. (2002). Microwave-assisted extraction of artemisinin from Artemisia annua L. *Separation and Purification Technology,* 28:191-196.

Henie, E.F. P., Zaiton, H., Suhaila, M. (2009). Bacterial membrane disruption in food pathogens by Psidium guajava leaf extracts. *International Food Research Journal* 16: 297-311.

Iwalokun, B. A., Ogunledun, A., Ogbolu, D. O., Bamiro, S. B., Jimi-Omojola, J. (2004). In vitro antimicrobial properties of aqueous garlic extract against multidrug-resistant bacteria and Candida species from Nigeria. *J. Med. Food,* 7: 327-333.

Jasso de Rodríguez, D., Hernández, C. D., Angulo, S. J. L., Rodríguez, G. R., Villarreal, Q. J. A., Lira, S. R. H. (2007). Antifungal activity in vitro of F. cernua extracts on Alternaria sp., Rhizoctonia solani, and Fusarium oxysporum. *Ind. Crops Prods.* 25:111-116.

Jasso de Rodríguez, D., Rodriguez, R., Hernandez, F. D., Aguilar, C. N., Saenz, A., Villareal, J. A., Moreno, L. E. (2011). In vitro antifungal activity of extracts of Mexican Chihuahuan desert plants against postharvest fruit fungi. *Industrial Crops and Products.* 34:960-966.

Kalemba D., Kunicka A. (2003). Antibacterial and antifungal properties of essential oils. *Curr. Med. Chem.,* 10: 813-829.

Kellner, P., Mermet, J., Otto, M., Valcarei, M., Widmer, M. (2004). *Analytical Chemistry.* 2nd edition. Wiley-VCH. Germany.

Khan, M. K., Abert-Vian, M., Fabiano-Tixier, A. S., Dangles, O., Chemat, F. (2010). Ultrasound-assisted extraction of polyphenols (flavanone glycosides) from orange (Citrus sinensis L.) peel. *Food Chemistry.* 119:851-858.

Kwon, J., Beälanger, J. M. R., Pare, J. R. J. (2003). Optimization of microwave-assisted extraction (map) for ginseng components by response surface methodology. *J. Agric. Food Chem.* 51:1807-1810.

Lenis, L., Peña-Salamanca E., Chito-Trujillo D. (2007). Extracción, separación y elucidación estructural de dos metabolitos secundarios de alga marina Bostrychia callipetera. *Scientia et Technica.* 13: 97-102.

Lira-Saldívar, R. H., Balvantín, G. F., Hernández-Castillo, F. D., Gamboa, A., Jasso de Rodriguez, D. and Jiménez, D. F., (2003). Evaluation of resin content and antifungal effect of Larrea tridentata (Sesse and Moc. Ex D. C. Coville) extracts against Pythium sp. Pringsh. *Rev. Mex. Fitopatol.* 21:97–101.

López, A., López, R. S., Vázquez, M. E., Rodríguez, A., Mendoza, M., Padrón, E. (2005). Inhibición del crecimiento micelial de Fusarium oxysporum Schlechtend. f. sp. lycopersici (Sacc.) Snyder and Hansen, Rhizoctonia solani Kuhn y Verticilllium dahliae Kleb. mediante extractos vegetales acuosos. *Revista Mexicana de Fitopatología.* 23(2):183-190.

Lozoya, X., About-Zaid, M.N., Nozzilillo, C., Arnason, J.T., (1990). *Spasmolytic effect of the methanolic extract of Psidium guajava.* Planta Medica 56:686–689.

Mabry, T. J., Hunkiker, J. H., DiFeo D. R. (1977). Creosote bush: biology and chemistry of Larrea in New World deserts. Stroudsburg, P. A.: Dowden, Hutchinson and Ross.

Martin-García, A.I., Moumen A., Yanez Ruiz D.R., Al-Caide M., (2003). Chemical composition and nutrients availability for goats and sheep of twostage olive cake and olive leaves. *Anim. Feed Sci. Tech.,* 107: 61-74.

Martínez, A., Contreras, J. C., Belmares, R. (2010). Extracción de polifenoles asistida por microondas a partir de Punica granatum L. Acta Química Mexicana. 2:4.

Martins, S., Aguilar, C. N., De la Garza, I., Mussato, S. I., Teixeira, J. A. (2010). Kinetic study of nordihydroguaiaretic acid recovery from Larrea tridentata by microwave-assisted extraction. *J. Chem. Technol. Biotechnol.* 85: 1142-1147.

Meckes, M., Calzada, F., Tortoriello, J., Gonzalez, J.L., Martinez, M., (1996). Terpenoids isolated from Psidium guajava hexane extract with depressant activity on central nervous system. *Phytotherapy Research* 10: 600–603.

Méndez, M., Rodríguez, R., Ruiz, J., Morales-Adame, D., Castillo, F., Hernández-Castillo, F.D., Aguilar, C.N. (2012). Antibacterial activity of plant extracts obtained with alternative organic solvents against food-borne pathogen bacteria. *Industrial Crops and Products.* 37: 445– 450.

Natta, L., Orapin, K., Krittika, N., Pantip, B. (2008). Essential oil from five Zingiberaceae for anti food-borne bacteria. *Int. Food Research J.,* 15: 337-346.

Osorio, E., Flores, M., Hernández, D., Ventura, J., Rodríguez, R., Aguilar, C. N. (2010). Biological efficiency of polyphenolic extracts from pecan nuts shell (Carya illinoensis), pomegranate husk (Punica granatum) and creosote bush leaves (Larrea tridentata Cov.) against plant pathogenic fungi. *Industrial Crops and Products.* 31: 153-157.

Pandey, A., Soccol, C., Nigam, P., Brand, D., Mohan, R., Roussos, S. (2000), Biotechnological potential of coffee pulp and coffee husk for bioprocesses. *Biochem. Eng. J..*, 6:153-162.

Peng, L., Jia, X., Wang, Y., Zhu, H., Chen, Q. (2010). Ultrasonically assisted extraction of rutin from Artemisia selengensis Turcz: *comparison with conventional extraction techniques Food Anal. Methods.* 3:261-268.

Pérez, C., Sáenz, A., Barajas, L. (2008). La química verde como herramienta indispensable en el aprovechamiento integral y sustentable de los recursos del semidesierto mexicano. In Fitoquímicos Sobresalientes del Semidesierto Mexicano: de la planta a los químicos naturales y a la biotecnología. Editorial: *Path Design* S. A. México, 67-82 pp.

Perino, S., Huma, Z., Abert, M., Chemat, F. (2011). Solvent free microwave-assisted extraction of antioxidants from sea buckthorn (Hippophae rhamnoides) food by-products. Food Bioprocess Technol. 4:1020-1028.

Peterson, J. W. (1996). Bacterial Pathogenesis. In: *Medical Microbiology.* Galveston University. Chapter 7.

Puangpronpitag, D., Sittiwet, C. (2009). Antimicrobial properties of Cinnamomum verum aqueous extracts. *Asian J. of Biological Sciences*, 2: 49-53.

Pundir, R. K., Jain, P., Sharma, C. (2010). Antimicrobial activity of ethanolic extracts of Syzygium aromaticum and Allium sativum against food associated bacteria and fungi. *Ethnobotanical Leaflets*, 14: 344-60.

Raissouni, T. (2005). Estudio comparativo de la eficacia de varios tratamientos tópicos de la estomatitis aftosa recurrente; Universidad de Granada. 1-46.

Raman, G., Gaikar, V. G. (2002). Microwave-assisted extraction of piperine from Piper nigrum. *Ind. Eng. Chem. Res.* 41:2521-2528.

Routray, W., Orsat, V. (2011). Microwave-assisted extraction of flavonoids: *A Review. Food and Bioprocess Technology*. 5 (2): 409-424.

Ruiz-Martínez, J., Ascacio, J. A., Rodríguez, R. Morales, D., Aguilar, C. N. (2011). Phytochemical screening of extracts from some Mexican plants used in traditional medicine. *Journal of Medicinal Plants Research.* 5 (13): 2791-2797.

Ryan, K. J., Ray, C. G. (2004). *Sherris Medical Microbiology*, 4th ed., McGraw Hill. 551–552.

Shao, P., He, J., Sun, P., Zhao, P. (2012). Analysis of conditions for microwave-assisted extraction of total water-soluble flavonoids from Perilla Frutescens leaves. *J. Food Sci. Technol.* 49(1): 66-73.

Sharma, K. D., Hashinga, F., (2004). Alpinia leaf extract: A prospective natural food preservative. *J. of Scientific and Industrial Research*, 63: 689-693.

Soares, M., Christen, P., Pandey, A., Soccol, C. R. (2000). Fruity flavour production by Ceratocystis fimbriata grown on coffee husk in solid-state fermentation. *Process Biochem.*, 35(8), 857-861.

Stead, D. (1993). The effect of hydroxycinnamic acids on the growth of wine-spoilage lactic acid bacteria. *J. Appl. Bacteriol.*, 75:135-141.

Tamalli, A., Arráez, D., Ibañez, E., Zarrouk, M., Segura, A., Fernández, A. (2012). Optimization of microwave-assisted extraction for the characterization of olive leaf phenolic compounds by using HPLCESI- TOF-MS/IT-MS2. *J. Agric. Food Chem.* 60:791-798.

Tayel, A. A., El-Tras, W. F. (2010). Anticandidal activity of pomegranate peel extract aerosol as an applicable sanitizing method. *Mycoses* 53:117–122.

Téllez, S. J., Ramírez, J. A., Pérez-Lamela, C., Vázquez, M., Simal-Gándara, J. (2001). Aplicación de alta presión hidrostática en la conservación de los alimentos. *Ciencia y Tecnología Alimentaria*. 3: 66-80.

Velasco, P.; Soengas, P.; Vilar, M.; Cartea, M. E.; del Rio, M., (2008) Comparison of glucosinolate profiles in leaf and seed tissues of different Brassica napus crops. *J. Am. Soc. Hort. Sci.* 133(4):551-558.

Vilkhu, K., Mawson, R., Simons, L., Bates, D. (2008). Applications and opportunities for ultrasound assisted extraction in the food industry-A review. *Innovative Food Science and Emerging Technologies.* 9:161-169.

Viveros-Folleco, J., Castaño-Zapata, J. (2006). Evaluation in vitro of vegetal extracts against Mycosphaerella fijiensis Morelet. *Agron.* 14(1): 37-50 (in Spanish).

Waller, C. W., Gisvold, O. (1945). A phytochemical investigation of Larrea divaricate Cav. *J. Am. Pharm. Assoc. Sci. Ed.* 34: 78-81.

Wang, S., Chen, F., Wu, J., Wang, Z., Liao, X., Hu, X. (2007). Optimization of pectin extraction assisted by microwave from apple pomace using response surface methodology. *Journal of Food Engineering*. 78:693-700.

Wang, L., Qin, P., Hu, Y. (2010). Study on the microwave-assisted extraction of polyphenols from tea. *Front. Chem. Eng. China.* 4(3):307-313.

Wollenweber, E., Dietz V.H. (1981). Occurrence and distribution of free flavonoid aglycones in plants. *Phytochemistry.* 20: 869-932.

Xia, E., Ai, X., Zang, S., Guan, T., Xu, X., Li, H. (2011). Ultrasound-assisted extraction of phillyrin from Forsythia suspensa. *Ultrasonics Sonochemistry.* 18:549-552.

Yan, M. M., Liu, W., Fu, Y. J., Zu, Y. G., Chen, C. Y., Luo, M. (2010). Optimization of the microwave-assisted extraction process for four main astragalosides in Radix astragali. *Food Chemistry.* 119:1663-1670.

Yemis, O., Mazza, G. (2012). Optimization of furfural and 5-hydroxymethylfurfural production from wheat straw by a microwave-assisted process. *Bioresource Technology.* 109:215-223.

Zarnowski, R., Suzuki, Y. (2004). Expedient soxhlet extraction of resorcinolic lipids from wheat grains. *J. Food Compos. Anal.* (doi: 10.1016/j.jfca.2003.09.007).

In: Plant Extracts
Editors: Adriana Giordano and Alfieri Costs
ISBN: 978-1-62417-534-3

Chapter 13

Plant Extracts as Antimicrobials: Prospects in Food Safety and Health Protection

Ahmed A. Tayel[1,*], Omnia A. Abdel-Monem[1], Shaaban H. Moussa[1,2] and Ahmed I. Al-Turki[2]

[1]Genetic Engineering and Biotechnology Research Institute,
Minoufiya University, Sadat City, Egypt
[2]College of Agriculture and Veterinary Medicine,
Al-Qassim University, Saudi Arabia

Abstract

The use of medicinal and aromatic plants has an aged history in many cultures; numerous varieties of these plants are listed in the historic manuscripts and religious books.

Many of the recurrent used spices and herbs were applied by the ancient nations throughout the world, and they were valued for their medicinal and preservative potentialities in addition to their flavor enhancement properties. The dependence on natural agent like plant extracts, as powerful alternatives to chemical and synthetic antimicrobials, is greatly desired from modern consumers.

While many ethno-pharmacological plants already exhibited promising characteristics for treating endemic and emergent diseases, however, the door is still opening for the discovery of other useful properties from phytomedicine.

The antimicrobial properties of plant extracts have been discussed by numerous investigators and everyday novel results are presented concerning their potentialities and prospects applications.

* For correspondence: Dr. Ahmed A. Tayel, PhD, Genetic Engineering and Biotechnology Research Institute, Minoufiya University, Sadat City, P. O. box 79/32897, Egypt; E-mail: tayel_ahmad@yahoo.com; Tel.: +20 10 196 9909; fax: +20 48 260 1266–8.

The present chapter will take us in a voyage, to indicate the apparent used antimicrobial plant extracts, in human therapy and as food biopreservatives, as well as the potential applications of these extracts to improve human health and safety.

1. Introduction: Historical Overview about Medicinal Herbs

Herbs are plants valued for their medicinal and aromatic properties and are often grown and harvested for these unique properties. The first gardens were herb gardens as some of the earliest of herb gardens were planted about 4000 years ago in Egypt. The present-day concept of an herb garden has developed largely from ancient Egyptian Christian and Islamic traditions. The knowledge of herbs has been handed down from generation to generation for thousands of years (Brown, 1995).

The word "herbs" comes from the Latin '*herba'*, which means a medicinal plant. The meaning of herb in a narrow sense is a non-lasting plant that weakens after blooming without its stems becoming woody. Perennial herbs are widely used for the purpose of dyeing and gardening utilized for medical purpose. Some edible herbs belong to the category of spice. Even some herbs containing poisonous components can be categorized as spices if the poisonous element can somehow be naturalized with heating or other cooking procedure. Herbs could be botanically classified as a perennial plant, but the meaning of spice comes from its use in cooking, not any plant classification. A medicinal plant should, therefore, be edible. In fact, no spice definitions clearly differentiate between a spice and an herb. The term spice can be broadly defined as a compound that has a strong flavor or coloring activity or one that increases appetite or enhances digestion. Herbs contain some complicated mixtures of organic chemicals that may vary depending upon many factors related to the growth, production, and processing of the herbal product (Tepsorn, 2009). Herbs could be obtained from seed, berries, buds, leaves, bark and roots of plants growing mainly in the tropical, the subtropical and the temperature zone. Culinary herbs and their essential oils have been used extensively for many years in food products, perfumery, dental and oral products due to their different medicinal properties (Suppakul et al., 2003a; 2003b).

Herbs have changed the course of history and in economic terms have greater importance as ingredients in food and medicine, perfumery, alcoholic beverages, cosmetics, coloring and garden plants. They are prepared in number of ways to extract their active ingredients for internal and external use. At home, dried herbs can be ingested as teas (steeped in hot water) or, rarely, tinctures (in alcoholic solutions) or inhaled via steam from boiling suspensions of the parts or powdered as food ingredients. Dried plant parts can be added to oils or petroleum jelly and applied externally. Poultices can also be made from concentrated teas or tinctures.

2. Bioactive Compounds from Plants

Herbs and spices are rich in volatile oils, which give pleasurable aromas. In addition, herbs may contain alkaloids and glycosides, which are of greater interest to pharmacologists. Some of the main active constituents in herbs and their tastes are as follows in Table 1.

Table 1. Main active constituents of herbs, their effects and tastes

Main active constituents		Effect	Taste
Volatile oils		Antiseptic, fungicidal, irritant and stimulant.	Aromatic
Acids		Antiseptic and cleansing	Sour
Alkaloids		Affect the central nervous system and many are very toxic and addictive.	Bitter
Glycosides	sulphur	Stimulant, antibiotic	Acrid
	mustard oil	Extremely irritant	Acrid
	synogenic	Entispasmodic sedative, affecting heart rate and respiration	Bitter
	cardiac	Affecting heart contractions	--
Resins		Acrid, astringen, antiseptic, healing	Acrid
Saponins		Stimulant hormonal, often anti-inflammatory, or diuretic.	Sweet
Coumarines		Antibacterial, anticoagulant	Smell of new-mown hay
Bitters		Increases and improves digestion.	Bitter
Flavones		Diuretic, antiseptic, antispasmodic and anti-inflammatory	Bitter or sweet
Gums and mucilages		Soothing and softening.	Bland, sticky or slimy
Anthraquinones		Irritant and laxative, acting also as dyes.	Bitter

Adopted from Brown (1995); de Guzman and Sienonsma (1999).

3. Plant Extracts as Antimicrobials

Since antiquity, humans have used plants to treat common infectious diseases, and some of these traditional medicines are still included as part of the habitual treatment of various diseases. The essential oil of coriander is reported to be analgesic, dill and anise oils as antipyretic, coriander, celery, parsley and cumin oils as anti-inflammatory. Recently, anticarcinogenic property has been reported for essential oils of cumin and basil and these can be used as protective agents against carcinogenesis. Also, methanol extracts of allspice, marjoram, tarragon and thyme strongly inhibited platelet aggregation induced by collagen in humans.

Long before the discovery of the existence of microbes, the idea that certain plants had healing potential, indeed, that they contained what we would currently characterize as antimicrobial principles was well accepted. Some of the spices essential oils are biocides against a broad range of organisms such as bacteria, fungi, protozoa, insects, plants, and viruses (Rosa et al., 2003; Esquenazi, et al., 2002); oregano and thyme show the highest antimicrobial activity. Carvacrol, present in the essential oils of oregano and thyme, has been proved to be the most important fungitoxic compound. Plant extracts or seed diffusates could be used for the control of seed-borne pathogens and can be a substitute for costly chemicals for seed treatment. Plant extracts of pimento can be used for controlling fungal growth during storage of wheat grains. Likewise, the seed diffusates of *Anthem graveolens* and *Coriandrum sativum* gave a high level of growth inhibition against seed-borne fungi such as *Alternaria alternata* and *Fusarium solani*.

Table 2. Some of the most economically used plants or their derivatives in phytomedicines

Plant names	Compounds	Therapeutic use
Apocyanaceae, Rubiaceae spp.	Yohimbine	Aphrodisiac
Artemisia annua L.	Artemisinin	Antimalarial
Camptotheca acuminata	Camptothecin	Antineoplastic
Capsicum spp.	Capsaicin	Topical analgesic
Cassia angustifolia	Sennosides	Laxatine
Catharanthus roseus	Vinblastin, vincristine	Antineoplastic
Cephaelis ipecacuanha	Ipecac	Emetic
	Emetine	Antiamoebic
Cinchona spp.	Quinine	Antimalarial
	Quinidine	Cardiac depressant
Colchium autumnale	Colchicine	Antigout
Digitalis spp.	Digoxin, digitoxin	Cardiotonic
Dioscorea spp.	Diosgenin, hecogenin, stigmasteroll	Oral contraceptives
Erythroxylum coca	Cocaine	Local anesthetic
Leucojumae stivum	Galanthemine	Cholinesterase inhibitors
Nicotiana spp.	Nicotine	Smoking cessation therapy
Papaver somniferum	Codeine, morphine	Analgesic, antitussive
Podophyllum peltatum	Podophyllotoxin	Antineoplastic
Rauwolfia serpentina	Reserpine	Antihypertensive, psychotropic
Taxus brevifolia	Taxol and other taxoids	Antineoplastic

Adopted from Balandrin, et al. (1993); Raskin, et al. (2002).

4. Modern Phytomedicine

Phytomedicines are defined as therapeutic agents derived from plants or parts of plants, or preparations made from them. The word's prefix `phyto' originates from the hellenephyton meaning plant. Phytomedicines are not isolated chemically-defined substances, they must therefore exhibit the totality of the medicinal plant or united of its parts (such as origin leaf, flower, fruit, etc.), rather than a single isolated chemical adjust.

In rational phytotherapy appropriate pharmacological investigations and clinical studies in patients have documented the efficacy of the products employed. In traditional phytotherapy, on the other hand, the efficacy of phytopharmaceuticals or herbal teas has not yet been established in that way. Phytomedicines have been valued in a traditional sensation for their therapeutic value. In other words, this value was known before modern scientific research studied them.

The European Medicine Licensing Agency (EMLA) established the term Herbal Medicinal Products (HMPs) in guidelines related to quality and specifications of products used in rational therapy. Phytomedicines are commonly found in as it was forms as teas, capsules, tablets, smoothed juices, tinctures, solvent extracted preparations, or combinations of these various yield forms, all of which are available in health meat stores. There is a great difference between phytomedicines in the European market, particularly Germany, compared with the American market is that, in Germany, they are considered "ethical drugs" prescribed by physicians and dispensed by pharmacists. In the United States, phytomedicines are sold in the form of dietary fill up products.

The US Food and Drug Administration (FDA) has published guidelines for standardized multi-component plant extracts referred as botanical drugs, thus making it possible to market these products under the New Drug Administration (NDA) approved process [*]. Some of the most economically plants or their derivatives used in phytomedicines in Table 2.

One of the major constraints in using plants in pharmaceutical discovery is the lack of reproducibility of activity for over 40% of plant extracts (Cordell, 2000). Reproducibility is the major problem, as the activities detected in screens often do not repeat when plants are re-sampled and re-extracted. This problem is largely due to differences in the biochemical profiles of plants harvested at different times and locations, differences in variety, and variation in the methods used for extraction and biological activity determination. It is important to identify the different agroclimatic or stress locations, climate, microenvironment, physical and chemical stimuli often called elicitors, which quantitatively and qualitatively alter the content of bioactive secondary metabolites.

5. Evaluation of Plants Antimicrobial Potency *In Vitro*

Novel compounds or agents from any origin, natural or synthetic, should be subjected to laboratory examination methods, i.e. different *in vitro* assays, before their application as a food preservative. Plant extracts should be also examined using the same methods.

The *in vitro* assays, to explore the antimicrobial potentiality of proposed extract as a food preservative, should be performed toward many pathogenic foodborne microorganisms and food spoilage microbial strains. *In vitro* assays could be divided into two groups [Figure 1], e.g. endpoint screening methods and descriptive screening tests.

For the endpoint tests, a microorganism is challenged for an arbitrary period. The results reveal the inhibitory activity of a specific agent only for the time specified. In descriptive test, the microorganism is also challenged, but periodic sampling is carried out to examine alterations in viable cell number over time.

Most of the *in vitro* techniques applied for the examination of preservative potentiality of natural compounds require the preparation of several concentrations from the natural agent, i.e. serial dilution.

A scheme to prepare serial dilutions from a potential natural biopreservative is shown in Figure 2, the compounds to be examined could be diluted with water or growth media, for water-soluble compounds, or using dimethyl sulfoxide (DMSO) or Tween 80 for water-insoluble agents.

6. Methodology for Applying Plants Extracts as Biopreservatives

An ideal approach was proposed in 1985, based on ethnomedical information, to experimentally examin plants as a source of drugs (Farnsworth et al., 1985). This approach was primarily planned for putting into practice by developing countries, where lack of hard currency often prevents sophisticated forms of research from being conducted. The possibility of drug development in the form of stable, standardized crude extracts and eventual development of the active principles from these plants was proposed. Figure 3 gives a brief overview of this approach.

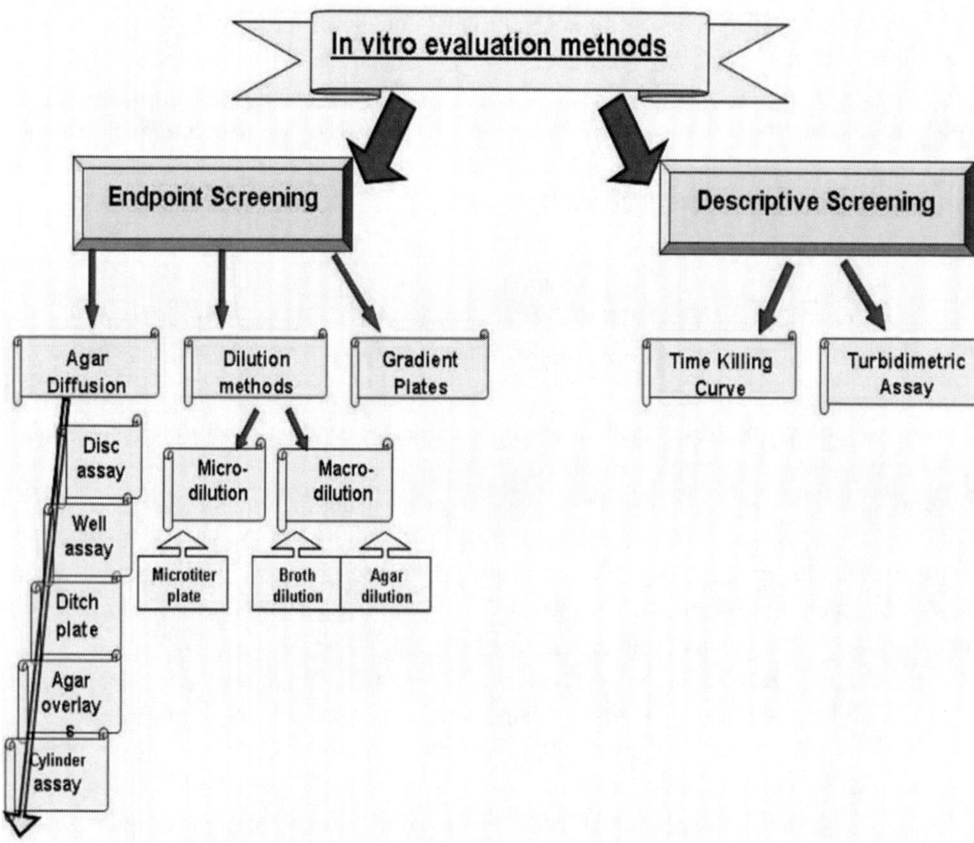

Figure 1. *In vitro* examination assays for evaluating the preservative activity of natural compounds.

7. Antifungal Activity of Plant Extracts

Phytopathogenic fungi can cause local or general symptoms on their hosts or can also cause excessive enlargement or induced growth and distortion of plant parts. The genera *Aspergillus* and *Fusarium* are examples of known phytopathogenic fungi reported to be widespread and harmful toxin-producers (Harris and Mantle, 2001; Harvey et al., 2001; Hussein and Brasel, 2001).

Aspergillus flavus is a common mold that grows on many substances, especially on grains and nuts, producing the dangerous aflatoxins (Vargas et al., 2001). *A. flavus* is also referred as being responsible for paranasal invasive aspergillosis, a rare but chronic diseases affecting man (Alrajhi et al., 2001).

Infections by *Fusarium* are responsible for destroying crops or dramatically reducing production yields. The economic importance of Fusarium species is enhanced by their ability to synthesize harmful mycotoxins, such as zearalenone, fusarins, fumonisins or trichothecenes (Koopmann et al., 1994; Blat et al., 1997; Desjardins et al., 2000).

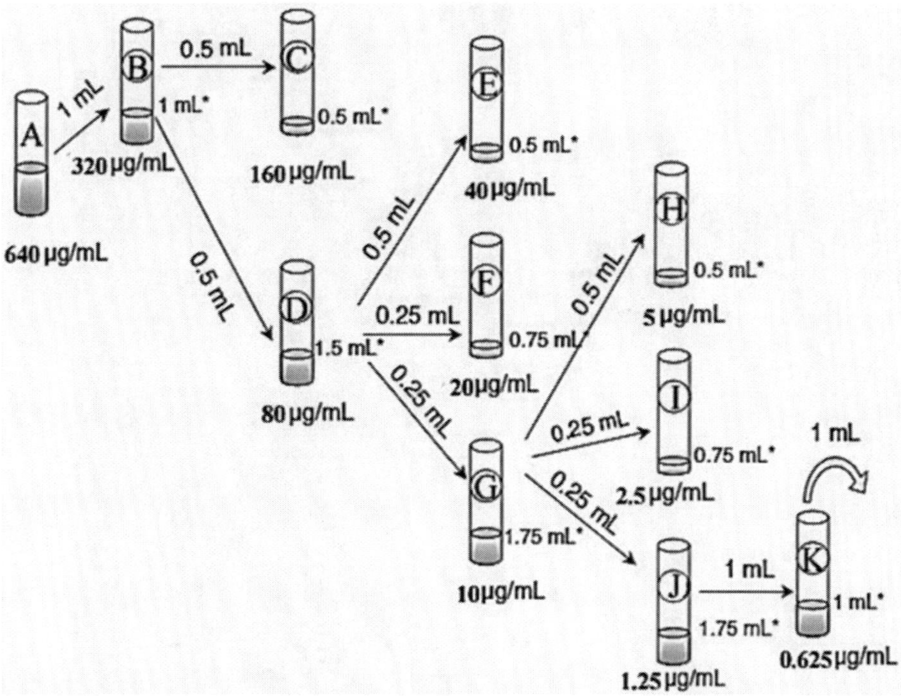

Figure 2. Scheme to perform the dilutions from natural compound in the range of 0.625 to 640 µg/ml. *Volume of DMSO or Tween 80 (for water- insoluble agents) or growth media (for water soluble compounds) before adding the antimicrobial dilutions. Adopted from Espinel-Ingroff and Cantón, 2007.

From the 1970s an increased interest on the study of the plants autoprotective capacity against pathogenic agents or adverse environmental conditions was noted (Bailey et al., 1992). The isolation of several substances synthesized by several plants as a response to biotic or abiotic stresses evidencing insecticidal, fungicidal, bactericidal or herbicidal action have been reported (Lamb and Lawton, 1989; Bouguerra, 1990; Keen, 1990; Matos et al., 1998).

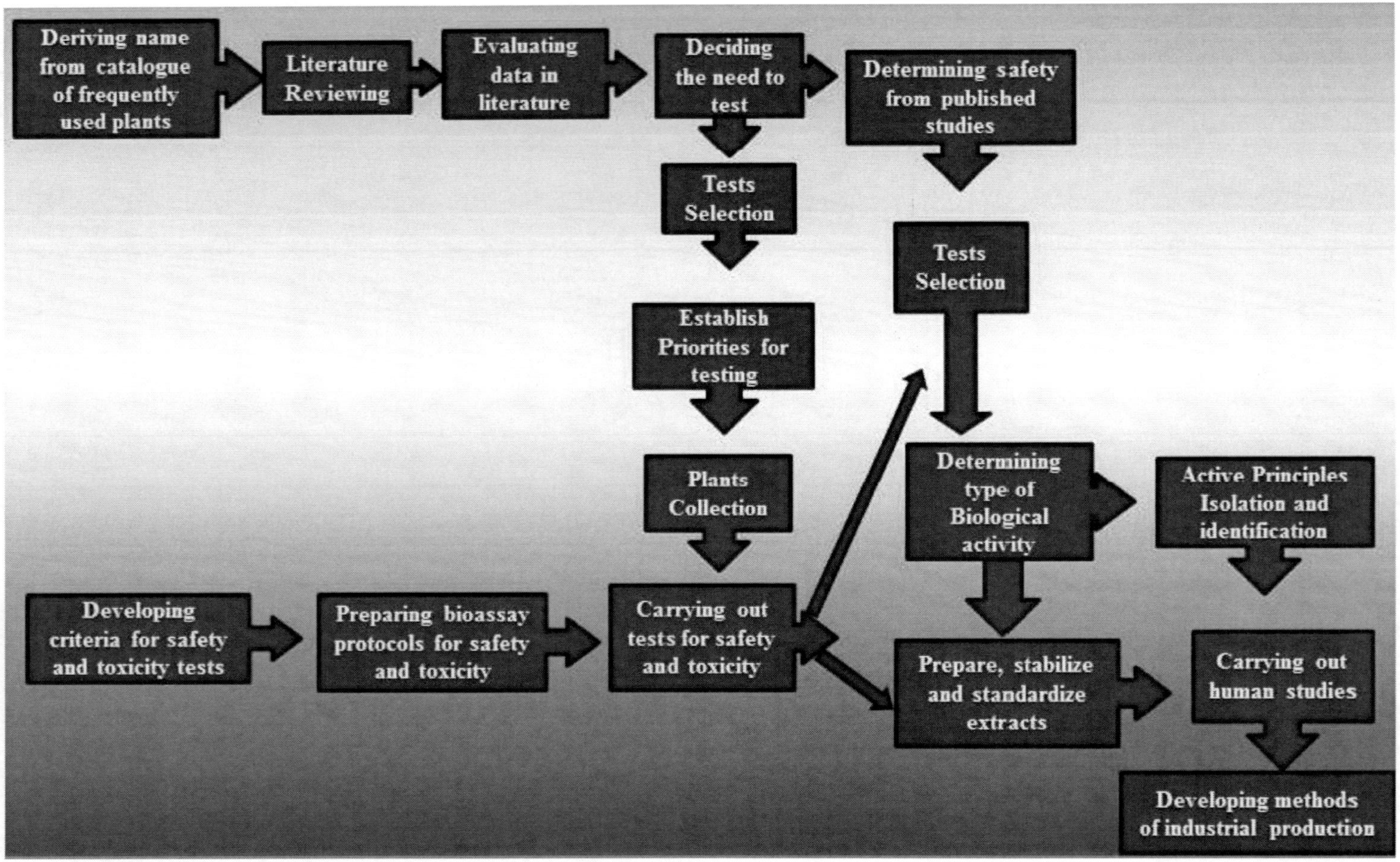

Figure 3. Systematic flow chart to study candidate plants for application as biopreservatives. Adopted from (Farnsworth et al., 1985).

The term natural fungitoxins has been sometimes used to refer substances of plant origin that cause a significant toxic or inhibitory action on the fungus growth at relatively low concentrations. Flavonoids, isoprenoids and alkaloids with these properties are essential for plant adaptation, participating in a broad range of physiological processes (Harborne, 1987; Mayer, 1989).

Roots are, in general, resistant to infection, which means that fungitoxic substances, present or exuded by the epidermis of the roots may play a protective role. In fact, more than twenty substances belonging to the group of isoflavones have been isolated from the roots of *Lupinus albus*. Borbonol-2, a hydrocarbonated lactone with antifungal activity, wasalso isolated from the roots of *Persea borbonica* (Harborne et al., 1976; Thara et al., 1984).

One of the most important defense systems of the plant consists in the synthesis of substances generally called phytoalexins, which are not present in healthy plant tissues but are synthesized when plant cells are exposed to microorganisms or other stressing stimulus (Darvill and Albersheim, 1984; Ebel, 1986; Fagboun et al., 1987; Adesanya et al., 1989; Osbourn, 1996). Phytoalexins have long been inferred to be important in the defense of plants against fungal infection (Brooks et al., 1986, 1987). Since the *in vitro* plant tissue cultures can synthesize substances different from those of the plant grown in natural conditions, the development of *in vitro* studies are also important. DiCosmo and Misawa (1985) referred several examples of phytoalexins produced in tissues of different plant species, elicited *in vitro* by some chemical compounds or extracts of fungi.

The novelty of applying pomegranate peel extract (PPE) as an alternative, reduced-risk antifungal agent for controlling citrus green mold invasion was investigated by Tayel et al. (2009). Their promising results indicated that the most potent antifungal PPEs against *Penicillium digitatum* isolates were those extracted with methanol, ethanol and water, respectively, using both agar diffusion assay and broth microdilution susceptibility test as qualitative and quantitative tests. Aerosolization with PPE confirmed its effectiveness as antifungal sanitizing agent against *P. digitatum. In vivo* prevention and control of *P. digitatum* invading harvested citrus fruit was successfully applied by immersing wounded fruit in PPE solutions. Therefore, we could recommend PPE application for the full control of citrus green mold via sanitation of storage rooms and treatment of harvested fruit.

Tayel et al. (2011) evaluated ten natural plant extracts, recommended in the Islamic medicine, as antifungal agents against mycotoxigenic Aspergilli, i.e. *A. flavus* and *A. ochracious*. Most of extracts had remarkable antifungal activities using both qualitative and quantitative evaluation methods. Cress (*Lepidium sativum*) seed extract was inventively proven to be the most powerful among examined plants. The most effective (garden cress seed, pomegranate peel and olive leaf extracts) were applied by blending with prepared organic maize silage, individually or as combinations, after inoculation of fungal spores. Treatment of silage with individual extracts, at their minimal fungicidal concentrations, resulted in the reduction of A. flavus colony counts by 99.9, 99.6 and 98.7 %, respectively, whereas silage blending with the combined extracts completely prohibited fungal growth up to 30 days of incubation under aerobic conditions.

It is not an easy process to select the appropriate preservation system for a particular food product. The target pathogen or spoilage microorganisms must be identified first, and then the possible preservation systems must be evaluated via model studies and studies in the food product in question.

8. Plant Antimicrobials Application in Food

Nowadays, food authorities continually search for "new" and unique medicinal plants because of the growing global demand for authentic ethnic and cross-cultural cuisines. Consumers are also seeking natural foods and natural preservatives for healthier lifestyles and natural ways of preventing ailments. So, medicinal plants are also being sought for their medicinal value, as antioxidants and as antimicrobials.

As well as their traditional role in flavoring and coloring foods, they have been increasingly used as natural preservatives and for their potential health-promoting properties, for example as antioxidants as they have been an essential factor in health care through the ages in all cultures and consumption of fresh herbs in the diet may therefore contribute to the daily antioxidant intake. Essential oils, oleoresins and even aqueous extracts of spices possess antioxidative properties. Phenolic compounds are the primary antioxidants present in spices and there is a linear relationship between the total phenolic content and the antioxidant properties of spices.

Consequences of microbial growth in foods include consumer hazards as a result of the presence of pathogenic organisms or microbial toxins, as well as economic losses as a result of spoilage (Davidson, 2001). Inactivation, growth delay, or growth prevention of spoilage and pathogenic microorganisms are main objectives of food preservation. Major antimicrobial systems naturally present in plants and animals or produced by microorganisms. Some of these natural antimicrobial systems are already used for food preservation, whereas many others are just being studied to be used in foods (Gould, 1995).

Plants, herbs, and spices as well as their derived essential oils and isolated compounds contain a large number of substances that are known to inhibit various metabolic activities of bacteria, yeast, and molds, although many of them are yet incompletely exploited. Wilkins and Board (1989) reported that more than 1340 plants are known to be potential sources of antimicrobial compounds. The antimicrobial compounds in plant materials are commonly contained in the essential oil fraction of leaves (rosemary, sage), flowers and flower buds (clove), bulbs (garlic, onion), rhizomes (asafoetida), fruit (pepper, cardamom), or other parts of the plant (Nychas, 1995; Shelef, 1983). Major components with antimicrobial activity found in plants, herbs, and spices are phenolic compounds, terpenes, aliphatic alcohols, aldehydes, ketones, acids, and isoflavonoids. As a rule it has been reported that antimicrobial activity of essential oils depends on the chemical structure of their components and on their concentration.

Cinnamon and clove were listed as exhibiting a strong inhibitory effect, and allspice was classified with a medium inhibitory effect (Zaika 1988). Also, Tayel and El-Tras (2009) investigated the antimicrobial potentialities of twenty five herbs and spices which are widely used in folk medicine by Egyptian housewives to treat gastrointestinal disorders against seven bacterial strains, mostly food borne including pathogens. They were tested by using paper disc diffusion technique as qualitative assay and agar dilution method for determining the minimum inhibitory concentration (MIC) of herbs extracts. Among screened plants, basil, cardamom, cinnamon, clove, lemon grass, mustard, oregano, rosemary, sage and thyme extracts exhibited notable antimicrobial activities against most of the tested strains. Cinnamon extract was the most inhibitor followed by clove, whereas extracts of chamomile, rose of Jericho, safflower and turmeric showed weak antibacterial activities against most of the tested strains. The most sensitive strain to plant extracts was *B. subtilis* and the most resistant strain was *Ps. fluorescens*. It was proved that herbs and spices extracts –used in Egyptian folk medicine for treating many gastrointestinal disorders - could be successfully applied as natural antimicrobials for elimination of food borne bacteria and pathogens growth.

Antimicrobial activity of cinnamon, allspice, and cloves is attributed to eugenol (2-methoxy-4-allyl phenol) and cinnamic aldehyde, which are major constituents of the volatile oils of these spices. Cinnamon contains 0.5% to 1.0% volatile oil, which contains 65% to 75% cinnamic aldehyde and 8% eugenol. Allspice contains up to 4.5% volatile oil, of which 80% is eugenol. Clove buds have an average essential oil content of 17% that is 93%to 95% eugenol (Bullerman et al.; 1977; Farrell, 1990).

Antimicrobial activity of essential oils and extracts from plants, herbs, and spices depends on the extraction method. Solvent selection for extraction including solvent ability to extract components is an important factor determining operation efficiency. Dry plants could be extracted with a variety of solvents and sometimes sequentially from low to high polarity. Polar solvents such as ethyl acetate or methanol are often used. For fresh plant materials (with high water content) a solvent mixture of dichloro-methane-methanol gave better results during extraction. Methanol separation and sample partition followed by ethyl acetate and butanol extraction help to separate lipophilic compounds from water-soluble materials.

Plants of the *Allium* species, namely garlic and onion, have been extensively used as antimicrobial plants. Garlic (*Allium sativum*) has been used medicinally for centuries. Many foodborne pathogenic bacteria are sensitive to onion and garlic extracts. *S. aureus, B. cereus, Clostridium botulinum, S.* Typhimurium, and *E. coli* have been adversely affected by garlic extracts (Beuchat and Golden, 1989). The antimicrobial effects reported for garlic and onion were attributed to the allicin concentrations and other sulfur compounds present in their essential oils. Allicin was described as a colorless oil, extremely pungent, that characterized the principle odor and taste of garlic and onion. It was reported that allicin in concentrations of 1:85,000 in broth was bactericidal to a wide variety of Gram-negative and Gram-positive organisms. Investigations have shown that extracts from *Allium* bulbs inhibit growth and respiration of pathogenic fungi and bacteria. Aqueous extracts from fresh garlic bulbs at levels of 3%, 5%, and 10% inhibited the growth of *B. cereus* on nutrient agar plates by 31.3%, 58.2%, and 100%, respectively (Saleem and Al-Delaimy, 1982).

Conner and Beuchat (1984a, b) suggested that antimicrobial activity of essential oils on yeasts could be the result of disturbance in several enzymatic systems involved in energy production and structural components synthesis. Once the phenolic compound crossed the cellular membrane, interactions with membrane enzymes and proteins would cause an opposite flow of protons, affecting cellular activity. Phenolic compounds could also denature enzymes responsible for spore germination or interfere with amino acids necessary in germination processes (Nychas, 1995). Rico Muñoz et al. (1987), after finding that various phenolic compounds have different effects on the membrane of *S. aureus,* concluded that phenolic compounds probably do not share a common mechanism of action and that there might be various targets associated with their antimicrobial effect.

The application of plant extract as biopreservatives, for meat products, was recommended by many researchers. Many plant extracts were evaluated as biopreservatives to decontaminate and maintain the quality of meat steaks (Tayel et al., 2012) and ground beef (Tayel and El-Tras, 2012); most of the examined extracts exhibited a remarkable antibacterial activity against antibiotic resistant strains from *Salmonella* and *Staphylococcus* Spp.

However, application of plant extracts was impressively recommended for comprehensive meat products decontamination and quality attributes enhancement.

Conclusion

Regarding their efficiency, biosafety, desirability and cost effectiveness, plant extracts could be highly recommended as the ideal biopreservative agents for the application in every food and feed processing sectors.

Acknowledgment

Always and foremost, Thanks are indebted to our god "ALLAH" forever for his mercy guiding and helping.

References

Adesanya, S. A.; Ogundana, S. K. & Roberts, M. F. (1989) Dihydrostilbene-phytoalexins from *Dioscorea bulbifera* and *D. dumentorum*. *Phytochemistry, 28*,773–774.

Alrajhi, A. A.; Enani, M.; Mahasin, Z. & Al-Omran, K. (2001) Chronic invasive aspergillosis of the paranasal sinuses in immunocompetent hosts from Saudi Arabia. *The American Journal of Tropical Medicine and Hygiene, 65*, 83–86.

Bailey, J. A.; O'Connell, R. J.; Pring, R. J. & Nash, C. (1992) Infection strategies of *Colletotrichum* species. In: Bailey J. A. & M. J. Jeger, (Eds.). *Colletotrichum: biology, pathology and control*. Wallingford CT: CAB International, pp. 88–120.

Balandrin, M. F., Kinghom, A. D. & Famsworth, N. R. (1993). Plant derived natural products in drug discovery and development. in *Human Medicinal Agents from Plants.* ACS Symposium Series 534, Washington, American Chemical Society, D.C. pp. 2-12.

Beuchat, L. R. & Golden, D. A. (1989). Antimicrobials occurring naturally in foods. *Food Technology,43*(1),134–142.

Blat, R. V.; Shetty, P. H.; Amruth, R. P. & Sudershan, R. V. (1997) A foodborne disease outbreak due to the consumption of moldy sorghum and maize containing fumonisin mycotoxins. *Clinical Toxicology, 35*(3), 249–255.

Bouguerra, M. L. (1990) Pesticides phototoxic, des armeslumineusecontre les nuisibles. *Recherche, 271*(21), 93–94.

Brooks C. I. W., Watson D. G. & Freerm I. M. (1986) Elicitation of capsidiol. Accumulation in suspended callus cultures of *Capsicum annuum*. *Phytochemistry, 25*(5), 1089–1092.

Brooks, C. I. W.; Watson, D. G.; Rycroft, D. & Freer, M. I. (1987) The biosynthesis of sesquiterpenoidphytoalexins in suspended callus cultures of Nicotianatabacum. *Phytochemistry, 26* (8), 2243–2245.

Brown, D. (1995). *The Royal Horticultural Society* – Encyclopedia of Herbs and Their Uses. Dorling Kindersley Limited, London.

Bullerman, L. B.; Lieu, F. Y. & Seier, S. A. (1977). Inhibition of growth and aflatoxin production by cinnamon and clove oils. Cinnamic aldehyde and eugenol. *Journal of Food Science, 42*, 1107–1109.

Conner, D. E. & Beuchat, L. R. (1984a). Effects of essential oils from plants on growth of food spoilage yeasts. *J. Food Sci.* 49:429–434.

Conner, D. E. &Beuchat, L. R. (1984b). Sensitivity of heat-stressed yeasts to essential oils of plants. *Applied and Environmental Microbiology, 47*, 229–233.

Cordell, G. A. (2000). Biodiversity and drug discovery- a symbiotic relation-ship. *Phytochemistry*, *55*, 463–480.

Darvill, A. G. & Albersheim, P. (1984) Phytoalexins and their elicitors – a defense against microbial infection in plants. *Annual Review of Plant Physiology, 35*, 243–275.

Davidson, P. M. (2001). Chemical preservatives and naturally antimicrobial compounds, In Doyle, M. P.; L. R. Beuchat; & T. J. Montville (Eds.). *Food Microbiology. Fundamentals and Frontiers*. 2nd ed. ASM Press, Washington, DC, pp. 593–628.

de Guzman, C. C. & Sienonsma, J. S. (1999), *Plant Resources of South East Asia*. No. 13. Spices. Backhuys Publishers, Leiden, The Netherlands.

Desjardins, A. E.; Manandhar, G.; Plattner, R. D.; Maragos, C. M.; Shrestha, K. & McCormick, S. P. (2000) Occurrence of *Fusarium* species and mycotoxins in Nepalese maize and wheat and the effect of traditional processing methods on mycotoxin levels. *Journal of Agricultural and Food Chemistry, 48*, 1377–1383.

di Cosmo F. & Misawa M. (1985) Eliciting secondary metabolism in plant cell culture. *Trends in Biotechnology, 3*(12), 318–322.

Ebel, J. (1986) Phytoalexin synthesis: the biochemical analysis of the induction process. *Annual Review of Phytopathology, 24*, 235–264.

Espinel-Ingroff, A. & Cantón, E. (2007). Antifungal Susceptibility Testing of Filamentous Fungi. In R. Schwalbe, L. Steele-Moore & A. C. Goodwin (Eds.), *Antimicrobial susceptibility testing protocols*. Boca Raton, FL: Taylor & Francis Group.

Esquenazi, D., Wigg, M., Miranda, M. M., Rodrigues, H., Tostes, J., Rozental, S., da Silva, A. J. & Alviano, C. (2002). Antimicrobial and antiviral activities of polyphenolics from *Cocosnucifera* Linn. (Palmae) husk fiber extract. *Research in Microbiology, 153*, 647-652.

Fagboun, D. E.; Ogundana, S. K.; Adesanya, A. S. A. & Roberts, M. F. (1987) Dihydrostilbenephytoalexins from *Dioscorearotundata*. *Phytochemistry, 26*(12):3187–3189.

Farnsworth, N. R., Akerele, O., Bingel, A. S., Soejarto, D. D. & Guo, Z. (1985). Medicinal plants in therapy. *Bulletin of the World Health Organization (WHO), 63*, 965–981.

Farrell, K. T. (1990). *Spices, Condiments and Seasonings*. 2nd ed. van Nostrand Reinhold, New York.

Gould, G. W. (1995a). Overview, In: *New Methods of Food Preservation.* Blackie Academic & Professional, Glasgow, pp. xv–xix.

Gould, G. W. (1995b). The microbe as a high pressure target. In: Ledward, D. A.; D. E. Johnston; R. G. Earnshaw; & A. P. Hasting, Eds.; *High Pressure Processing of Foods.* Nottingham University Press, Nottingham, U. K.

Harborne, B. J. (1987) Natural fungitoxins. Biologically active natural products. *Proceedings of the Phytochemical Society of Europe, 27,*195–211.

Harborne, J. B.; Ingham, J. L.; King, L. & Payne M. (1976) Isopentenyl-isoflavoneluteone as a preinfeccional antifungal agent in the genus Lupinus. *Phytochemistry, 15*, 1485–1487.

Harris, J. P. & Mantle, P. G. (2001) Biosynthesis of ochratoxins by *Aspergillus ochraceus. Phytochemistry, 58*,709–716.

Harvey, R. B.; Edrington, T. S.; Kubena, L. F.; Rottinghaus, G. E.; Turk, J. R.; Genovese, K. J. & Nisbet, D. J. (2001) Toxicity of moniliformin from *Fusarium fujikuroi* culture material to growing barrows. *Journal of food protection, 64*, 1780–1784.

Hussein, H. S. & Brasel, J. M. (2001) Toxicity, metabolism, and impact of mycotoxins on humans and animals. *Toxicology, 167*,101–134.

Keen, N. T. (1990) Microbe and microbial products as herbicides–phytoalexins and their elicitors. *ACS Symposium Series, 439*(6), 114–129.

Koopmann, B.; Karlovsky, P. & Wolf, G. (1994) Differentiation between *Fusarium culmorum* and *Fusarium graminiarum* by RFLP and with species-specific DNA probes. In: Schots A.; F. M. Dewey & R. Olivier, (Eds.) *Modern assays for plant pathogenic fungi: identification, detection and quantification.* CAB International Wallingford, Oxfordshire, pp. 37-46.

Lamb, C. J. & Lawton, M. A. (1989) Signals and transduction mechanisms for activation of plant defense against microbial attack. *Cell, 56*, 215–224.

Matos, O. C.; Baptista, S.; Passarinho, J. A.; Delgado, F.; Ricardo, C. P. & Vilas-Boas, L. (1998) Watering and mineral supply effects on plant production, on hydrossoluble metabolites content and on the antifungal properties of three aromatic plants used as spices. *1st International Meeting of Aromatic and Medicinal Mediterranean Plants, Proceedings*, pp. 158–163.

Mayer, A. M. (1989) Plant–fungal interactions: a plant physiologist's viewpoint. *Phytochemistry, 28*, (2):311–317.

Nychas, G. J. E. (1995). Natural antimicrobials from plants, In: Gould, G. W. (Ed.) *New Methods of Food Preservation.* Blackie Academic and Professional, Glasgow, pp. 58–89.

Osbourn, A. E. (1996) Preformed antimicrobial compounds and plant defense against fungal attack. *Plant Cell, 8*, 1821–1831.

Raskin, I.; Ribnicky, D. M.; Komarnystky, S.; Ilic, N.; Poulev, A.; Borisjuk, N.; Brinker, A.; Moreno, D. A.; Ripoll, C.; Yakoby, N.; O'Niel, J. M.; Cornwell, T.; Pastor, I. &Fridlender, B. (2002). Plants and human health in the twenty-first century. *Trends in Biotechnology, 20*, 1–10.

Rico-Muñoz, E.; Bargiota, E.; and Davidson, P. M. (1987). Effect of selected phenolic compounds on the membrane-bound adenosine triphosphate of *Staphylococcus aureus. Food Microbiology, 4*, 239–249.

Rosa, M. S. S.; MendoncaFilho, R. R.; Bizzo, H. R.; Rodrigues, I. A.; Soares, R. M. A.; Souto-Padron, T.; Alviano, C. S. & Lopes, A. H. C. (2003). Antileishmanial activity of a linalool-Rich 2 essential oil from *Croton cajucara. Antimicrobial Agents and Chemotherapy*, *47*, 1895–1901.

Saleem, Z. M. & Al-Delaimy, K. S. (1982). Inhibition of *Bacillus cereus* by garlic extracts. *Journal of food protection, 45*, 1007–1011.

Shelef, L. A. (1983). Antimicrobial effects of spices. *Journal of food Safety, 6*, 29–44.

Suppakul, P., Miltz, J., Sonneveld, K. & Bigger, S. W. (2003b). Antimicrobial properties of basil and its possible application in food packaging. *Journal of Agricultural and Food Chemistry, 51*, 3197-3207.

Suppakul, P., Miltz, J., Sonneveld, K. & Bigger, S. W. (2003a). Active packaging technologies with an emphasis on antimicrobial packaging and its application. *Journal of food Science, 68*, 408-420.

Tayel A. A., Salem M. F., El-Tras, W. F. & Brimer L. (2011) Exploration of Islamic Medicine Plant Extracts as Powerful Antifungals for the Prevention of Mycotoxigenic Aspergilli Growth in Organic Silage. *Journal of the Science of Food and Agriculture, 91*, 2160–2165.

Tayel A. A., El-Baz A. F., Salem M. F. & El-Hadary M. H. (2009) Potential Applications of Pomegranate Peel Extract for the Control of Citrus Green Mold. Journal of Plant Diseases and Protection, 116 (6) 252-256.

Tayel A. A., El-Tras W. F., Moussa S. H. & El-Sabbagh S. M. (2012). Surface Decontamination and Quality Enhancement in Meat Steaks Using Plant Extracts as Natural Biopreservatives. *Foodborne Pathogens and Disease, 9*(8), 755-761. doi:10. 1089/fpd.2012.1203.

Tayel, A. A. & El-Tras W. F. (2012). Plant Extracts as potent Biopreservatives for *Salmonella* Typhimurium Control and Quality Enhancement in Ground Beef. *Journal of Food Safety, 32* (1), 115-121.DOI:10.1111/j. 1745-4565.2011.00357.x.

Tayel, A. A. & El-Tras, W. F. (2009). Possibility of Fighting Food Borne Bacteria with Egyptian Folk Medicinal Herbs and Spices Extracts. *The Journal of the Egyptian Public Health Association, 84*(1&2), 21-32.

Tepsorn, R. (2009). *Antimicrobial Activity of Thai Traditional Medicinal Plants Extract Incorporated Alginate-Tapioca Starch Based Edible Films against Food Related Bacteria Including Foodborne Pathogens*. Ph. D. Dissertation, Faculty of Agricultural Sciences, University of Hohenheim, Pattani, Thailand.

Thara, S.; Ingham, J. L.; Nakahara, S.; Mizutani, J. & Harborne, J. B. (1984) Fungi-toxicdihyd-rofuranoisoflavones and related compounds in white lupine, *Lupinusalbus. Phytochemis-try, 23*, 1889-1893.

U. S. Department of Health and Human Services, Food and Drug Administration for Drug Evaluation and Research. Guidance for Industry Botanical Drug Products.http://www. Fda.gov/cder/guidance/1221 dft.htm# P131_3293.

Vargas, E. A.; Preis, R. A. Castro, L. & Silva, C. M. G. (2001) Co-ocurrence of aflotoxins B1, B2, G1, G2, zearalenone and fumisonin B1 in Brasilian corn. *Food Additives and Contaminants, 18*, 981–986.

Wilkins, K. M. & Board, R. G. (1989). Natural antimicrobial systems, In Gould, G. W. (Ed.), *Mechanisms of Action of Food Preservation Procedures*. Elsevier Science Publishers Ltd.; New York, pp. 285–362.

Zaika, L. L. (1988). Spices and herbs: their antimicrobial activity and its determination. *Journal of Food Safety, 9*, 97–118.

In: Plant Extracts
Editors: Adriana Giordano and Alfieri Costs

ISBN: 978-1-62417-534-3

Chapter 14

Microwave Extraction of Antioxidants from Rosemary Leaves: A Green Approach

***Meziane Assami Dalila*[1,2], *Valérie Tomao*[3], *Karine Ruiz*[3], *Brahim Youcef Meklati*[1,2] *and Farid Chemat*[3*]**

[1] Centre de Recherche Scientifique et Technique en Analyses Physico-Chimiques, Alger, Algérie

[2] Laboratoire d'Analyse Organique Fonctionnelle, Faculté de Chimie, USTHB, Alger, Algérie

[3] Université d'Avignon et des Pays de Vaucluse, INRA, Avignon, France

Abstract

This chapter reports the application of microwave technology for extraction of antioxidants compounds from natural product. This novel method based on an electromagnetic heating was named by our research group as "Microwave Integrated Soxhlet" abbreviated as the MIS method. This technique was compared to the traditional method, Soxhlet extraction, which has been used as the reference method for evaluating the performance of the proposed method. Rosemary (*Rosmarinus officinalis L.*) was selected as natural source used for extracting antioxidants compounds by MIS process, owing to its antioxidants properties, as well as it is easily grown in the Mediterranean area. Response Surface Methodology (RSM), was applied to the MIS extraction of the antioxidants components of the rosemary plant, to determine the most influential factors in the process in order to optimize it and therefore, find the combination of factors which would result in obtaining the extract with the highest antioxidant content. After executing a set of designed experiments the MIS extraction parameters including extraction time, leaching time and microwave irradiation power were optimized.

* E-mail: farid.chemat@univ-avignon.fr.

1. Antioxidants: Composition, Properties, and Applications

Antioxidants (AOs) are defined as chemical substances that prevent the oxidation of molecule that is found in foods or biological materials, including carbohydrates, DNA, lipids and proteins [1]. They protect cells from being damaged by unstable molecules called free radicals, those molecules are normally present in the body in small numbers. They are created during the body's basic metabolic functions, such as digestion and breathing. Indeed, not all free radicals are harmful. Some free radicals help destroy virus and bacteria, other are also involved in producing some very necessary hormones and enzymes that are vital to life. Many different factors can lead to an excess of free radicals, including radiation exposure (from the sun, medical x-rays, etc.), environmental pollutants (tobacco smoke, automobile exhaust, household cleaners, harsh chemicals), diet that is high in fat (fried foods, fatty foods...) [2]. It should be pointed out, that if there is excessive free radical formation, damage to cells and tissues can occur. Free-radicals mechanisms have been implicated in the pathology of several human diseases, including cancer, atherosclerosis, malaria, and rheumatoid arthritis and neurodegenerative diseases [3]. These diseases may be prevented or treated by antioxidants [4-6]. Scientists attach great importance on the effect of those molecules against damage caused by free radicals [7-9]. Antioxidants are added in food during processing to improve food quality and stability, they may also act as nutraceuticals to terminate free radical chain reactions in biological systems and therefore may provide additional health benefits to consumers [10]. AOs can be divided into natural and synthetic groups (figure 1). They can differ in their composition, in their physical and chemical properties, in their mechanism and in their site of action.

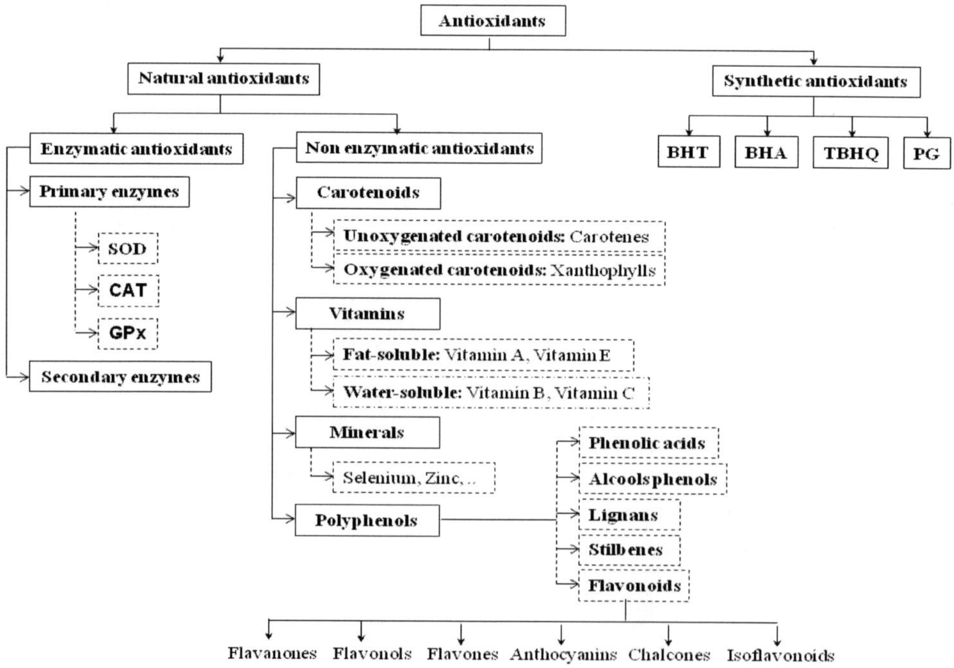

Figure 1. Some of the most important antioxidants.

Synthetic antioxidants are manmade chemical molecules. The principal synthetic antioxidants (table 1) are butylated hydroxyanisole (BHA), butylated hydroxytoluene (BHT), propyl gallate (PG), and tert-butylhydroquinone (TBHQ) [11-13]. Natural antioxidants are grouped into enzymatic and non enzymatic antioxidants. Enzymatic antioxidants include the primary enzymes, superoxide dismutase (SOD), catalase (CAT) and glutathione peroxidase (GPx), the secondary enzymes, glutathione reductase and glucose-6-phosphate dehydrogenase [14]. As presented in figure 1, non enzymatic antioxidants can be classified into various classes of which polyphenols are the largest class. Synthetic antioxidants cause a wide range of health problems. A lot of published findings agree with the fact that BHA and BHT are tumour promoters [15]. In that respect, the demand for natural antioxidants is going up, they can be used to replace their synthetic counterparts. The sources of natural antioxidants include a variety of fruits, vegetables, culinary herbs and medicinal herbs. Investigations by the researchers [16-19] have revealed that rosemary provides compounds capable of inhibiting lipid oxidation. It was found that numerous polyphenolic compounds which possess high antioxidant activity were present in rosemary extract, including the three main antioxidants compounds (table 2) rosmarinic acid, carnosol and carnosic acid [20-23].

Table 1. Principal synthetic antioxidants

	SYNTHETIC ANTIOXIDANTS			
Name	Butylated hydroxyanisole (BHA)	Butylated hydroxytoluene (BHT)	Tert-butylhydroquinone (TBHQ)	Propyl gallate (PG)
Synonyms	E320	E321	E319	E310
	4-Methoxy-2-(2-methyl-2-propanyl) phenol	4-Methyl-2,6-bis (2-methyl-2-propanyl) phenol	2-(2-Methyl-2-propanyl)-1,4-benzenediol	Propyl 3,4,5-trihydroxy-benzoate
Formula	$C_{11}H_{16}O_2$	$C_{15}H_{24}O$	$C_{10}H_{14}O_2$	$C_{10}H_{12}O_5$
Structure	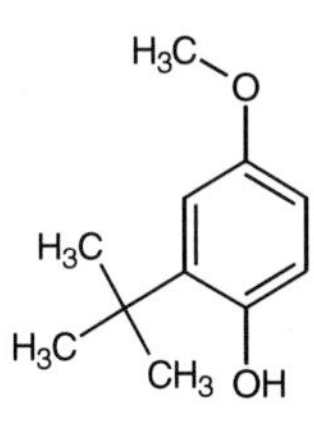	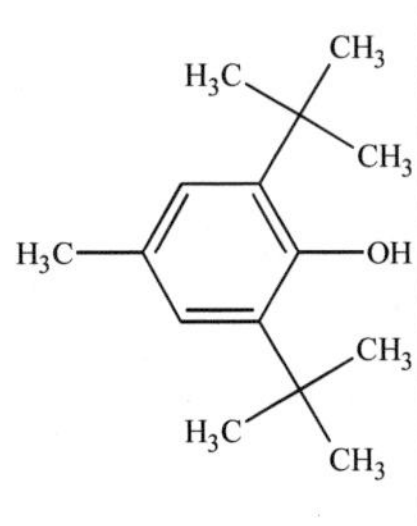	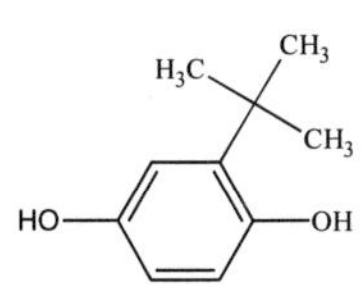	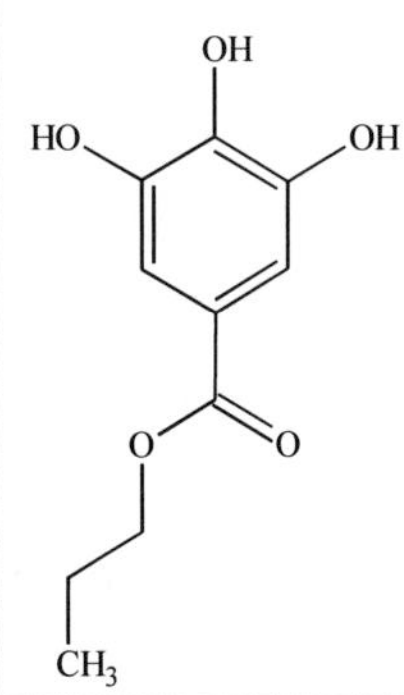
Molecular weight [g/mol]	180.24	220.35	166.21	212.06
Persistence Time [hours]	1.07e+003 hr	1.5e+003 hr	1.76e+003 hr	687 hr

Table 2. The main rosemary antioxidants

Name	NATURAL ANTIOXIDANTS		
	Carnosol	Carnosic acid	Rosmarinic acid
Synonyms	(5β,7α)-11,12-Dihydroxy-7,20-epoxyabieta-8,11,13-trien-20-one	11,12-Dihydroxyabieta-8(14),9(11),12-trien-20-oic acid	**(2R)-3-(3,4-Dihydroxyphenyl)-2-{[(2E)-3-(3,4-dihydroxyphenyl) prop-2-enoyl]oxy}propanoic acid**
Formula	$C_{20}H_{26}O_4$	$C_{20}H_{28}O_4$	$C_{18}H_{16}O_8$
Structure	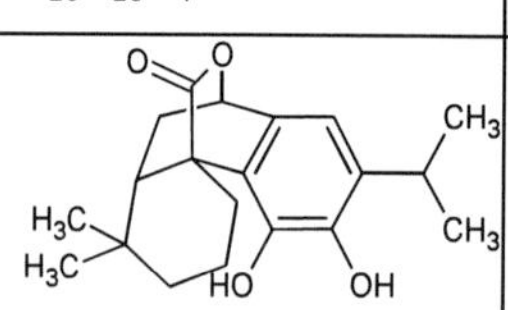	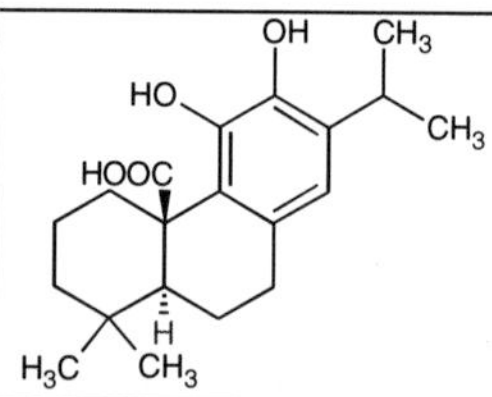	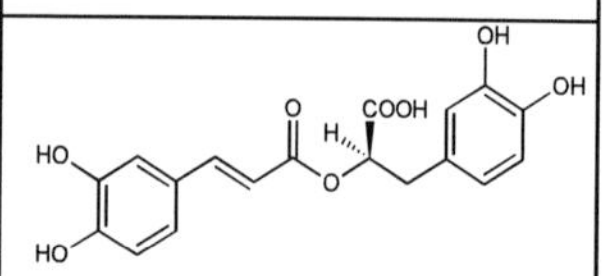
Molecular weight [g/mol]	330.41	332.19	360.31
Persistence Time [hours]	2.94e+003	2.42e+003	689

Carnosic acid and carnosol are phenolic diterpenes of abietane type, and rosmarinic acid is an ester of caffeic acid and 3,4-dihydroxyphenyllactate (phenolic depside) [23].

This chapter provides an overview about the isolation using novel technology and analyses of the major natural antioxidants in rosemary leaves. Rosemary (*Rosmarinus officinalis L.*), belongs to the Lamiaceae plant family. A native of the Mediterranean region, its name means "dew of the sea". It is one of the oldest and still most popular plants used since ancient times for a variety of purposes. The ancient Egyptian civilization used it as incense and as strong smelting herb in mummification process and the Romans used it in religious ceremonies.

Nowadays, there has been an increasing interest in the use of this plant, especially its secondary metabolites, such as antioxidants which are used in the food industry, cosmetic product and some pharmaceutical product [24].

2. Materials and Methods

2.1. Raw Material

Rosemary leaves (*Rosmarinus officinalis L.*, Lamiaceae) were collected in March 2010 with pruning shears in their highest state of greenness, from the National Institute of Agronomy (INA - El Harrach - Algiers - Algeria), where plant growth is uniform over the entire area. INA was chosen as a sampling site for the current study, because the best yield of extraction of secondary metabolites was found in rosemary harvested from this site [25]. It is not polluted and no pesticides are used in the selected area. The climate of the study site is humid Mediterranean according to the Emberger bioclimatic classification [26], characterized by a relatively high mean yearly precipitation (640 mm/year) and an average annual temperature vary from 12.2°C to 25.5°C. Sufficient quantities necessary to perform several extractions of rosemary were collected and dried in the air before utilisation. The initial moisture content of fresh rosemary leaves was 61.54%, it was reduced to a final moisture content of 13 %. Dried leaves were finely ground using an electrical mill to gain smaller particles, which can be more easily penetrated by the solvent. Finally, the dried and crushed leaves were subjected to extractions.

2.2. Chemicals

Standards of rosmarinic acid, carnosol and carnosic acid were purchased from Sigma Aldrich (St. Louis, USA). Solvents and reagents were all of analytical grade from Merck (Darmstadt, Germany). Water used in the mobile phase was deionised and filtered through a 0.45 μm membrane.

2.3. Extraction Procedures

With the growing demand for natural products, the extraction of the active components using a variety of methods from simple traditional technologies to advanced extraction techniques is a current research topic aimed to improve extraction technologies most suitable for the production of natural extracts of good qualities while minimizing the extraction time, energy consumption and the amount of solvents used. Driven by these goals, advances in microwave extraction have given rise to apply a novel technique, microwave integrated Soxhlet (MIS) for isolate the antioxidant molecules from rosemary leaves. The MIS technique was also compared with hot continuous extraction (Soxhlet) which is the traditional way of isolating antioxidants compounds. The MIS method has been done for extracting lipids from food products [27] but to our knowledge, at present it has never been applied for isolation of antioxidants from natural products.

2.3.1. MIS Apparatus and Procedure

The obtention of antioxidant extracts by Microwave Integrated Soxhlet extraction was carried out using the experimental apparatus illustrated in figure 2, it was largely described in a Virot et al [27] study.

MIS apparatus is composed of a milestone ETHOS microwave oven (12) which is a multimode microwave reactor (2.45 GHz) with a maximum delivered power of 1000 W variable in 10 W increments. Time, temperature, pressure and power were controlled with the "easy WAVE" soft ware package during experiments.

Beside this oven, the used apparatus include three other major elements:

- The flask (1) suited for microwave reaction, for containing the plant material.
- The extraction tube (5) which is placed on top of the flask.
- A condenser (7) which is placed on top of the extraction tube in order to allow the solvent present in base vessel to reflux upon microwave irradiation.

In microwave extraction, the polarity of the solvent is very important. Polar solvents are usually believed better than non-polar ones [28, 29]. It has been reported that by addition of water to the solvent, increased yields are obtained [30]. Ethanol/water was the solvent mixture selected for all extraction.

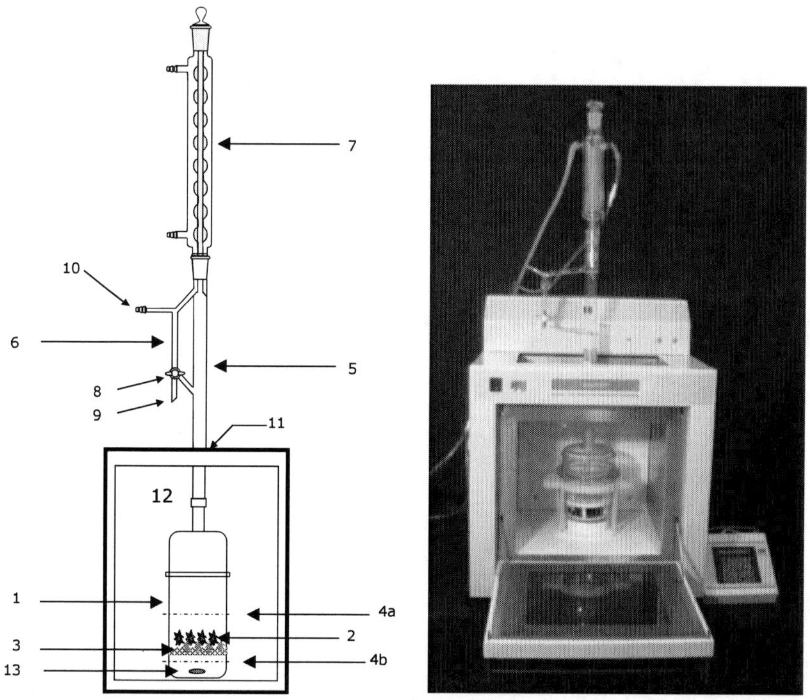

Figure 2. Experimental apparatus for extraction of rosemary antioxidants by MIS. 1: base vessel; 2: solid material; 3: support; 4a: (ethanol/water) level immersing the sample; 4b: (ethanol/water) level below the sample; 5: extraction tube; 6: side arm; 7: condenser; 8: 3-way valve; 9: side arm opening to collect (ethanol/water) ; 10: side arm opening to pull a vacuum in the system; 11: opening on upper surface of the microwave oven; 12: microwave oven.

Briefly, dried and ground INA rosemary leaves (2) were loaded onto the support (3) and ethanolic solution [ethanol/water (80:20, v/v)] (4a) was added in order to immerse the sample. Then, the base vessel was placed in the microwave oven (12) and screwed together with the extraction tube (5). The condenser (7) was placed on the extraction tube and the system was started. Extractions were performed in triplicate and the mean values were reported.

2.3.2. Soxhlet Extraction Procedure

Soxhlet extraction technique was used for a long time, it was developed in 1879 by the German agricultural chemist, Franz Ritter von Soxhlet for the determination of fat in the milk. Nowadays, it is a standard technique and the main reference for evaluating the performance of other methods of solid-liquid extraction. In that respect, this method was carried out and compared with MIS process. The dried and ground rosemary leaves were put in a cellulose thimble clogged with cotton and inserted into the Soxhlet apparatus, which was fitted with a reflux condenser and placed on a distillation flask containing ethanolic solution [ethanol/water (80:20, v/v)] and boiling glass regulator. Samples were thus, extracted under reflux with ethanolic solution during 10 h of continuous extraction. At the end of this period, the solvent was removed from the extract in a vacuum rotary evaporator, yielding the extracted compound.

2.4. Statistical Analysis

The efficiency of the MIS extraction is dependent on several factors such as the extraction time, microwave power and the leaching time. Thus, it is imperative to optimize the extraction conditions in order to maximize antioxidant capacity of the extract. The optimization of the main variables involved in the MIS process has been performed by means of a statistical method, Response Surface Methodology (RSM), which allows not only the identification of the optimum conditions but also the determination of the factors that influence the extraction process, while minimising the number of experiment required. The RSM was achieved with a central composite design (CCD), using the software Statgraphics Plus (Rockville, USA, 2000).

It was used to investigate the effects of the three controlled factors (namely extraction time, leaching time and irradiation power) on the dependent variable (extraction yield). The CCD application comprises a two-level full factorial design (coded ±1), superimposed by centre points (coded 0) and "star points" (coded $\pm\alpha$). The three key variables studied were pointed at five separate coded levels: $-\alpha$ (=−1.68), −1, 0, +1, $+\alpha$ (=1.68) and their values were selected on the basis of previous experiments. The natural values and coded levels are given in table 3 and involved twenty experiments with six replications of the centre points to take the experimental error of the measurements into account.

The data obtained from the CCD design was expressed through the following second-order polynomial equation:

$$y = \beta_0 + \sum_{i=1}^{3} \beta_i x_i + \sum_{i=1}^{3} \beta_{ii} x_i^2 + \sum_{i=1}^{2} \sum_{j=i+1}^{3} \beta_{ij} x_i x_j \quad (1)$$

where y is the predicted response (extraction yield); β_0 is a constant; β_i, β_{ii} and β_{ij} are regression coefficients; x_i and x_j are independent variables. The adequacy of the model was determined by evaluating the coefficient of regression (R^2) and the fisher test value (F-value) obtained from the analysis of variance (ANOVA).

Table 3. Coded levels and natural values used in the experimental design

	Coded levels				
	$-\alpha$ (-1.6818)	−1	0	+1	$+\alpha$ (1.6818)
Extraction time (min)	5	8	12	16	19
Leaching time (min)	5	8	12	16	19
Microwave power (W)	430	500	600	700	770

2.5. Extraction Kinetics

A comparative study was conducted between the Soxhlet and MIS techniques after the optimisation of conditions. The variation of the extraction yield according to the extraction time was followed.

2.6. Chemical Analyses

Identification of phenolic compounds of extracts obtained by MIS and soxhlet, was performed using a high performance liquid chromatography system (HPLC) Waters (Milford, MA, USA) consisting of a Model 600 pump and a Model 600 gradient controller, to which were connected a Model 717 auto sampler and a Model 996 photodiode-array detector. HPLC data were acquired and processed by Empower software (Waters). The separations of the analytes were obtained through a Purospher Star C18e column (250 mm x 4.6 mm i.d., 5 μm particle size), provided by VWR International. The binary mobile phase consisted of solvent (A) water/formic acid (99.5/0.5) and solvent (B) acetonitrile. The flow rate of the mobile phase was 1.5 mL/min. The injection volume was 20 μL. Detection was carried out at 284 nm for carnosic acid and carnosol, and at 330 nm for rosmarinic acid. Quantification was carried out by using the external standard method. All analyses were made in triplicate. The analytes were identified by comparing the retention time and UV- spectra of samples to the standard analysed under the same chromatographic conditions.

3. Results and Discussion

3.1. Central Composite Design results

Table 4 reports the three controlled variables studied and the yield obtained in the multivariate study. The yield of antioxidants compounds obtained from rosemary leaves by MIS ranged from 8.90 to 14.21 %. The results were fitted with the following second order polynomial equation:

$$\text{Yield of extracted antioxidants (\%)}: -39.4117 + 0.6340E + 1.9062L + 0.1362P - 0.0421E^2 + 0.0075EL + 0.0005EP - 0.0699L^2 - 0.0002LP - 0.0001P^2 \quad (2)$$

where E denotes extraction time (min), L the leaching time (min) and P is the microwave irradiation power (W). The responses surfaces obtained from this equation were plotted and presented in Figure 3, in order to show the effects of parameters and their interactions on antioxidants yield. These types of plots show effects of two factors on the response at a time and the other factor was kept at level zero.

Statistical testing of the model was performed in the form of analysis of variance (ANOVA). The ANOVA for the fitted polynomial model of extraction of antioxidants were shown in Table 5.

Table 4. Fully coded Central Composite Design and responses obtained

Experiment	Coded variable level			Response
	Extraction time	Leaching time	MW power	(%)
1	-1	-1	-1	12,46
2	1	-1	-1	11,05
3	-1	1	-1	12,50
4	1	1	-1	12,26
5	-1	-1	1	8,90
6	1	-1	1	9,05
7	-1	1	1	9,26
8	1	1	1	9,21
9	$-\alpha$	0	0	11,53
10	α	0	0	13,81
11	0	$-\alpha$	0	9,00
12	0	α	0	13,82
13	0	0	$-\alpha$	12,86
14	0	0	α	9,06
15	0	0	0	14,01
16	0	0	0	13,92
17	0	0	0	14,21
18	0	0	0	14,17
19	0	0	0	13,84
20	0	0	0	13,90

Table 5. Summary of the ANOVA model statistics

Source	SS[a]	d.f[b]	MS[d]	*F*-Ratio	*P*-Value
E:Extraction time	0,382157	1	0,382157	0,35	0,5693
L:Leachintg time	7,1423	1	7,1423	6,47	0,0292
P:Microwave power	24,3634	1	24,3634	22,07	0,0008
EE	6,55006	1	6,55006	5,93	0,0351
EL	0,117612	1	0,117612	0,11	0,7508
EP	0,382813	1	0,382813	0,35	0,5690
LL	18,0662	1	18,0662	16,37	0,0023
LP	0,0666125	1	0,0666125	0,06	0,8109
PP	23,5655	1	23,5655	21,35	0,0010
Total error	11,0394	10	1,10394		
Total (corr.)	84,474	19			

R^2 = 86,93 %, R^2 (adjusted for d.f.) = 75,17 %.
[a] Sums of squares. [b] Degree freedom. [c] Mean square.

In this table, the quadratic regression model showed the value of the determination coefficient (R^2) was 0.8693, which implied that 86.93 % of the variations could be explained by the fitted model. R^2 adj was 0.7517, which implied that 75.17 % of the total variations were explained by the model. It also indicated that there was a high degree of correlation between observed and predicted data from the regression model. The statistical significance of the regression model was checked by *F*-ratio and *p*-value. The corresponding variables would be more significant if the *F*-value becomes greater and the *p*-value becomes smaller [31]. Values of *p*-value less than 0.05 showed model terms were significant. From the results given in Table 5, it can be seen that microwave power (P) has the most important influence on the antioxidants yield ($p < 0.05$), followed by squared terms of power and leaching time (PP and LL), leaching time (L) and quadratic term (EE). Results also showed that the *F*-ratio of the independent variable (microwave power (P)) and of the quadratic term (PP) was greater (22.07 and 21.35 respectively), which implied that the independent variable (microwave power (P)) was the most significant factor on the experimental yield of antioxidants.

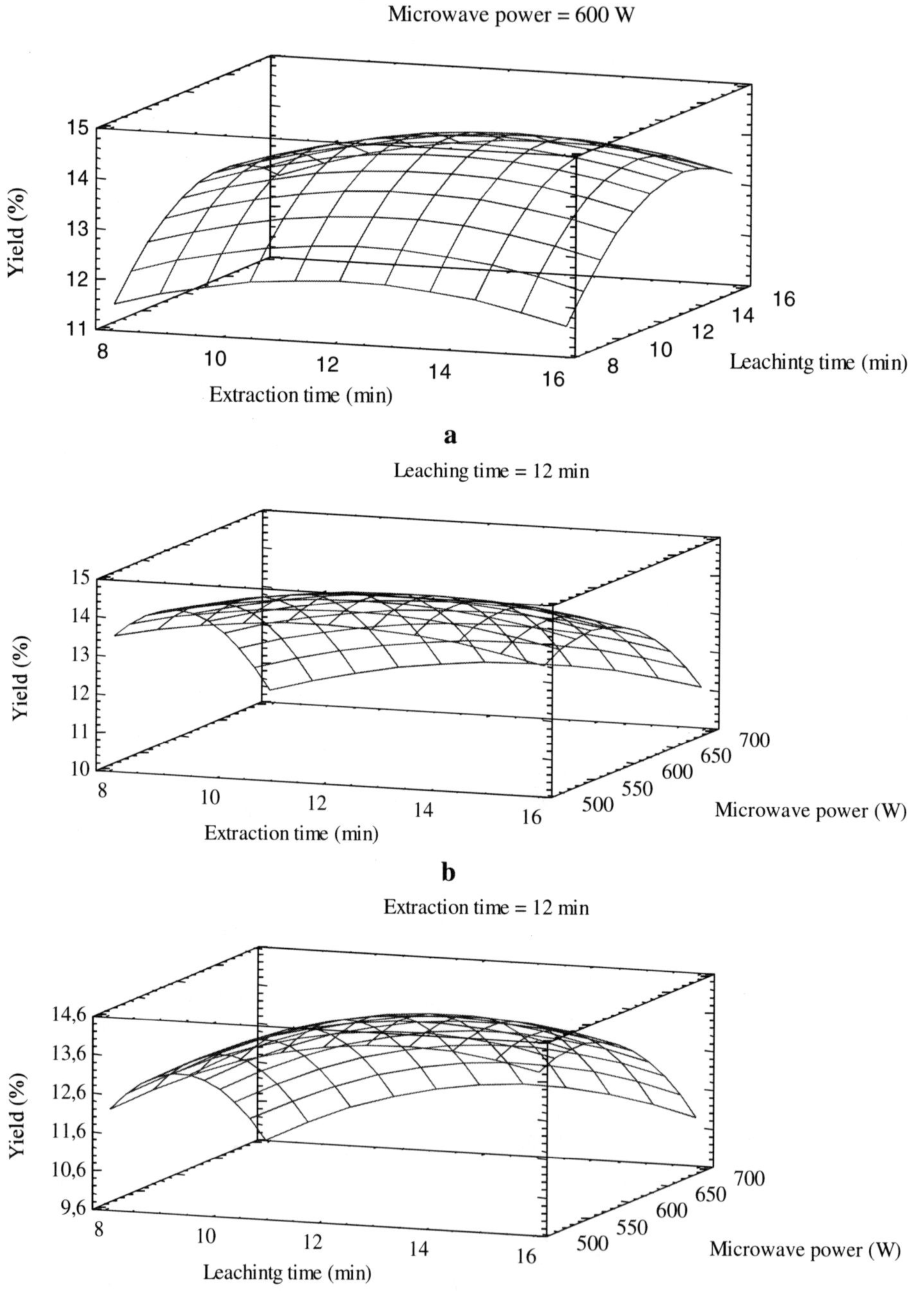

Figure 3. Surfaces obtained with the multivariate study: (a) estimated yield, extraction time, leaching time response surface (b) estimated yield, extraction time, power response surface and (c) estimated yield, leaching time, power response surface.

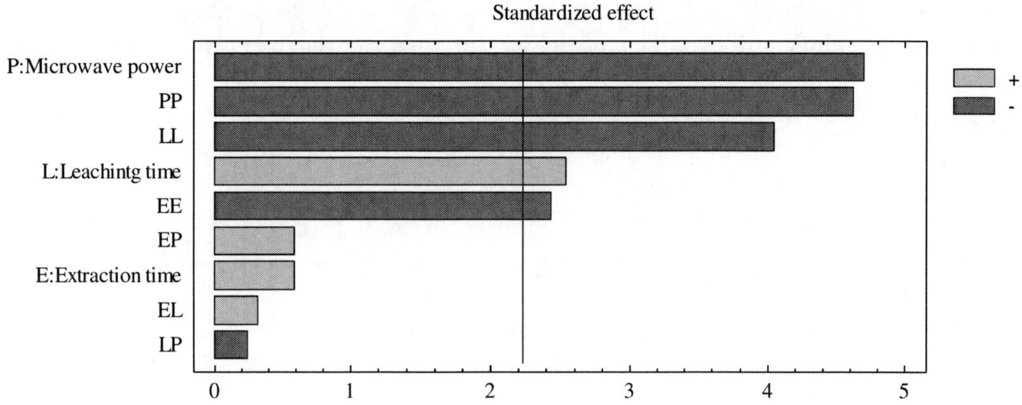

Figure 4. Standardized Pareto chart of rosemary showing the effect of different factor terms on extraction rate. Bar exceeding the vertical line on the graph indicate that the corresponding factor terms are significant ($p<0.05$).

Pareto chart illustrated by figure 4 showed the most important influence of the three studied variables as well as their interactions and the quadratic effects in the MIS extraction process. The vertical line on the Pareto chart determines the effects that are statistically significant. The order in which the bars are displayed corresponds to the order of the size of the effect. The standardized effect is the estimated effect divided by its standard error. Hence a low standardized effect can mean either a low effect of the parameter or a large experimental error. Pareto chart confirmed that the microwave power has the most important influence on antioxidants contents, whereas the cross product terms (EP, EL, LP) and the extraction time show no significant effects. However, the quadratic effect of the three variables seemed to be significant with negative sign which indicated that the extraction rate increased with increasing the three variables up to a certain level then begins to decrease after.

3.2. Optimal Conditions

Response surface optimization can be calculated depending on the three key variables, namely irradiation power, extraction and leaching times. The maximum yield computed by the software is calculated by the maximum value of the surface response for a set of variables lying between the minimum and maximum value of the CCD plan. The optimized conditions obtained using the model for each parameter were: extraction time 12 min; leaching time 13 min and irradiation power 550 W. Using those optimized conditions, the model predicted a maximum response of 14.5%.

3.3. Kinetics of Extraction: Comparison of MIS vs Soxhlet

The rosemary antioxidants extraction yield obtained by optimized MIS extraction (figure 5a) proved the efficiency of this process when compared with the conventional Soxhlet extraction (figure 5b). The duration of extraction was clearly improved, as the extraction of antioxidants was found twenty times as fast under microwaves (30 min) as in the conventional method (10 h) which could be attributed to the microwave energy which increase the mass transfer of the analyte from the sample into the solvent. As can be seen in fig. 5a, five phases are observed in the process of the MIS extraction. The first step (0 to 2 min) corresponds to the time necessary for heating the solvent to the boiling point by microwave. The step 2 (2 to 14 min) corresponds to the extraction phase, it characterizes the rapid increase in the yield. This step is followed by the third part, where the level of the solvent is lowered below the sample, approximately 66 % of the totally yield were obtained from the beginning of the extraction. The fourth step (realised into 13 min) is the stage corresponding to the leaching time, the maximum of the yield was obtained (14.16 %) in the end of this step. The last stage (step 5) corresponds to a horizontal line which marks the end of the extraction process.

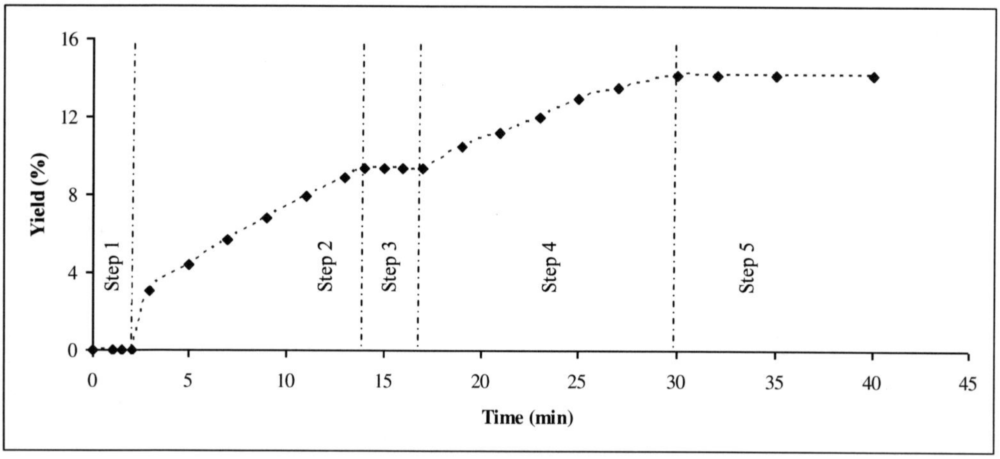

a

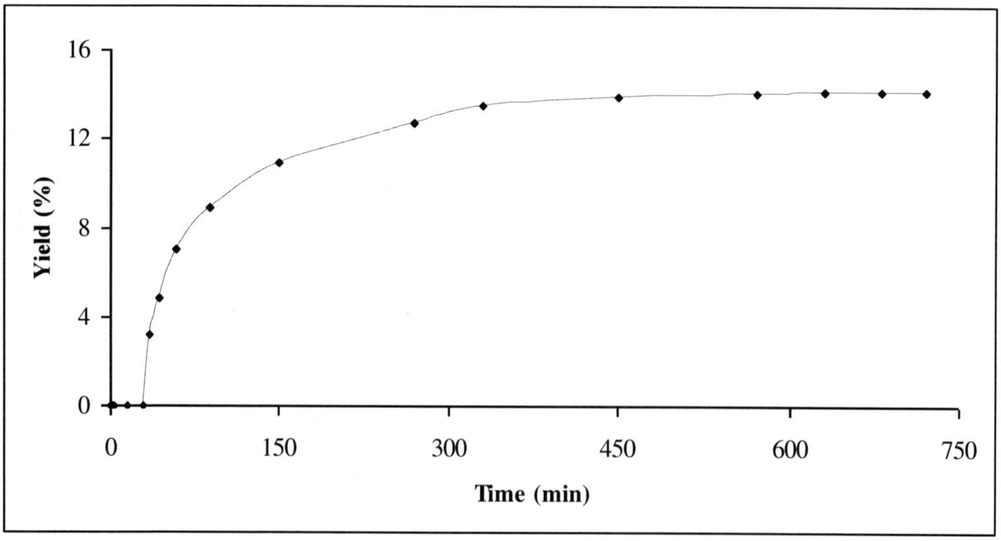

b

Figure 5. Kinetics of extraction realized by MIS (a) and by conventional Soxhlet (b).

3.4. Antioxidants Quantification

The identification and the quantification of the major antioxidants phenolic compounds in the rosemary extract were accomplished by HPLC analysis. The calibration curve equations, concentration range used for each compound are presented in Table 6. The concentration ranges used were selected according to the relative amounts of each compound found in the extracts.

Table 6. Calibration curves and concentration ranges employed for the quantification

Compound	Tr (min)	Concentration range (μg/mL)	Calibration curve	R^2
Rosmarinic acid	12.115	17.7 – 60	y = 3E+07x	0.9953
Carnosol	24.959	100 – 1000	y = 4E+06x + 55331	0.9985
Carnosic acid	28.052	136 - 1360	y = 3E+06x-106223	0.9964

Table 7. Concentration of carnosol, rosmarinic and carnosic acids (mg/g) obtained from rosemary leaves

Extraction method	Major phenolic compounds of Rosmarinus			Yield (%)
	Rosmarinic acid	Carnosol	Carnosic	
Soxhlet	1.83	0.49	10.85	14.13
MIS	1.44	0.53	11.12	14.16

The linearity of these curves was always good, with R^2 values higher than 0.99 for all the studied compounds. The results (Table 7) showed that the rosemary extracts obtained either by MIS (12 min, 13 min, 550 W) or Soxhlet (10 h) contain the same dominant components (carnosic acid, carnosol and rosmarinic acid). Extract composition was equivalent in term of major antioxidants concentration for both extraction methods and also in term of yield. As it was reported by Kuhlmann, A et al [22], Carnosic acid represents the main constituent of the phenolic diterpenes in rosemary extracts obtained by MIS and Soxhlet methods.

Conclusion

In this work, it has been shown the usefulness of the MIS process for extraction of natural antioxidants from a low cost material. This original device employing the microwaves as heating source for Soxhlet device provides a valuable alternative for antioxidants extraction. Compared to Soxhlet, MIS could reach the same antioxidants yield in 30 min instead of 10 h. RSM was used to estimate and optimize the experimental parameters: extraction time (min), leaching time (min) and microwave power (Watt). The results showed that the independent variable (microwave power (P)) and the quadratic term (PP) had highly significant effects on the experimental yield of antioxidants ($p < 0.05$). The optimum conditions for extraction rate of antioxidants selected with the Response Surface Methodology were as follows: extraction time, 12 min; leaching time 13 min and irradiation power 550 W. Using those optimized conditions, the model predicted a maximum response of 14.5%. The experimental yield of rosemary antioxidants was 14.16 %, which was close with the predicted yield value.

The proposed extraction procedure, Microwave-Integrated Soxhlet extraction, provides an extract of rosemary with similar antioxidant content as those produced by Soxhlet. More over the proposed process takes short time, below 30 min and shows a good yield.

References

[1] Wanasundara, P. K. J. P. D; Shahidi, F. Antioxidants: Science, Technology, and Applications. Bailey's Industrial Oil and Fat Products, Sixth Edition, Six Volume Set. Edited by Fereidoon Shahidi. Copyright 2005. John Wiley and Sons, Inc.

[2] Phyllis A. Balch. Prescription for Nutritional Healing. A practical A to Z reference to drug free remedies using vitamins, minerals, herbs and food supplements. Fourth edition. Published by the penguin group, USA, 2006.

[3] Aruoma, O. I. Free Radicals, Oxidative Stress, and Antioxidants in Human Health and Disease. *JAOCS*. 1998, 75, 199-212.

[4] Neff, J. Big companies take nutraceuticals to heart. *Food Process*. 1997, *58*, 37- 42.

[5] Merken, H. M; Beecher, G. R. Measurement of food flavonoids by high-performance liquid chromatography - A review. *J. Agric. Food Chem*. 2000, *48*, 577-599.

[6] Espin, J. C; Soler-Rivas, C; Wichers, H. J. Characterization of the total free radical scavenger capacity of vegetable oils and oil fractions using 2,2-diphenyl-1-picryhydrazyl radical. *J. Agric. Food Chem*. 2000, 48, 648-656.

[7] Duthie; S.J, Ma; A, Ross; M.A, Collins; A.R. Antioxidant Supplementation Decreases Oxidative DNA Damage in Human Lymphocytes. *Cancer Res*. 1996, *56,* 1291–1295.

[8] Pezzuto; J.M. Plant-Derived Anticancer Agents. *Biochem. Pharmacol.* 1997, *53,*121–133.

[9] Geronikaki; A.A, Gavalas; A.M. Antioxidants and inflammatory disease: synthetic and natural antioxidants with anti-inflammatory activity. Combinatorial chemistry and high throughput screening. 2006, 9, 425-442.

[10] Yu, L; Haley, S; Perret, J; Harris, M; Wilson, J; Qian, M. Free Radical Scavenging Properties of Wheat Extracts. *J. Agric. Food Chem*. 2002, 50, 1619-1624.

[11] Buck, D. F; Edwards, M. K. Antioxidants to prolong shelf-life. *Food Tech. Int.* 1997, 29-33.

[12] Shahidi, F. Antioxidants in food and food antioxidants. *Nahrung - Food.* 2000, 44, 158-163.

[13] Shahidi, F. Zhong, Y. Antioxidants: regulatory status. Ed. Bailey's Industrial Oil and Fat Products. John Wiley and Sons. 2005, 491–512.

[14] Venkat Ratnam, D; Ankola, D.D; Bhardwaj, V; Sahana, D.K. Ravi Kumar. M.N.V. Role of antioxidants in prophylaxis and therapy: A pharmaceutical perspective. Review. *Journal of Controlled Release*. 2006, 113, 189–207.

[15] Kahl; R, Kappus; H. Toxicology of the synthetic antioxidants BHA and BHT in comparison with the natural antioxidant vitamin E. *Z Lebensm Unters Forsch.* 1993, 196, 329-38.

[16] Aruoma, O.I; Spencer, J.P; Rossi, R; Aeschbach, R; Khan, A; Mahmood, N; Munoz, A; Murcia; A, Butler J, Halliwell B. An evaluation of the antioxidant and antiviral action of extracts of rosemary and Provençal herbs. *Food Chem Toxicol.* 1996, 34, 449-56.

[17] Martínez-Tomé, M; Jiménez, A.M; Ruggieri, S; Frega, N; Strabbioli, R; Murcia, M.A.

[18] Antioxidant Properties of Mediterranean Spices Compared with Common Food Additives. *Journal of Food Protection.* 2001, 64, 1412-1419.

[19] Djenane, D; Sánchez-Escalante, A; Beltrán, J.A; Roncalés, P. Ability of α-tocopherol, taurine and Rosemary, in combination with vitamin C, to increase the oxidative stability of beef steaks packaged in modified atmosphere. *Food Chemistry.* 2002, 76, 407-415.

[20] Wijeratne, S. S. K; Cuppett, S.L. Potential of Rosemary (Rosemarinus officinalis L.) Diterpenes in Preventing Lipid Hydroperoxide-Mediated Oxidative Stress in Caco-2 Cells. *J. Agric. Food Chem.* 2007, 55, 1193–1199.

[21] Schwarz, K; Ternes, W. Antioxidative constituents of *Rosmarinus officinalis* and *Salvia officinalis*. II. Isolation of carnosic acid and formation of other phenolic diterpenes. *Z. Lebens-unters Forsch.* 1992, 195, 99–103.

[22] Cuvelier, M.E; Richard, H; Berset, C. Antioxidative activity and phenolic composition of pilot-plant and commercial extracts of sage and rosemary. *J. Am. Oil Chem. Soc.* 1996, 73, 645–652.

[23] Shetty, K. Biotechnology to harness the benefits of dietary phenolics; Focus on Lamiaceae. *Asia Pacific J. Clin. Nutr.* 1997, 6, 162–171.

[24] Kuhlmann, A; Rohl, C. Phenolic Antioxidant Compounds Produced by in Vitro Cultures of Rosemary (Rosmarinus officinalis) and Their Anti-inflammatory Effect on Lipopolysaccharide-Activated Microglia. *Pharmaceutical Biology.* 2006, 44, 401–410.

[25] Navarrete; A, Herrero; M, Martin; A, Cocero; M.J, Ibáñez; E. Valorization of solid wastes from essential oil industry. *Journal of food Engineering.* 2011, 104, 196-201.

[26] Meziane-Assami; D, Tomao; V, Ruiz; K, Meklati; B.Y, Chemat; F. Geographical Differentiation of Rosemary Based on GC/MS and Fast HPLC Analyses. *Food Analytical Methods.* 2012, DOI 10.1007/s12161-012-9430-6.

[27] Emberger; L. Une classification biogéographique des climats. Recueil des travaux des laboratoires de botaniques, de géologie et de zoologie de la faculté des sciences de l'université de Montpellier. 1955, 7, 3–43.

[28] Virot; M, Tomao; V, Colnagui; G, Visinoni; F, Chemat; F. New microwave-integrated Soxhlet extraction. An advantageous tool for the extraction of lipids from food products. *Journal of Chromatography A.* 2007, 1174, 138–144.

[29] Wang; L, Weller; C. L. Recent advances in extraction of nutraceuticals from plants. *Trends in Food Science and Technology.* 2006, 17, 300-312.

[30] Proestos; C, Komaitis; M. Application of microwave-assisted extraction to the fast extraction of plant phenolic compounds. *LWT.* 2008, 41, 652–659.

[31] Pan; X, Niu; G, Liu; H. Microwave-assisted extraction of tea polyphenols and tea caffeine from green tea leaves. *Chemical Engineering and Processing.* 2003, 42, 129-133.

[32] Atkinson; A.C, Donev; A. N. Optimum experimental designs. Oxford: Oxford University Press. 1992, 132–189.

Index

B

C

D

E

F

G

H

I

N

Q

R

S

T

U

V

W

X

Y

Z